OUTPOST IN ORBIT

A PICTORIAL & VERBAL HISTORY OF THE INTERNATIONAL SPACE STATION

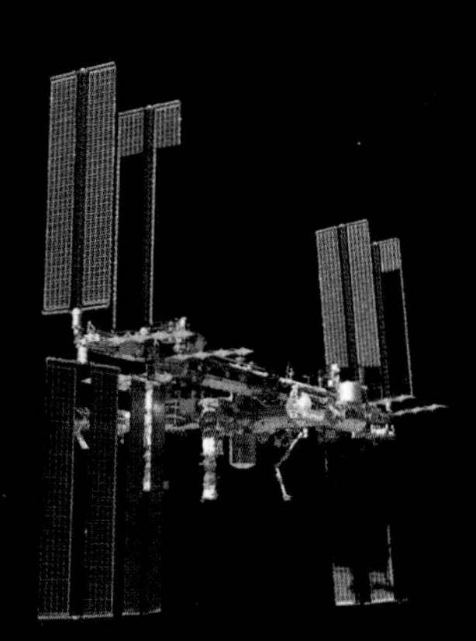

DAVID J. SHAYLER + ROBERT GODWIN
EXECUTIVE EDITOR – DR. GARY KITMACHER

AN APOGEE BOOKS PUBLICATION

Outpost in Orbit: A Pictorial & Verbal History of the International Space Station
by David J. Shayler & Robert Godwin
Edited by Dr. Gary Kitmacher
ISBN 9781-989044-03-2 - ISSN 1496-6921 ©2018 Apogee Books

Published by Apogee Books under a Space Act Agreement with the National Aeronautics and Space Administration
Images provided by NASA except Pages 5-13, 19, 22-24, 62, 65-66 © Reserved or fair usage, 16 (top), 17-18 US Army, 54-55 Energia
Further details of the many space stations in this book can be found at www.thespacelibrary.com
http://www.apogeebooks.com Apogee Books, Burlington, Ontario, Canada, L7R 1Y9
Cover and Design: Robert Godwin, Printed and bound in South Korea

Special thanks to:
Shannon Abbott, Hans Bolender, Tim Braithwaite, David Christensen, Michael Ciancone, Robert C. Dempsey, Cayce C Dowell, Bill Gerstenmaier, Reagan Grimsley, Chris Hadfield, Fred Haise, Bob Jacobs, Juan Jaime, Mark Kahn, the Koelle family (Elisabeth, Karin, Ingrid, Patricia and Dietrich), John Lange, Colleen Lapp, Conrad Lau Jr., Carolyn Lawson, Isabelle Lord, Charles Lundquist, Stephen McNeilly, Paul MacNaughton, Alexander Nitsch, Jean Olivier, Fred Ordway IV, Gwen Pitman, Hans Schlegel, Sarah Schwaner, Robert Thirsk, Sarah Waddington, Mary Wilkerson, Dave Williams, Frank Williams, Frank Winter, Jordan Whetstone, Mitch Youts, Brent Ziarnick

ABOUT THE AUTHORS

Spaceflight historian David J. Shayler, FBIS was born 1955. His lifelong interest in space exploration began by drawing rockets aged 5 but it was not until the launch of Apollo 8 to the moon in December 1968 that the interest for human space exploration became a passion. Dave joined the BIS in January 1976, became an Associate Fellow in January 1983 and Fellow in January 1984 and was elected to the Council in September 2013. Since 2012 he has coordinated the annual BIS Russian/Sino Technical Forum at the BIS Headquarters in London. His first articles were published by the British Interplanetary Society in the late 1970s and in 1982 he created Astro Info Service (www.astroinfoservice.co.uk) to focus his research efforts. His first book was published in 1987 and he now has over 28 titles to his name including works on the American and Russian space programmes, on the topics of space stations, space walking, women in space, and the human exploration of Mars. Dave's authorised biography of Skylab 4 astronaut Jerry Carr was published in 2008. Since 1969 he has had a keen interest in EVA and spacesuits, which is reflected in many of Dave's presentations and published works.

Robert Godwin is the owner and founder of Apogee Books. He was the Space Curator at the Canadian Air & Space Museum in Toronto and is a member of the International Astronautical Federation and American Astronautical Society History Committees. He is also a director of the General Astronautics Research Corporation. He has written or edited over 150 books including the award-winning series "The NASA Mission Reports". He has appeared on dozens of radio and television programs in Canada, the USA and England as an expert on space exploration. His books have been discussed on CNN, the CBC, the BBC and CBS 60 Minutes. He produced the first virtual reality panoramas of the Apollo lunar surface photography and the first time-synchronized multi-camera angle movie of the Apollo 11 moonwalk and moon landing. He consulted on the Discovery Channel TV series "Rocket Science" and "Mars Rising" and the science fiction web series "Deep Six". In 2002 Godwin's Mission Reports series won the Space Frontier Foundation's Best Presentation of Space Award. In 2007 the International Astronomical Union's Committee for Small Body Nomenclature approved the naming of a main belt asteroid after Robert for his efforts in documenting space history and raising public awareness about Near Earth Objects. "4252 Godwin" is an absolute magnitude 12.7 minor planet. He has staged space exhibitions in the USA and Canada and has spoken at countless venues including the Canadian Embassy in Washington DC. In 2013 he co-authored a biography of Arthur C. Clarke with Fred Clarke, brother of Arthur. In 2014 he co-authored *2001 The Heritage and Legacy of the Space Odyssey* with Frederick I. Ordway III, who was technical adviser to Stanley Kubrick for the film *2001 A Space Odyssey*. In October 2015, his essay *The First Scientific Concept of Rockets for Space Travel* contributed a previously unknown chapter to early space history by revealing that a Scottish Presbyterian Minister named William Leitch was the first scientist to determine, for the correct reasons, that rockets were the best method for powering space flight.

Dr. Gary H. Kitmacher has worked at Johnson Space Center since 1981. He has worked on the ISS, Freedom, Mir and several other development efforts over the course of his career. For the space station, Kitmacher was the Architectural Manager for Man-Systems and was instrumental in establishing the ISS lab module, node, Cupola, and standard rack design requirements and configuration, and he established the human-systems habitability subsystems. Later he served as the subsystem manager for the Crew Health Care System. As the Mir/Priroda manager and later the manager for Mir operations and integration, he negotiated the contracts, and established the requirements and processes for the integration of US hardware and payloads on Mir. During this time he invented and developed the soft stowage system (CTBs), Crew On Orbit Support (COSS) computer system, and US GFE mechanical, electrical and data systems for Mir. He worked extensively in Moscow and Baikonur. Previously, Kitmacher was the mission manager for the first Spacehab commercial missions. He began his work for NASA as the stowage and crew equipment subsystem manager for Shuttle. Before joining NASA, he worked for Rockwell in the Shuttle Commercialization Office, streamlining the payload integration process and establishing agreements for the first Shuttle commercial payloads. Kitmacher has an extensive portfolio of books and technical papers about ISS, Mir, future exploration, spacecraft design and space commercialization. He recently wrote the chapter for the Standard Handbook for Aerospace Engineers, about spacecraft design for human habitation and operations. He wrote the NASA/Apogee Books ISS Reference Guide and developed the ISS interactive website which won the Adobe Max international award. He has been awarded the NASA Exceptional Achievement Medal, 2 JSC Commendations, and the Silver Snoopy.

CONTENTS

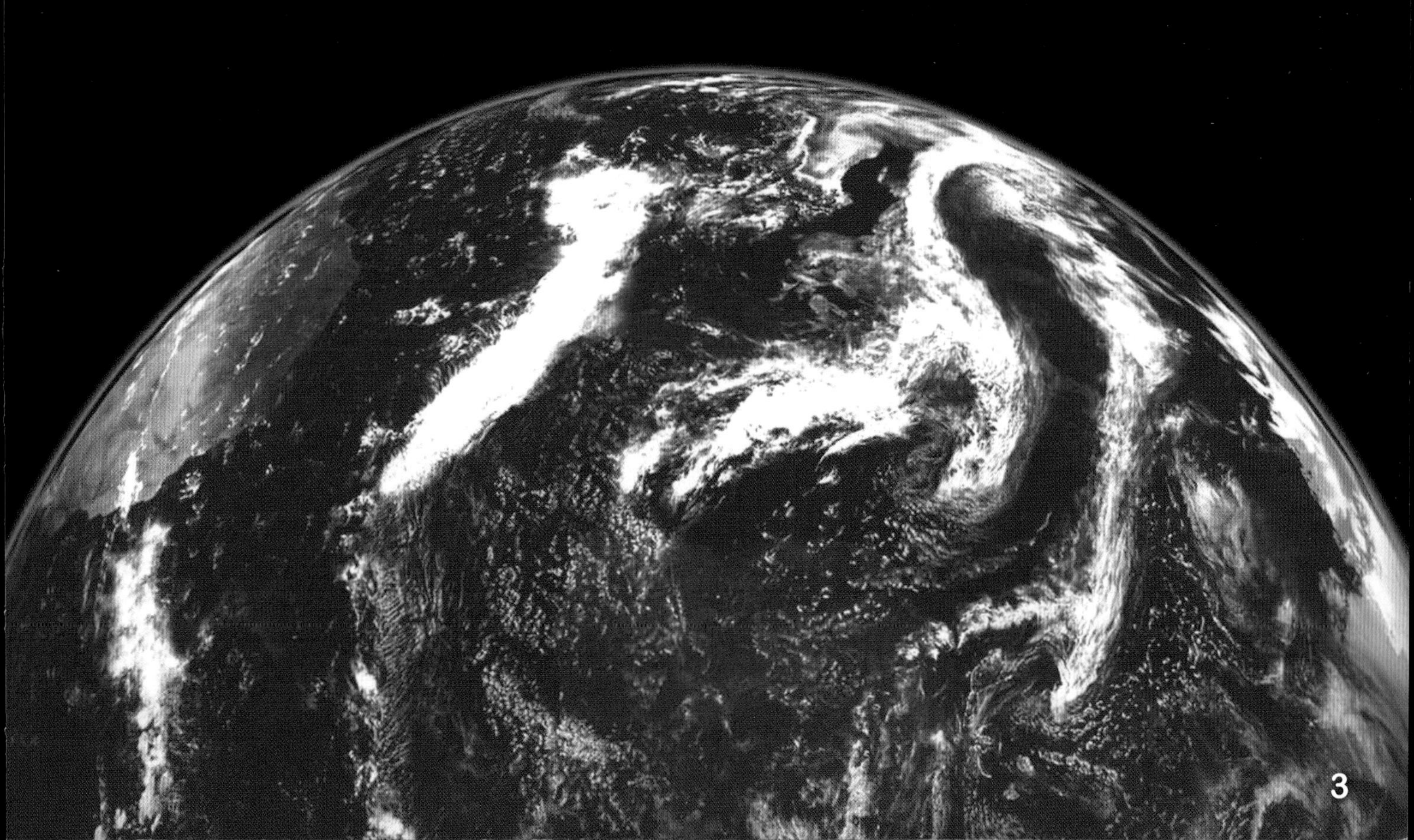

INTRODUCTIONS

"Over thirty years ago, when the idea of an International Space Station was seriously being considered by NASA I, like many others, wondered if we would ever see the result. Today, twenty years after construction started and seven years after the major assembly was completed we can now look back at a very special time in human spaceflight history, when sixteen countries worked together for a common goal in the creation of a large international research facility in Earth orbit. The resulting ISS has focused not only in the fields of space science and exploration but shared, via the recent expansion of social media, the experiences of orbital flight to a global audience. In researching this book one common factor links crewmembers, engineers, scientists, managers and investigators and that is the ISS is an International space station for all. Clearly this is the most rewarding achievement for those who have been involved in the program. This cannot be the definitive story of ISS, as the station is still operating for hopefully another decade, but this book offers a visual and vocal celebration of what the past two decades have achieved. Building upon over a century of dreams, unrealised plans and early station operations across Salyut, Skylab, Almaz, Spacelab, Mir, and Tiangong ISS is our stepping stone to the next phase of human space exploration. This then is the story of ISS up to the 20th anniversary of its creation from the people who helped build and operate it, the crews who lived and worked on it and their hopes for the future. Research conducted specifically for this project has spanned more than two decades. In this time there have been a large number of people who have helped bring this book to its conclusion, most notably Rob Godwin of Apogee books for keeping the faith in the project over many years, and Gary Kitmacher at NASA. There have been numerous people, too many to list here, at NASA, JAXA, CSA, RSA and ESA I would like to thank, especially those involved in Public Affairs, and those individuals who agreed to provide interviews cited in the text. This participation added a personal insight into developing and building the ISS, operating and controlling it and living and working on board. Finally thanks to my brother Michael D. Shayler for his early research efforts and wordsmith skills at editing the original draft document." - David Shayler

"In 1998 I quite unexpectedly found myself sitting down to dinner with Buzz Aldrin. Of course everyone remembers Buzz for his epic journey to Tranquility Base, but we shouldn't forget that his EVA on Gemini 12 is what really proved that the space station was more than just a fanciful dream. After finishing dinner that night Buzz asked me to put together a book to honour the crew of Apollo 8. I agreed without hesitation. On November 20th 1998 the first module of the ISS took flight and just twelve days later Apogee Books was also launched - at the Apollo 8 dinner in Chicago. Thanks to that chance meeting - two decades ago - I now find myself celebrating two anniversaries. Of course back then I had no idea that Apogee would still be around in 20 years to produce this volume about one of the truly astonishing feats of human ingenuity. In 2019 as the world celebrates the first voyage to the moon, let's not forget that it will also be the 150th anniversary of the idea of the space station. Edward Hale's "Brick Moon" may not have looked much like the ISS, but just like the spectators in his story, anyone with keen eyesight can see it sprinting across the night sky just after sunset, or just before sunrise. It is literally a glowing testament to the genius and imagination of our species, and to its best and brightest, many of whom have also helped to make this book better and brighter than it might have been. Over the last two decades I have been able to interact with many of the people who contributed to this book. It has been my singular honour to meet, interview and work with the people from all walks of life who have made the ISS possible and to now celebrate Apogee Books 20th anniversary in the company of Dr. Gary Kitmacher, whose direct contributions to ISS can be found in amongst these pages, and Mr. David Shayler of the British Interplanetary Society whose exemplary contributions to space history continue herein." - Robert Godwin

"People have dreamed of living off of the Earth for centuries. Today the space station is a reality. Men and women are living and working on the International Space Station. We have not known a time that men and women have not been living in space for more than three decades. Hundreds of years ago civilization's greatest achievement were the great cathedrals. They often would take decades and even centuries to complete and they required the work of the greatest organizers, artisans, architects, craftsmen and laborers of the day. Today that multi-generational life-affirming project is the space station. ISS is a laboratory, an observation post and a testbed for future trips to other worlds. It has been a home for hundreds of visitors. It has been the dream, occupation, vocation and focus of attention of tens of thousands of dreamers, planners, scientists, architects and engineers over several generations. One day the idea of a space station was only a dream and a challenge, but today it is a functioning reality. ISS has been a tremendous achievement and it is an important step to the next frontier. ISS contains all of the elements that will be needed for future life beyond Earth. One day in the future more people will probably live in space and off the Earth than live on the Earth today. When that day arrives, it will be due in large measure to the men and women who conceived, designed, developed and who today operate, live and work on the ISS. The ISS is a critical step towards that future. The desire to explore the universe beyond Earth has been the dream of generations and has stirred hearts and minds. The International Space Station has happened because of human curiosity, science, technology, politics and cultural advancement. ISS has advanced the cause of peace and goodwill with sixteen countries working closely together in an integral partnership. It has established a new model for cooperation in peaceful technological projects. ISS has demonstrated what can be accomplished when different countries and different cultures work together over a long period for a common cause. What we are learning today will be critical for civilization in the future." - Dr. Gary Kitmacher

THE BEGINNING – A FOUNDATION

It was 1687 when Isaac Newton revealed the nature of gravity to the world. In 1835, using Newton's insights, astronomer Joseph von Littrow explained that if a sufficient propulsive force could be found, artificial moons could be launched into orbit. Then in 1869 *Atlantic Monthly* magazine published Edwards Everett Hale's story, *The Brick Moon*. Hale's moon was a manned space station 200 feet (60.96 meters) in diameter which flew in a polar orbit 4000 miles (6436 km) above the Earth. Hale reasoned that because of the extreme heat resulting from its exposure to the Sun above the atmosphere, the satellite should be made from brick. The brick moon had a crew of 37 engineers and in Hale's story was *accidentally* launched into space. Hale realized that his artificial moon would have an extremely low gravitational field, allowing its inhabitants to enjoy super-human athletic feats. He also realized that they would have an unprecedented view of the Earth below, due to the fact that they would be looking down through only a few miles of dense atmosphere.

◀ Hale's station was to be launched by gigantic flywheels which were powered by a waterfall. More than one 'Brick Moon' was envisaged – multiple "moons" would assist vessels with navigation. Made of 12 million bricks, it would have cost $60,000 ($1M in 2018$) to fabricate the satellite and a further $214,729 to fabricate the launching system, build the monorail and launch it. By reusing the launching system, a second moon could be launched for another $159,732.

Edward Everett Hale - 1869

Konstantin Tsiolkovsky - 1883 & 1933

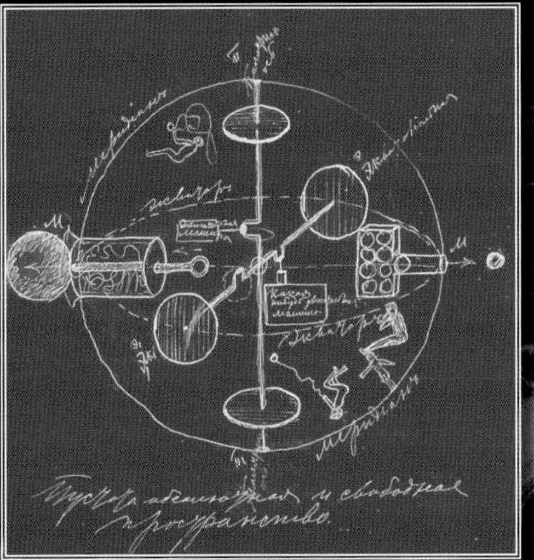

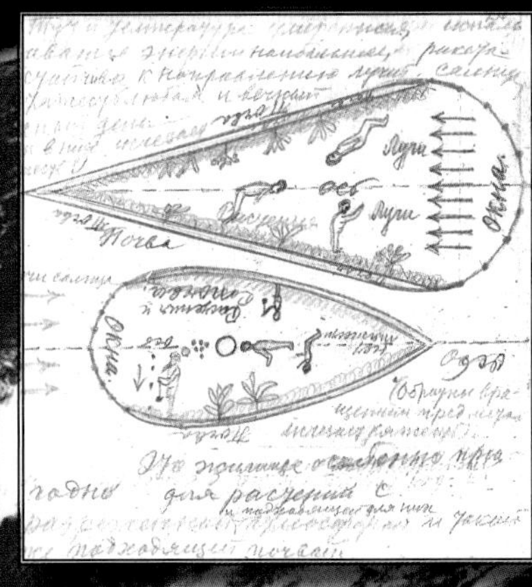

Russian scientist Konstantin Tsiolkovsky drew the diagram seen here (above left) in 1883. In his essay *Svobodnoe Prostranstvo* (Free Space) he first described how humans might live in space. Then in his 1896 novel *Vne Zemli* (Beyond the Planet Earth) he described gardens and habitats in space; but it would not be until his *Al'bom kosmicheskikh puteshestviy* (Album of Space Voyages) in 1933 that a sketch of his proposed habitat would be seen, complete with plants growing on the walls and a large window to allow sunlight into the space station (above centre, colour added). At bottom left of the lower spacecraft in this image appears to be what might be a self portrait of the author wielding a shovel. ▶

Hermann Oberth, a German-Rumanian, began publishing his ideas for space flight in 1923. His book *Die Rakete zuden Planetenräumen* influenced an entire generation of German and British engineers to contemplate space stations. At this time the term "space station" was generally considered to mean any artificial object placed into orbit, whether manned or unmanned. The distinction between a manned orbiting spacecraft and an unmanned one would not come until much later. In some circles the term "space station" was still in common use to describe any satellite until well into the 1950s. The impact of Oberth's book would be far-reaching and would trigger a revolution in rocketry. At first Oberth concentrated his ideas on placing large mirrors in orbit to bring light and heat to remote places on earth. One of the very first to be excited by the concepts described by Oberth was an electrical engineer and physicist named Otto Willi Gail. In 1925 and 1926 he would write two novels about spaceflight, and in the second of these novels, which was entitled *Der Stein vom Mond* (The Stone from the Moon) he described in detail what it might be like to visit a space station such as that described by Oberth. In Gail's story the space station is named the *Astropol* and the enormous habitat is under construction by spacewalking astronauts. The station itself is a disc-like oblate sphere with an arm attached to a pear-shaped habitat. The arm is an elevator shaft which takes the crew from the core to the living quarters and is in perpetual motion, spinning in orbit around the central core to provide artificial gravity. Alongside the *Astropol* the construction workers are also building Oberth's massive space mirror for beaming heat to the earth. Although Gail's novels are science fiction they had a profound impact on the fledgling astronautical community in Germany. He soon became aquainted with Oberth and the year after his novel appeared the first rocket society, the *Verein für Raumschiffahrt* (VfR), was formed in Germany. Gail then met the rocket pioneers of that society, including a young man named Wernher von Braun. Over the next 30 years Gail would write many non-fiction books on astronautics.

Hermann Oberth - 1923

"The traffic between (the space station) and the earth can be maintained by smaller rockets so that the space station will always remain aloft for its particular mission.The space station could also be a refueling station for interplanetary operations since hydrogen and oxygen could be stored here over extended periods in a solid state. A rocket which is refuelled here and leaves from the station does not encounter air resistance and suffers only a small impulse loss from gravitational effects." - Hermann Oberth

Otto Willi Gail - 1926

Gail's 1926 *Astropol,* as depicted in 1930 in an English edition. This image shows the spacewalking astronaut construction workers. The spinning habitat, seen at right, was too close to be a viable source of artificial gravity. In the 1929 edition of Oberth's book "Ways to Spaceflight" he suggested that for a space station to be given artificial gravity two modules would have to be on opposite ends of a cable as much as 20km in length. In Oberth's 1954 book *Menschen im Weltraum* he would fully describe these ideas for a manned "springboard station," (below left) which might be many kilometers across and would include "watchdog bombs" to protect it from hostile intruders while his "fixed-orbit station" (below right) would have a space telescope mirror 21 km across.

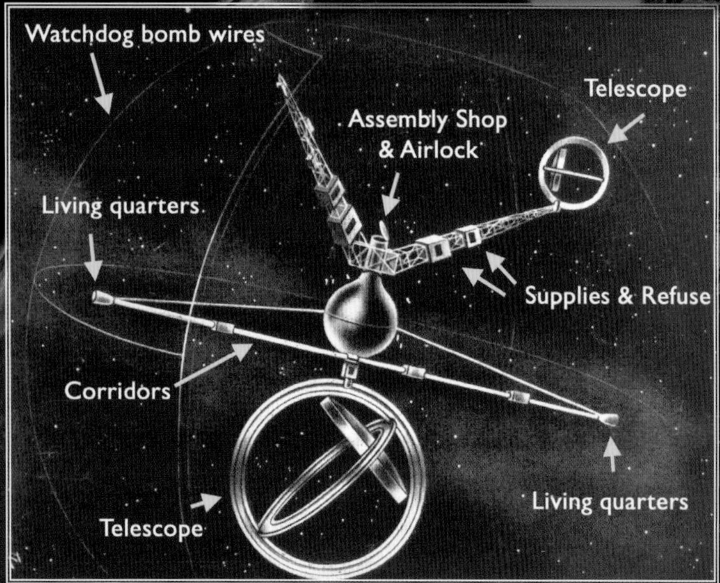

In 1928 Baron Guido von Pirquet, devised a method to overcome the obstacle of reaching orbital velocity by proposing a system of three space stations that could be used for refuelling and could expedite travel to the planets. His *Aussenstations* would orbit at different altitudes; the lowest would be in a 100 minute orbit, and could be used for earth observation, and the highest would be in a 200 minute orbit and could be used for launching an escape vehicle to the moon or planets. The middle station, in a 150 minute orbit, would be used for transferring crew and equipment. Another engineer who read Oberth, was Herman Potočnik, born in Slovenia in 1892. Writing under the pseudonym of Hermann Noordung, Potočnik further refined Oberth's ideas and proposed a space station in his 1929 book *Problem Vožnje Po Vesolju* (The Problem of Space Travel). The *Wohnrad* living quarters spun to create artificial gravity but unlike Gail's unstable arm, the habitat was a balanced toroid. A large mirror was also used to generate steam from solar power and a separate observatory module was connected by cables to the main space station. Potočnik's station would have been placed in a 24-hour orbit which would have only allowed it to observe one half of the Earth's surface. Oberth criticised Potočnik for not realising that the radius of his space station would not be large enough to create viable artificial gravity. In the same year J.D. Bernal, in his book *The World, the Flesh and the Devil,* described a space station built almost entirely from space resources. His description still resonates today for its uncanny accuracy.

Modern renderings of the Potočnik *Wohnrad (Living Wheel)* (centre) and the Bernal sphere (below), the latter was to be constructed from non-terrestrial resources. Oberth had also postulated carving out an asteroid to use as a space station; an idea which dated back at least as far as Scottish minister William Leitch's 1861 book *God's Glory in the Heavens* in which he suggested *"The very globe itself might be tunnelled and split up so that contending parties might have little worlds of their own to live in."*

Guido von Pirquet - "Aussenstation" - 1928

"From a few important figures we can gauge the tremendous significance of the Aussenstation for the purposes of cosmonautics. This is not only of physiological importance, but also quite specifically because it can significantly reduce the amount of fuel required. Thus we find that the situation and the prospects for the realization of space travel with the technical means available today are as follows: The journey from the Aussenstation to the planets is easier to cope with than the trip to the Aussenstation or its construction." - Guido von Pirquet

Herman Potočnik - "Wohnrad" - 1929

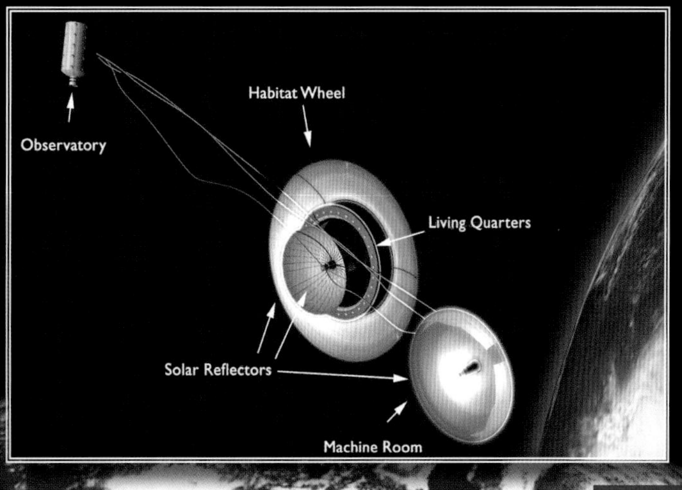

John Desmond Bernal - 1929

"When the technicalities of space navigation are fully understood there will, from desire or necessity, come the idea of building a permanent home for men in space. The ease of actual navigation in space together with the difficulties of taking-off from or landing on planets like the earth with considerable gravitational fields will in the first place lead to the necessity for bases for repairs and supplies not involving these difficulties. A damaged space vessel would, for instance, almost be bound to be destroyed in attempting earth landing. At first, space navigators, and then scientists whose observations would be best conducted outside the earth, and then finally those who for any reason were dissatisfied with earthly conditions would come to inhabit these bases and found permanent spatial colonies. Even with our present primitive knowledge we can plan out such a celestial station in considerable detail. Imagine a spherical shell ten miles or so in diameter, made of the lightest materials and mostly hollow. Owing to the absence of gravitation its construction would not be an engineering feat of any magnitude. The source of the material out of which this would be made would only be in small part drawn from the earth; for the great bulk of the structure would be made out of the substance of one or more smaller asteroids, rings of Saturn or other planetary detritus. The outer shell would be hard, transparent and thin. Its chief function would be to prevent the escape of gases from the interior, to preserve the rigidity of the structure, and to allow the free access of radiant energy. Immediately underneath this epidermis would be the apparatus for utilizing this energy either in the form of a network carrying a chlorophyll-like fluid capable of re-synthesizing carbohydrate bodies from carbon dioxide, or some purely electrical contrivance for the absorption of radiant energy. In the latter case the globe would almost certainly be supplied with vast, tenuous, membranous wings which would increase its area of utilization of sunlight." - John Desmond Bernal

World War II had proven that the dreams of reaching space were now possible. In the Spring of 1945 as the war came to an end Arthur C. Clarke, one of the early members of the British Interplanetary Society (BIS) wrote a paper called *The Space Station* (below) in which he explained how three satellites in a 24-hour orbit could be used as radio relays to provide communications across the globe. At the time Clarke assumed his relays would be manned and the crew would also provide meteorological reports, zero-g research and, in the distant future, traffic control for ships coming and going from Earth. Clarke cited Noordung's book as a reference. While Clarke was busy reforming the BIS in the rubble of post-war England, many of the pre-war German rocket pioneers were now in captivity. One of them was an engineer named Heinz Gartmann who had worked on rockets for BMW during the war. Gartmann was sent to the Wright Patterson USAF base in Ohio where he consulted with the Air Force on rockets. A massive stockpile of paperwork captured at the end of the war had also been sent to Wright Patterson to be analysed. A young USAF engineer named Donald J. Ritchie was given access to this paperwork and quickly used his new-found knowledge to design a space station. Learning from Oberth's objections to a small radius for artificial gravity, Ritchie, in a paper he wrote in May 1946, conceived of a 16,000' diameter *Space Terminal*. His solution was to have two rings, the inner one static, while the outer spun for artificial gravity. In 1947 Ritchie went to work at *Bell Aircraft* and then became president of the *Detroit Rocket Society* (DRS). In January 1948 on the other side of the Atlantic a group of aerospace enthusiasts had reformed a society in Stuttgart called the *Gesellschaft für Weltraumforschung* (Society for Space Research) (GfW). During the war the group's president had been a rocket engineer named Krafft Ehricke. It was now led by Gartmann and a 24 year-old engineer named Heinz Hermann Koelle. When the U.S. government first announced plans for an artificial satellite in late 1948 Ritchie's space station plans hit the front page of the newspapers across the USA. Almost immediately the DRS and GfW made contact. They shared memberships with Koelle, Gartmann, Ehricke and Oberth joining the DRS and Ritchie and his colleagues joining the GfW. To promote their renewed efforts, Koelle edited a magazine named *Weltluftfahrt* (Air World) which would feature short articles on space flight. In the second issue of Koelle's magazine Gartmann reintroduced readers to the *Aussenstation* concept. The DRS journal also featured numerous articles including the first English translation of Oberth's book (accompanied by Ritchie's space station paintings) and the first articles by Ehricke in English.

Arthur C. Clarke - 1945

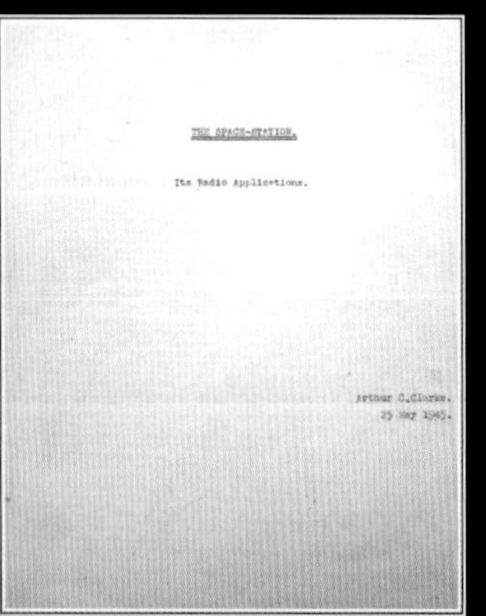

THE SPACE-STATION.

Its Radio Applications.

Arthur C.Clarke.
25 May 1945.

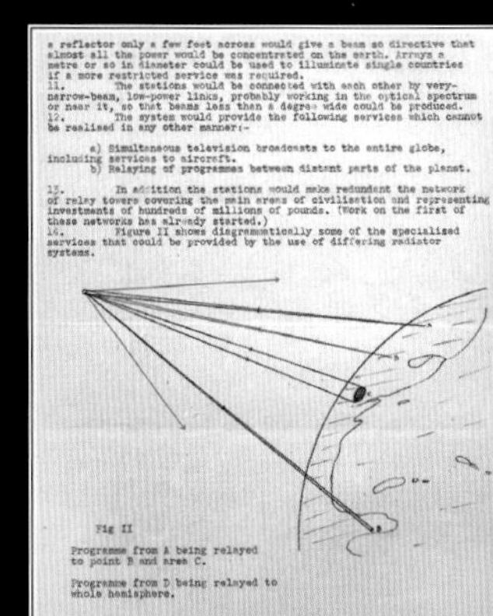

a reflector only a few feet across would give a beam so directive that almost all the power would be concentrated on the earth. Arrays a metre or so in diameter could be used to illuminate single countries if a more restricted service was required.
11. The stations would be connected with each other by very-narrow-beam, low-power links, probably working in the optical spectrum or near it, so that beams less than a degree wide could be produced.
12. The system would provide the following services which cannot be realised in any other manner:-

a) Simultaneous television broadcasts to the entire globe, including services to aircraft.
b) Relaying of programmes between distant parts of the planet.

13. In addition the stations would make redundant the network of relay towers covering the main areas of civilisation and representing investments of hundreds of millions of pounds. (Work on the first of these networks has already started.)
14. Figure II shows diagrammatically some of the specialised services that could be provided by the use of differing radiator systems.

Fig II

Programme from A being relayed to point B and area C.

Programme from D being relayed to whole hemisphere.

Clarke's original *Space Station* paper (left) was subsequently renamed *Extraterrestrial Relays* by the editor of *Wireless World* magazine and published in October 1945.

Donald Ritchie - "Space Terminal" - 1946

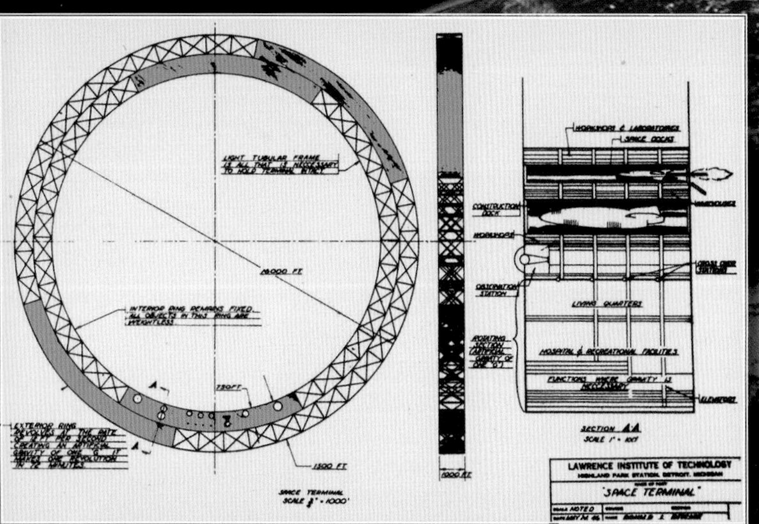

▲ Ritchie's elaborate drawings were completed in May of 1946. His airbrush painting of the massive Space Terminal would appear in newspapers around the world. A transport shuttle can be seen departing from one of the hangars. Critically it can be seen that the station's rotating ring is mostly symmetrical to eliminate oscillations.

Hermann Oberth's most famous pupil was Wernher von Braun. In the 1920s von Braun had been a member of the *Verein für Raumschiffahrt*, the German Society for Space Travel (VfR). As a young man he had also been captivated by a science fiction novel named *Auf Zwei Planeten* (On Two Planets) written in 1898 by Kurd Lasswitz. The novel described a toroidal space station motionless above the Earth's North Pole. At the end of World War II von Braun was captured and brought to America to work on missiles for the U.S. Army. In January 1947, during his "rehabilitation", he was invited to speak to the Rotary Club in El Paso Texas. In this short speech, his first in America, (later published in *The Voice of von Braun*) he described his design for a space station. It would be an inflatable spinning wheel with a solar dynamic power supply, 150' in diameter, to be used as a permanent harbour for ships coming and going to and from the moon and planets. By 1949 von Braun had refined these ideas and included them in a novel which later became known as *Das Marsprojekt* (The Mars Project). He described a manned mission to Mars staged from the Earth-orbiting space station which he called the *Lunetta*. Von Braun offered his novel to many publishing companies in the United States but was turned down. In March of 1950 he was invited to speak at the University of Chicago conference *Biological Aspects of Manned Space Flight*. His paper that day (later published in *Space Medicine*) was called *Multi-Stage Rockets and Artificial Satellites* and it would be the first time that his ideas for a space station drew national attention. Von Braun suggested that such a station would have both peaceful and military uses. The space station he described was larger than the rockets that could launch it and so, just as Gail and others had suggested, it would have to be assembled in orbit. In April 1950 von Braun relocated to the Army Ballistic Missile Agency (ABMA) in Huntsville Alabama.

"...the major part of construction would be performed by men...specially trained and equipped to operate in space. Automated construction alone would not be possible. The individual foldable rubberized segments of the rim would be carried into orbit by rocket; bolted together and then inflated" - Wernher von Braun

Wernher von Braun/ US Army Fort Bliss and ABMA - 1947 to 1950

The *Lunetta* described in von Braun's Mars novel only had ten sides. But by March of 1950 the first version of his space station which he showed to the public (right) had 20 sides and a large circular mirror to direct the sunlight to a central spherical boiler. This was called a *point-focus system;* similar to that proposed by Potočnik.

Harold Ross & Ralph Smith/BIS - 1948

In 1948 BIS engineer Harold Ross and BIS artist Ralph Smith designed a large manned space station which, like Potočnik, also derived its power from the sun and steam turbines but used a "line focus" trough heat capture system. Although Smith and Ross hadn't read Potočnik, they had seen his illustrations. Their station would have been 200' in diameter and gyro-stabilised. Their rotating habitat was more like Potočnik's Wonhrad than Ritchie's Space Terminal. The station was designed with Clarke's orbital relay in mind and included a radio mast.

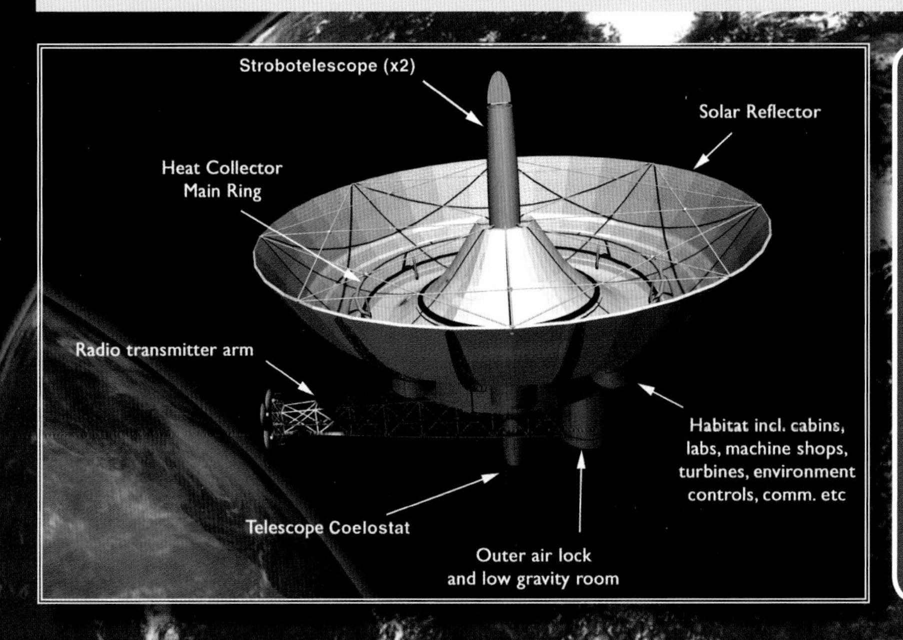

"The station consists of three principal parts; the bowl, the bun and the arm. The bowl is a 200' mirror used to collect and concentrate the sun's rays. The bun behind the mirror is the living quarters, etc. The whole of this structure is made to rotate to provide a pseudo-gravitational effect. The arm behind the mirror does not rotate and serves the dual purpose of no-gravity laboratory and means of entrance and exit to the space station." - Harold Ross

The term *Aussenstation* had first been coined by Pirquet and for a time it became the most popular phrase in Germany to describe an orbiting spacecraft. Although Pirquet had expressed his concern that a spinning space station would induce motion sickness in the crew, many scientists believed that some form of artificial gravity would be necessary. Some doctors thought the first astronauts might not be able to eat without choking but others pointed out that you could swallow food lying on your back or even upside down, therefore gravity was unnecessary. Gartmann's 1949 article in *Weltluftfahrt* illustrated a manned station with a rotating arm that was almost identical to that proposed by Otto Willi Gail in the 1920s. Gail, who was now a respected expert on astronautics, was enlisted that same year to write a screenplay for a documentary film which was to have been titled *Raketen im Weltraum* (Rocket into Space). An animation was created for this film showing that Gail had also listened to Oberth's concerns and his rotating arm was now over 500m long. Gail favoured a disc-shaped space station and his connecting arm was to be pneumatically inflated. Gail's cabin was called a *Wohnbirne* (residential pear) due to its unusual shape. Gartmann preferred a spherical station core (image below). The sphere was considered the optimal shape not only due to its inherent strength as a pressure vessel but because it was the most efficient use of the limited lifting power that was likely to be available, and the ratio of volume and surface is proportional to the mass ratio, a point Ehricke reiterated in a paper in 1950. This design for a spherical space station with one long spinning arm became popular in Germany and appeared on the covers of Gartmann's astronautics books for several years. In an apparent compromise the disc and sphere were later blended together; but with only one arm the station would have been subject to inconvenient oscillations. By 1949 the GfW's ideas were being illustrated by artist Klaus Bürgle who seems to have been the first to illustrate a one-man space taxi using robotic grapples to aid in construction. It is unclear whether Gartmann, Koelle or Bürgle were the first to suggest this. Koelle would make Wernher von Braun an honorary member of the GfW in 1949 and by April 1950 word of von Braun's lecture in Chicago and his grand plans for a space station had already reached Germany.

Heinz Gartmann/Otto Willi Gail - "Aussenstation/Wohnbirne" - 1949

Heinz Gartmann, was a member of the post-war German *Society for Space Research (GfW)* and later a co-founder of the *International Astronautical Federation (IAF).* After returning to Germany at the end of the war he joined forces with Hermann Koelle and reintroduced the Aussenstation to the German public.

After World War II a wall of secrecy was placed around the Soviet Union and very little about space research in Tsiolkovsky's homeland was revealed to the West. However, for more than 20 years a flourishing community of space writers and enthusiasts had done their very best to excite the Russian people about spaceflight. One of these was a young engineer from Poland named Ary Sternfeld. In 1949 Sternfeld's book entitled *Полёт в мировое пространство (Flight into Planetary Space)* outlined how a rotating spacecraft could be disorientating, but this could be overcome by increasing the radius of spin. Later in his book Sternfeld showed a large toroidal space station very similar to the one designed by Ritchie three years earlier. Sternfeld and most of his predecessors believed that it would be necessary to build an earth-orbiting space station before moving further into deep space. Sternfeld was familiar with the notion of rotational artificial gravity, and the problems associated with it. He wrote many books on astronautics.

Ary Sternfeld - "Orbital Hub" - 1949

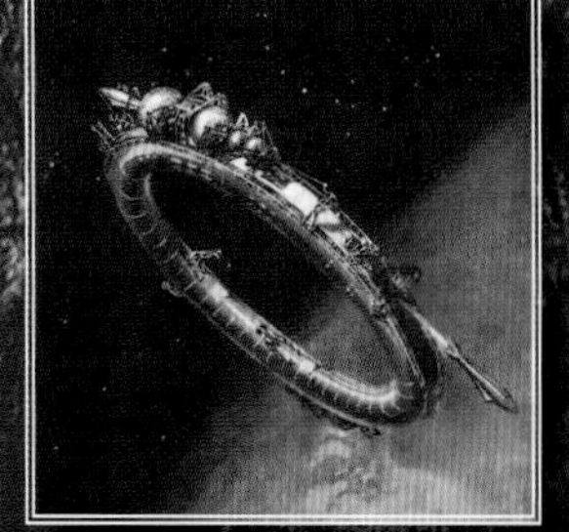

A shuttle departs from Sternfeld's rotating station (above) while his large rotating orbital hub would have been used to stage interplanetary expeditions (right).

"At first glance it is easiest to create an artificial "gravitational field", by the continuous operation of the engine, at least at a reduced power. However, such a method would require an exorbitant fuel consumption. Meanwhile, there is an extremely simple way to create artificial gravity, namely, the rotation of a spacecraft. At the end of the last century (Hermann) Ganswindt published a spacecraft design. In this project, the rotation of the passenger cabin provides the creation of an artificial gravity. The drawback of the (Ganswindt) project is that due to the small size of the cabin it would have to be rotated very quickly, and this could entail all kinds of physiological disorders. However, even earlier than Ganswindt, K.E. Tsiolkovsky (in 1895) expressed the idea of a more perfect method of creating artificial gravity. According to his idea, developed subsequently by other researchers, the apparatus must consist of two interconnected parts, which take off as one unit; at the right moment, these parts are separated from each other, remaining, however, connected with each other by cables, and then with the help of small rocket engines are driven in a circular motion near their common center of gravity. It is obvious that after the system reaches the required rotation speed in an airless medium, the rotation will continue by inertia without the participation of the engines." - Ary Sternfeld

By the spring of 1950 Wernher von Braun had all but given up trying to find an English language publisher for the story which featured his toroidal *Lunetta* space station. Hermann Koelle wrote to von Braun to suggest that he should try to publish a German edition. So von Braun sent the manuscript to Koelle in August of 1950. Several months earlier Koelle had published a new method for calculating launch loads which gave the exact specifications for getting to and from a space station. He sent this paper to von Braun, followed by an eight-page critique of the manuscript in which he explained how his calculations for what he called the "optimum rocket" could accomplish many of von Braun's goals. Von Braun encouraged Koelle to apply his methods to the ideas in *The Mars Project* so Koelle teamed up with Helmut Hoeppner, an engineer who had worked at the Peenemünde rocket center, and together they began to formulate exactly what would be required to launch and build a large toroidal Aussenstation. At the 1st International Astronautical Congress (IAC), in Paris in September 1950, Koelle called for a *Commission on Space Stations* which would present its findings at the 2nd IAC in London. This was agreed to, and a year later, on September 6th 1951, the proceedings for what would be the first space station conference opened in Westminster. Pirquet delivered a paper with no less than 14 reasons for why zero-g was more desirable aboard a space station. IAF chairman Eugen Sänger, who during the war had designed the first winged space glider, summarized the multitude of new disciplines that the international partners would need to master, in order to build a space station. Arthur Clarke co-chaired the conference and Ralph Smith spoke about the problems of rendezvous. Von Braun, unable to attend, sent a proxy to read a summary of his Mars Project which now addressed some of Koelle's concerns. But it was Koelle and Hoeppner's presentations that cut to the heart of the problem. To illustrate their plans for building a toroidal station, like that currently in favour with von Braun, Koelle had built display models of the "Optimum Rocket" and Aussenstation (which his wife declared to British customs as a somewhat peculiar new kind of hat.) Their calculations proved that it would require a lot of expensive launches, and so to optimize this procedure Koelle designed a space station built of standardized modules brought to orbit by Hoeppner's space shuttle - which had been specifically designed for the task. Their shuttle could then either be re-cycled back on Earth or integrated into the final station. They combined robot aided construction, modular design, large winged boosters and Sänger's winged re-entry concepts into a holistic approach to space station assembly. Koelle and Hoeppner had invented a bespoke spacecraft, with the singular purpose of delivering space station modules to space in the most economical method possible. In this respect their system anticipated many of the space station construction proposals of the next 20 years.

Heinz Hermann Koelle & Helmut Hoeppner - "Aussenstation" & "Optimale Raketen" - 1951

"The construction of the satellite is very closely connected with the construction of the optimum cargo rocket, hence the designs cannot be separated from each other." - Hermann Koelle

The nickname for Hermann Koelle's (far left) station was *The Rosary*, because it resembled a string of beads. Co-author Helmut Hoeppner (near left) would later become Hermann Oberth's assistant in Huntsville Alabama before becoming chief aerospace scientist for Chrysler in Detroit. The Koelle Aussenstation (lower left) was a 200' diameter ring of 36 spheres, each 16 feet in diameter, connected to a hub by eight supporting condenser tubes, four of which had elevator shafts. The hub would be constructed from two of the third stages of the cargo vehicle connected via two more of the spherical modules, bringing the total number of the spheres to 38. This station would accommodate up to 65 people and would weigh 150 metric tons. Koelle and Hoeppner realised that the vehicle used to carry the components to orbit would have to be designed in tandem with the space station. They named their shuttle the *Optimale Raketen* (Optimum Rocket 51 or SR51) (near left). The assembly process was to be assisted by using a one-man *Raumtrecker* (Space Tractor) which Koelle adapted from an idea by von Braun. Some of the underlying concepts for their design also came from studying Smith and Ross' work and the Sänger *Silverbird* space glider. Koelle anticipated recovering at least 50% of each launch vehicle for reuse.

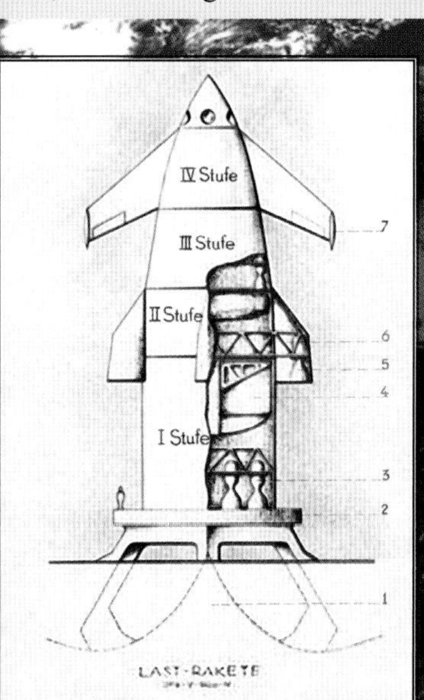

Illustrations from 1951 show Koelle and Hoeppner's plans for building their modular space station. The Satellite Rocket is shown with the third stage being returned to earth by parachute. The one-man *Raumtrecker* removes the 16' sphere from the shuttle and delivers it for installation by spacewalking construction workers.

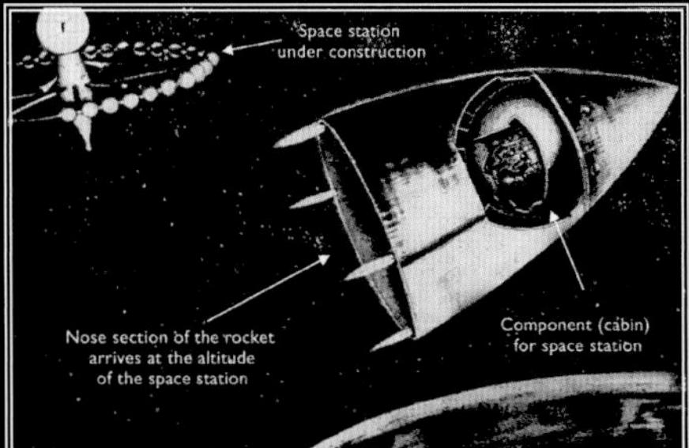

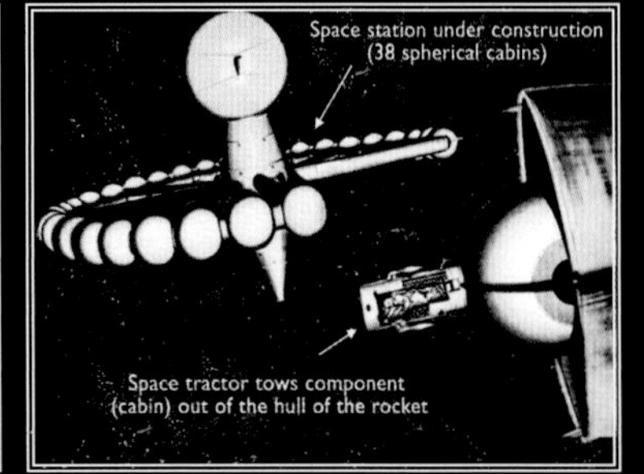

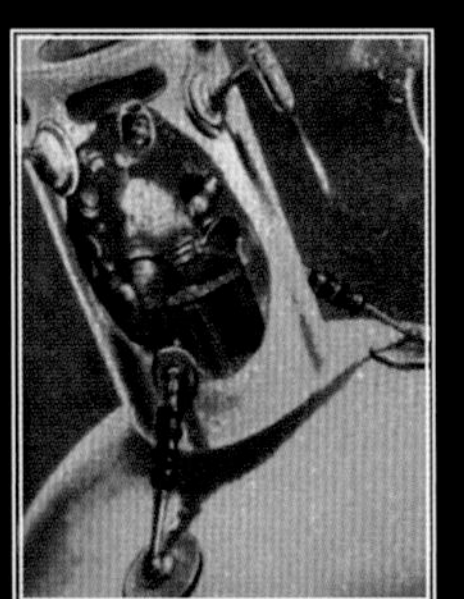

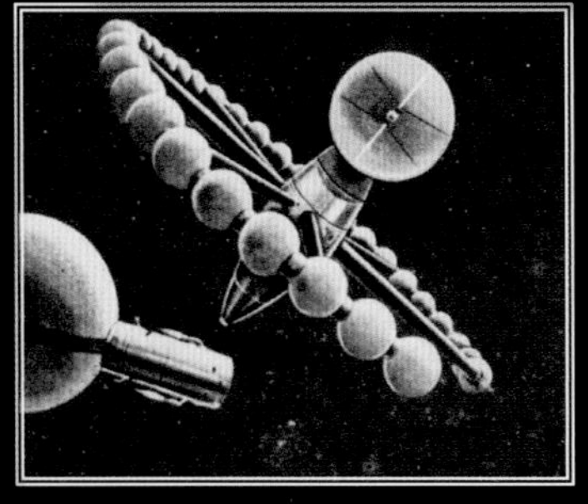

Wernher von Braun/ ABMA - 1952

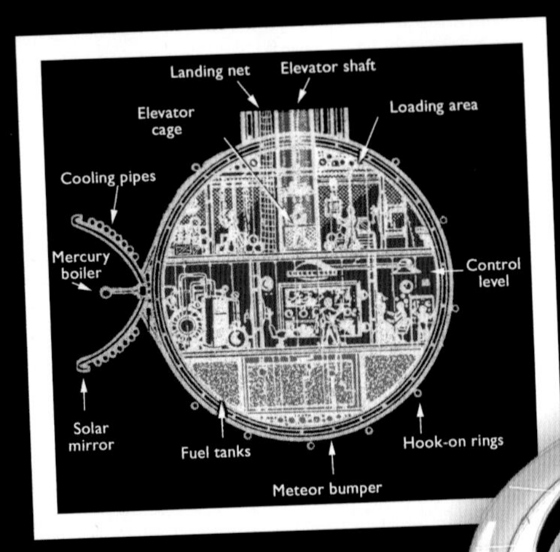

By 1952 Wernher von Braun had adopted the Ross and Smith toroidal trough-shaped mirror and boiler, (a line-focus system), to harness the sun's power. His station was also made much larger than his *Lunetta* to allow for artificial gravity without the problems associated with the Coriolis effect. The inflatable segments were now replaced with a multi-decked hard fuselage (above). In February 1952 his book *Das Marsprojekt* was finally published. First in Germany, with help from Gartmann and Koelle, and later in Chicago. However, only the technical appendices were printed and the full unaltered novel would not be published until 2006.

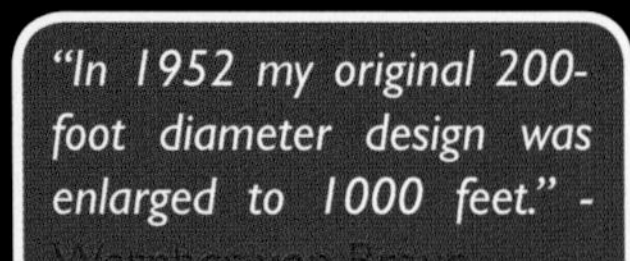

"In 1952 my original 200-foot diameter design was enlarged to 1000 feet." - Wernher von Braun

Helmut Hoeppner - "Astropol" and "Satellite Rocket" - 1952

In 1952 Hoeppner updated the plans for his and Koelle's reusable space shuttle which he now called *Satellite Rocket 52*. It still consisted of three booster stages and a winged orbiter. Hoeppner suggested that if four of the SR52s were to meet in orbit, the third stages of all four could be docked together to create a space station. He gave it the same name Otto Gail had used, *The Astropol*. The station (right) was to be used as a habitat and as a workshop for engineers to build a lunar exploration vehicle. Although *The Astropol* used expended stages like the *Aussenstation* it dispensed with the toroidal ring. Shortly after proposing this design Hoeppner and Koelle both accepted employment in America at the Army Ballistic Missile Agency in Huntsville Alabama and went to work for Wernher von Braun.

After a period also working with von Braun for the Army, Krafft Ehricke left the ABMA to take a position at the Bell Aerospace Preliminary Design Department in Buffalo, New York. In February 1954 he prepared his paper *Analysis of Orbital Systems* for the International Astronautical Congress in Innsbruck Austria. Ehricke had concluded that a spinning toroidal station would require constant attention to keep it balanced as it rotated. His counter proposal included a design for a space station which could house a crew of four and would, like Hoeppner's *Astropol*, be built mainly from large tanks left over from the stages of his proposed launch system. His paper also included plans for a flyback shuttle, similar to Koelle and Hoeppner's, which could be used to build the station.

Ehricke's station would be 400' long with most of the mass accumulated at the center of a bar-bell configuration. He called it a *Manned Observational Satellite*. The station would have weighed in at 500,000 lbs with a volume of 20,000 cu ft. In late 1954 Ehricke left Bell for a position at Convair. While at Convair he presented his design to the US Government using a large demonstration model as a talking point. ▲

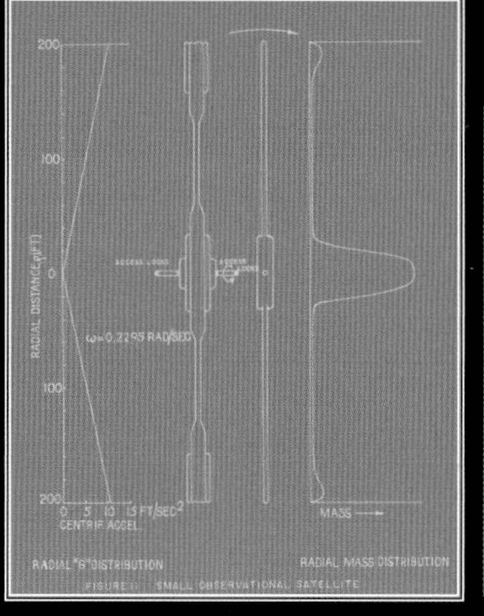

Krafft Ehricke/ Bell Aerospace - "Observation Satellite" - 1954

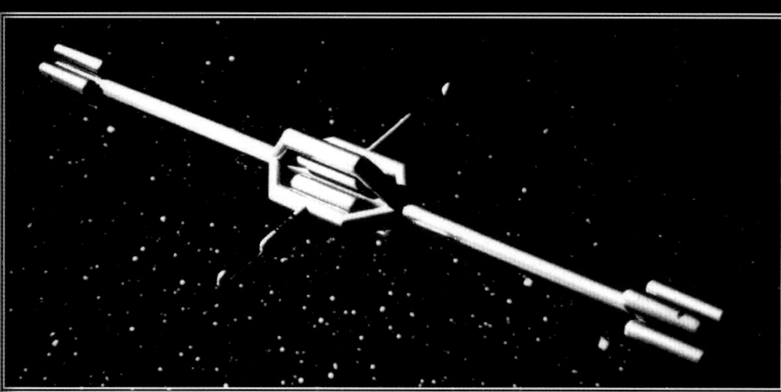

"In order to produce a functional satellite which combines the advantage of apparent gravity with smaller size and a less precarious balancing situation, it is proposed to concentrate most of the mass in a center body from which two extensions, oppositely directed, lead to the crew space. Most of the satellite's mass is concentrated in the center which contains all stowage, reserve parts, emergency equipment, radioactive power supply with shielding material, earth scanning equipment, purifiers, attitude control and, of course the entrance tubes on the hub with the double air locks. In the peripheral sections are located the living and working quarters (for 2 persons on each side), control motors and actuators." - Krafft Ehricke

In 1954, in the Soviet magazine *Knowledge is Power* B. Lyapunov described two adaptations of Tsiolkovsky's original design. Illustrated by A. Orlov and I. Fridman they showed a hybrid of Tsiolkovsky's orbiting conical greenhouse and Noordung's toroidal solar-powered habitat. (*Key:- 1. Observatory and lab; 2. Greenhouse; 3. Mirror for solar plant; 4 & 5 Auxiliary rooms; 6. Radar; 7. Elevator; 8. Workshop and power plant; 9. Hangar; 10. Radio antenna; 11. Observatory; 12. Storerooms; 13. Turbogenerator; 14. Solar boiler*) ▶

Lyubanov/Tsiolkovsky station - 1954

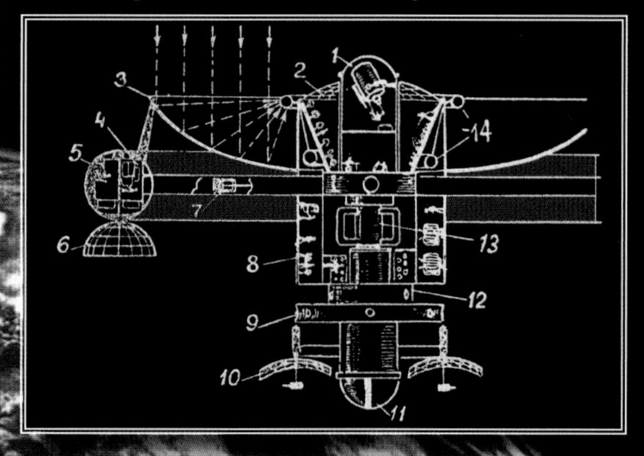

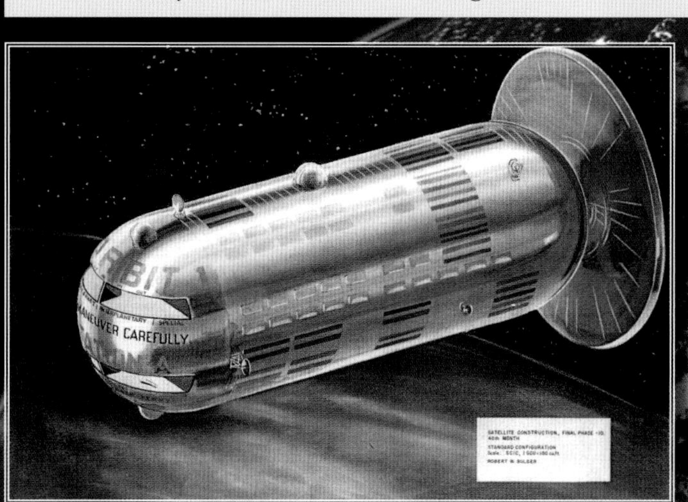

Darrell C. Romick/Goodyear - "Meteor Station" - 1955

"(It will)...end up as a small city in the sky, rivaling any on earth of comparable population, and dedicated to fulfillment of a special mission - serving as an outpost on a new frontier." - Darrell C. Romick

▲ At the IAC in Rome in 1956 Goodyear's Darrell C. Romick took the concept of building a space station using the spent-stages from launch vehicles to its extreme limit. His METEOR (*Manned Earth Satellite Terminal Evolving from Earth-to-Orbit Ferry Rockets*) station was to be built from dozens of used stages. It was to be 3000' feet long and 1000'-1500' in diameter. The wide section in the image (above left) would rotate and provide artificial gravity for the living quarters, gymnasiums, theatres and churches on board. As ambitious as this appeared it was just one part of a more elaborate plan by Romick to colonise the moon.

With the control of nuclear energy for peaceful purposes, engineers at the Army Ballistic Missile Agency in Alabama envisioned a new kind of space station. By using an onboard fission reactor it became possible to dispense with the large solar mirrors needed to supply power to the earlier designs. A more advanced kind of propulsion also became possible, the electric ion-engine. Wernher von Braun, Ernst Stuhlinger and Hermann Koelle at the ABMA began to see the possibility of building enormous space stations that, in addition to orbiting the Earth, could also be used in orbit around Mars or on flyby missions to Venus. In his speech in Chicago von Braun had outlined his idea for orbital construction which he said would be done by astronauts in spacesuits, perhaps assisted by small space taxis he called *Bienen* (Bees). Building a space station in orbit required the invention of a whole series of procedures, such as rendezvous and docking. A nuclear powered space station also required new safeguards to protect the crew from radiation. In 1955, in the Disney movie *Man and the Moon*, von Braun finally introduced his one-man spacecraft armed with robotic grapplers. It was similar to Koelle's *Raumtrecker* but von Braun called his "the bottle suit." In 1957 all of these ideas came together when the ABMA team demonstrated an entire architecture for putting humans on Mars in the Disney film *Mars & Beyond*. A nuclear powered toroidal space station was to be used as a base for constructing a fleet of nuclear powered ion propelled Mars vessels which would also use spin-up artificial gravity for the long voyage. Years later many of the requirements of a planetary vessel and an orbiting space station were found to be much alike.

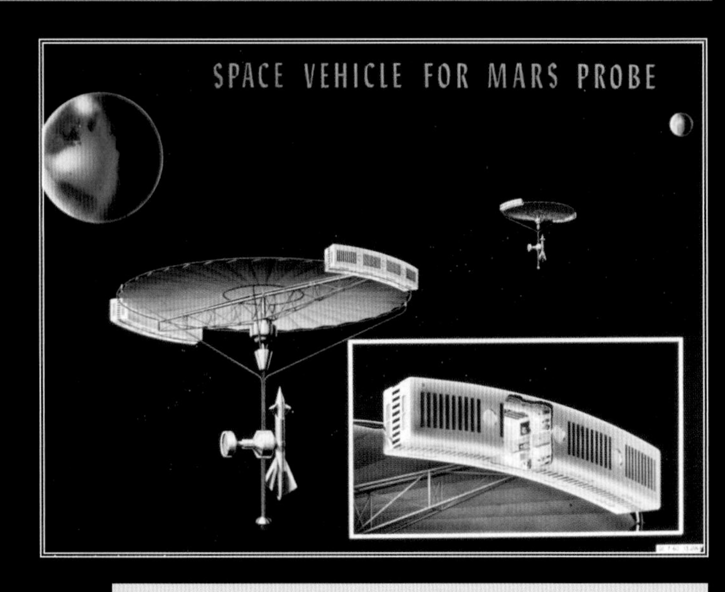

▲ Free-floating astronauts doing orbital construction (above) was an idea that was briefly replaced by one-man spacecraft with robotic grapplers (below). The original concept was advanced by Koelle and Hoeppner before von Braun publicized it. With the advent of on-board nuclear reactors it also offered a way of protecting the crew from harmful radiation. In the image below a bottle suit is used to service a satellite. The reactor is the red striped cylinder. ▼

▲ Ernst Stuhlinger's vehicle (above) would have been left in orbit around Mars. Two bottle suits would have been aboard to prepare the lander and to service the reactor core. In the image below "bottle-suit" servicing vehicles are seen providing maintenance to a Mars orbiting space station. The image shows the nuclear core radiating heat into space while the maintenance vehicle returns to the main orbiting habitat.

The idea of small one-man repair craft would be central to space station planning for years. ▼

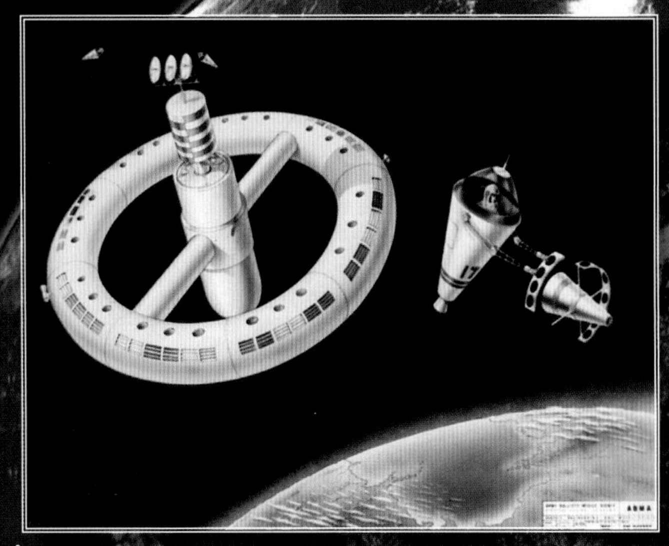

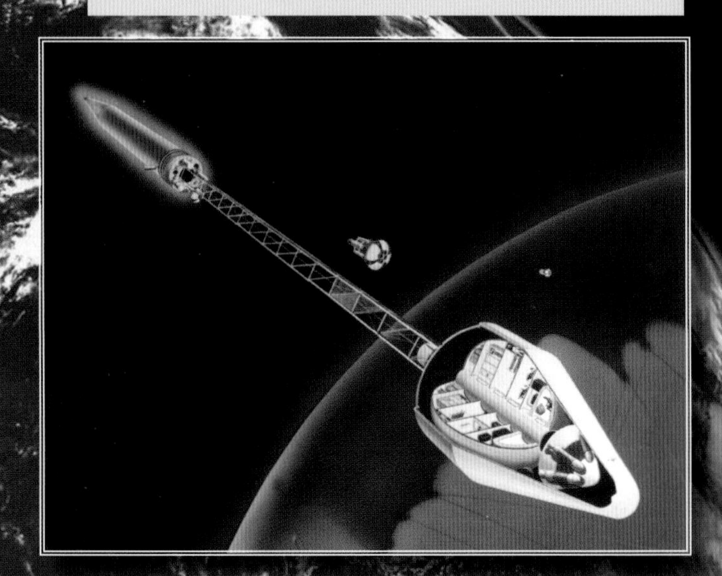

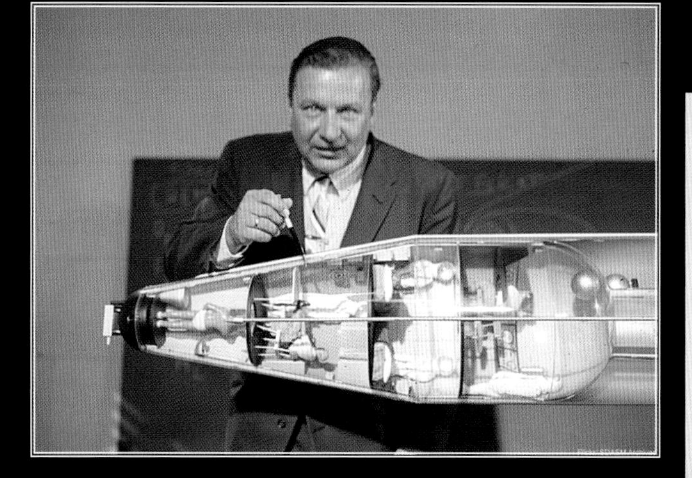

In early 1958 Krafft Ehricke, now at Convair Astronautics in San Diego, presented his study for what he called the *Experimental Manned Space Station* (*EMSS*). In the image at left Ehricke can be seen in April 1958 standing with a model of his "outpost" Model I. It would have used an upgraded ICBM-class booster to place it into low earth orbit. A single Atlas would be launched and the empty tank later equipped with the necessary equipment. The volume of habitation space was to be made of a rubber/nylon and would have been inflated in orbit. A crew would then install the decks and furnishings later. Ehricke placed the crew quarters at one end so that the station could still be rotated if necessary. The crew return gliders would attach at the opposite end to shift the mass load.

"The principal purpose of Outposts are: 1) To serve as a test bed for selection and training of crews in the actual environment of space, who will later operate space vehicles on long-term missions. 2) To serve as a test bed for development of gear and equipment to be used in subsequent manned and unmanned space vehicles. 3) To develop techniques of earth-orbit operations such as rendezvous maneuvers, rotation of crews etc. 4) To serve as a simulator for conditioning space crews to special missions. This includes training for emergencies, living at low gravity levels, etc." - Krafft Ehricke

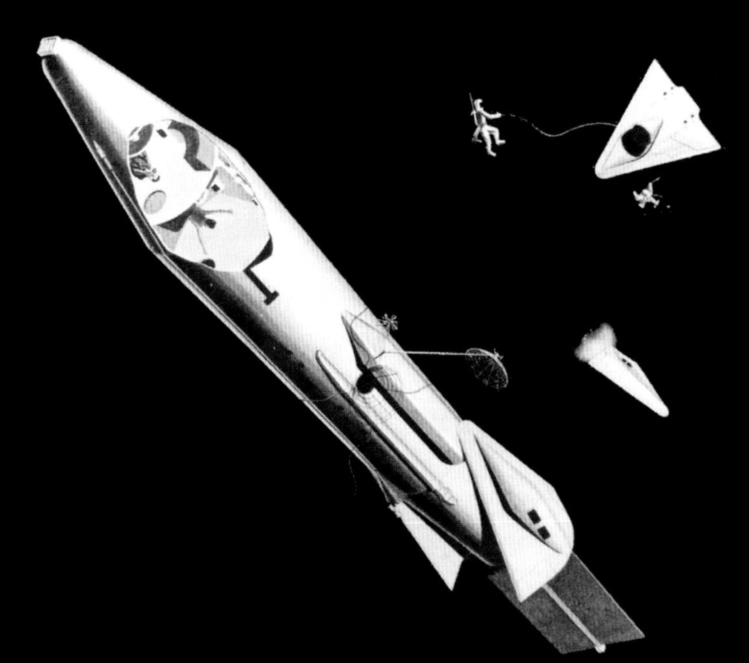

At right is an artist's rendering showing Ehricke's Atlas-launched outpost Model I station accompanied by emergency gliders. A nuclear reactor at lower right would have increased the shielding requirements and weight. In his Model II (below) Ehricke added telescoping booms to increase the distance from the reactor while simultaneously increasing the radius of rotation.

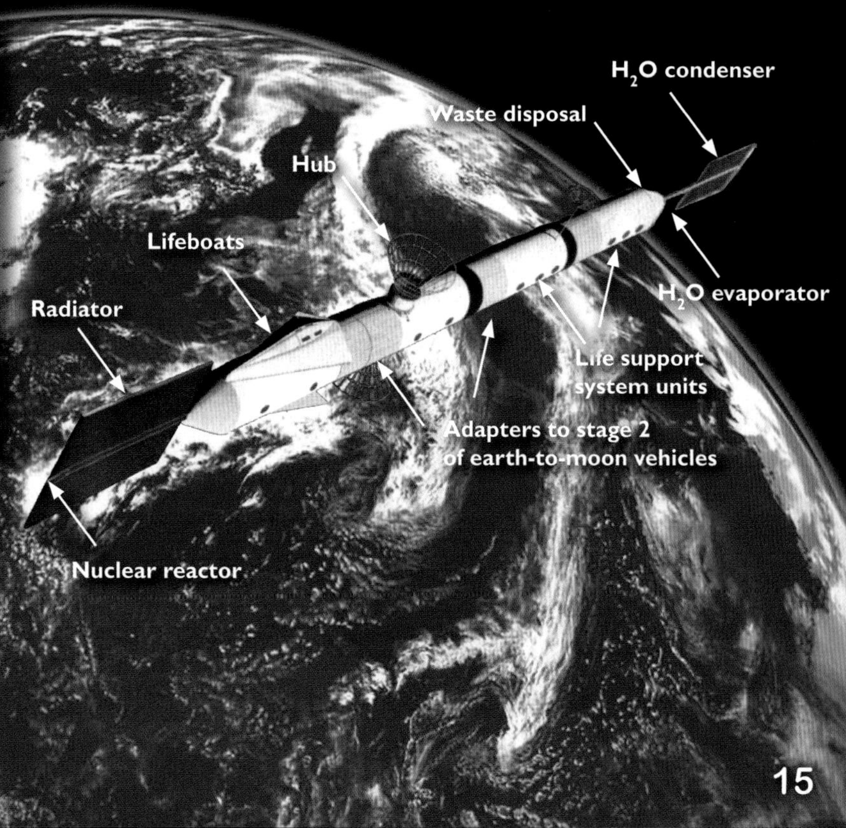

Ehricke introduced his Model III in 1960. Four Model II's would have been combined to create Model III which was a four-compartment station to be assembled using Centaur and Saturn launch vehicles. Model III was to train crews for deep space missions. It would have weighed 50,000 lbs, was ten feet in diameter and 140 feet long not including the cooling radiators. The station also had a "storm shelter" which was heavily shielded against solar flares.

ABMA - "Terra" - 1958

By July of 1958 the ABMA team proposed a series of space stations each one bigger than the last, to be launched incrementally. They called them *Terra.* Von Braun had used the name *Supraterra* in his earlier writings and had used the *Terra* name for the Disney movies. By this time the concept of using spent-stages was well entrenched. The *Terra I* and *II* were variations of Ehricke's *EMSS* and his *Observation Satellite*. The *Terra III* was von Braun's toroid, capable of housing 50 people, and reduced in size back to only a 240' diameter.

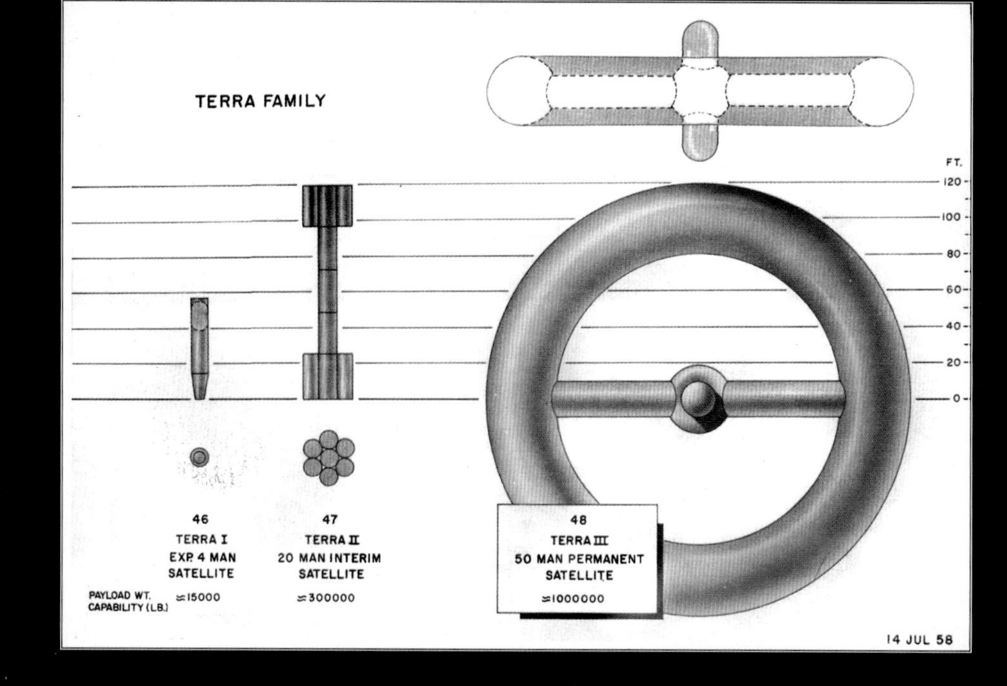

Saunders Kramer & Richard Byers/ Lockheed - 1958

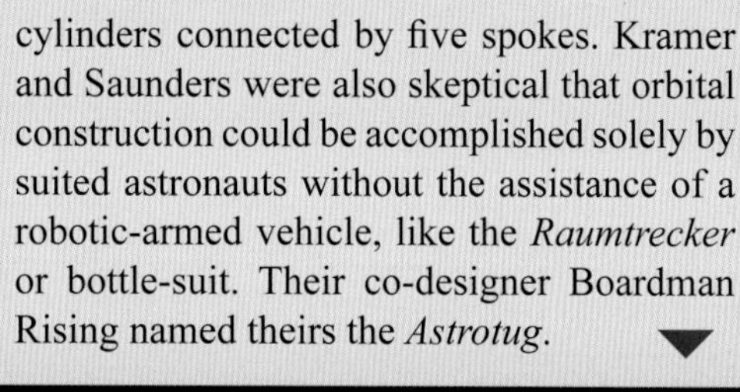

Late in 1958 Lockheed recruited a team of 30 engineers under the guidance of Saunders Kramer and Richard Byers to develop the methods for building a space station. On December 29th of that year their concept for a toroidal flywheel (below) was presented at the American Astronautical Society conference. Their design seemed to revert back to von Braun's *Lunetta*. It was made of 15 cylinders connected by five spokes. Kramer and Saunders were also skeptical that orbital construction could be accomplished solely by suited astronauts without the assistance of a robotic-armed vehicle, like the *Raumtrecker* or bottle-suit. Their co-designer Boardman Rising named theirs the *Astrotug*.

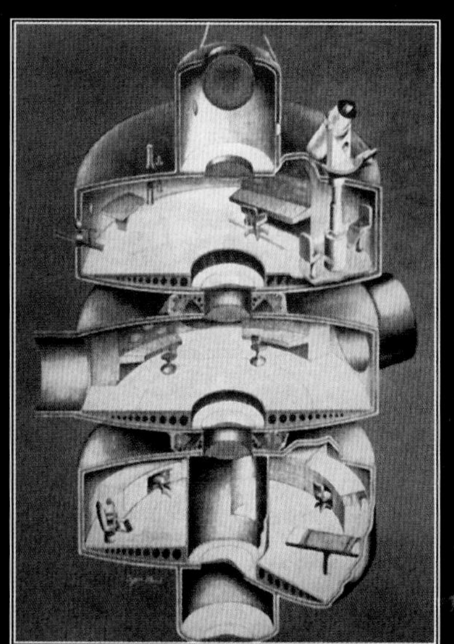

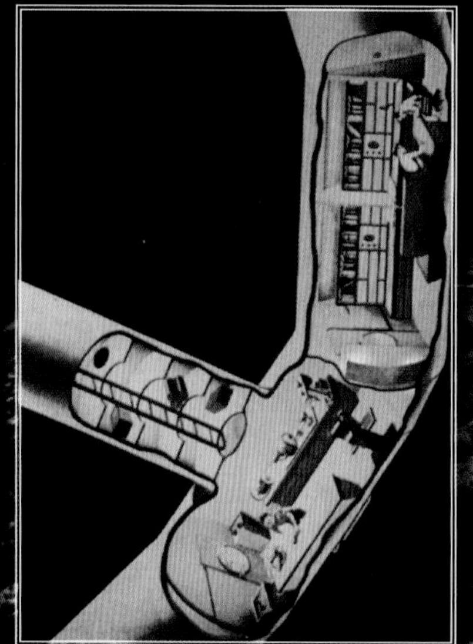

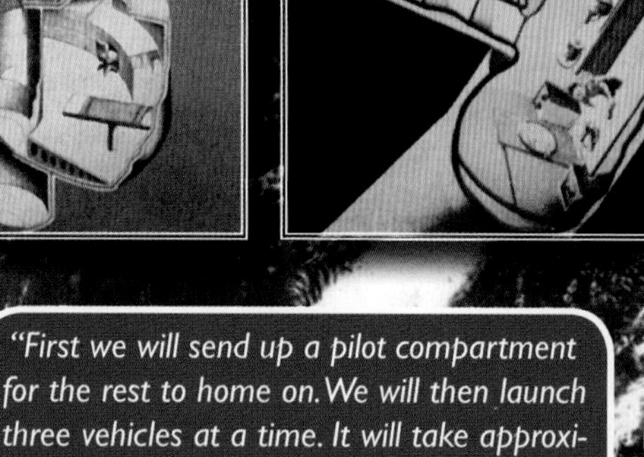

"First we will send up a pilot compartment for the rest to home on. We will then launch three vehicles at a time. It will take approximately 35 launchings and it will take about 30 days." - Saunders Kramer (left)

Key: 1. Zero-g labs; 2. Slipring (rack and pinion); 3. Command center; 4. Food, O_2 and H_2O; 5. Spare parts; 6. Thrusters; 7. Chemical supplies.

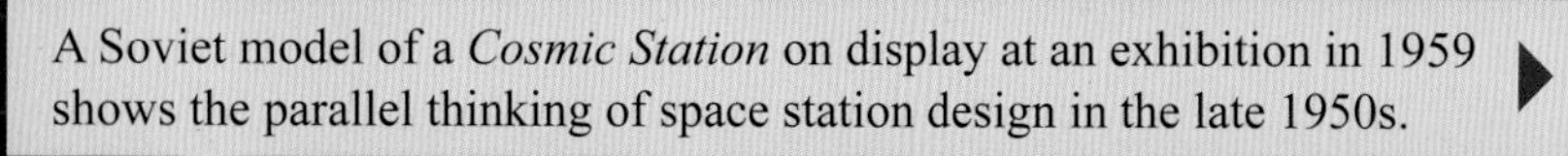

A Soviet model of a *Cosmic Station* on display at an exhibition in 1959 shows the parallel thinking of space station design in the late 1950s.

At the end of 1958 von Braun, Koelle and Stuhlinger delivered their suggestions for future space exploration to the newly formed National Aeronautics and Space Administration. In their presentation they informed the NASA management of their plans for a new large super booster which would ultimately become known as *Saturn*. Despite Ehricke's ambitious schemes for an ICBM launched space station even he had realised that the proposed new heavy-lifter would make his outposts easier to build. Up to this point the large space station, like the *Terra III*, had only one real purpose, to serve as a staging post for missions to the moon and beyond. That meant that it would be used as an assembly point for building lunar landers and interplanetary vehicles. The new heavy lifter presented the opportunity to build even larger spent-stage space stations, but could also be used to create a refuelling depot to prepare for the assault on the moon. The best documented early example of this kind of staging post is the ABMA's study *Project Horizon*. By 1959 the US Army was reluctantly being phased out of the space business and so a "requirement" for a military presence on the moon was proposed by a team of Army engineers in Huntsville, Alabama. The task fell to Koelle who was now in charge of advanced projects.

"The non rotating earth-assembled manned laboratory is the next logical and decisive step in the development of orbital operations. It will be technically feasible as soon as carrier vehicles become available which can lift loads in the order of 40,000 lbs. into a 96-min. orbit." - Heinz Hermann Koelle

Heinz Hermann Koelle & Frank Williams/ABMA - "Project Horizon" - 1959

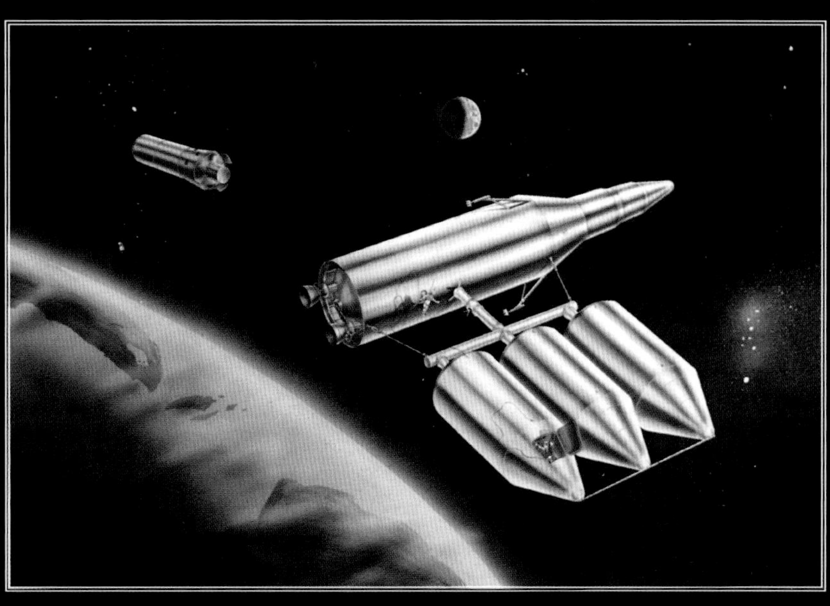

Project Horizon was proposed by Frank Williams and supervised by Koelle, with encouragement from von Braun. It would have used dozens of heavy lift launch vehicles to create a 12 man permanent base on the moon. To accomplish this, at that time, would have required a low earth orbiting space station and fuelling depot (right). The image below shows the proposed structure of a *Project Horizon* toroidal station made from spent stages.

"Wernher wanted me to get help from somebody named Hermann Koelle. Wonderful guy. He was my boss. I briefed General Trudeau and Colonel Nels Parsons who was my interface with the Army and the Pentagon. So we set up a meeting and we met with General Taylor and told him our story and he said, 'OK I want a study done. A report about going to the moon and returning. How long will it take to do that?' I told him that with carte blanche from the Dept of Army to get the Signal Corps, Engineers, Ordnance everybody together we could do it in a year. He asked if we could do it in 30 days. I said, 'Sir we couldn't get the people all together in 30 days.' He said, '90 days.' I said, 'Six months.' He said, '90 days is a long time you ought to be able to do that.' So I said, 'Yessir we'll do it.' and ninety days later, to the day, we presented Project Horizon." - Frank Williams

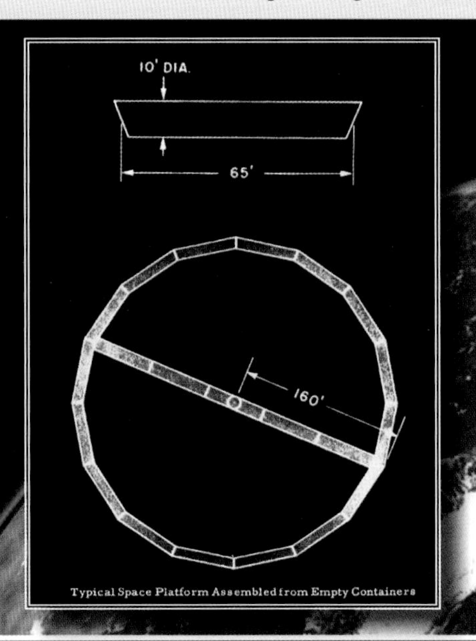

Typical Space Platform Assembled from Empty Containers

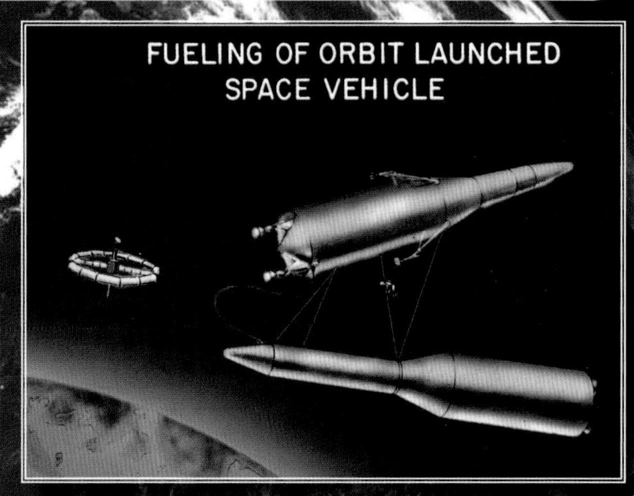

"Through the use of the tankage of the Saturn I high-energy last stage such additional space can be made available with minimum orbital assembly and with no additional vehicle requirements. Although special tanks for conversion to habitable quarters will weigh slightly more than the standard tankage, the useful weight in orbit will be greater. Tankage conversion for this dual purpose role will be necessary only for the initial manned orbital fueling missions. When rotation of the fueling crew takes place, the arriving crew and capsule will utilize the same tank converted to quarters." - Project Horizon Report

The concept of using an orbiting space station for staging expeditions to the moon, and beyond, goes back at least as far as Hermann Oberth in the 1920s. Until rocket engines could be refined and perfected, engineers anticipated the need to first launch a multitude of boosters carrying fuel and other supplies to low earth orbit. Crews would then use these supply depots to enable their missions into deep space. The value of the space station as an observation post and unique laboratory was also recognised immediately by the earliest advocates of space exploration. Certainly there were other uses for a space station, but many designs, beginning as early as 1959, were solely to be used to study the earth; or to look outwards; or simply to conduct experiments in the unique environment of space. One of the first companies to realise that you might be able to create custom designed hardware to use for both a space station and a deep space vehicle was the *Chance-Vought Company*. In early 1959 an entire Vought Astronautics Division was formed with John Russell Clark as its general manager. Clark was keenly aware that all the chat around the aerospace community was about spaceflight and, more specifically, going to the moon. Working under Clark in the role of Chief of Advanced Projects was a Trinidadian-Chinese immigrant named Conrad Lau, who had been a member of the NACA's Subcommittee on Aerodynamic Stability and Control. Lau's team had already been working on the multi-stage *Scout* missile for NASA and the *X-20 Dyna-Soar* military space shuttle for the USAF. Following the directive from the upper management Lau began working on a plan for going to the moon. His final summary was completed in late 1959 and sent to Abe Silverstein at NASA by John Russell Clark, on January 12th 1960. The 190 page report included a complete architecture for a space station and lunar landing system using a series of partly interchangeable components. Lau's team envisioned two versions of their manned modular space system. One was an earth-orbiting laboratory that used two modules; a command/re-entry module and a habitat which they called *Satellab*. The second version used a three module system, a command/re-entry module, a lunar mission module,

Chance-Vought - "Satellab" - 1959

The *Satellab* was a low earth orbit laboratory with a space telescope, which would have weighed 10,090 lbs with an entry vehicle weighing 5,750 lbs. Satellab took advantage of the newly proven technology of photovoltaic power with its conspicuous fan of solar panels.

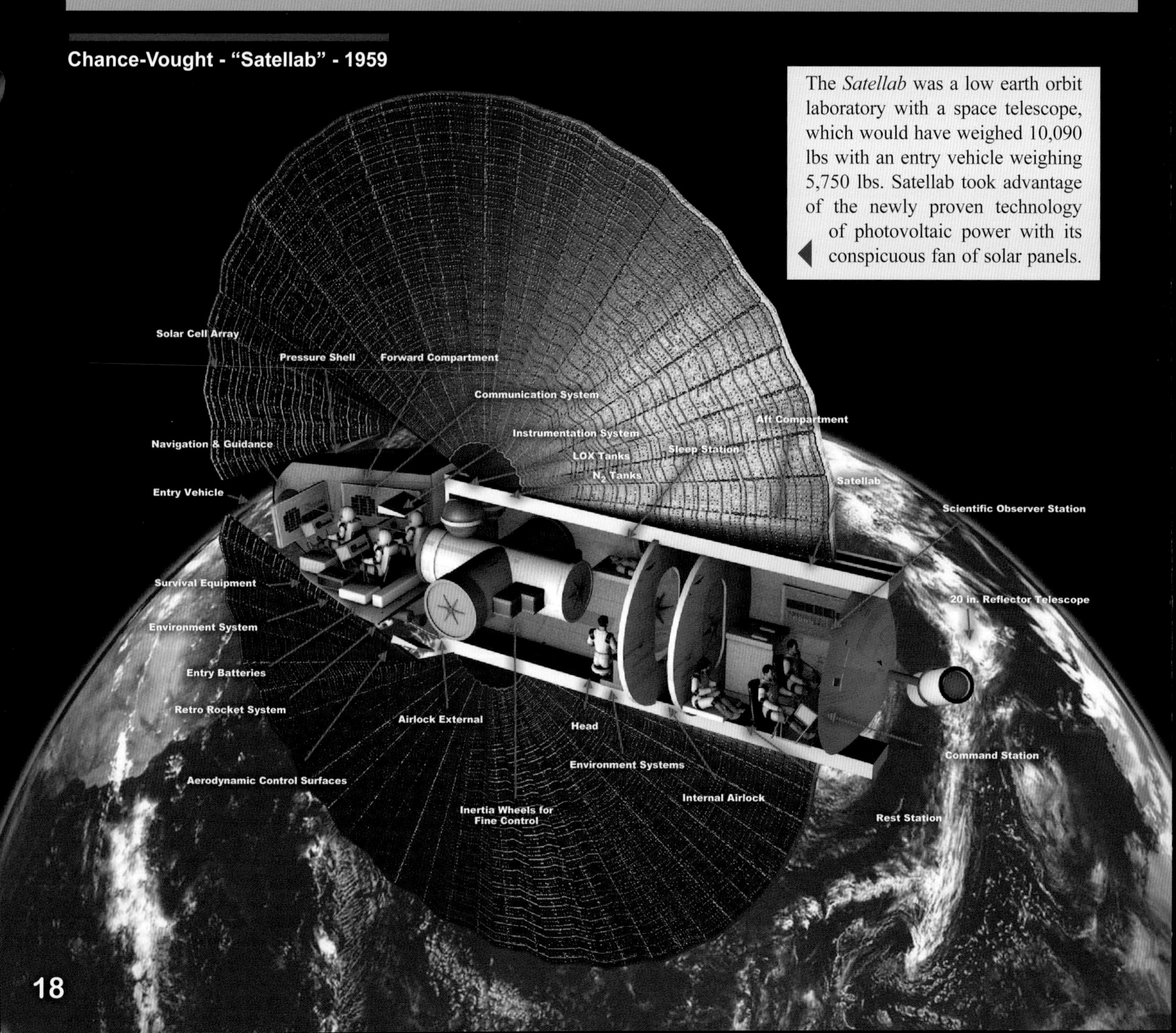

and a lunar landing module. The lunar version was to use something called "lunar orbit rendezvous" (LOR). This was a method for getting to the moon and back. At the moment that Lau's report arrived on Silverstein's desk no one at NASA had seriously considered LOR. Inexplicably the Chance-Vought lunar vehicle almost immediately vanished into history, but the earth-orbiting version, the *Satellab*, became the first serious design for an earth-orbiting laboratory, complete with a space telescope. Chance-Vought sent Thomas Dolan, one of Lau's team, out on the road to try to sell the *Satellab* to NASA and the aerospace industry. Dolan would be unsuccessful, but in the process he would put together a complete analysis of how much a space station might cost. His budgets became a template for future efforts. He would later be appointed to the President's Space Council. Conrad Lau would not live to see Lunar Orbit Rendezvous or the Apollo, he died in 1964.

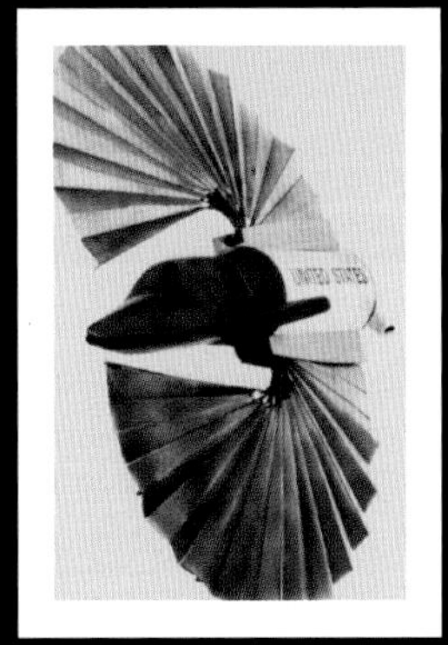

Chance-Vought was also a sub-contractor on the USAF X-20 program and the image at left shows a later version of Satellab using *Dyna-Soar* as the main re-entry vehicle. The team at Chance-Vought (below) John Clark, Conrad Lau and Tom Dolan.

"A 300 mile orbit has been selected to provide a true space environment while still close enough to afford economical proximity to the earth and also avoid the radiation hazards of the Van Allen belt." - Tom Dolan

In November 1959 this *Manned Space Laboratory* was designed by Koelle. It proposed using the same inverted cone-shaped reentry vehicle that had been part of the *Horizon* lunar lander. Refuelling of the lander would be handled by the space station crew. 25kW of power came from the two petals which contained a Mercury Rankine line-focus solar plant. The proposed upper stage Centaur would be replaced by the space station. The second stage would sit on the basic Saturn first stage and would use four cryogenic Pratt and Whitney engines.

Hermann Koelle/ ABMA - 1959

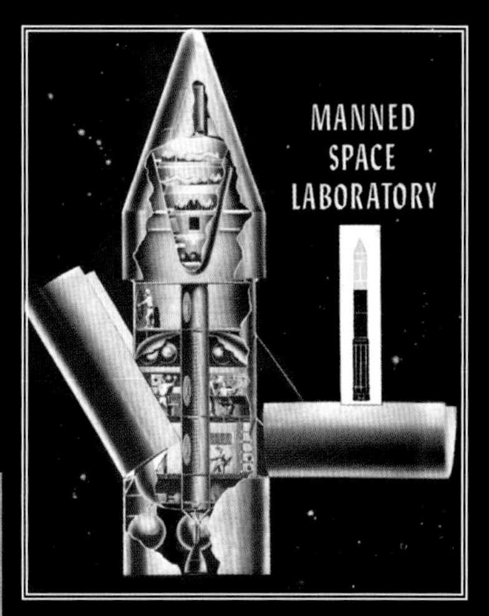

On July 8th 1960 Koelle, Frank Williams, Frederick Ordway and Harry Lange of MSFC met to talk over what von Braun should reveal about space station plans at the forthcoming IAC in Stockholm. Later that day Ordway met with Tom Dolan to discuss Vought's plans for their space station before meeting with von Braun. On August 16th von Braun delivered the speech written by Ordway. For the first time the public got to see the plans for the fourth generation launch vehicle, the Saturn C1 and its upgrade the C2. Von Braun planned to use multiple C2s to stage the lunar trip.

MSFC - "C2 Station" - 1960

The four-stage Saturn C2 would carry at least four different payloads (below), including cargo, fuel, a space station, or a smaller, more compact version of the station, which could be used to fly around the moon. Until the fifth generation launcher (the Saturn V) became available, a manned space station used for orbital refuelling was considered the best way to get to the moon.

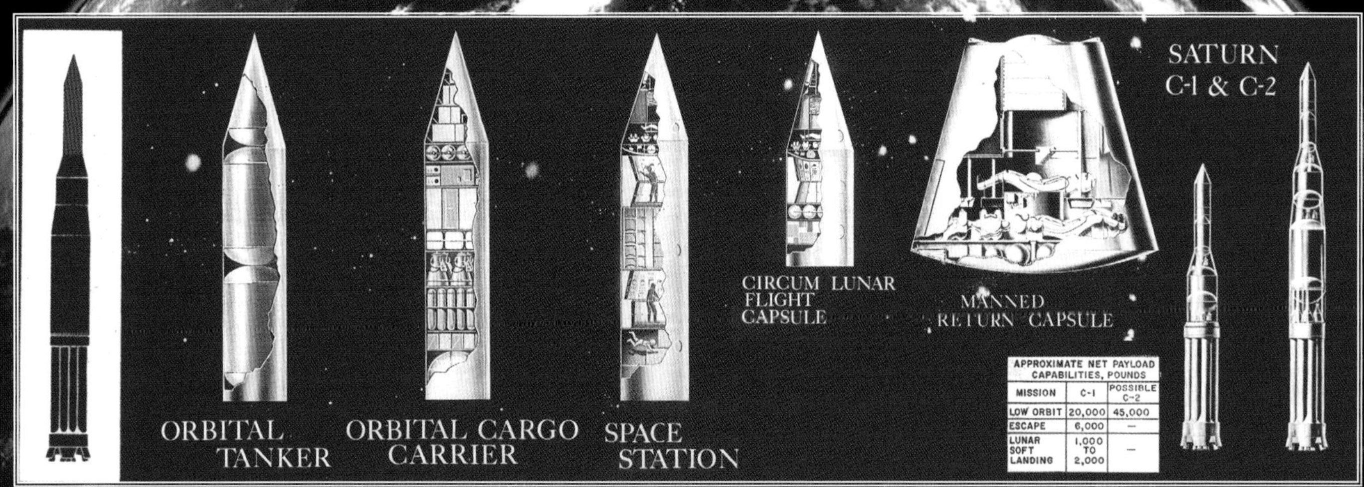

APPROXIMATE NET PAYLOAD CAPABILITIES, POUNDS

MISSION	C-1	POSSIBLE C-2
LOW ORBIT	20,000	45,000
ESCAPE	6,000	—
LUNAR SOFT LANDING	1,000 TO 2,000	—

Saunders Kramer & Richard Byers/ Lockheed - 1960

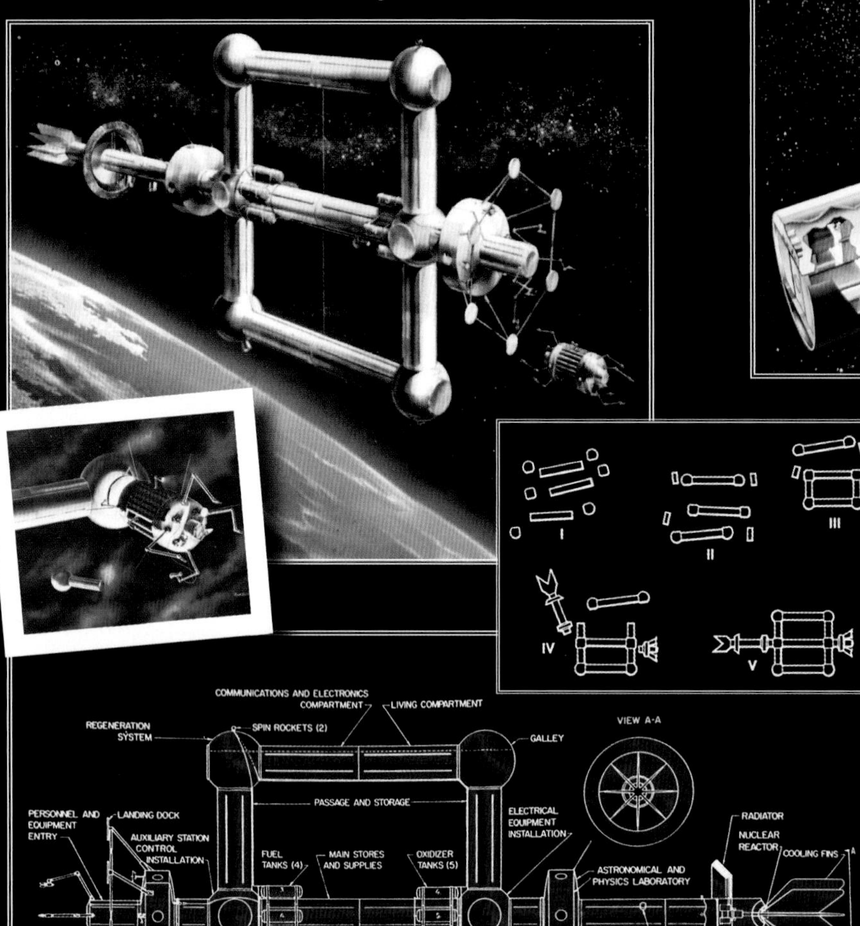

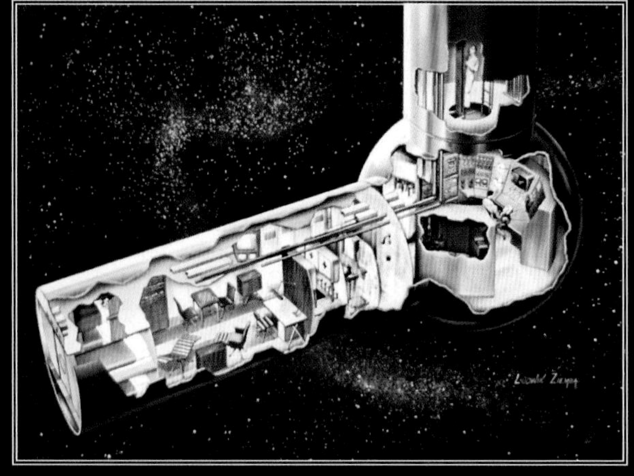

By 1960 Kramer and Byers at Lockheed had introduced a modular station adopting some of the design principles first suggested by Koelle back in 1951. It was also to be built in orbit (left). Their new design was 94 feet wide and 108 feet long. The spherical connecting nodes were 18 feet in diameter. The components were to be shipped to orbit already pressurized. The assembly sequence seen at left would be accomplished using two Astrotugs (seen inset at far left). The prime power source was a fast breeder nuclear reactor cooled by liquid sodium and shielded by lithium hydride and lead. This design's main attraction was that it could always be made larger by adding more nodes and modules. It was also designed so that it could rotate around its long axis to provide artificial gravity. This configuration was considered stable because the vehicle turned about its minimum axis of inertia.

Dandridge Cole/Martin Company - "Macro-life" - 1960

In June 1960, Dandridge Cole, the senior advanced planning specialist at the *Martin Co.*, delivered a lecture called *Extraterrestrial Colonies* to the USAF Academy. Cole had discussed *Project Horizon* and the latest details of the Saturn rocket with Koelle. Cole revealed his ambitious plans for a series of space colonies, some built using the raw materials from asteroids. He coined the expression *Macro-life* to describe these enormous space stations which were illustrated by Roy Kerswill, a British-born artist working at Martin who also aided with the ambitious design for a unique solar-dynamic power supply (bottom). Cole further developed these ideas while working at the *General Electric Space and Missile Division* at Valley Forge in Pennsylvania. In the summer of 1961 he discussed these ideas with Frederick Ordway of NASA MSFC. By 1964 Ordway's company *General Astronautics* partnered with GE on a contract at MSFC for interplanetary mission studies. A few months later Ordway introduced Cole's work to Stanley Kubrick when the director hired *General Astronautics* to consult on *2001: A Space Odyssey*. In 1975 Cole's work was also referenced by Gerard O'Neill's team at NASA Ames during a protracted study on space colonization.

Gerd de Beek/ABMA - "Space Village" - 1960

1960 was the year which produced more space station studies than any other. Five months after Cole's *Macro-life* presentation, Gerd de Beek, who had been the main conceptual artist for the ABMA and had provided the artwork for *Project Horizon*, presented his own ambitious design for a *Space Village* which would have included orbital apartments, a hospital and a deep space staging post for building and servicing planetary vehicles. The painting seen here was created by Harry Lange, a partner at Ordway's *General Astronautics*, who also worked for de Beek and Koelle. Lange went on to do most of the design work for *2001: A Space Odyssey.*

Emanuel Schnitzer & Paul Hill/ NASA Langley/Goodyear - 1960

"A rotating station in space must have a large diameter but it must also fit into a small-diameter booster payload stage. Consequently, the station must be erected in space in some manner. Automatic erection would save time over piece-by-piece assembly. It is possible to erect a complete rigid station of 150-ft diam. automatically from a reasonable payload package." - Emanuel Schnitzer

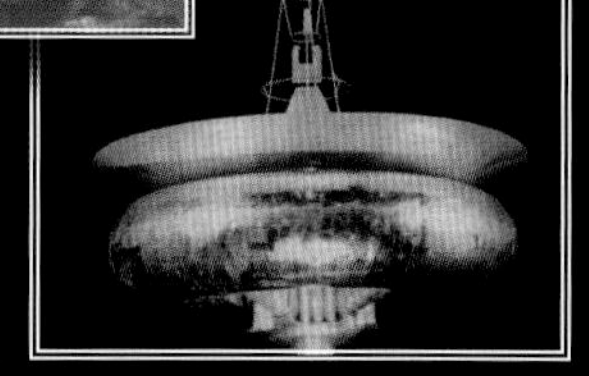

INVENTOR
EMANUEL SCHNITZER

Schnitzer and Hill of NASA Langley filed a patent for an auto-inflatable deployment method. Schnitzer then demonstrated the design to the American Rocket Society (above). The living quarters are in the inflatable toroid. It had an erectable solar collector as the source of heat for the power plant. The station would have rotated at 6rpm to provide some artificial gravity and stabilization and a Dynasoar escape taxi (not shown) would have attached at the bottom.

Gene Konecci & Neal Wood - Douglas Aircraft - 1960

Konecci and Wood's inflatable space station for use in lunar orbit. Each module would be manned by three to five men; a total capability of up to 15 men at one time with a min. of 350 cu.ft. allocated per man. Completely fabricated on earth it would be launched folded in a pie-shaped wedge. In orbit the station unfolded and formed a 36-ft.-dia. 5,300-cu.ft.-volume, disc-shaped vehicle. There were three large compartments and a central air-lock (far left).

Goodyear Inflatables - 1961

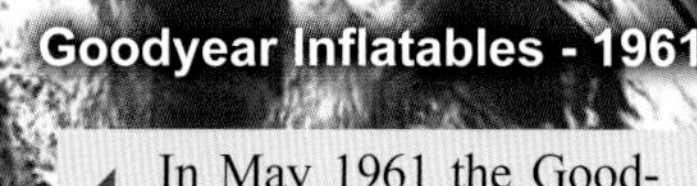

In May 1961 the Goodyear company's Robert Madden presented this design to the US Congress, for a one-man 24 ft. inflatable toroid for launch on an Atlas. It was to be mounted directly below a Mercury capsule and would have weighed 3,432 lbs. A three-man 50 ft version (far left) was also considered. A model showing the interior of the later version (above left) shows the Mercury replaced with a larger ballistic re-entry capsule at the axis.

In 1961 Milton Rosen, head of the Viking rocket and Vanguard satellite programs and one of the top astronautical engineers in the United States, testified in front of the Committee on Science and Astronautics at the U.S. House of Representatives. In his testimony Rosen outlined the long list of new procedures that would be required to put America ahead in the space race. Foremost on his list was the ability to conduct what became known as *Orbital Launch Operations* (OLO), principally consisting of rendezvous, docking and orbital assembly. The ABMA team in Huntsville had moved over to NASA in late 1960 and Koelle was now working as head of the Future Projects Office (FPO) at the Marshall Space Flight Center. He was aware of Lau and Dolan's *Satellab* so one of the first things he did was to issue a large research contract for OLO to the Chance-Vought Company to compile a complete study covering all aspects of working in orbit. Vought subsequently issued sub-contracts to many other companies. While Vought and Douglas concentrated on OLO, Martin, Northrop and STL created a flight performance manual; AMF worked on assembly; Convair studied orbital launched vehicles (OLV); Lockheed concentrated on orbital docking; North American looked at rendezvous, and Grumman worked with Chrysler and various universities to work on orbital transfers and guidance. In November 1961 Koelle explained the acronyms to von Braun; OLV was an orbital launch vehicle and OLF was an orbital launch facilty. In 1961 the FPO spent 21% of its manpower on the Saturn V, 20% on OLO, 14% on the Nova superbooster and 12% on Apollo. The resulting OLO report from Vought was filed in January 1962 and would be cited and referenced for years to come.

NASA MSFC, Vought, Lockheed, Douglas - "OLO & OLF" - 1960 to 62

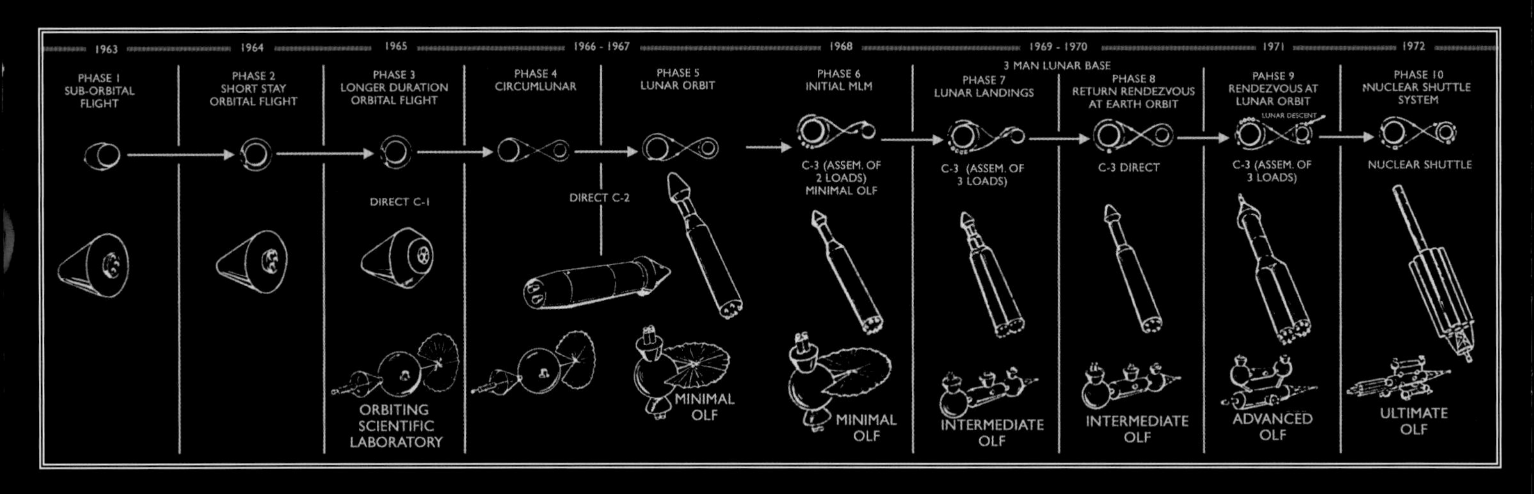

▲ Vought's final report presented an entire multi-year architecture for space stations based around what was then called an OLF, or *Orbital Launch Facility*. The ultimate goal of these stations was to provide an entire infrastructure to support missions to the moon and beyond. The main component was the 18' diameter sphere from Kramer and Byers, similar to that proposed by Koelle back in 1951. This size was chosen to be suitable for a Saturn C-2 launch vehicle. In its first iteration it was to be used for a 3-man orbiting science laboratory (left). This small space station would have weighed 15,000 lbs and was powered by Vought's ingenious fan of photovoltaic solar panels. The crew would arrive via Saturn rocket aboard an Apollo capsule.

The *Minimal OLF* came in two versions, both designed to fit into a single Saturn C-2 launch vehicle. This version (right), did not use the spherical component but instead used a cylindrical module that could house six crew.

The preferred *Minimal OLF* seen here, weighed 45,000 lbs and could house nine crew. It could also accommodate two Apollo capsules. Escape provisions were only needed for six people because the OLF would include unspecified escape provisions for its own crew.

The *Intermediate OLF* (center left) was powered by a 30 kilowatt nuclear reactor seen at the end of the long boom in the image. It housed eight crew and was comprised of two OLF spheres connected by a cylindrical module the size and shape of a *Saturn S-V* propellant tank. It would have weighed 95,000 lbs at launch and provided three Apollo spacecraft for escape. This was one of the earliest designs to propose using a tank from a Saturn rocket as a habitation module; although the Saturn S-V would never fly.

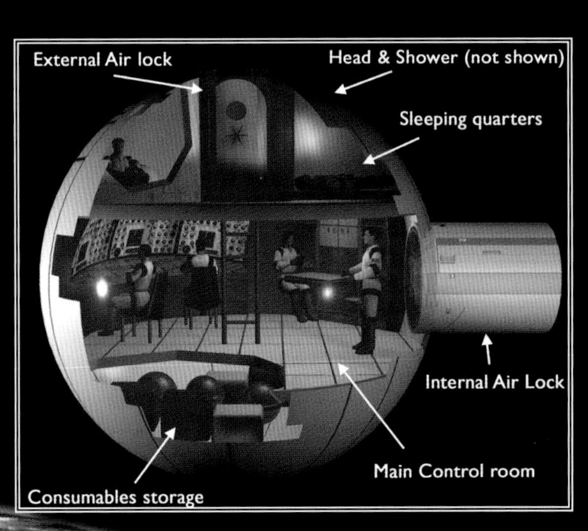

A modern computer rendering of an OLF node configuration as detailed in Vought's report.

The next step in this evolution was the nuclear powered *Advanced OLF* which would have weighed 115,000 lbs. It could have housed 18 crew and included a small hangar to allow for a shirt-sleeve environment for checking out an *Orbital Launch Vehicle* (OLV). It would also have included an *Astrotug* for moving the OLV around in the vicinity of the OLF.

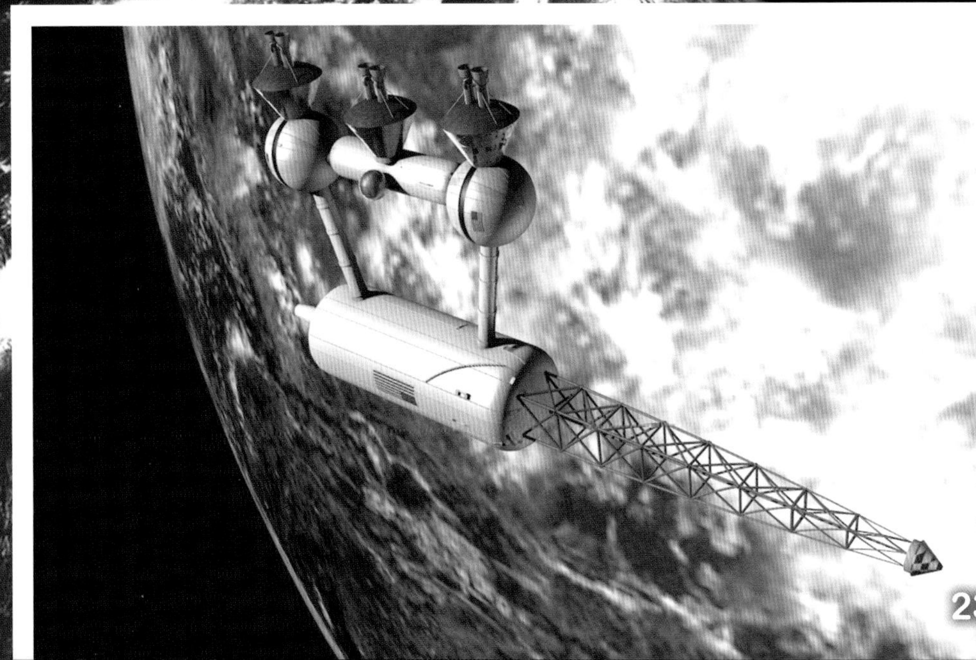

The final step in this ambitious architecture was the *Ultimate OLF* which was devised in two different configurations. Since Lockheed worked with Vought on the OLO study, both of these designs look very similar to those proposed by Saunders and Byers. The UOLF might have consisted of multiple Intermediate OLF's arranged symmetrically around a central partial space hangar for the OLV. Onboard power would again be derived from a nuclear source.

By the end of 1963 the OLO studies had produced a wide array of potential designs for space stations. This *Orbiting Space Platform* concept (below) was presented to the US Congress in 1964 and featured two very obvious features from OLO. A tank suitably equipped with life-support systems and a nuclear power generation system installed on a tower at some distance from the tank to reduce shielding requirements. The estimated electrical power requirement of 30 to 60 kilowatts from on-board power to operate the platform was within the capability of the SNAP-8 electrical generating system, including shielding of some 15,000-30,000 pounds. It also featured a space telescope similar to Vought's *Satellab*. The entire structure was within the allowable 200,000 pound gross weight capability of a Saturn V launch vehicle.

The *Ultimate OLF* could house 36 crew with 100% escape vehicle capability. As it would have been used to support OLVs it would have a fuel transfer capability. This UOLF would have weighed 225,000 lbs and in its wheel configuration (below) could have provided artificial gravity.

Martin Co. - Semi Inflatable - 1960

◀ Engineers still pursued more economical single-launch space stations, which included this ingenious design for a partially inflatable space station from the Martin Company (left).

Key: Inflatable-structure space vehicle: (1) escape and resupply vehicle, (2) airlock, (3) 16-ft.-dia. inflatable sphere, (4) stowable light weight compartmentizer, (5) pressure bulkhead, (6) middle bay, (7) main rigging cables-equipment deployment, (8) stowed inflatable sphere, (9) container section, (10) stowed annular equipment, (11) booster adapter, (12) booster, (13) airlock stowed equipment, (14) retracted airlock hatch.

Martin Co. - "MSOL" - 1960

Stoiko, Kayten and Dorsey of the Martin Co. (right) then designed the *Manned Scientific Orbital Laboratory* (MSOL) a fully integrated space station which could be launched on a single Saturn rocket. It would have housed between four and six people. Crews would have been rotated every month and resupplied by an advanced ICBM-class booster. A wedge-shaped escape vehicle would provide emergency egress capability. The entire structure including crew and supplies for 30 days would have weighed 31,806 lbs. ▶

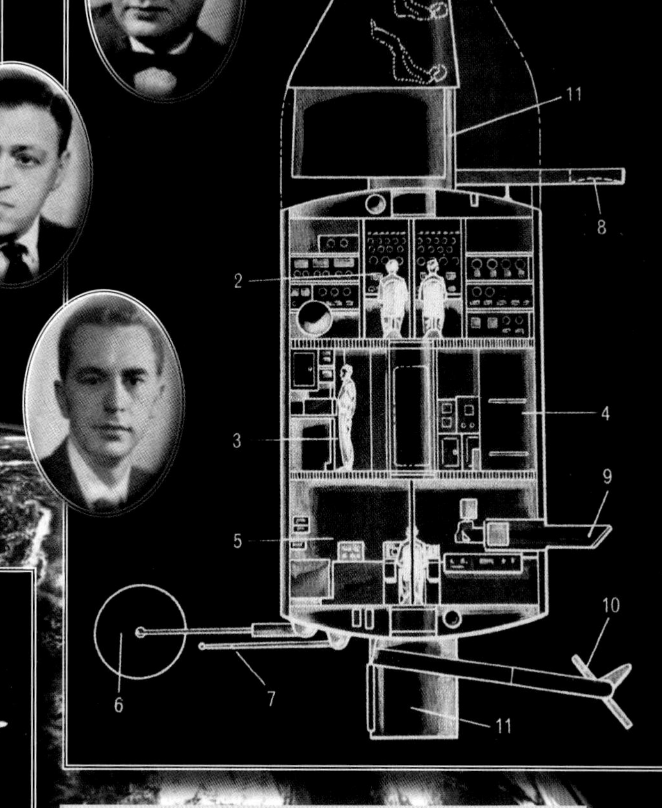

▲ The MSOL (above) was one of the first station designs to seriously incorporate all of the functions which would later appear aboard systems like *Skylab* and the *ISS*.
Key: 1. Escape and re-entry vehicle; 2. control centre; 3. biological and chemical laboratory; 4. medical laboratory; 5. astronomy and geophysical laboratory; 6. antenna; 7. magnetic sensor; 8. parabolic mirror; 9. telescope, focal length 28 ft; 10. telescope reflector; 11. air locks, which could be used as a vacuum laboratory.

Freeman D'Vincent/Convair - "TASSEL" - "MARS" - 1960 to 1962

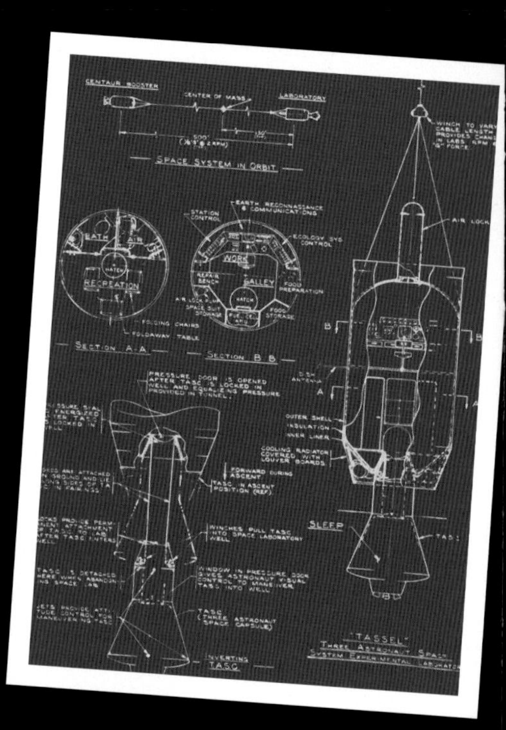

Driven by current booster capability, Convair engineers Freeman D'Vincent and Krafft Ehricke proposed a three-man space lab named *TASSEL* (*Three Astronaut Space System Experimental Laboratory*). TASSEL was to have been launched by a Convair Atlas-Centaur with an enlarged three-seat Mercury capsule. Once in orbit the TASSEL was released from the Centaur stage with a tether and the Mercury would rotate and dock with the lab. Using the idea proposed by Oberth and Tsiolkovsky decades earlier, the Centaur would provide a counter-balance for spin-up artificial gravity. TASSEL eventually evolved into a full size mockup named *MARS* (*Manned Astronautical Research Station*) which was built at the Convair factory (left). TASSEL was the first program to simulate conditions for a three-man spacecraft and would set some of the parameters later used for *Apollo*. The illustration at lower right shows a two-person version of TASSEL with the Centaur stage on the far end of the tether. In this version the return capsule remained connected to the lab.

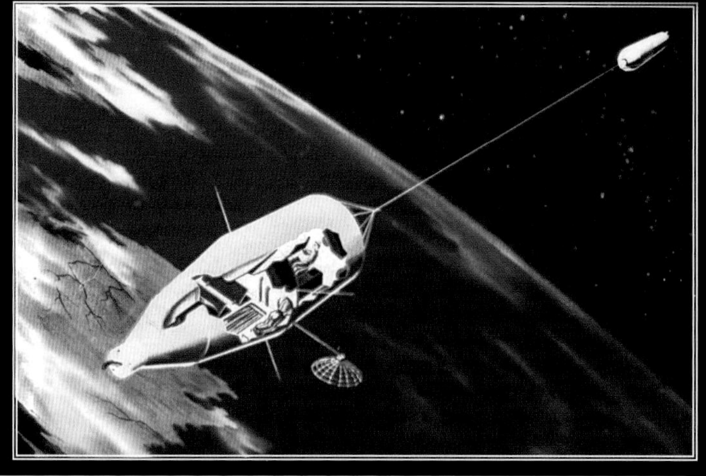

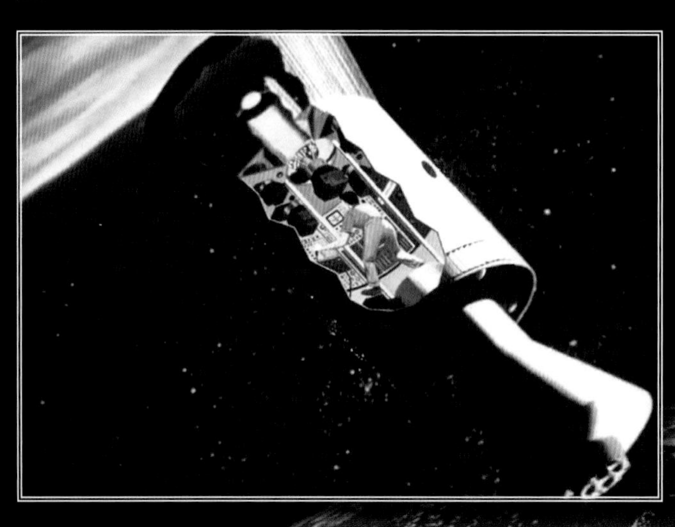

"TASSEL possesses certain unique advantages that permit it to become operational early in the history of manned space flight. No technological breakthroughs are required before it can perform a successful mission. For example, it does not require a solution to the rendezvous problem as its cabin, cargo, earth return vehicle and the astronauts are delivered into orbit in one package." - Freeman D'Vincent

Aeronutronic - "Mercury Space Trainer" - 1961

The Aeronutronic Company proposed this Mercury *Space Trainer* (right), a one-man inflatable space station. This 15' x 7' reinforced module which could be used to build a space station or as the central core of a reentry glider (centre) was part of an 18-month study for the US Air Force. At far right is their *Space Train.* The command module is up front, in the middle is the propulsion unit, with engine systems and a retro-pack for returning to Earth. In the back is a laboratory "car" with maneuvering thrusters. The boost section at far right prepares to dock.

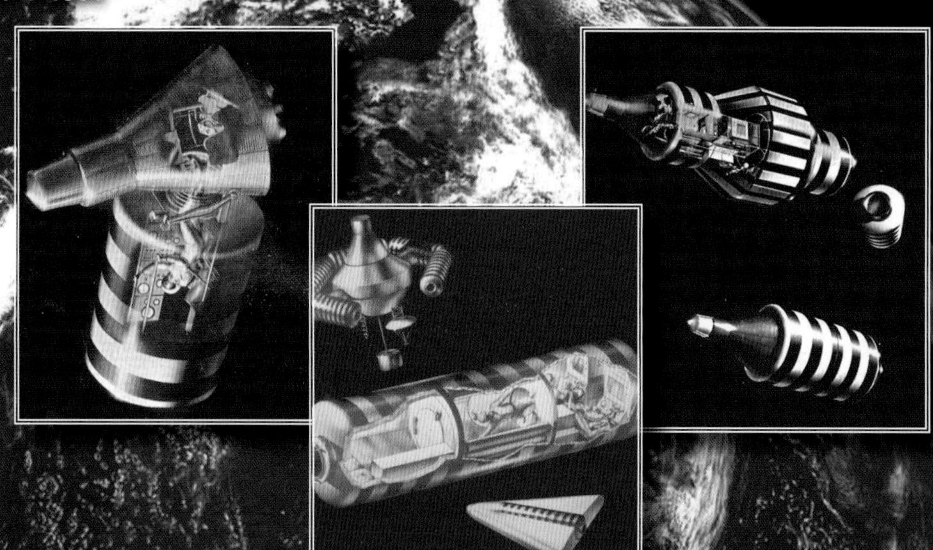

NASA Langley Research Center/Goodyear - 1962

While attempting to overcome the limitations of the available launch vehicles, in 1962 Schnitzer's team of engineers at NASA's Langley Research Center were still working with Goodyear to build mockups of an inflatable space station module. The spoked toroidal shape was chosen to test the different materials in this early study.

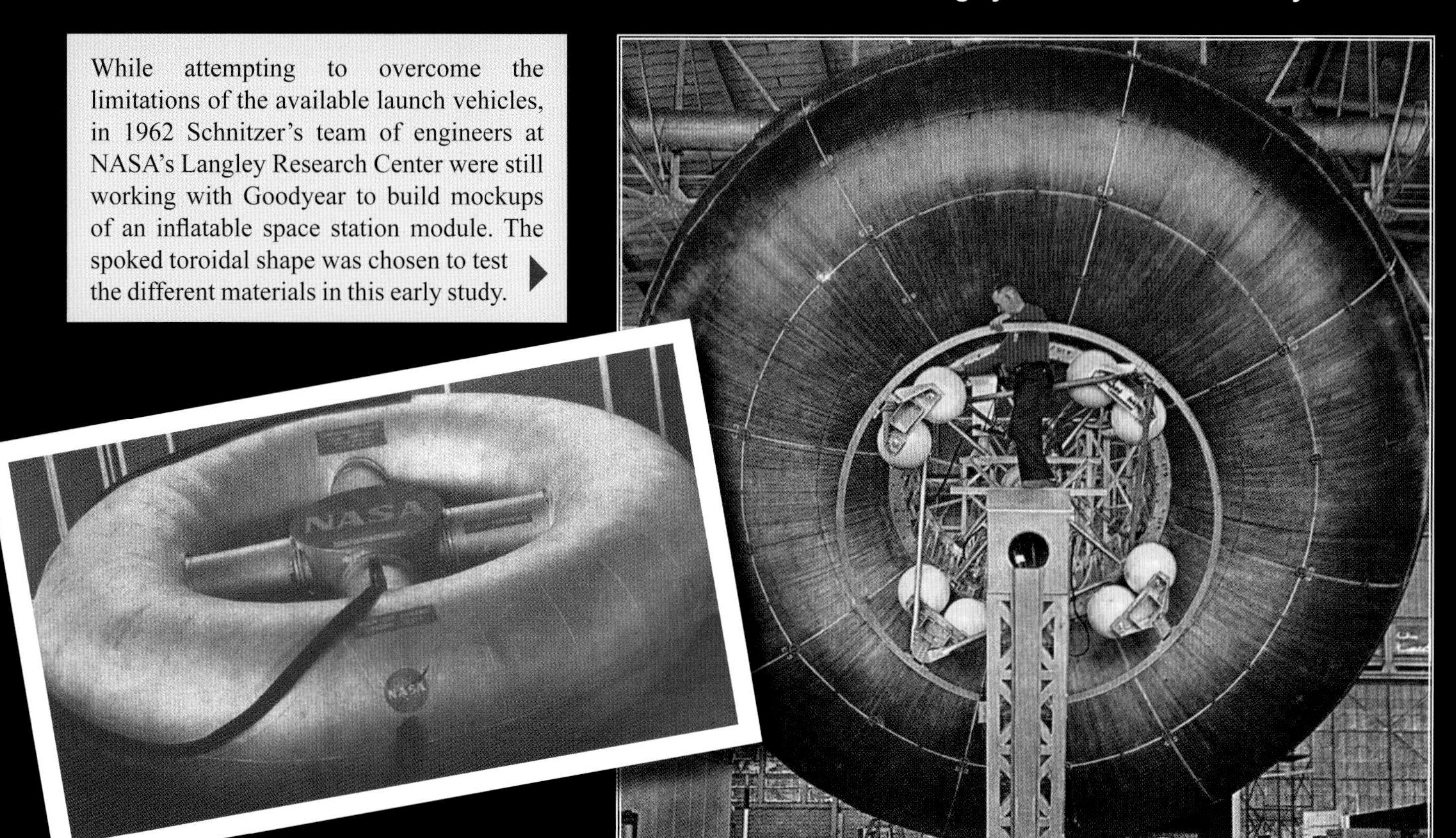

NASA MSFC/North American Aviation - 1962

Koelle's team at Marshall Space Flight Center worked with Schnitzer and North American Aviation to engineer methods for launching a self-deploying inflatable toroid from a single launch. In the image seen below the structure is folded into one stage of the proposed Saturn C-2 booster (10% scale model pictured at left). The 50' diameter of this station was later considered too small for spin-up and would have caused crew disorientation. Another undesirable feature of this concept was that equipment could not be installed in the proper places before launch.

Since 1960 engineers at NASA's Langley Research Center had been conducting a general study of space stations. A radial "Y' shaped station was proposed by Kurt Strass in November 1961. By February 1962 the Manned Spaceflight Center (MSC) established the *Space Station Study Office* in the Engineering and Development Directorate. The first report from that office, called *Project Olympus,* was filed three months later. Owen Maynard at MSC sketched something similar to Strass' proposal in June but the report recommended that NASA pursue as large a combined artificial- and- zero-gravity rotating space station as could be orbited by a two-stage Saturn V launch vehicle. In June Koelle told von Braun that his FPO team had begun working on a space station made from a modified Saturn fuel tank. Two months later Edward Olling of MSC visited Koelle and told him that over 60 people in Houston were now working full time on space stations. He told Koelle to inform von Braun that he would be submitting an official request to MSFC for Saturn V engineering support for the MSC space station project. By October of 1962 Koelle informed von Braun that the space station was now considered an urgent priority at the executive level of NASA.

O. Maynard & E. Olling/NASA MSC - "Olympus" - 1962

In 1962 Owen Maynard and Edward Olling of NASA's Manned Spacecraft Center created a new spoked design for an inflatable space station which used solar panels for power. The central hub, which was 33' in diameter would provide zero-g, while each of the three habitat spokes were 15' in diameter. Over 4000 sq ft of solar panels provided more than enough power for 24 crew.

"I was at a meeting where contractors were presenting what they thought the space station should be. I was very negative. That was when I got the assignment: "Well, you go figure it out by yourself." I got the assignment to write a paper on space stations. I had to invent the concept. On the airplane home, I made a sketch of what I thought the space station ought to be. When I got home I walked into my office and I gave the picture to Will Taub. I said, "We've got to write a paper to talk about space stations." Will took my terrible sketch; he's artistic; he made that into a picture that looked rational. We didn't know what President Kennedy was going to say at that time. We had to define enough about the station to tell NASA Headquarters what they could tell the President so he could use it as a policy statement. The Space Station was a big cylinder that sat on top of the lower stages of a Saturn launch vehicle. Several of these station modules could orbit Earth. At the central part, you fly in a mating spacecraft, the Apollo command module with a short little service module on it because it's not going to the moon. That was our space station. We invented the concept. NASA decided to patent it, and the patent covered lots of other kinds of space stations as well as this, so that when other people wanted to propose something, they wouldn't get the patent rights; NASA already had them." - Owen E. Maynard

"The technology to develop the space station is in existence, but it is still necessary to exercise the utmost ingenuity and intelligence in planning and designing the space station to achieve maximum results." - Edward Olling

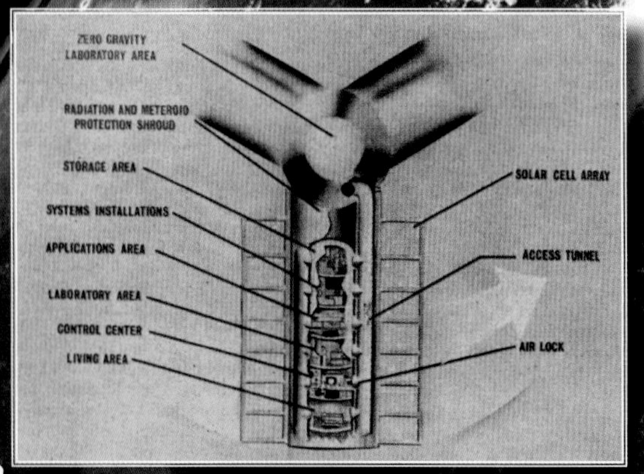

One of the main problems with this "spoke" design was that crew members moving from the tip of one spoke to another would have to move in and out in a radial direction, which would subject them to the disorienting Coriolis force.

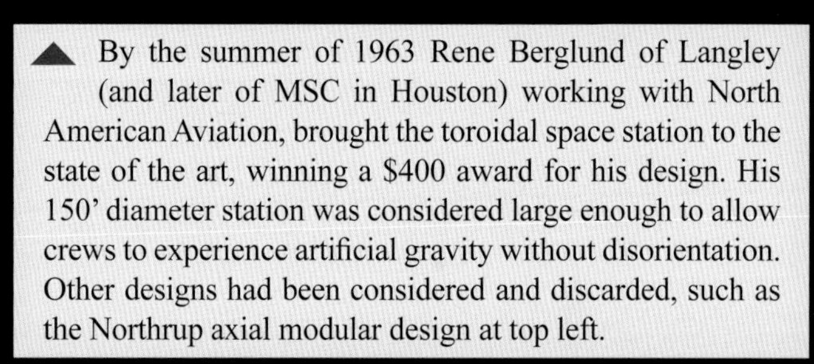

▲ By the summer of 1963 Rene Berglund of Langley (and later of MSC in Houston) working with North American Aviation, brought the toroidal space station to the state of the art, winning a $400 award for his design. His 150' diameter station was considered large enough to allow crews to experience artificial gravity without disorientation. Other designs had been considered and discarded, such as the Northrup axial modular design at top left.

"Emphasis was placed on obtaining the largest volume and largest diameter station without the necessity of the crew having to assemble a number of small units. Studies under structural analysis indicated that six sides was optimum since weight trade-offs showed that a greater or lesser number of modules resulted in a greater total weight for the space station." - Rene Berglund

Seen here with a model of his team's design, Berglund had perfected a method for deploying a toroidal station, with solar panels, from a single launch of the upcoming Saturn V lunar booster. Initially the six hexagonal sections would have been connected by inflatable joints. This was soon deemed unnecessary, which left only the spokes as inflatables. Later even the inflatable spokes were replaced with telescoping tubes. This concept combined the best features of the inflatable and rigid space station concepts; that-is, the compactness of inflatable designs and the prelaunch equipment installation features of the rigid designs. ▶

1963 would prove to be a busy year for space station planning. Koelle was asked to investigate advanced 5-man and 12-man Apollo spacecraft as crew transports. Then, based on a preference at NASA HQ to use Apollo hardware, Langley asked MSFC to look at launching a small station on a Saturn IB. This would require a trans-stage to be fitted to the Saturn IB. By the end of June Koelle told von Braun that such a station would be marginal and that he recommended one Saturn V and one Saturn IB. In April MSFC had its first formal space station committee meeting. After completion of the initial contract given to Chance-Vought for *Orbital Launch Operations*, a further study named *Advanced Orbital Launch Operations* (AOLO) was commissioned by Koelle at Marshall Space Flight Center. This again included Douglas, Boeing, Lockheed and LTV (the new name of Chance-Vought). Heading this project for NASA was the Launch Operations Center (LOC) Future Studies Branch at MSFC in Huntsville Alabama. The LOC was technically part of the NASA facility at Cape Canaveral in Florida but at this time was based in Huntsville Alabama. Its job was to predict future launch needs and so if someone wanted to launch to the moon or planets from Earth orbit, the team at LOC considered it part of their assignment. The department was run by Georg von Tiesenhausen and staff engineer Jean R. Olivier who had both worked with Koelle on *Project Horizon*. The final AOLO contractor reports would be filed in October of 1965 and included an array of advanced Orbital Launch Facility concepts. Each of these designs were orbital staging posts for planetary missions or regular lunar ferry missions. This would require several classes of launch vehicles, including the Saturn IB, Saturn V, a reusable shuttle and a Nova class booster. The Phase A of AOLO recommended three different orbital labs. They were a small station called *Apollo X*; a *Manned Orbiting Laboratory* (MORL) proposed by Langley as a single module lab; and a multi-module 12-18 man OLF by Boeing. A sophisticated computerized simulator was created named the *Orbital Launch Operations Simulation Model* to try and keep pace with the rapid change of development of the various OLV and ELVs. The AOLO studies would be one of the last things Koelle did for NASA before accepting a prestigious post back in Germany.

Georg von Tiesenhausen/NASA LOC - "OLF" - 1963

In March 1963 von Tiesenhausen unveiled the new advanced OLF (below) to a meeting of the American Rocket Society. The Future Studies Branch of NASA's Launch Operations Center's large orbiting launch facility would have been launched by two Saturn V's and could have housed a crew of 25. It would have been used to stage launches to the moon and the planets and was capable of receiving and preparing Saturn C-1-class low earth orbit shuttles, Saturn C-5-class lunar ferries and Nova-class interplanetary orbital launch vehicles. The laboratory would handle research into materials and the whole thing would be powered by a SNAP nuclear reactor.

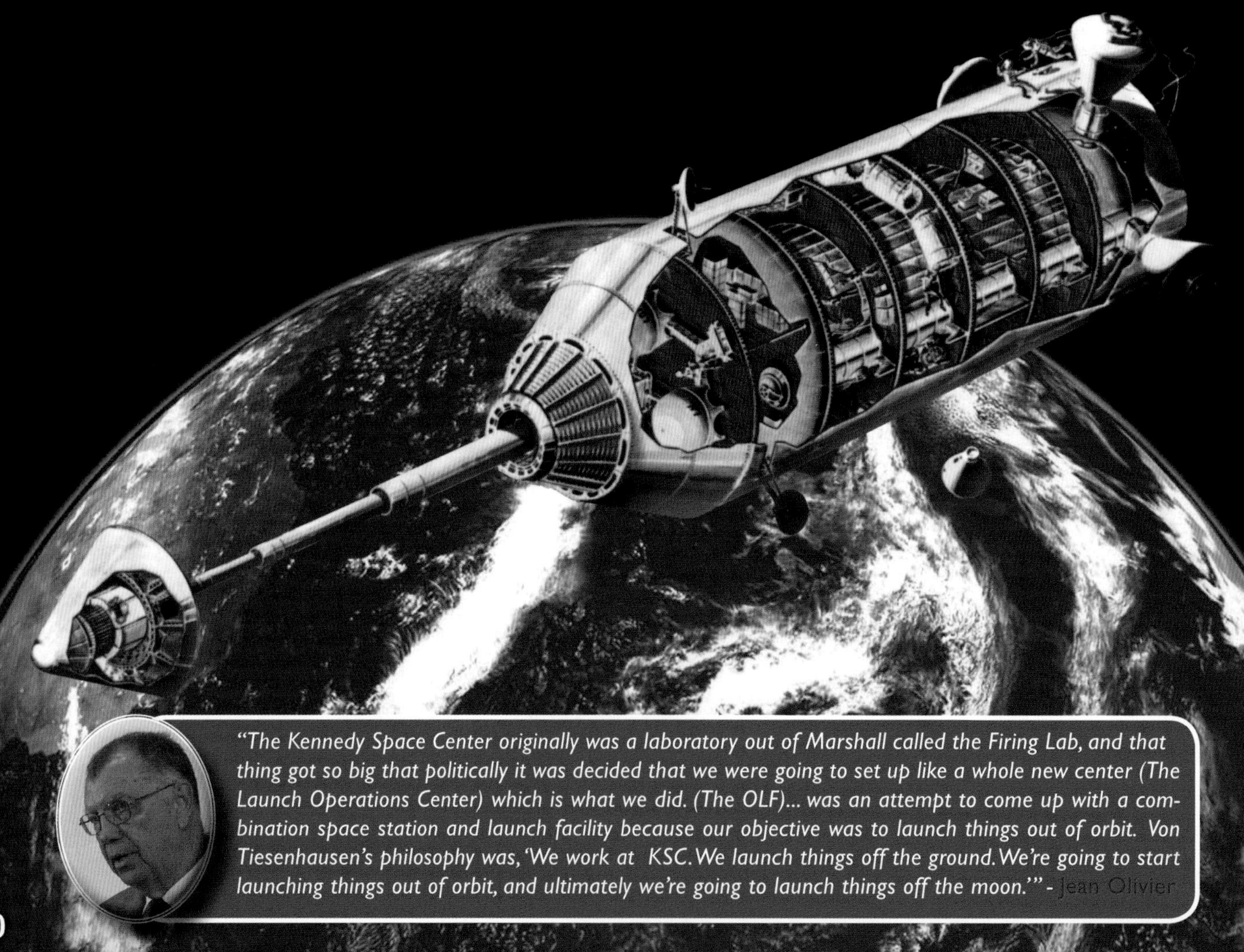

"The Kennedy Space Center originally was a laboratory out of Marshall called the Firing Lab, and that thing got so big that politically it was decided that we were going to set up like a whole new center (The Launch Operations Center) which is what we did. (The OLF)... was an attempt to come up with a combination space station and launch facility because our objective was to launch things out of orbit. Von Tiesenhausen's philosophy was, 'We work at KSC. We launch things off the ground. We're going to start launching things out of orbit, and ultimately we're going to launch things off the moon.'" - Jean Olivier

In June 1963 NASA Langley had organised the Manned Orbiting Research Laboratory (MORL) Studies Office and Steering Committee. In October of that year the first MORL contract - to see if the Air Force Titan IIIC launch vehicle could be used for MORL - was concluded by the Martin Company. Titan was rejected because it would be incompatible with Apollo systems. Boeing and Douglas then competed for further contracts, with Douglas winning out. The original design was for a station which could be launched on a single Saturn IB and would be powered by solar panels. This MORL would have stayed in orbit for up to 5 years with six or nine crew. As plans for MORL slowly developed the enormous Saturn V offered bigger options for larger stations.

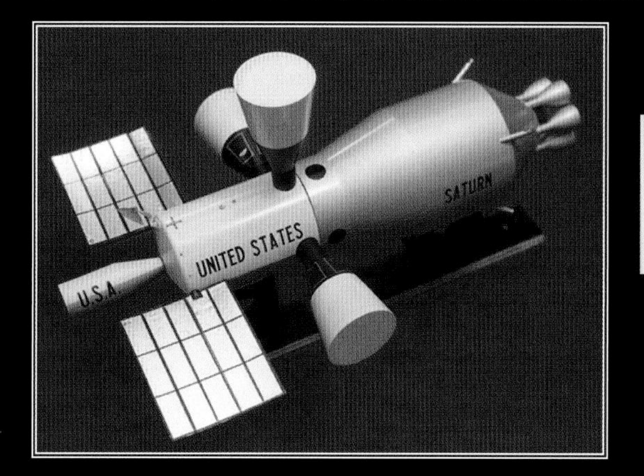

NASA Langley/Douglas - "MORL"- 1963 to 67

A model from July 1963 of the earliest design of the solar powered Orbiting Research Laboratory with its three Gemini escape vehicles and an Agena-D supply vehicle. The multi-engine spent-stage Saturn IV could have been used as a counterbalance for artificial gravity.

Cutaway image of MORL interior (right) shows an Apollo with a small service module docked at the bottom. A model of MORL seen at bottom right shows the solar panels folded up inside the upper part of the station. It also shows two Gemini escape craft attached to the docking ports. Later a radioisotope/ Brayton power system was considered instead of the solar panels. This system used thermal energy from radioactive decay to power a turbine to create electricity. With the arrival of the more powerful Saturn rockets MORL could be launched on an S-IVB with its single engine (below). MORL would have had a volume of 4000 cu ft and would have housed a crew of six. The docking hangar can be seen at the bottom of all three images.

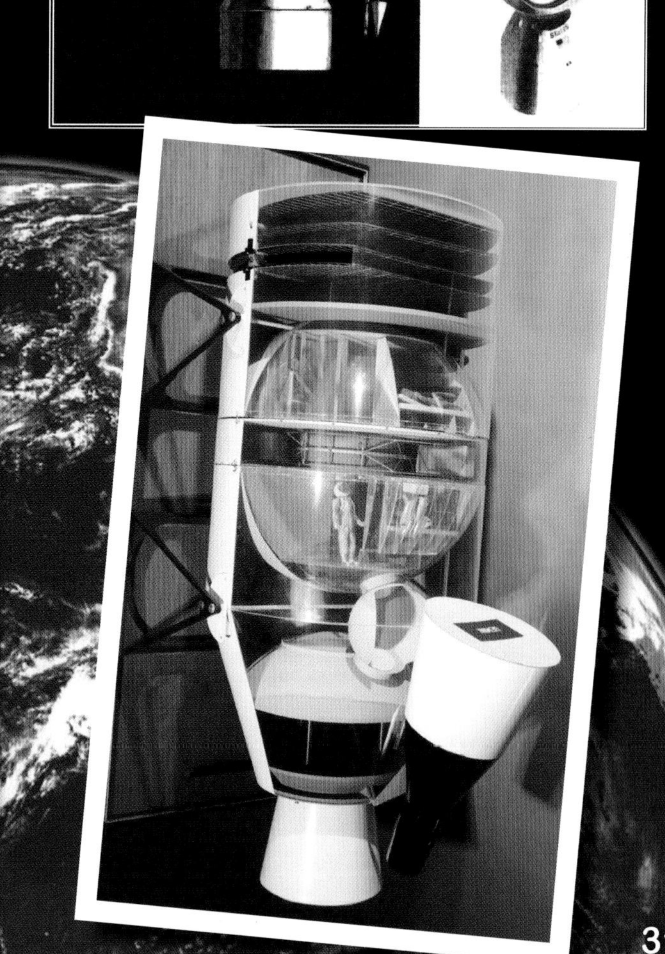

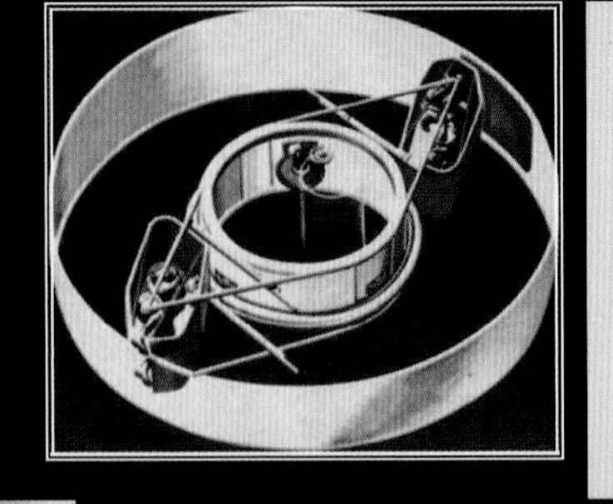

Many of the early station concepts also assumed that a telescope would be aboard for either looking down at the earth or out into space. One version of MORL (below) would have had a large aperture space scope attached.

MORL decks changed over time (above left) and although MORL was originally to have been spun on tethers (above), once it was decided that it would not spin in space a mid-deck centrifuge (above right) became a permanent fixture.

Boeing - "MORL/OLF" - 1963-65

This "Permanent Facility" was to be a large OLF with fuel tankers attached. It was decided that the optimum approach was to use two modified MORL units connected by a cylindrical adapter which housed docking collars and a Gemini emergency-escape vehicle. The OLF would use a radioisotope/ Brayton system located at the hub to power a turbine with the heat differential. By using an ingenious telescoping mechanism, similar to that suggested by Ehricke in his EMSS, the two MORLs would have moved apart like a telescope and created a large hangar space in which service work for the OLV could be accomplished. This adaptation of the MORL would house 12 permanent crew and its standard operating mode would be zero-g, with an option for spin-up. Several modified Apollo spacecraft or reusable shuttles would carry crew to the OLF and large LOX tankers developed by von Tiesenhausen and Olivier at LOC, (below) would arrive to fuel the OLV.

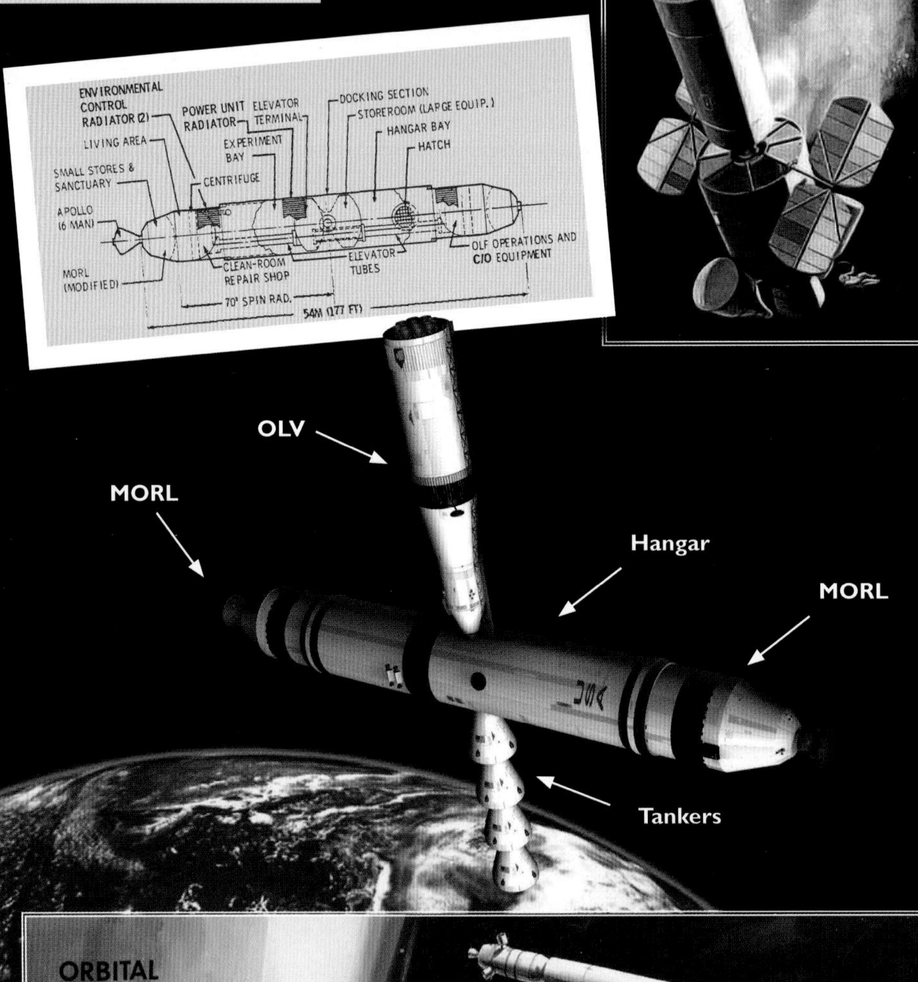

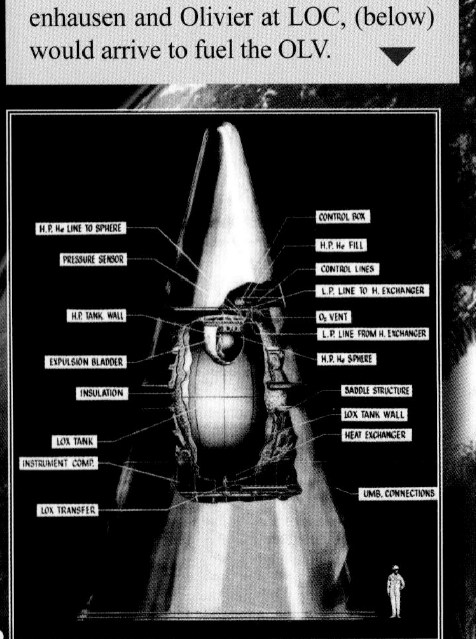

ORBITAL LAUNCH OPERATIONS

MARS ORBITAL LAUNCH VEHICLE

ORBITAL LAUNCH FACILITY

DEPART STAGE

ORBITAL SUPPORT ASSEMBLY VEHICLE

LOGISTIC SPACECRAFT

The first USAF space station project was called the *Military Test Space Station* (MTSS). In 1960 Col. Lowell Smith of the USAF stipulated that the MTSS would be a platform for testing weapons in space and where advanced technologies like nuclear engines could be fired without polluting the Earth. It would also conduct science experiments in conjunction with NASA.

Douglas/General Electric/USAF - "MTSS-MODS-MOL-DORIAN" - 1960 to 69

Powered by two large solar dynamic arrays and built from expended launch vehicle stages, contractor General Electric equipped MTSS with a downward looking telescope and a lifting body type reentry vehicle. It aimed to be self-sustaining for a year at a time in a 500 mile orbit using spin-up for artificial gravity. At the end of 1963 after more than six years of discussion the USAF *Manned Orbiting Laboratory* (MOL) was formally approved. Replacing MTSS it would launch a two-man crew in a modified Gemini, attached to a cylindrical space laboratory, atop a Titan III booster. Essentially a manned spy satellite, MOL/DORIAN consisted of a Laboratory Module built by the Douglas Aircraft Company, a Mission Module built by General Electric and a modified Gemini spacecraft built by McDonnell. MOL began when the US government had cancelled the Air Force's plans for a military space shuttle - the X-20. A requirement had been issued for what was originally called *Manned Orbital Development Station* (MODS). General Electric responded with two designs, both using the same station module. One was for a zero-g environment (right center) and the other used a large boom for spin-up artificial gravity (below). *Air Force Digest* also showed a design to the press in July 1963 which claimed to fulfill all of the specifications of MODS and MOL, a shirt-sleeve environment for four crew in a 200-250 mile orbit (bottom left).

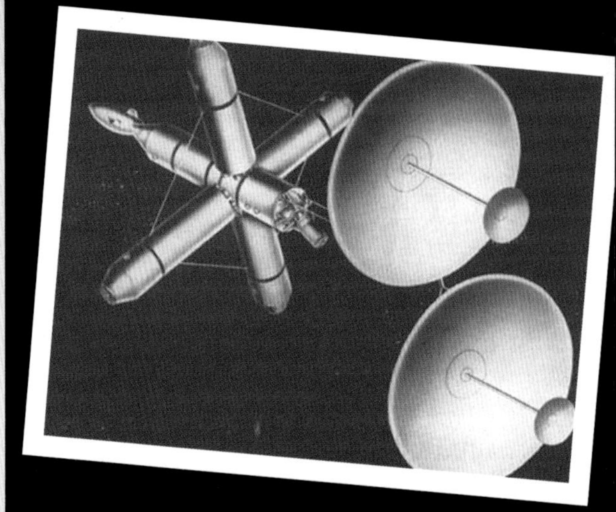

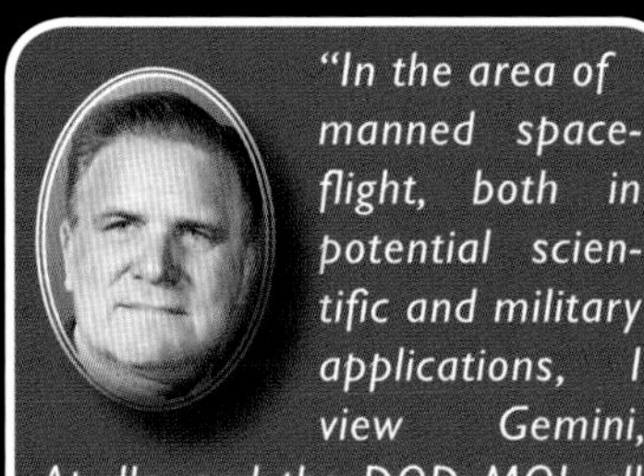

"In the area of manned spaceflight, both in potential scientific and military applications, I view Gemini, Apollo, and the DOD MOL all as important contributors to the ultimate justification and definition of a national space station." - James Webb

The early work done on MORL at NASA Langley and Douglas impacted the final version of MOL which looked much like the image seen here.

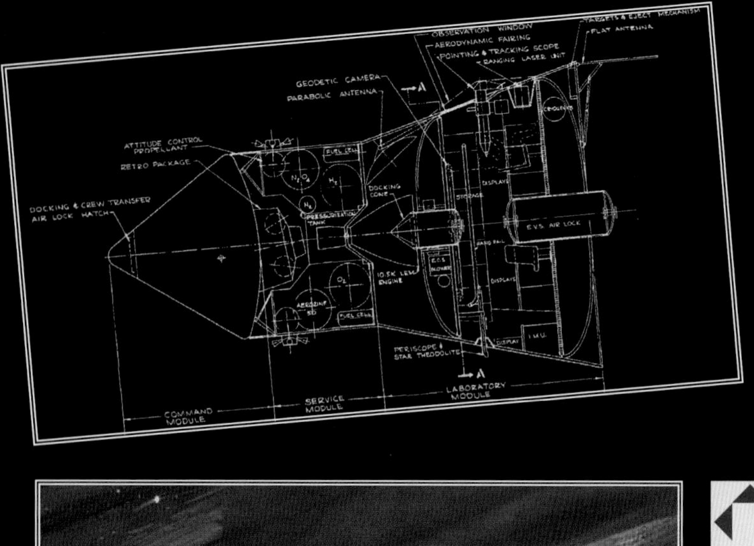

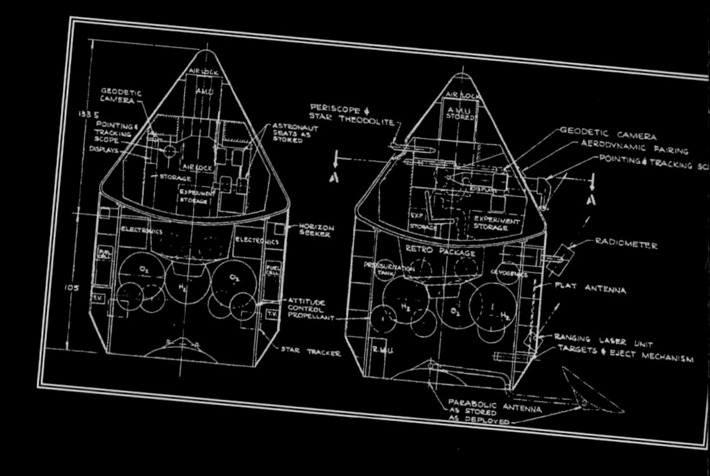

As the Apollo system evolved serious consideration was again given to creating MOL using Apollo and Saturn equipment (above left) or even a much smaller version of Apollo/MOL launched on a Titan III. (above right) These ideas were related to the *Apollo X* studies done for OLO which Koelle gave to the USAF in October 1963. In 1964 a version of the MOL (left) was considered by Douglas Aircraft as a 3 to 6-man lab for exploring the moon from lunar orbit. It would still have had an internal centrifuge for studying artificial gravity.

USAF astronauts were to have reached MOL aboard a modified Gemini spacecraft called "Gemini B". There were at least four different methods of getting the crew from the Gemini into the MOL laboratory. A hatch in the Gemini heatshield, an inflatable tunnel between the outer hatches, a mechanism which could rotate the Gemini to the MOL's side hatch or a simple spacewalk. There were many design changes for MOL, usually precipitated by changes in the capabilities of the booster technology. The version seen below had two Geminis, one at each end and a large outboard telescope for ground surveillance. The final configuration (bottom) was launched in full-size mockup on a Titan III in November of 1966 (left). MOL was cancelled in 1969, when the Nixon administration concluded that unmanned satellites could do the job of orbital surveillance for less money.

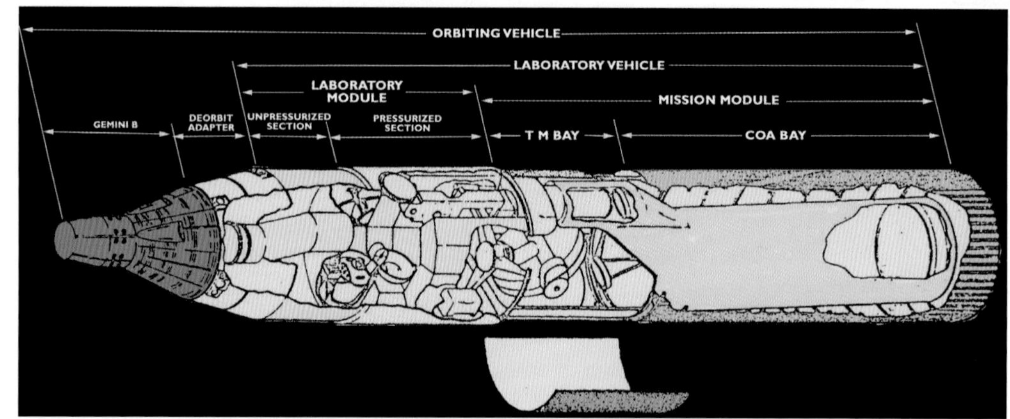

Grumman - "Spin-Up" Stations - 1963-1969

A tethered spin-up configuration, such as Ehricke's TASSEL, would have to be reeled in and simultaneously despun for resupply or crew rotation. This showed that tethered designs would have minimal capacities. Since May 1963 Koelle's FPO had been investigating what was called a "minimum space laboratory" which had a spin-up radius of at least 100'. This design required a constant need for counterbalance, combined with the requirement for a docking port at the center of the spin-axis or it would be prone to rolling.

Grumman Aircraft Corporation began touting an assortment of laboratories which, rather than using a tether, spun the whole spacecraft, exactly as Krafft Ehricke had suggested in 1954 with his *Manned Observational Satellite*. As can be seen in both of these images, above and right, spacecraft are seen docking on the ends and at the center.

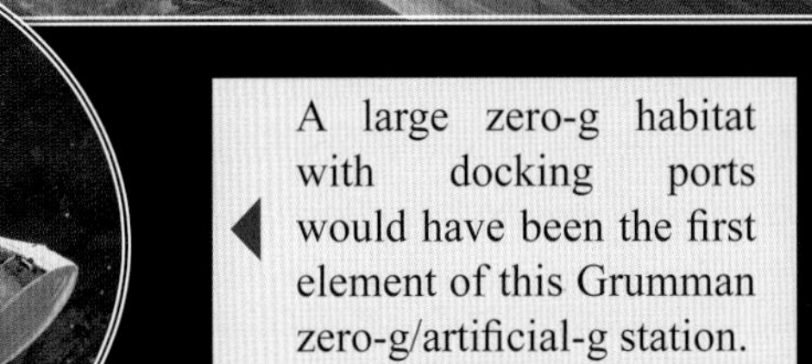

A large zero-g habitat with docking ports would have been the first element of this Grumman zero-g/artificial-g station.

Further designs from the late 1960s by Grumman (above and right) used rotating sections to provide artificial gravity. The station seen above left was to be a single launch station with two swing-out arms, while the one at right and above would have required several launches and shows the Grumman Lunar Module Lab being used for *semi-detached* and *free-flying* experiments such as space telescopes. Habitats were then mounted on the rotating arms.

Atomics International - Nuclear Orbiting Space Station - 1964

The desire to fly nuclear reactors to provide long-term power in space was still being studied by such companies as Atomics International. This image shows a proposed station similar to the Douglas MORL or the USAF MOL which had been under consideration since at least 1960. It has a Zirconium Hydride Mercury Rankine power conversion system to provide up to several hundred kilowatts of power. The reactor is used as the counterbalance in spin-up mode. In this instance having the reactor a long way from the crew made it the obvious candidate to be used as a counter-balance.

Hermann Koelle and his assistant Harry Ruppe issued contracts to Lockheed, General Dynamics (GD) and Ford/Aeronutronic in May of 1962 to study manned planetary missions. These studies became known by the acronym EMPIRE (Early Manned Planetary-Interplanetary Roundtrip Expedition). The resulting reports were filed in 1963 and 1964 and included in-depth analysis of whether a crew needed artificial gravity to survive long-term exposure to the space environment. Most of the contracting engineers, which included Krafft Ehricke's team at GD and Saunders Kramer at Lockheed, concluded that designing an interplanetary vehicle would simultaneously deliver the basic requirements for a manned space station; or a surface habitat.

General Dynamics/Lockheed/MSFC - EMPIRE studies - 1962-64

The General Dynamics EMPIRE station would have been deployed using fold-out arms with modules attached. This station could then have been sent to Mars or simply left in Earth orbit.

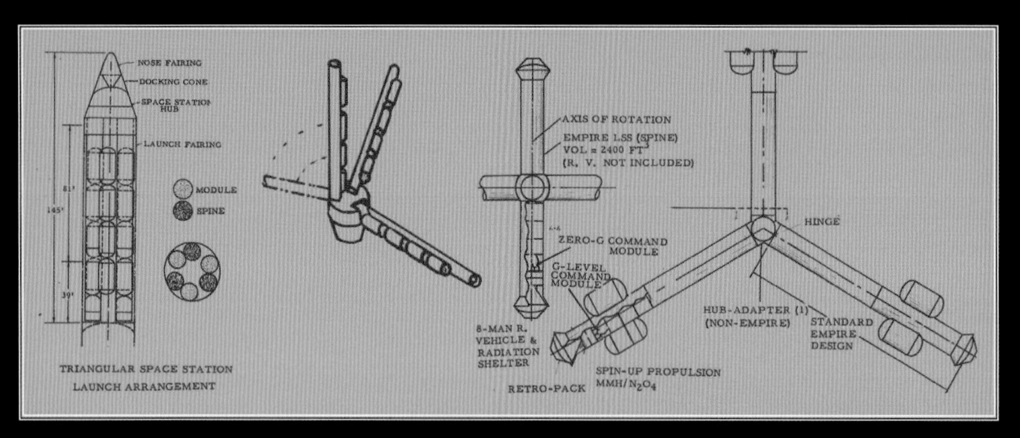

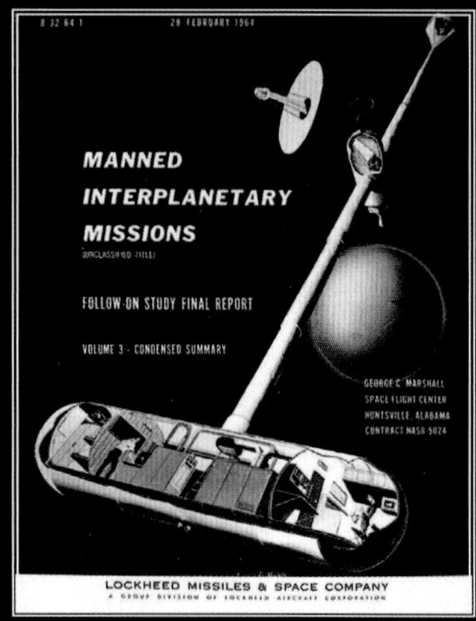

NASA - "Apollo Labs" - 1964

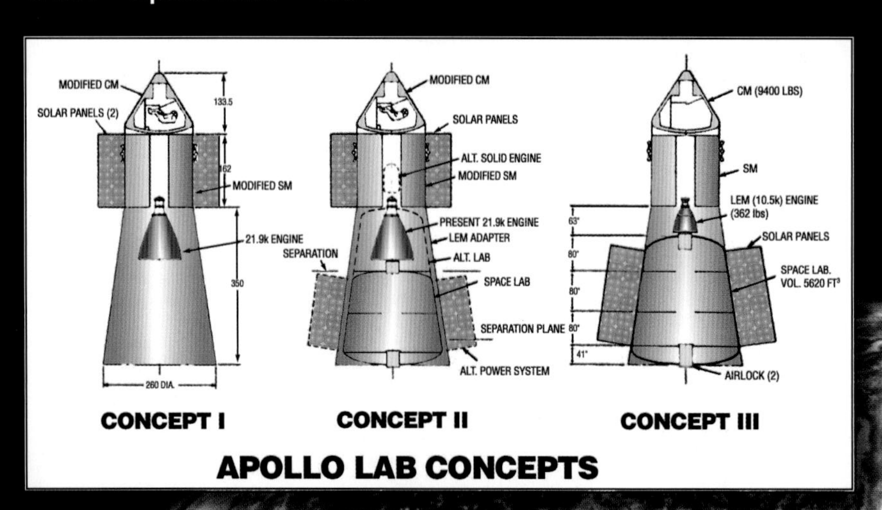

In a presentation to the Congressional Committee on Astronautics NASA offered several concepts for advanced Apollo laboratories with broader Earth orbit capabilities. The minimum extended Apollo consisted of a modified command and service module to be launched on a Saturn IB (Concept I). The SM would be modified by off-loading propellants and providing more life support consumables. This configuration would have allowed 2 crew to stay in orbit for 100 days. Concept II had a lab module inside the LEM adapter. This lab would use the extra consumables in the SM for life support. Concept III would have carried its own expendables and would only rely on the CM/SM for communication, guidance and navigation. The Apollo Labs main benefit was its simplicity and early availability.

NASA Langley/NAA - "Apollo X" - 1961 to 65

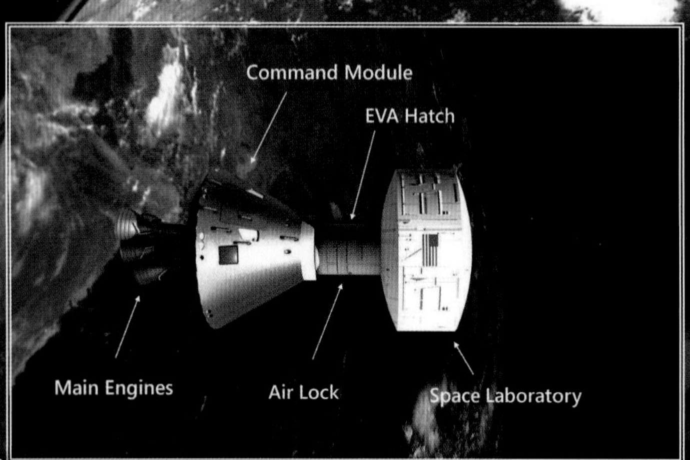

Proposed by Emanuel Schnitzer at Langley in 1961 the *Apollo X* small laboratory was to originally have been inflatable. Later designs by North American Aviation became part of the same studies which led to MORL and the large OLF. Like the other Apollo Labs it would be launched inside the lunar module adapter of a Saturn rocket. The *Apollo X* had a small centrifuge inside, similar to the one used in the MORL. A three-person crew could remain in orbit for up to 45 days inside the 1300 cubic foot laboratory. In theory a second identical lab could be stacked inside the same launch vehicle and connected to the first.

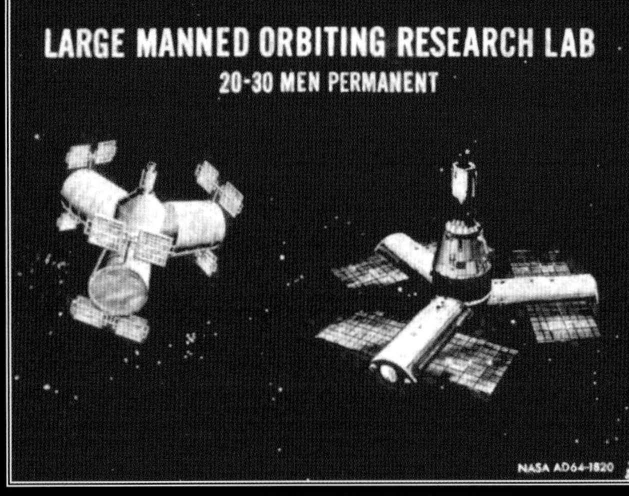

Lockheed - Large Rotating Manned Orbital Laboratory - 1964

In June of 1963 Lockheed was awarded a contract to investigate different rotating configurations for space stations. Working with MSC they concluded that the three arm station was more efficient than other designs. Some consideration was given to using four of the Douglas MORLs to create the preferred configuration (far left) But by June of 1964 the original design by Maynard and Olling's team had been improved with larger solar panels and pressurized arms which were now cylindrical. Each arm used a stack of pressurized modules that were 180" in diameter. This size was specifically chosen to fit inside the Saturn launch vehicle and would precipitate an entire family of possible space stations derived from this building-block module. Maynard and Olling's patent with the 180" modules was approved in 1967. (center left) Inset below is a cutaway showing the internal hangar.

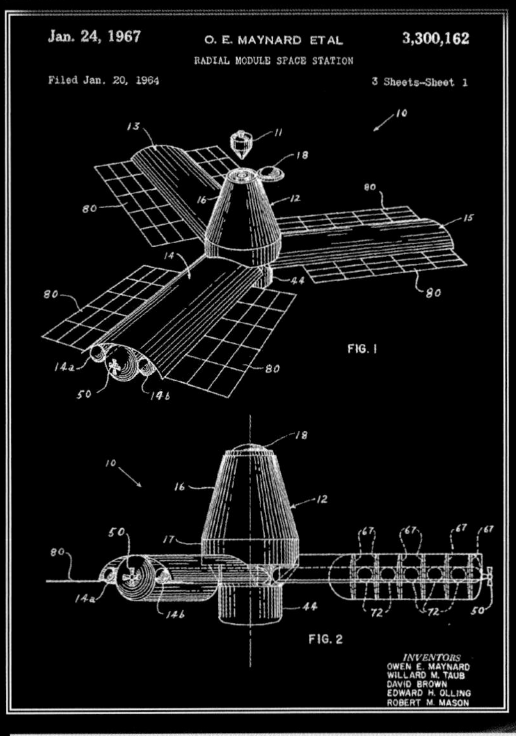

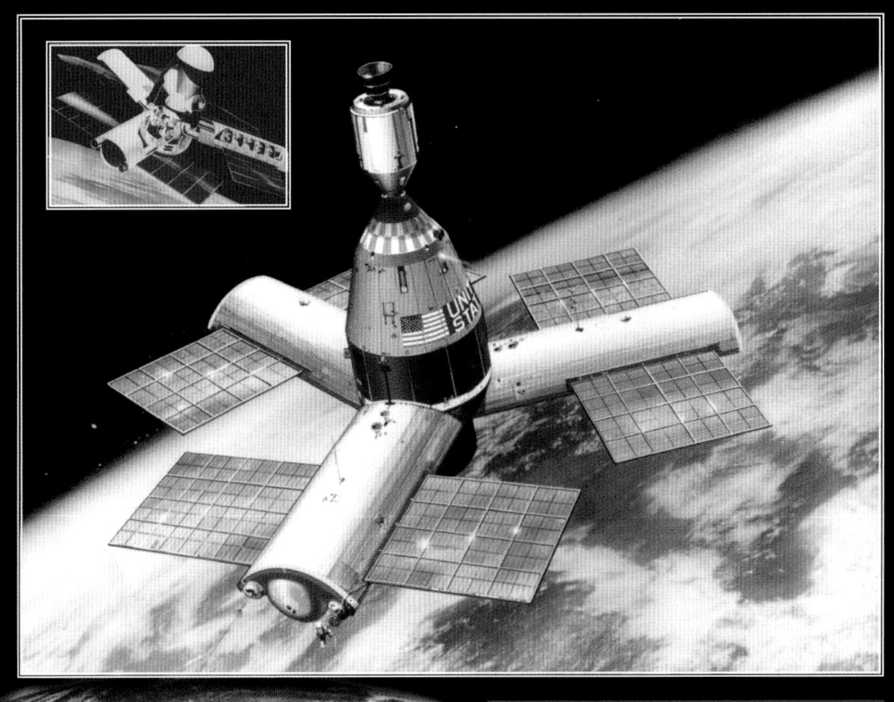

Lockheed - NASA - MPMM - 1964

Between September 1964 and July 1965 Lockheed fulfilled their NASA MSC contract to study how to improve what was now called the *Large Orbiting Research Lab* (LORL), by using the basic building-blocks concept, i.e. standardized modules. Module diameters of two sizes - 183 inches and 260 inches - were evaluated. The 183-inch diameter was favored because one or two modules conveniently fit into the spacecraft lunar-module adapter (SLA), and three such diameters were precisely circumscribed by the 33-foot diameter of the second stage of the Saturn V, which was the launch vehicle assigned to the LORL. Lockheed labelled the 183-inch-diameter configuration the *Multi-Purpose Mission Module* (MPMM), and it was to be used as the basic modular diameter for the six space-station configuration seen here (below right). The basic single module version (below left) was extremely similar to the North American *Apollo X*.

Boeing - NASA - MPMM/CAN - 1964

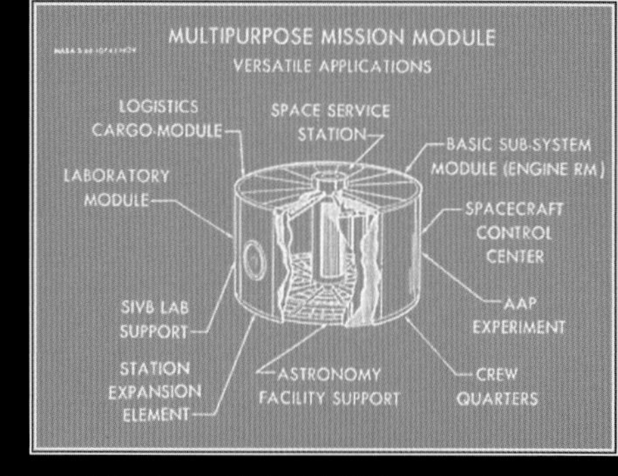

Running concurrently with the Lockheed study the Boeing Company was contracted to develop the *Multi-Purpose Mission Module* (also known as the Universal Dependent CAN). Their study established the design requirements and definition of the constraints of the module, and development of the preliminary design of a laboratory module including its structure. Boeing prepared mockup drawings and prepared program plans and cost data for the basic laboratory module. The MPMM was so versatile that a version of it was considered by Bendix as a lunar roving laboratory (below). The large volume of the CAN meant that it could be used for space stations, lunar habitats or even planetary voyages.

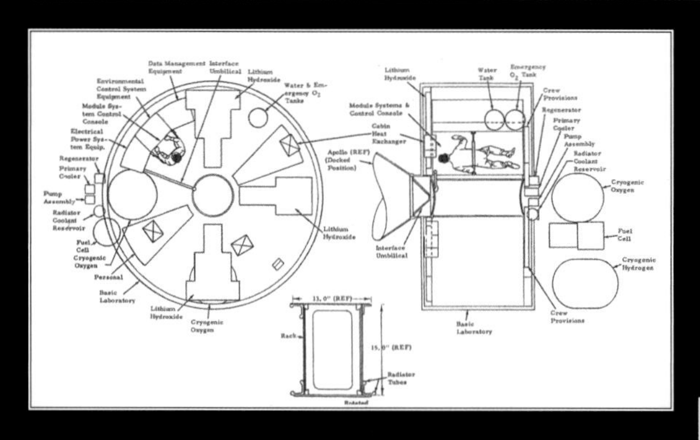

Saunders Kramer of Lockheed submitted a 100-ton axial "Y" shaped station with three 18' x 40' double-decked parallel modules. He called it *Bucephalus*. This axial configuration was soon discarded as it was too easy to shift masses a considerable distance out from the plane of rotation; such a shift could cause substantial dynamic imbalance and wobble. Kramer also introduced the *Astrocommuter* a Saturn-launched shuttle to service the Bucephalus.

Lockheed - "Bucephalus" - 1964

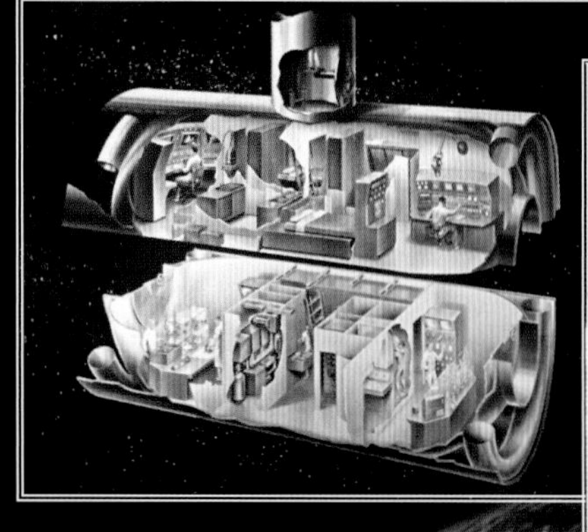

NASA/Douglas - MOSS - 1964

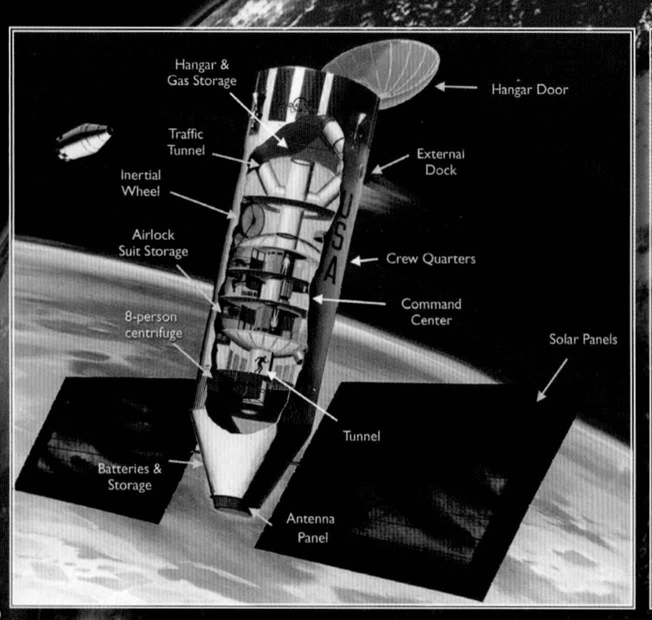

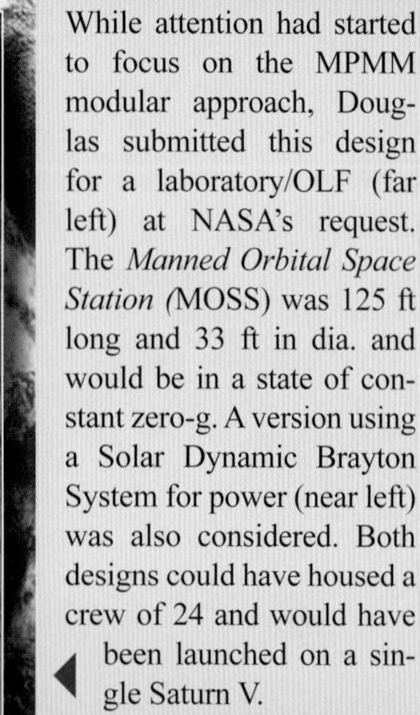

While attention had started to focus on the MPMM modular approach, Douglas submitted this design for a laboratory/OLF (far left) at NASA's request. The *Manned Orbital Space Station (*MOSS) was 125 ft long and 33 ft in dia. and would be in a state of constant zero-g. A version using a Solar Dynamic Brayton System for power (near left) was also considered. Both designs could have housed a crew of 24 and would have been launched on a single Saturn V.

Sergei Korolev, the world's foremost space pioneer, suggested the need for a space station to his superiors as early as 1954. Korolev said that solving the problem of rendezvous and orbital assembly was critical to further success in space. In March 1962 he proposed launching multiple Vostok spacecraft and having them rendezvous and dock in orbit. Six months later he recommended that research should begin on Tsiolkovsky's space garden. In 1963 his solution was to have his proposed larger lunar spacecraft dock with a "space tug" and a tanker. These three modules would all be called *Soyuz,* which meant "Union". This space station could meet up with as many tankers and spacecraft as necessary and in modified form could be used to go to the moon. Vladmir Chelomei was one of Korolev's chief competitors and the designer of the Proton launch vehicle. When the USAF announced the MOL military observation platform, Chelomei immediately began designing a Soviet counterpart which he named *Almaz*. Built with a two meter camera aboard Almaz was to be a manned spy satellite like MOL. The station was basically two cylindrical sections connected together and was to be powered by solar panels. Chelomei's station included a re-entry vehicle known as the VA (Return Craft) which could be launched along with a laboratory module known as the FGB (Functional Cargo Block), the two units combined were known as TKS (Transport Supply Ship). After Korolev's death a team of designers at his OKB-1 factory, anxious to move ahead with a space station, realised that although Chelomei's Almaz still lacked the basic internal structures to make it functional it would give them a headstart. So political pressure was applied to Chelomei to relinquish his Almaz to the greater goal of getting a space station into space before the USA.

Sergei Korolev/OKB-1 - Soyuz-A Station - Vladimir Chelomei/ OKB-52 - "Almaz" - 1964 to 67

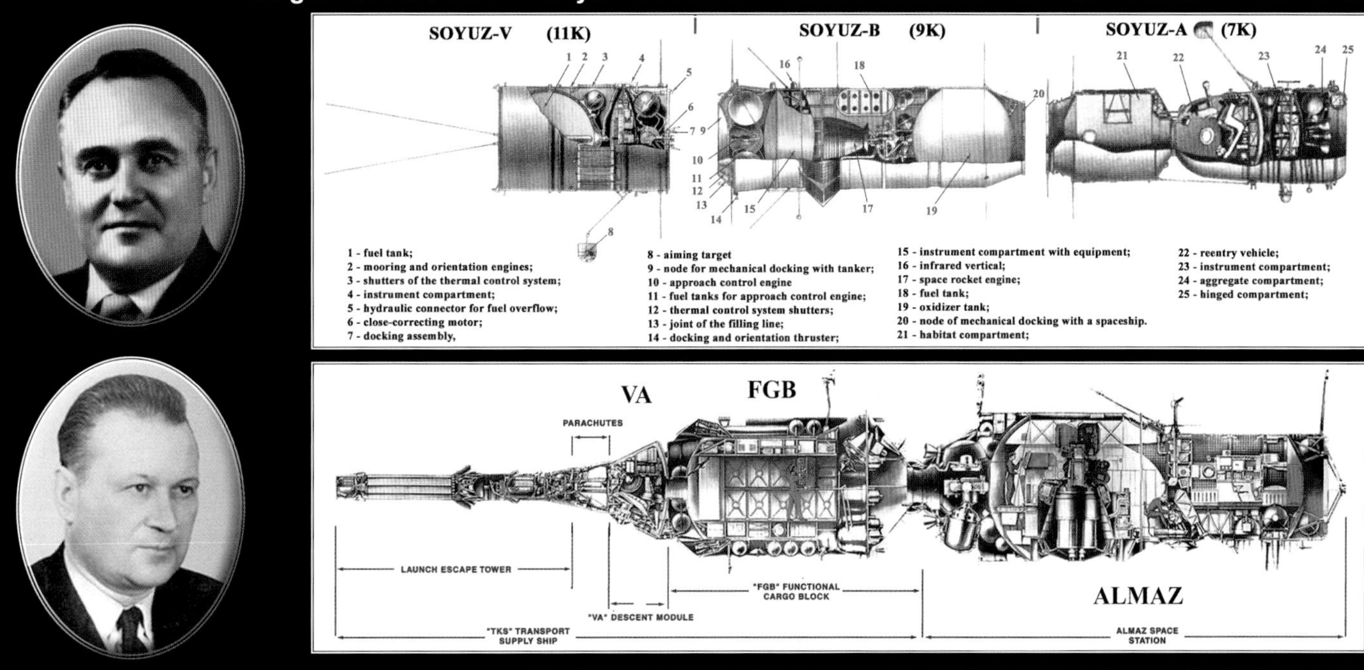

NASA MSFC/Douglas - "Apollo Applications Program" - 1965 to 1970

Since June 1962 plans had been underway to build a space station which could be launched on a Saturn rocket. Koelle's FPO had first suggested using an S-IC LOX tank, then at the end of 1964 von Braun asked Koelle to look at the S-II. Eventually an *S-IVB Orbital Experiment Module* (below left) or "wet" workshop was considered, in which part of the Saturn IV-B would still have an engine and be full of fuel (below). During its long development this station was just one element in the *Apollo Applications Program* (AAP). By 1966 this station became known as the *Orbital Workshop* and included a separate airlock as seen at lower right.

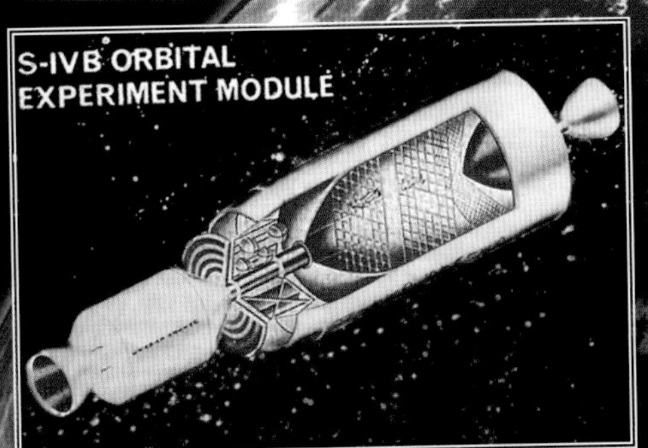

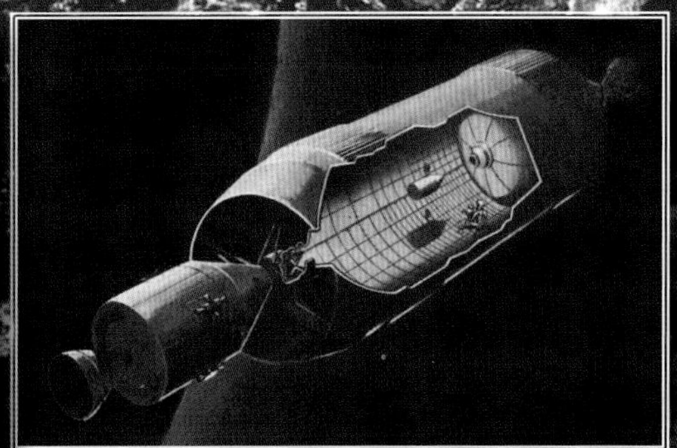

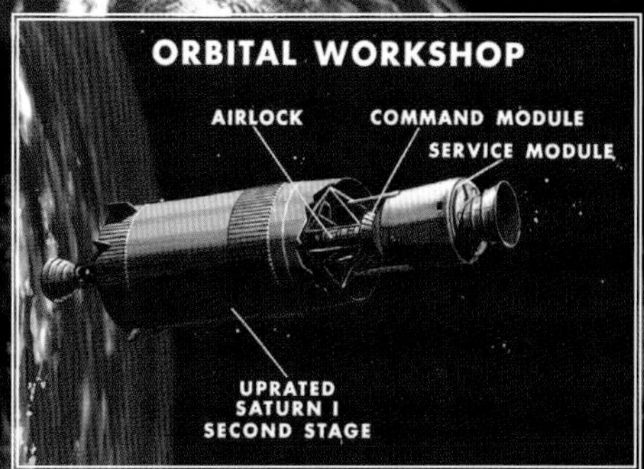

"The Orbital Workshop starts as the 2nd stage of the uprated Saturn I. A rigid metal grating-type floor, which also allows easy flow of the fuel, will be placed inside the stage's liquid hydrogen tank. Interior walls and ceiling support structures will also be preinstalled, as well as handholds, ladders, electrical wiring for lights, fan mounts, experiment brackets, and so forth. When other payloads are joined to the Multiple Docking Adapter, we have an Orbiting Workshop Cluster." - Wernher von Braun

AAP-1 THRU 4 ORBITAL CONFIGURATION

LM-A/ATM
S-IVB WORKSHOP
AIRLOCK
M&SS
DOCKING ADAPTER
CSM

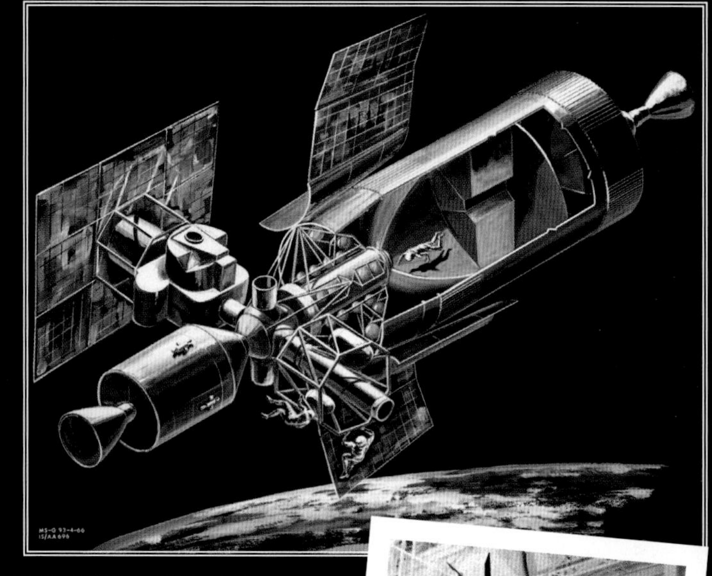

The AAP station had a multiple docking adapter attached to the airlock, this configuration became known as the "cluster". Later a Mapping and Survey System Module (red above left) and a modified Apollo lunar module (LM-A) pressurized observatory would be added. The Grumman LM went through many modifications in the hope that it might be used as a free flying lab or an observatory module. The LM-A/Apollo Telescope Mount (ATM) (right) was to be launched separately and docked using a combination of remote control and crew assisted guidance from the station. As the AAP design evolved the wet workshop was replaced with a ground prepared dry workshop (below) with no large engine at the base of the stage, and no LM-A.

"A US Federal agency gives a five-year budget projection. So in 1966, we were looking at 1971 and new activities beyond Apollo. Johnson Space Center set up the Apollo Applications Program Office. I became the Program Manager and began to sort out what the program should be, mainly centered on using Apollo hardware in Earth orbit." - Robert F Thompson

Martin/General Dynamics/NASA MSC - STB and BSM - 1965-1968

Over a period of more than three years, from May 1965 to September 1968, NASA MSC awarded the Martin Company $436,000 to develop a full scale mockup of the standard space station module which they called the *Subsystem Test Bed* (STB) (right). At the same time General Dynamics were awarded a $500,000 contract to develop the *Basic Subsystem Module* (BSM), an engine room designed for a space station which would maintain the station and life contained therein. The five basic engine-room subsystems included the structural shell, electrical power, environmental control, life support, and stabilization and control. The subsystems were further modularized to support a three-, six-, or nine-man crew. The end result of this mid-1960s design program was the creation of a standardized template for building space station modules suited for specific needs, that could be used both in orbit or on the surface of the moon. Most importantly, the modules were all 180" in diameter so that they would fit snugly inside the Saturn V launch vehicle.

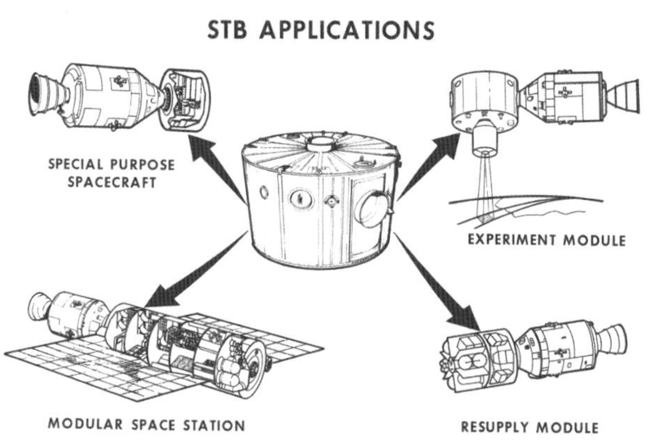

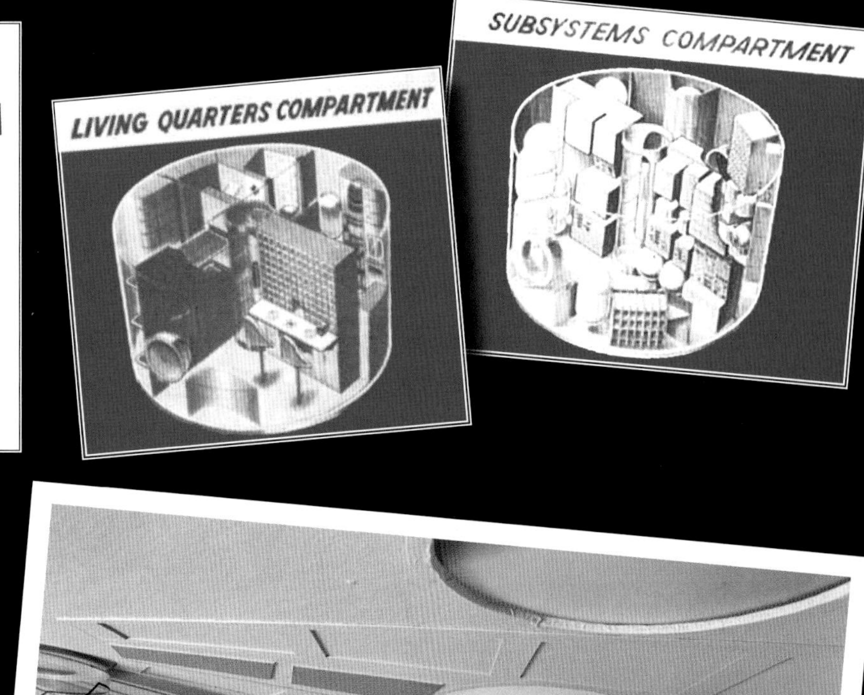

The 180" diameter module became the template for much of the work being done by the United States' contractors during the peak period of NASA funding, from 1965-1968. Almost all of the major players had a hand in designing and perfecting this one-size-fits-all approach to space station building. A single module could be used to fly alongside Apollo or several could be stacked like a layer cake to establish larger and larger stations (above). Each module could be equipped as needed (top right) and the large fifteen foot diameter interior would allow for relatively comfortable surroundings for the crew. The image at center right shows a model of the interior of one of the modules in a spin-up artificial gravity environment, which would come later as many other designs for this kind of module were postulated. One of the first station designs to be presented using this approach was the image at bottom right which shows a stack of four modules connected to an observation platform protruding at right angles. The internal configuration is shown below. Two early space shuttles are seen servicing the station and a free-flying module is seen at far right. Inset is an *Earth Resources Module* attached to the docking port. One of the principal flaws with this stacked approach was the lack of multiple avenues of escape for the crew in the event of an emergency.

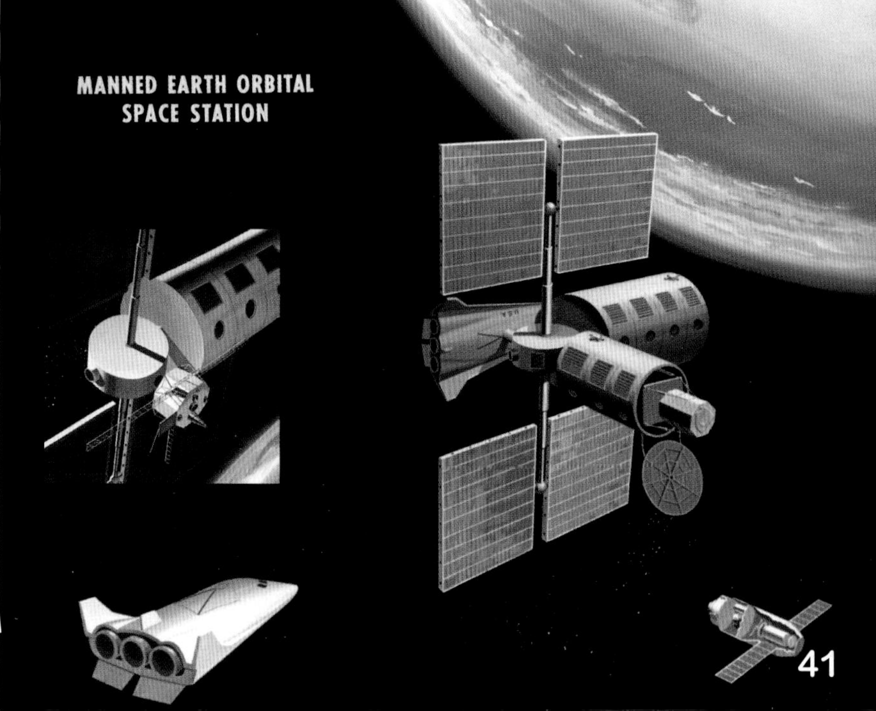

MSC NASA Planning & Steering Group - "I" Type Rotating Station - 1966

From June 1966 to November 1966 a study was conducted under the auspices of the Planning Coordination Steering Group of NASA to examine the needs for, the requirements of, and constraints to, a manned space station capable of supporting broad-based programs in science and technology, including earth-oriented applications. It was decided that most of the requirements for such a manned orbiting space laboratory could be satisfied by a space station operating in a zero-gravity mode, in a low-altitude orbit (200 miles), and at an inclination of 50° to 70°. Such a space station would be capable of accommodating 8 to 12 people. The study reviewed and analyzed the various types of zero-gravity and artificial gravity space stations, of both the Saturn I and Saturn V sizes. The analyses showed that the larger Saturn V stations were much more cost effective with respect to working space. A stack of the Boeing MPMMs was chosen to be telescoped out as the counterbalance for the so-called "I" type station seen at left.

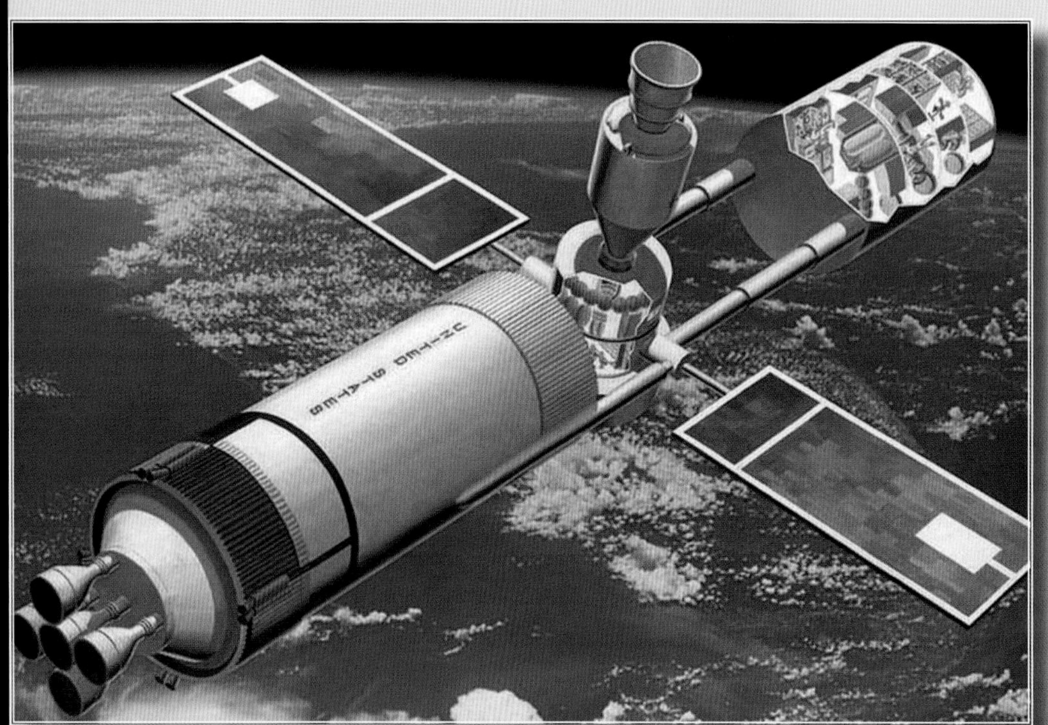

NASA MSFC/MSC/Langley/KSC Donlan Group - LDSS - 1966

A study group headed by NASA's Charles Donlan was conducted in-house from April to August of 1966 at MSFC with assistance from teams at MSC, Langley and the Kennedy Space Center. The result was the *Long Duration Space Station* (LDSS). Built from a modified S-IVB it was expected it could have been launched by 1975 and could have housed a crew of from nine to twenty four. It would have been launched on a Saturn V and some consideration was still given to providing artificial gravity in some areas. It was expected to last in orbit for five years with all of the scientific experiments mounted inside before launch. Although not shown in this image it would have been powered by 15kW solar cells. The instruments at the top of this image include a Schmidt Telescope, a Ritchey Chretien Multi Purpose Telescope, MM wave parabolic dish, Imaging X-Ray, Spectroheliograph, and a Coronograph. Although it used the same layer-cake approach as the stacked MPMM, the LDSS would have been integrated on the ground before launch. Donlan's committee delivered an eight volume report on the LDSS on November 15th 1966.

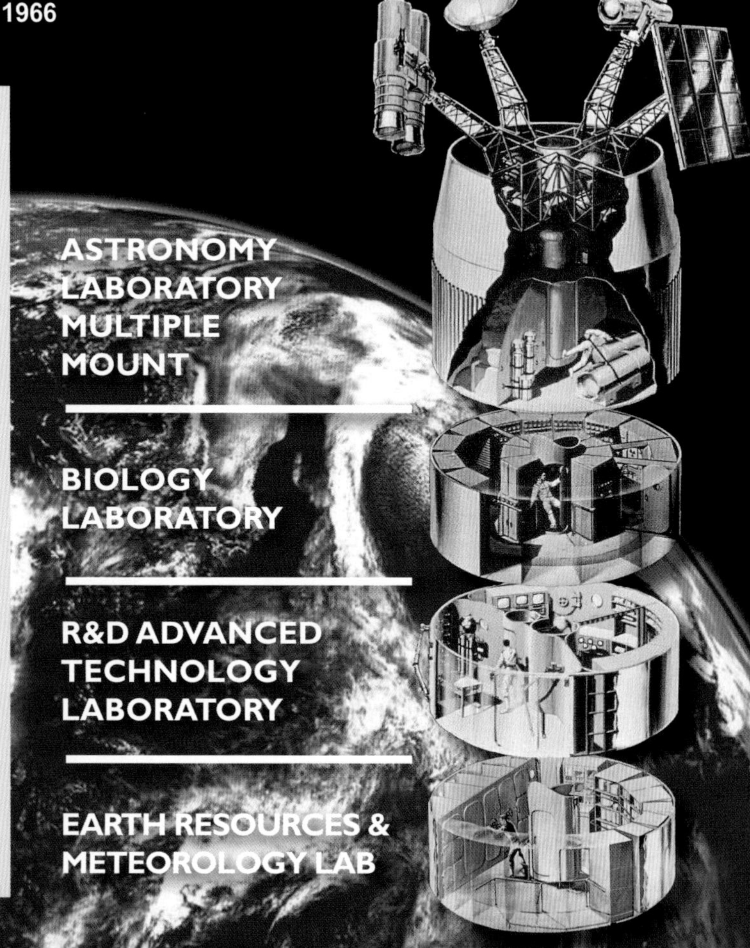

Krafft Ehricke/Convair - Space Hospital - 1966

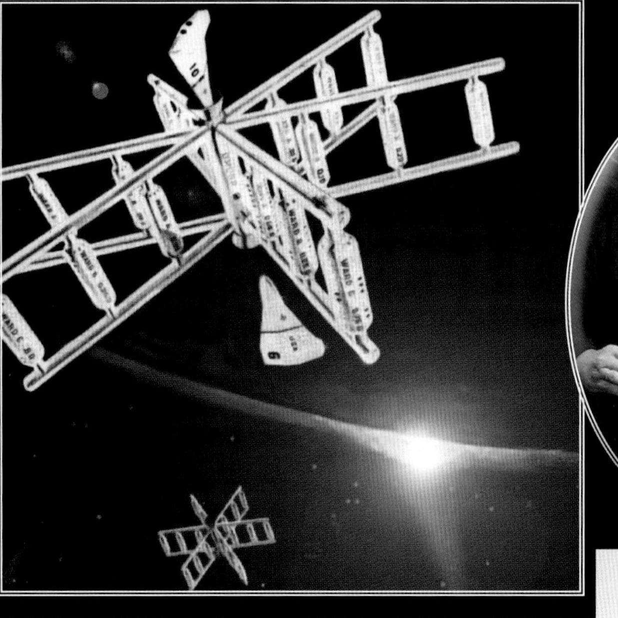

In 1966 Krafft Ehricke (center) demonstrated his plans for an orbiting space hospital (top left). His idea clearly influenced engineers at Douglas who responded with an MPMM based space hotel (top right). The idea for a large spinning hotel in orbit would resonate for years in the minds of film makers such as Stanley Kubrick, and the Hilton family, who began to seriously consider the viability of a commercial operation in space. Many years later hotelier Robert Bigelow began plans for an orbiting hotel which continues today.

Douglas - "EOSS" - 1967

From January to December of 1967 the Douglas Aircraft Company created an elaborate plan for their *Early Orbital Space Station* (EOSS). This multi-decked tiered space station was to have been based on a modified S-IVB and could accommodate a crew of six. The EOSS still contained many of the design elements from the earlier MORL station, including the onboard centrifuge, but the larger size allowed for multiple docking adapters and more expendibles. With its large supply of solar power and a design which allowed it to fit neatly on top of the Saturn V's large second stage, the EOSS was a scaled up version of the AAP station being proposed at the same time. Getting an early start on space stations was the main goal, with the original design (below left) to be in space by 1969-1971 and the advanced version with its modular companions by 1975-1977. Douglas worked closely with Jean Olivier at MSFC's Advanced Systems Office to best determine how the EOSS and MORL could be used as an *Orbital Astronomy Support Facility* (OASF). Eight different configurations and dozens of different instruments were considered. *Integrated* EOSS (far right), had all the instruments inboard, the *semi-detached* version (lower right) had modules which could come and go, whereas the "*detached*" had modules which might be launched with the station but would then fly in tandem; known as *free-flyers.*

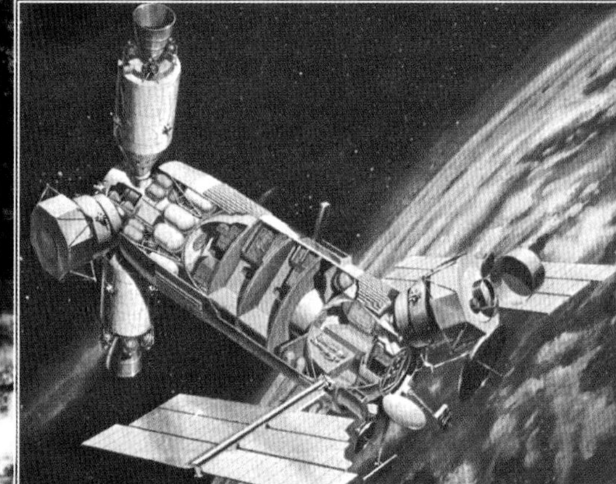

"This was the first time that we began discussing the possibility of separate modules which could operate independently of the space station, but dock periodically for resupply and data retrieval." - William R. Lucas

Douglas - "Commercial Space Station" - 1967

▲ The enormous lifting power of the Saturn V emboldened contractors like Douglas to also envision a time when ▲ this 130 ft long, 33 ft diameter, 285,000 lb space laboratory (left), powered by a Uranium-Zirconium Hydride nuclear reactor, could welcome four privately owned commercial modular labs at its multiple docking ports.The Douglas commercial space lab would accommodate a crew of 32 people and was supplied with ample nuclear power from the reactor at the end of a 100 ft boom. Modified Apollo command modules could be used as laboratories. Douglas also believed it would be possible to build a much larger second generation version capable of housing 400 crew, by the early 1990s (right).

Boeing - NASA - Saturn V Single Launch Space Station - 1967

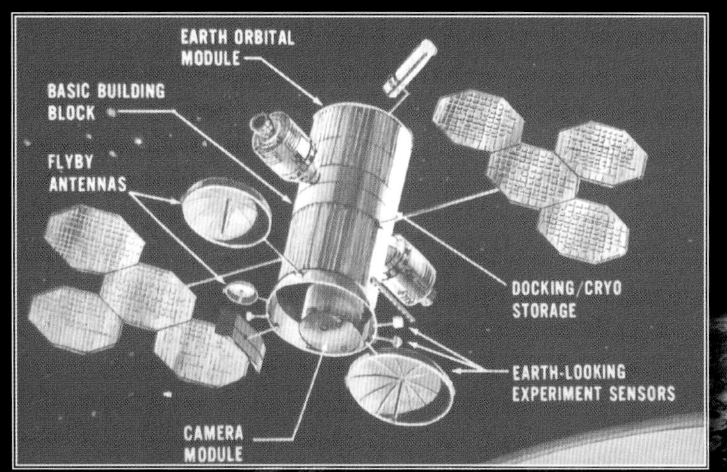

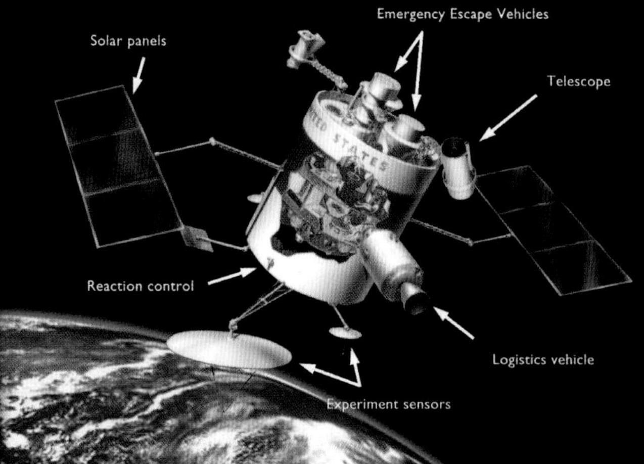

▲ From March to November of 1967 the Boeing company worked on what was known as the *Saturn V Single Launch Space Station.* The NASA contract was divided into two parts with the basic study covering the conceptual design of an earth-orbital space station launched on a Saturn V booster. At least five configurations were considered, with anything from two to six decks (above). The baseline was a heavily modified S-IVB at 189,000 lbs, with four decks, that was just over 21' in diameter and similar to the EOSS. An addendum to the contract covered the feasibility of using a Mars flyby spacecraft in low earth orbit, as described in NASA's *Joint Action Group (JAG) Planetary Missions Study*. The final preferred version (right) was a 33' diameter, twin deck, 232,000 lb variant of this planetary vehicle (very similar to the image seen at above right.) However, the study showed that it definitely was not feasible (based upon projected or anticipated funding levels) to consider such a space station in the stated time period. The addendum however, determined that it was technically feasible to use a major portion of a Mars flyby spacecraft as an earth-orbital space station.

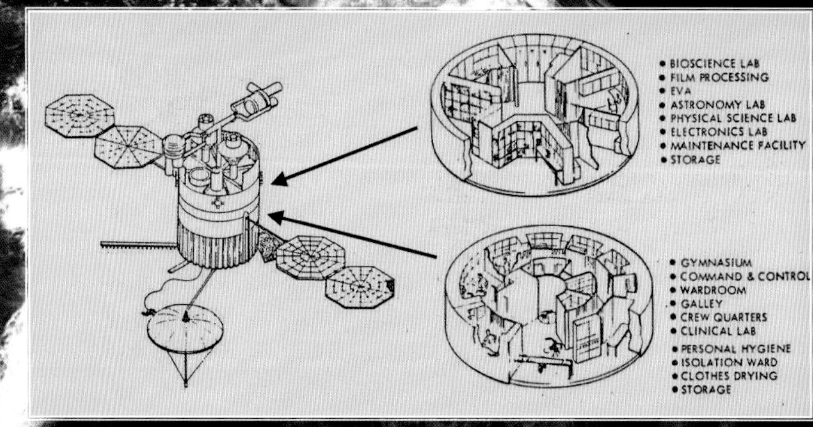

"A space station is a "base" in space, equivalent in function to those used in many forms of terrestrial exploration, for example, those in the Antarctic. The base has six important characteristics: it is a central location for power, volume, logistics, experimental equipment, communications, and data reduction. Such a definition encompasses schemes in which all of these things might be located within a single volume, or alternately where there may be several devices which are orbiting the Earth in close proximity with one another. An example of this latter form would be a station which consists of a large living center surrounded by free modules containing telescopes, Earth Sensors, and specialized scientific satellites orbiting close enough so that they could be tended or serviced intermittently by personnel from the living quarters." - Robert Gilruth

Robert Gilruth - NASA MSC - "S-II Station" - 1968

In the summer of 1968 NASA's Robert Gilruth, who was running the Manned Spacecraft Center, made a presentation in San Antonio Texas in which he made his case for the space station. Gilruth showed his audience a 1,000,000 lb space station that could house a dozen crew (center left). The team at the Manned Spaceflight Center in Houston created the design between April and July of 1968. This concept allowed for many more modules to be added onto the main module, or on to the axis core, while still adding S-II counterbalances at the other end. The space station would have included a power base, a volume base, a logistics base, an experiment equipment base, a communications base, and a data-reduction base. But this permanently rotating station was soon replaced. It was decided that a 12-man-capable core module would be put through a temporary artificial gravity phase to conduct experiments (below). Once these studies were concluded the S-II would be deorbited and space shuttles would begin to deliver an assortment of modules to attach to the core or to fly in formation with the main station.

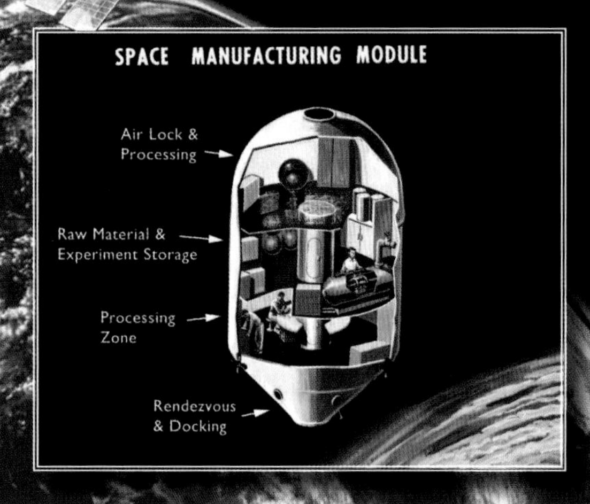

An early version of the 12-man station (left) would have used solar panels attached to the main fuselage. Key characteristics of its proposed companion a free-flying space manufacturing module (above) were deemed to be:- ready access to vacuum; zero gravity; raw material and experiment storage; and processing chambers. A proposed Earth Resources Module would need sensors pointing at the Earth; astronaut access to equipment for calibration; film retrieval and analysis; and a suitable control station.

In February 1969 Dr. William R. Lucas the Director of Program Development at the Marshall Space Flight Center spoke before the House Subcommittee on Manned Space Flight. In his presentation he broke down all of the work on space stations undertaken at Marshall, Langley, Lewis and at MSC in Houston during the previous four years. He explained how NASA's teams had adapted to the evolution of the launch vehicles and had gradually moved from the Saturn IB launched MORL class of station, to the LDSS, the EOSS, the Saturn V Workshop and the "B" series of workshops before arriving at the IOWS (bottom). That same year Dr John D. Hodge was transferred from Chief Flight Control Division of the MSC Lunar Exploration Group to become Director of Advanced Manned Missions. In this role he would lead the planning in Houston for a large manned space station. In March of 1969 Hodge informed the Congressional Committee on Science and Astronautics that his superiors had asked for suggestions for what to do with the unused Apollo hardware. Hodge took the opportunity to show the committee the plans for a large nuclear powered space station to be launched by up to six Saturn V rockets. In his presentation Hodge used the Martin STB as an example of how the budget for advanced projects was being spent. This began what became known as the Phase B studies.

NASA MSFC/Langley/MSC/KSC Saturn V Workshop - 1968

From January to March 1968 in-house teams at NASA's Marshall Space Flight Center, Langley, MSC and KSC designed what became known as the Saturn V Workshop "B" Configuration. Once again it was to have been based on a modified S-IVB. It was similar in many ways to the Skylab "dry" workshop in that it would be fully equipped on the ground prior to launch. Configuration B-1 would have used more solar panels than the proposed wet workshop, providing up to 15kW. The Apollo Telescope Mount would have been placed where the J-2 engine would normally have been fitted. Launch date was predicted as 1971 with a crew of three. A more advanced version known as Configuration B-2 incorporated individual floors and a pressure tunnel running from one end of the station to the other for safety purposes. An ATM-type astronomical experiment module was added that could be pressurized so that the equipment could be worked on in an IVA fashion. The configuration could also operate with remote experiment/application modules. All the configurations carried their expendables on the initial launch, and crew rotation varied from 60 to 90 days.

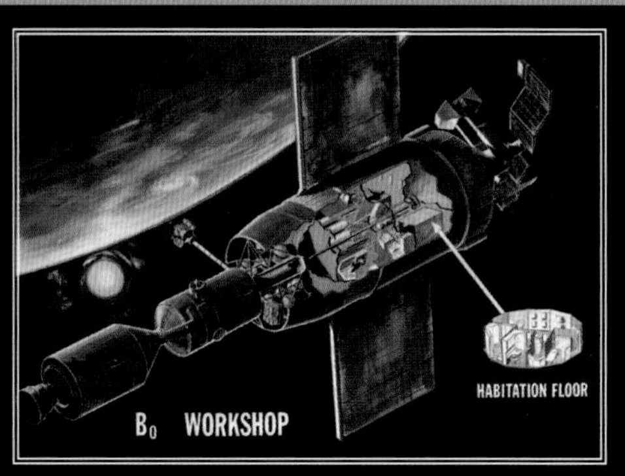

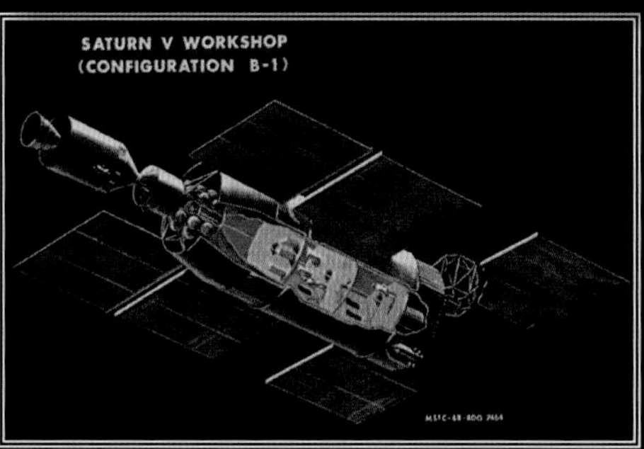

"Again the S-IVB hardware would be utilized and launched on the Saturn V, that is, the first two stages of Saturn V, the S-IC and the S-II stages. This was to have been an extension of our AAP program, but it was determined that this program would not be a sufficient advance and therefore, we proceeded into the next study, the Intermediate Orbital Workshop." - William R. Lucas

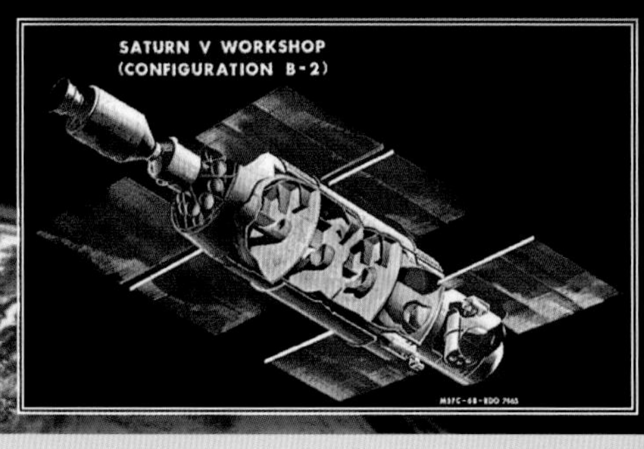

NASA MSFC/Langley/MSC/KSC IOW Modular - 1968

The *Intermediate Orbital Workshop* (IOW) was designed at NASA in three configurations between May and July of 1968 and could house a crew of six with a 90-180 day rotation. Powered by 15kW of solar panels or a nuclear reactor, the modular configurations considered were all well below the Saturn V orbital capability. The basic 260" station module consisted of subsystems, crew quarters, and the integral experiment laboratory. This wider module was favoured because it reduced the number of "Category 1" interfaces (involving crew safety) between station compartments and allowed for a central tunnel. The 6-man core of the hybrid version would be considered for the Phase B studies to accommodate up to 12 crew. In order to accomplish operation with the larger crew sizes in later stages of the program, a 6 or 9-man "Big Gemini" logistic spacecraft was also considered.

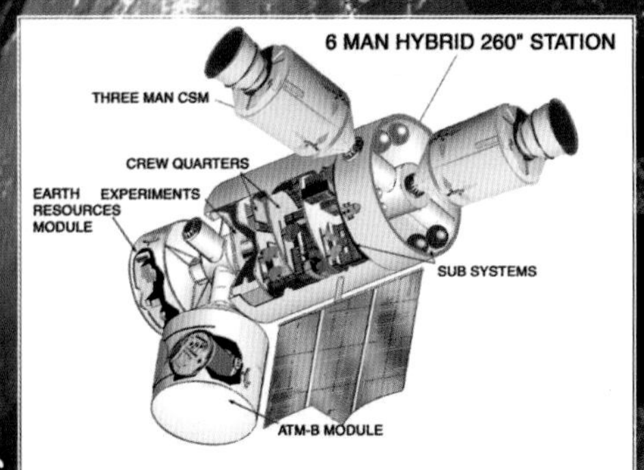

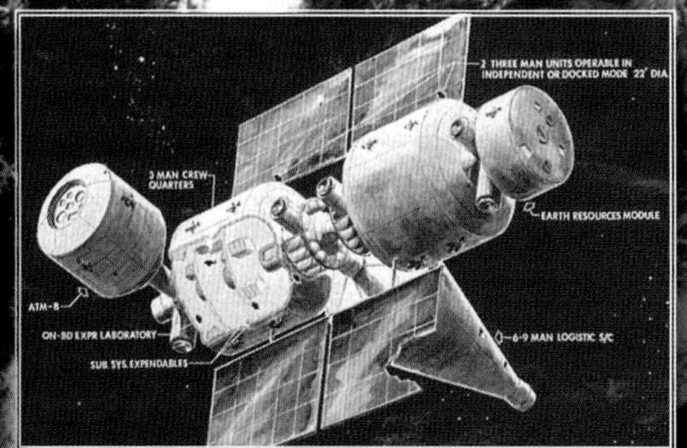

"This was the time that we began consideration of a modular build-up concept. The modules would accommodate six men and additional ones could be flown up as the situation changed." - William R. Lucas

In 1969 NASA administrator Thomas Paine approved the contracts for the *Phase B* space station studies. The Manned Spacecraft Center in Houston partnered with North American, while the Marshall Space Flight Center in Huntsville partnered with McDonnell Douglas. The goal of the studies was space station program definition, logistics system interfaces for resupply, and growth to a permanent space base or planetary module. Almost 150 man-years went into the studies. Some of the goals were to investigate compatibility with a synchronous station, a lunar station or a manned Mars mission. Contractors were asked to consider a 12-man station with an eye to ending up with a 50-man crew. The initial station was to be operational by 1977 with the large *Space Base* operational in the early1980s. The station was to be launched on two Saturn V's, have a ten year lifespan, a maximum diameter of 33 ft, capable of both zero G and spin-up, and a closed loop environmental system. It was to then be capable of being incorporated into a larger Space Base. The Space Base was to be able to accommodate zero-G and artificial gravity simultaneously. NASA issued a *Blue Book* in May of 1969 which would allow the contractors and agency centers to synchronize their efforts in designing the space base which could remain useful for at least a decade. A full-size mockup was built by North American Aviation in California.

NASA /McDonnell Douglas/Martin/North American Aviation/GE - Space Station "Phase B" Studies 1969 - 72

At the heart of this program was the central core which both North American (right) and Douglas (below) envisioned to be much the same. Beyond the basic 12-person core module this broad plan included definitions and outlines for an assortment of accompanying modules which would be brought to the station by a space shuttle. These experiment modules could be attached to the core docking ports or allowed to fly in formation and occasionally be brought back to dock for refurbishment or resupply.

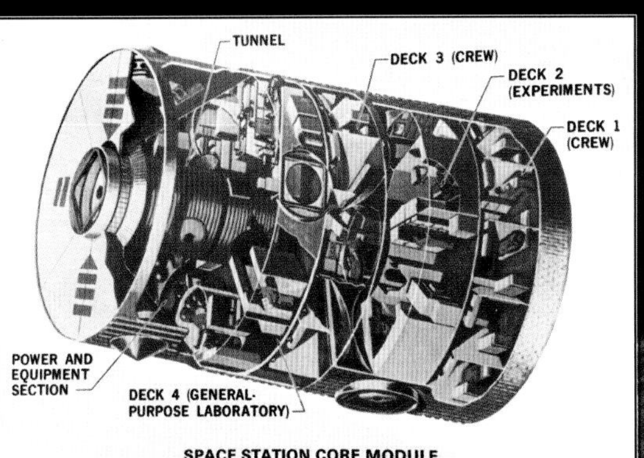

SPACE STATION CORE MODULE

The Douglas version seen being assembled (below left) used an S-II stage counterbalance and a telescoping turret (below). This would have still allowed artificial gravity experiments to be conducted. These core modules would be serviced by supply ships and space "tugs" would help effect close-up docking procedures for the modules. A central tunnel was one of the main safety features (right). An expanded version (bottom center) featured a large nuclear reactor on the boom.

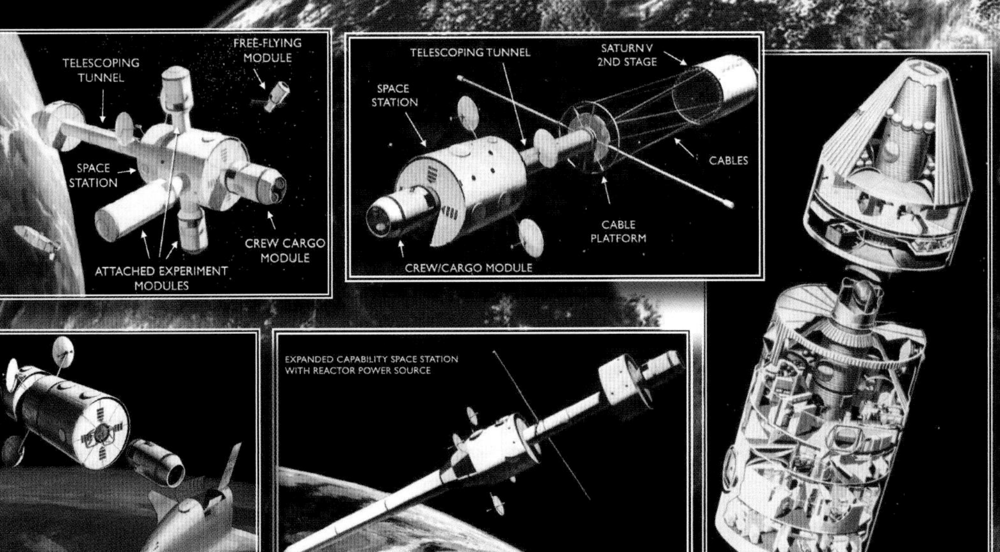

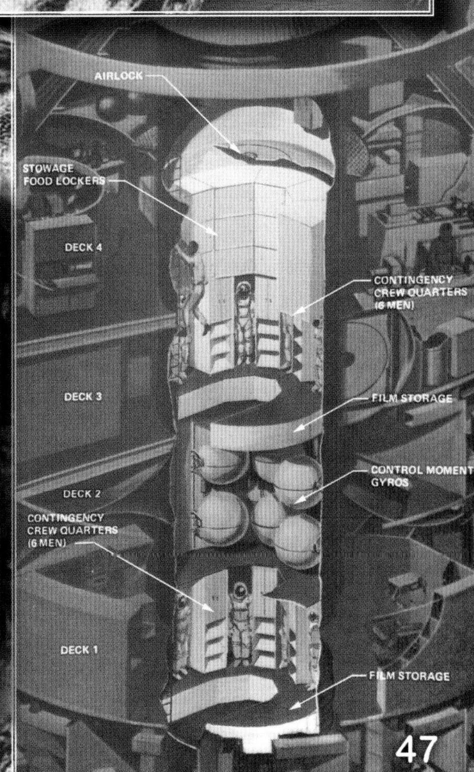

Accommodation aboard the Space Base would have been generous, including individual quarters that allowed the international crew to sleep in bunks, as well as command centers with artificial gravity (right).

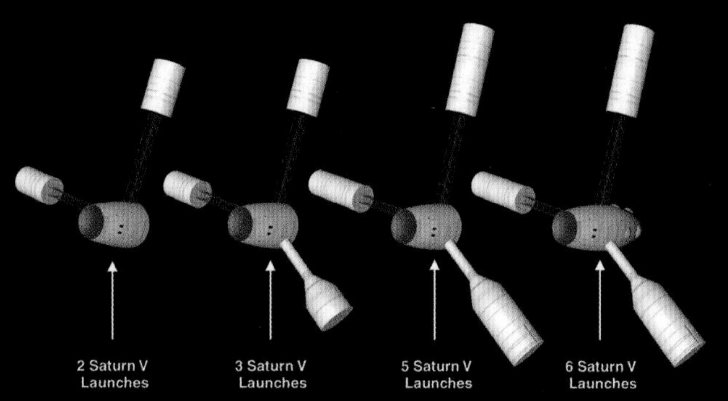

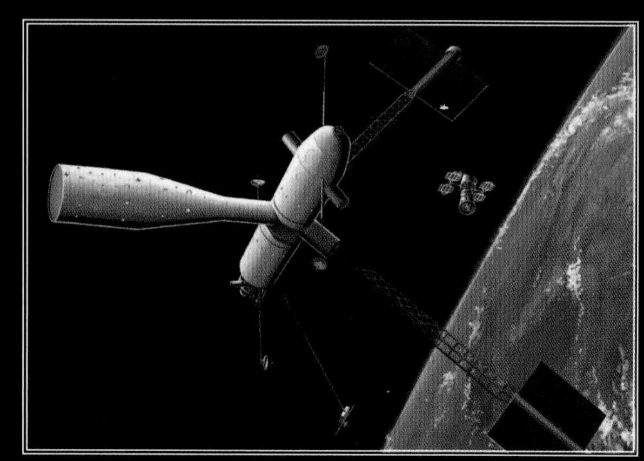

The designs for the Space Base came from NASA and were given to the contractors to study. Using the station as the core it would have been enlarged with each Saturn V launch (above). The "Y" configuration (above right) has the 12-man core at left and two nuclear reactors on the other arms. Crew quarters would be situated on the end of the 150' habitat arm and they would be rotated to provide from 0.3 to 0.7g. The two off-axis arms in the Douglas "hub" version (below) and its North American equivalent (right) are two nuclear reactors. In these the 12-man core is integrated into the non-rotating hub. The basic Space Base would later have modular labs attached to the central axis (bottom left & below right). Eventually the 12-man space station would evolve into the 50-man or even 100-man Space Base.

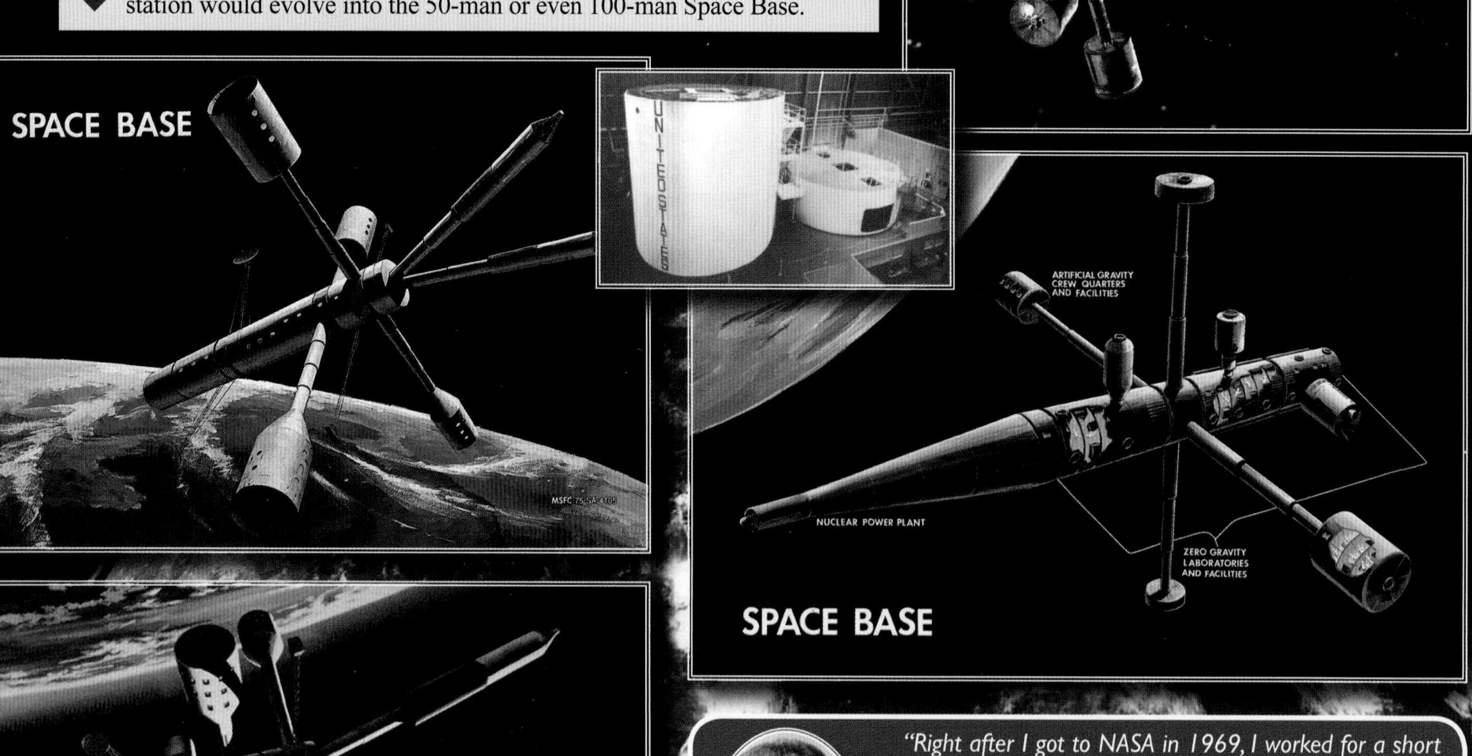

"Right after I got to NASA in 1969, I worked for a short time on the space station program. I remember going out to the contractor at Seal Beach, California. They had a model of the station based on an S-II, the Saturn second stage, which was huge. (inset above) That stage is thirty-three feet in diameter and 80 feet long; a big diameter and four decks; it was a fantastic station." - Henry Hartsfield

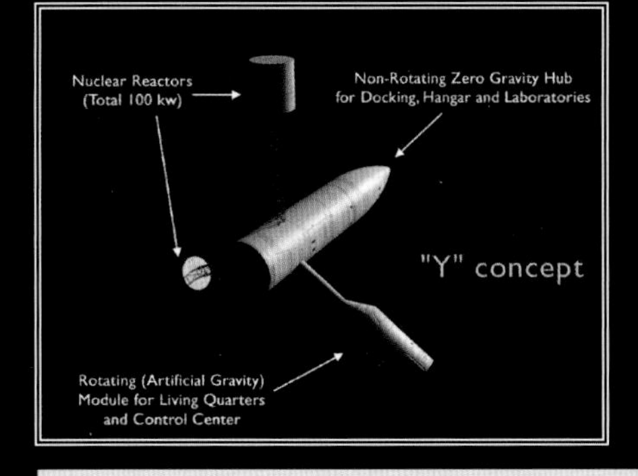

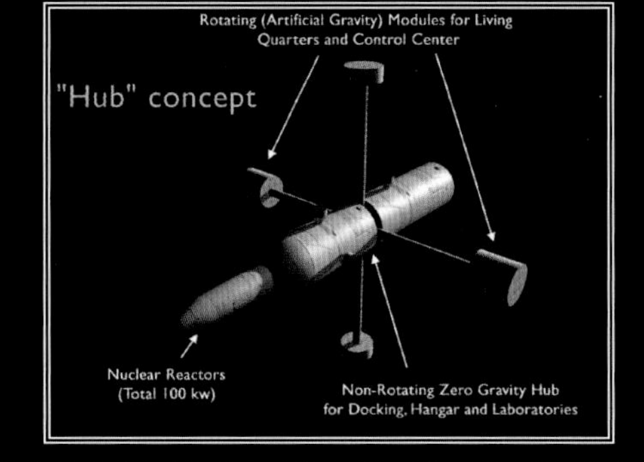

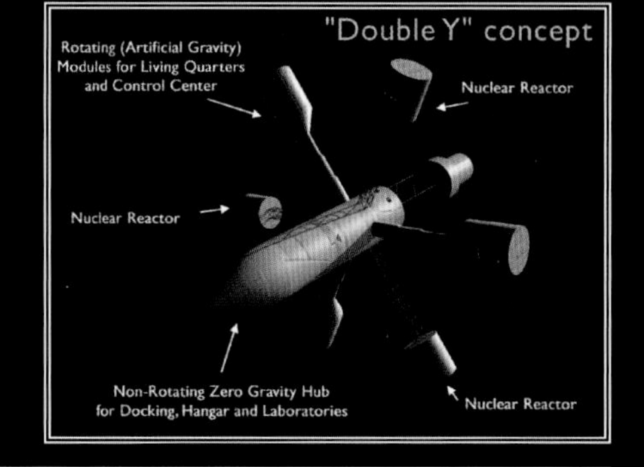

▲ Plans for the Space Base proposed three different configurations, launched atop as many as seven Saturn V's. At left the "Y" configuration provided ample power through four nuclear reactors. In the center image the zero-G lab volume is maximized. The right image shows three habitation modules at 120° intervals and more nuclear reactors to meet the additional power requirements.

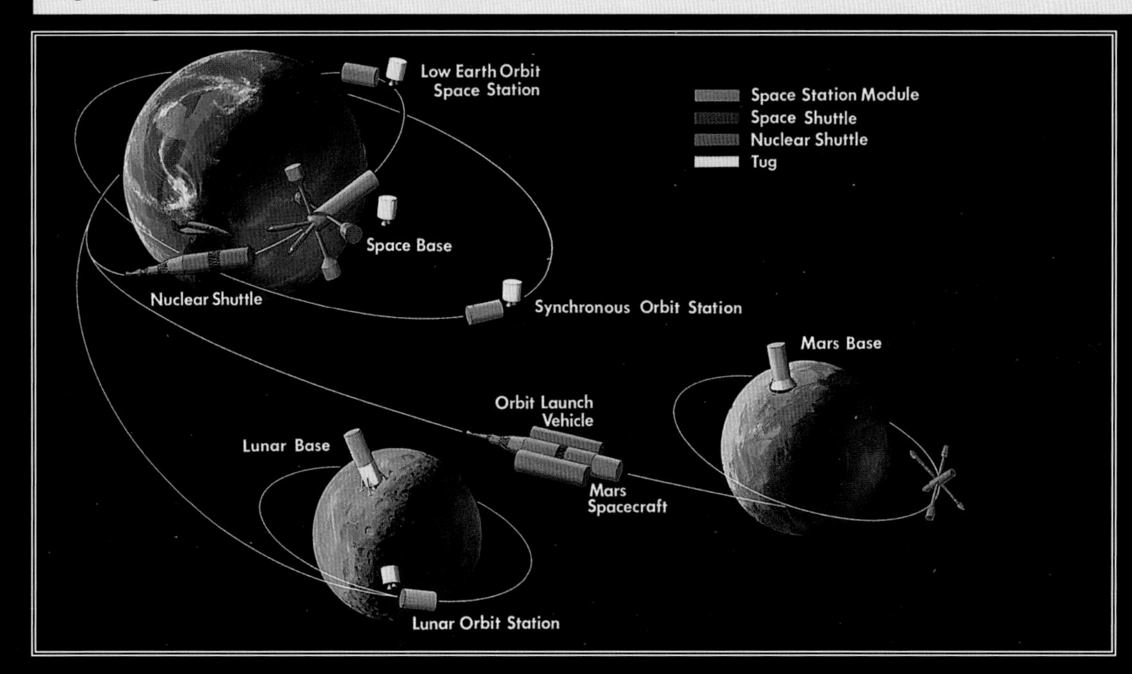

▲ The grand plans for using the Saturn V for launching the Space Base would have allowed the same equipment to be used on the Moon and Mars (left and above).

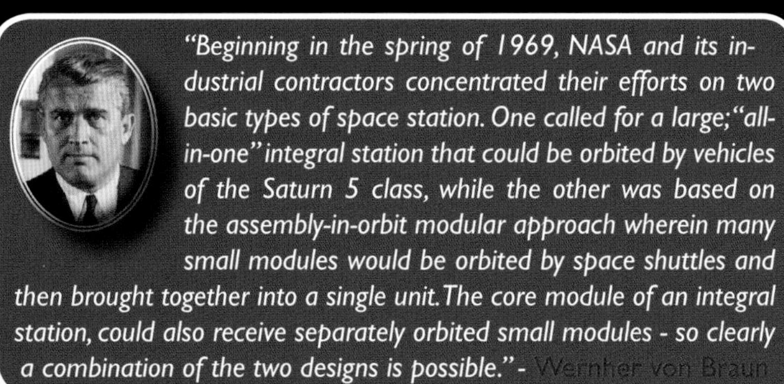

"Beginning in the spring of 1969, NASA and its industrial contractors concentrated their efforts on two basic types of space station. One called for a large; "all-in-one" integral station that could be orbited by vehicles of the Saturn 5 class, while the other was based on the assembly-in-orbit modular approach wherein many small modules would be orbited by space shuttles and then brought together into a single unit. The core module of an integral station, could also receive separately orbited small modules - so clearly a combination of the two designs is possible." - Wernher von Braun

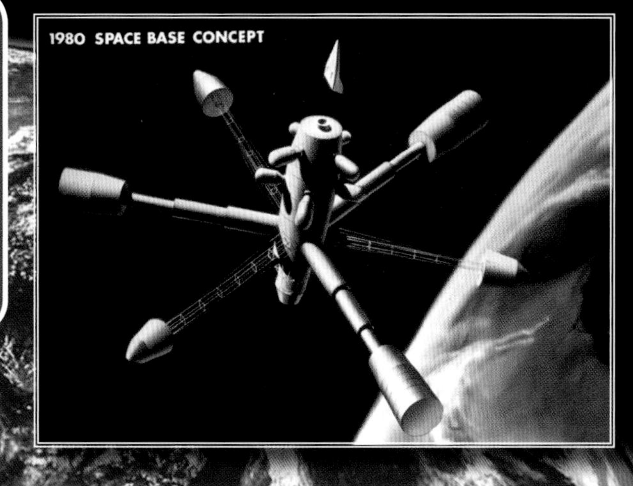

◀ Two more images from immediately after the moon landing when NASA was lobbying to keep the Saturn V, while also developing a reusable shuttle. The picture at left shows two space shuttles arriving at a different configuration of the Space Base which would still have required the Saturn V to launch it. This base has four arms, two for reactors and two for habitats. By 1980 (above) the "Double Y", powered by multiple nuclear reactors, would have served as an OLF to launch planetary and lunar missions. However, no new contracts for Saturn V rockets had been issued since 1967. ▲

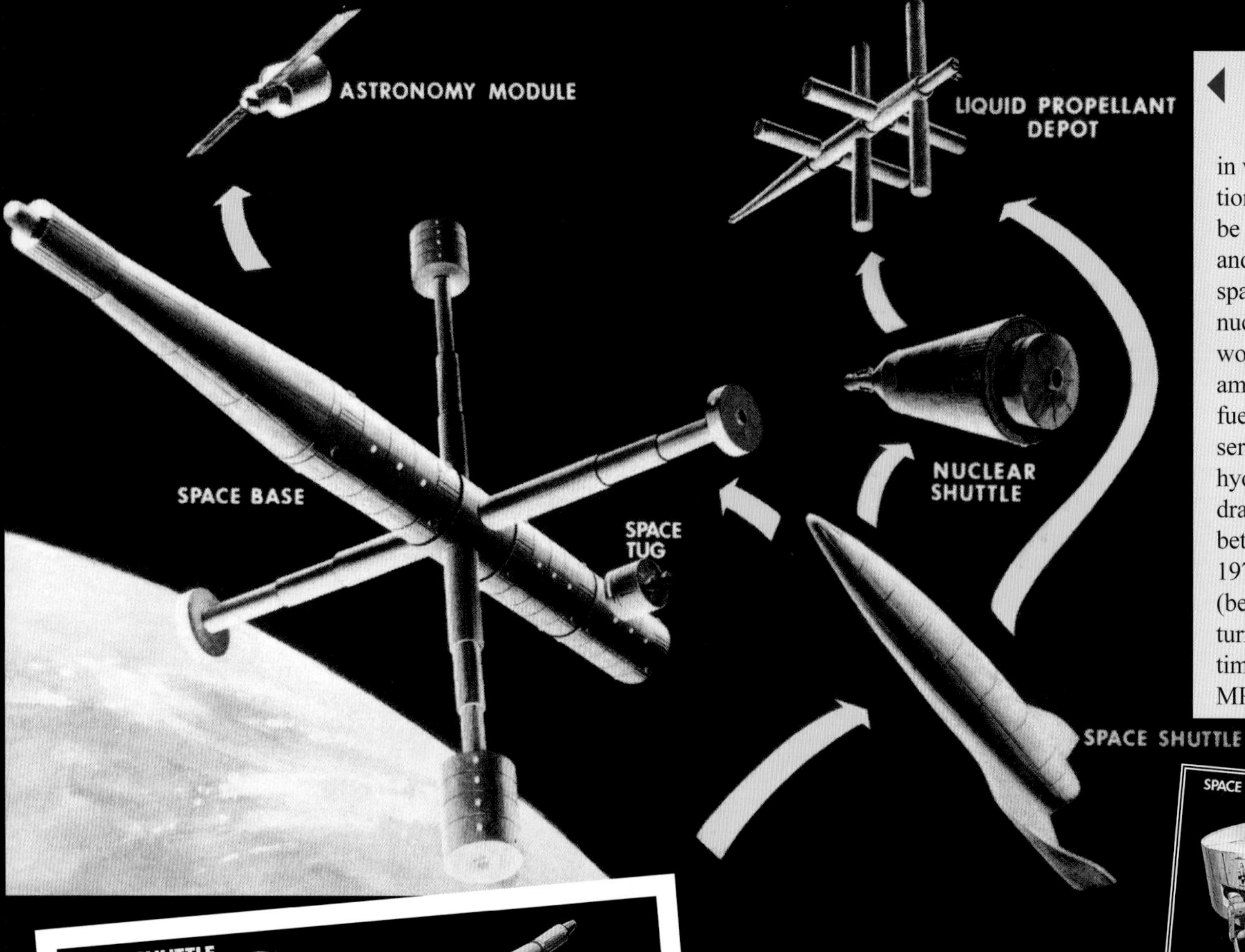

This proposed Phase B infrastructure led to studies in which the space station was once again to be used as a refuelling and transit point. Deep space missions using nuclear propulsion would require large amounts of hydrogen fuel, consequently a series of plans for huge hydrogen stations were drawn up by NASA between 1969 and 1971. The Space Tug (below) once more returned to the table, this time derived from the MPMM.

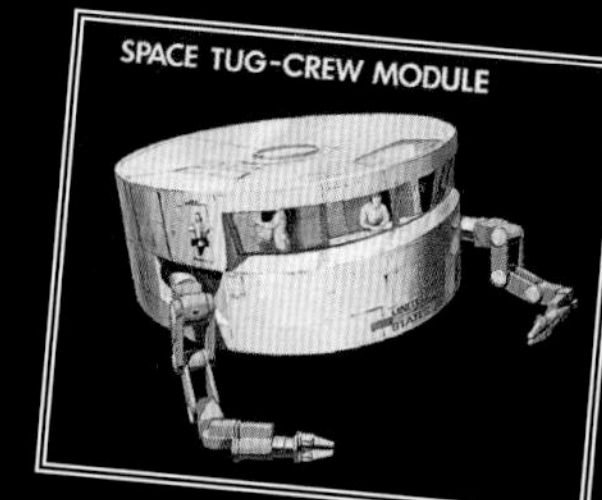

SPACE SHUTTLE MISSIONS
ORBITAL PROPELLANT STORAGE FACILITY
CAPACITY PER SPHERE 17,000 ft^3 LH^2 (75,000 Lbs)
CAPACITY TOTAL 120,000 ft^3 LH^2 (525,000 Lbs)
POWER 100 KW

The image above shows a low earth orbit space shuttle being used to service the massive propellant depot for a nuclear lunar and planetary shuttle.

A typical nuclear orbital launch vehicle is seen refuelling at the orbital hydrogen depot. It could then be used to take the large station/habitats to the moon or Mars.

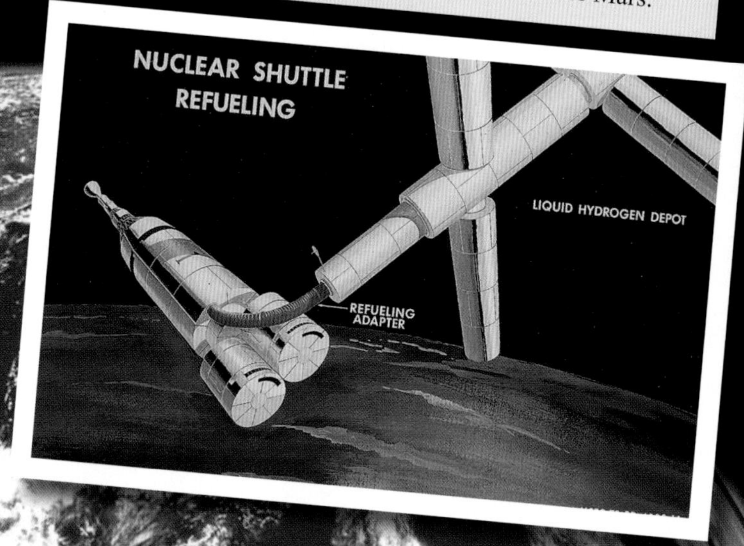

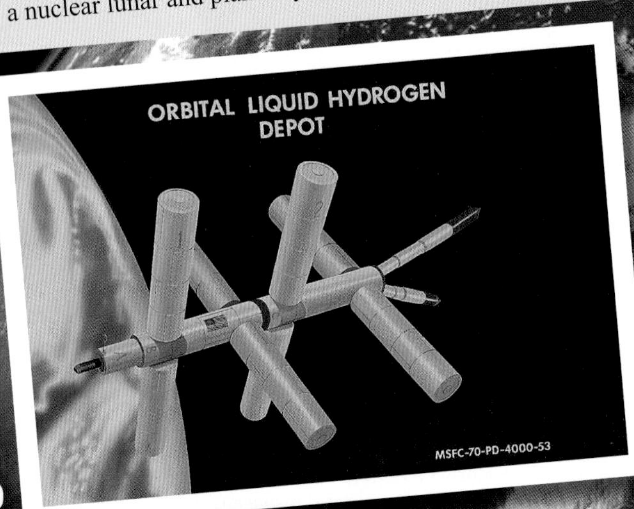

Orbital refuelling depots and storing large quantities of propellant in orbit had been studied by people like Jean Olivier since *Project Horizon* in 1959, but as the Apollo program came to an end in 1972, the NERVA nuclear engine program was also cancelled. This brought a long hiatus to the era of planning for orbital fuel depots and manned deep space missions.

Grumman - “Space Base” - 1968

Although not selected as a prime contractor for the Phase B space station studies, Grumman would receive a contract for Crew Operations. In 1968 the company submitted this axial modular design for consideration. Built around a group of large Saturn V launched modules the whole Grumman station would have rotated to provide artificial gravity in the three outlying four-storey habitation modules. The Grumman presentation model (above) was reminiscent of the futuristic *Star Trek* television show sets.

Following on from the Douglas OASF studies done in 1967 and 1968, the next logical step was to consider the various options for free-flying or semi-detached modules which could produce science onboard a space station. These studies which suggested an entire array of dozens of possible astronomical instruments to fly in space were conducted by Douglas and overseen for NASA by Jean Olivier. Originally these modules would have flown in conjunction with MORL or EOSS, but now the possibilities offered by the enormous Space Base concepts allowed for more, bigger and specialized modules, including a free-flying space telescope. In the two years following the start of the Phase B Studies the *Blue Book* would be updated several times to accommodate changes in expectations.

Convair/General Dynamics - “Research Modules” - 1970

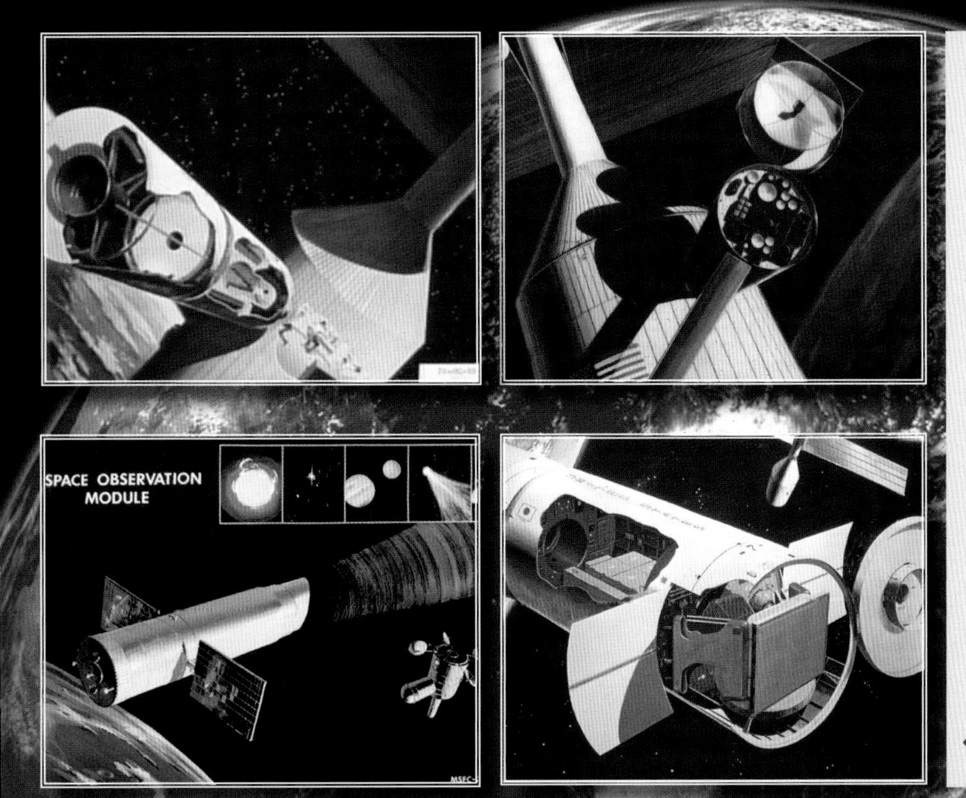

Convair/General Dynamics experiment modules were overseen by Jean Olivier at MSFC. In October 1970 Convair identified 17 candidates for stand-alone modules, a Grazing Incidence X-Ray Telescope, a Stellar Astronomy (top right), Solar Astronomy, High Energy Stellar Astronomy, Plasma Physics & Environmental Perturbations, Cosmic Ray Physics Laboratory (bottom right), Primates (Bio A), Small Vertebrates (Bio D), Plant Specimens (Bio E), Earth Surveys (top left), Remote Maneuvering Subsatellite, Materials Science & Processing, Contamination Measurements, Exposure Experiments, Fluid Physics in Microgravity, Component Test & Sensor Calibration, and a Physics & Chemistry Laboratory. Some of these would be man-tended and some automatic free flyers (bottom left). Considerable thought was also given to using the Apollo CM and LM, as well as the MPMM as semi-detached stand-alone laboratories

By the beginning of 1970 the integral 12-man station had been pushed back to 1977 and the larger *Space Base* was not expected to be flying before 1986. But none of this would happen because by July of 1970 it became clear that the massive launching power of the Saturn V would not be available. In January 1971 with budget constraints and the imminent demise of NASA's heavy lifter looming on the horizon, parallel studies were begun at both MSFC in Huntsville and at the MSC in Houston to see if a space station constructed entirely from space shuttle compatible modules might make more sense. The Integral Station would have been based around the 33 ft diameter core module launched by the Saturn V, but a modular station would be entirely launched by space shuttles and assembled in orbit. The key to this was to try and find a commonality in the structure of the modules which could be used for different purposes.

NASA/MSFC/MSC/NAR/Douglas - "Shuttle Launched Modular" - 1971 to 72

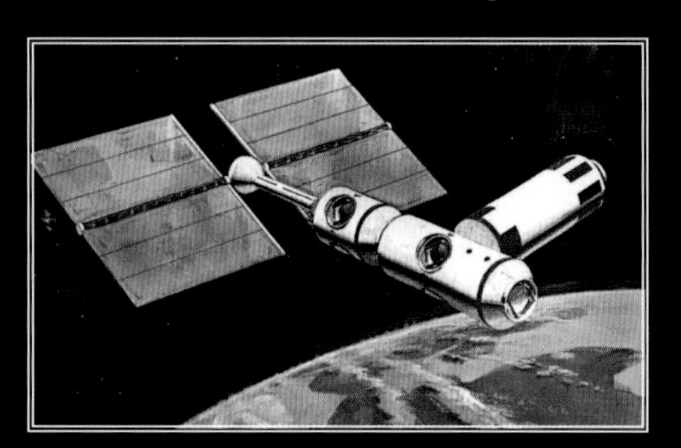

Up to seven modules could be attached to the core of the Integral Station but in the case of a Modular Station the entire station would be constructed over 17 flights. An assortment of configurations was considered, almost all of which would be launched and assembled in a manner reminiscent of the very earliest *Orbital Launch Operations* plans made by Vought and Koelle's team in 1960 - by connecting multiple nodes and modules in an incremental fashion. The first step would be to launch what was known as the *Initial Space Station*. This consisted of a power subsystems module, a crew operations module and a general purpose laboratory (left).

Taking the *Initial Space Station* to what was known as "maximum cluster" (bottom right) was done by attaching four *Research Application Modules* (RAM) which would be configured according to experiment needs. Each RAM would arrive via space shuttle (right). RAM modules came in an assortment of configurations (below). Many of the scientific payloads were configured to study the earth and the local space environment since the modular space station would be a laboratory and not an OLF.

RAM

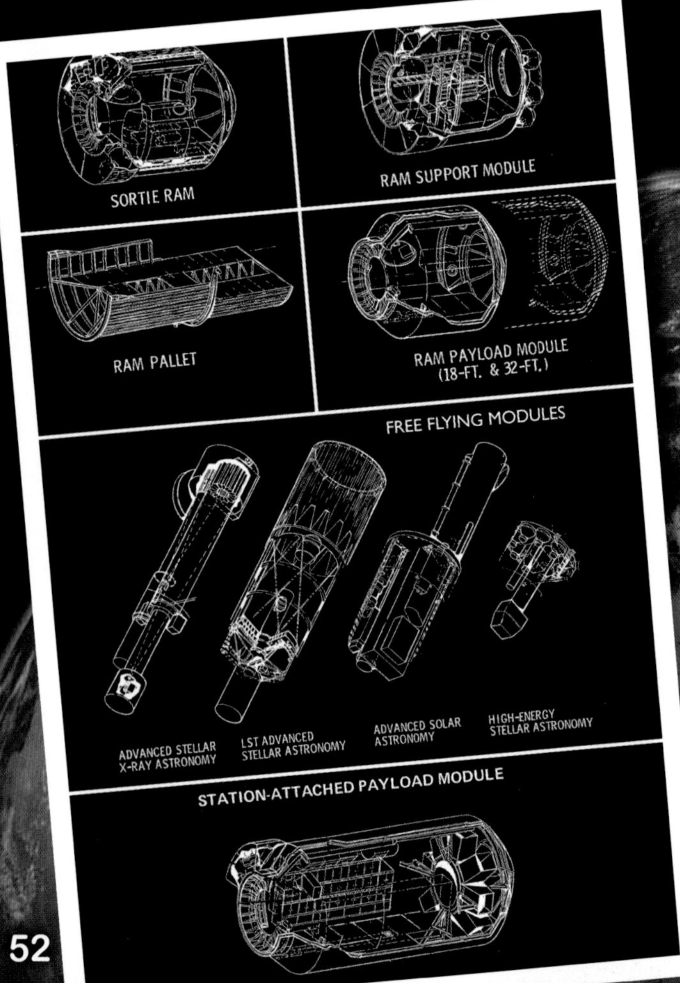

A General Dynamics/TRW cutaway of a maximum cluster configuration for the modular station.

Finally the station would become known as *Growth Space Station* (GSS) by adding duplicate crew operations and power subsystems (right). This would then house a crew of 12 and match the capabilities of the Saturn V launched Integral Station. Full-size mockups were created of the laboratory module (below).

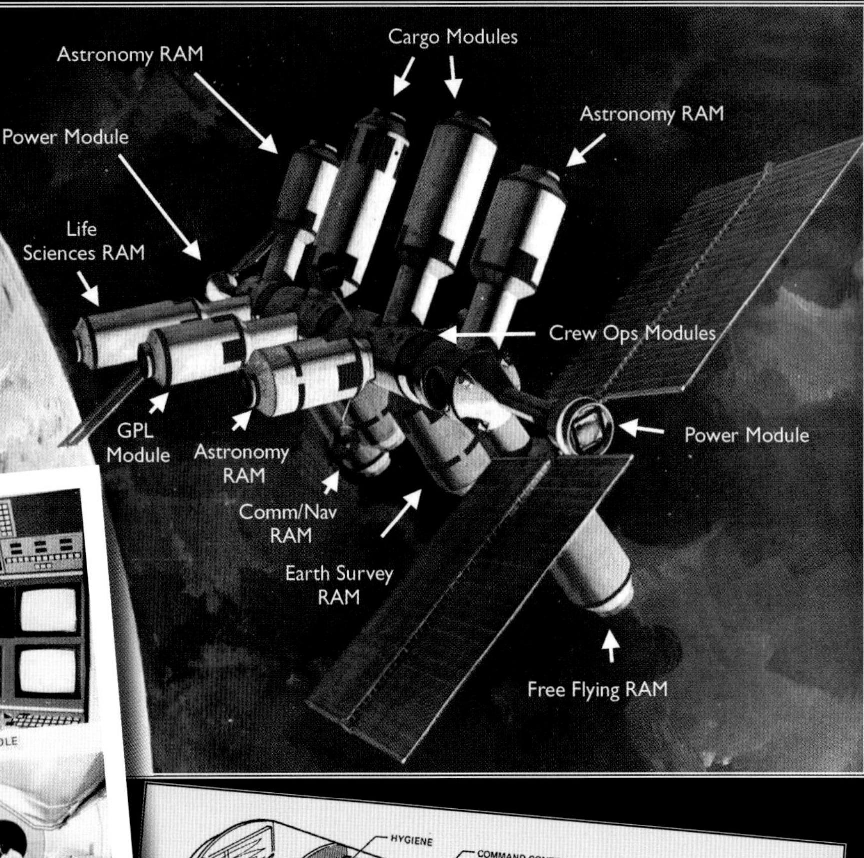

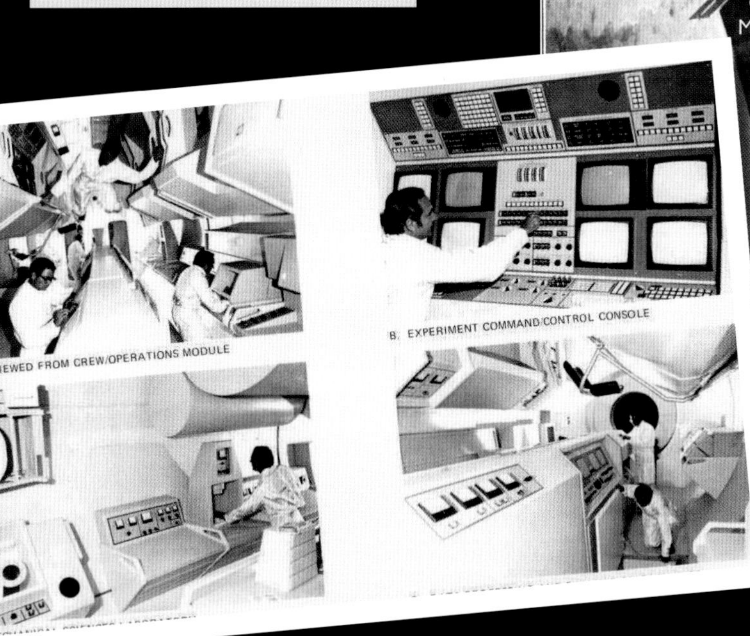

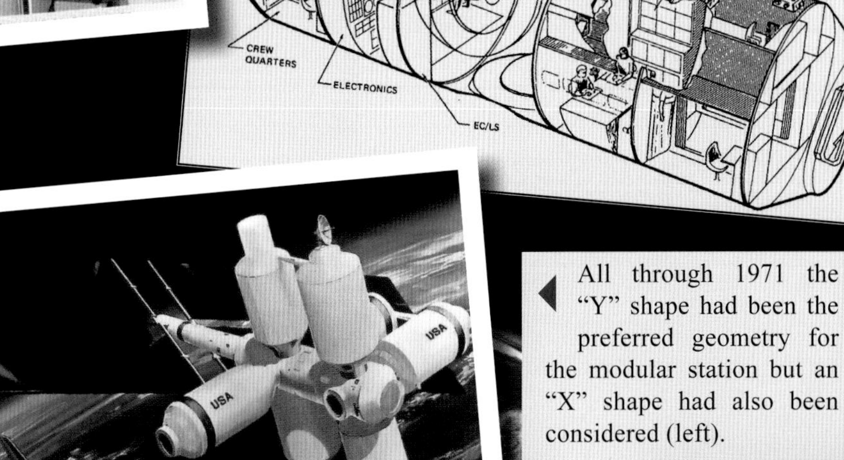

Extensive design architecture was created for the entire station as evidenced by this schematic showing the Crew Operations Module (far right). The 1971 image (below) was created to demonstrate how the Shuttle could be used to sustain the station. An injured crewman is transferred from a disabled Space Station. Upon safe recovery of the injured crewman, the remaining station personnel would evacuate, using a lifeline as a tether.

All through 1971 the "Y" shape had been the preferred geometry for the modular station but an "X" shape had also been considered (left).

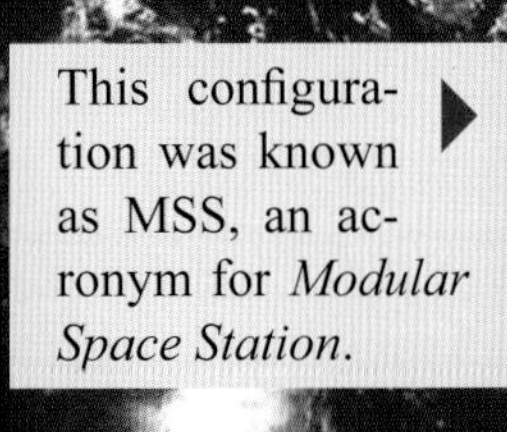

This configuration was known as MSS, an acronym for *Modular Space Station.*

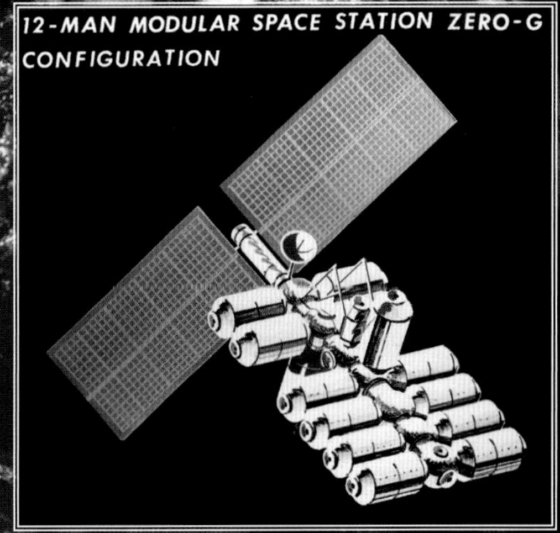

The Nixon administration would authorise NASA to move ahead with its plans for a reusable space shuttle. This would replace the traditional disposable boosters and was expected to lower the cost of access to space. If the space shuttle could live up to expectations huge amounts of payload could be lifted to low earth orbit and plans could begin anew for a long term presence for Americans in space. On the other side of the world after the failure of the N-1 rocket and the race to place a man on the Moon, the Soviets turned their attention to trying to beat the US in placing the first manned space station in orbit. Korolev's old bureau OKB-1 and Vladimir Chelomei's OKB-52 had been working on competing civilian and military space station designs since 1964. The decision was made to consolidate these efforts to rush the first Soviet space station to completion. In 1971 they launched the first in a series of seven stations, the *Salyut 1*. The first Salyut took the core of Chelomei's *Almaz* military platform and combined it with many of the flight and control systems of the newly finished Soyuz spacecraft. Versions of the central core of Salyut would prove to be versatile and adaptable enough to ultimately make its way into all of the subsequent Russian space stations. Despite encountering many problems the first Salyut still marked the beginning of the manned space station era - which continues to this day.

OKB-52 /OKB-1 - "Salyut" - 1971 to 1986

A rare image of the historic Salyut 1 in space. In April of 1971 the Salyut 1 space station was moved from the assembly building and was launched by a three stage version of Chelomei's Proton launch vehicle. The station weighed almost 19,000 kg and it was 15.8 m long and as much as 4.1 m in diameter. It had four large solar arrays for power and one docking port. The first Salyut was plagued with problems with life support systems, docking systems and other scientific equipment. The first crew docked, but could not enter the station; whereas the second crew (above right), Georgy Dobrovolsky, Viktor Patsayev and Vladislav Volkov conducted the longest space mission on record up to that time before perishing in a tragic re-entry accident that was unrelated to the Salyut.

"Chelomei's design bureau had already begun work on the military orbital station "Almaz", it had reached the stage of experimental development of the structural elements. The hulls were already manufactured, but there were no on-board systems or equipment yet. In this area we (OKB-1) were significantly ahead of them, since we could use almost the entire airborne apparatus, units and engines from the Soyuz spacecraft for the orbital station. At the beginning of December 1969, when Mishin (head of OKB-1) was somewhere in the south on vacation, and Chelomei was resting, I consulted with Chertok and called on Ustinov and asked for an appointment. He offered to drop by at 5 o'clock in the evening. I told him that if we used the on-board systems of Soyuz as a basis, the cylindrical part of the fuselage from Chelomei's orbital station, and the propulsion system, solar batteries and the docking unit also from the Soyuz spacecraft; in a short time, about a year, we could create an orbital station. Chelomei quite reasonably considered this as a pirate raid on his domain by our side. Of course, there was an element of piracy, but our conscience was clear: according to the laws of the socialist system, everything belonged to the state, and so consequently, to us. We acted in the interests of the cause." - Konstantin Feoktistov

Salyut 7 with Soyuz docked in orbit.

Seven *Salyut* space stations were built and, although some failed on launch, crews manned the stations, on and off, for fifteen years before Salyut was replaced by the next generation of Soviet space station, the *Mir*. The last mission to Salyut was the Soyuz T-15 in late 1985. Cosmonauts Leonid Kizim and Vladimir Solovyev spent several weeks aboard both Salyut 7 and Mir and even shuttled equipment back and forth between the two space stations; the only time this has ever happened.

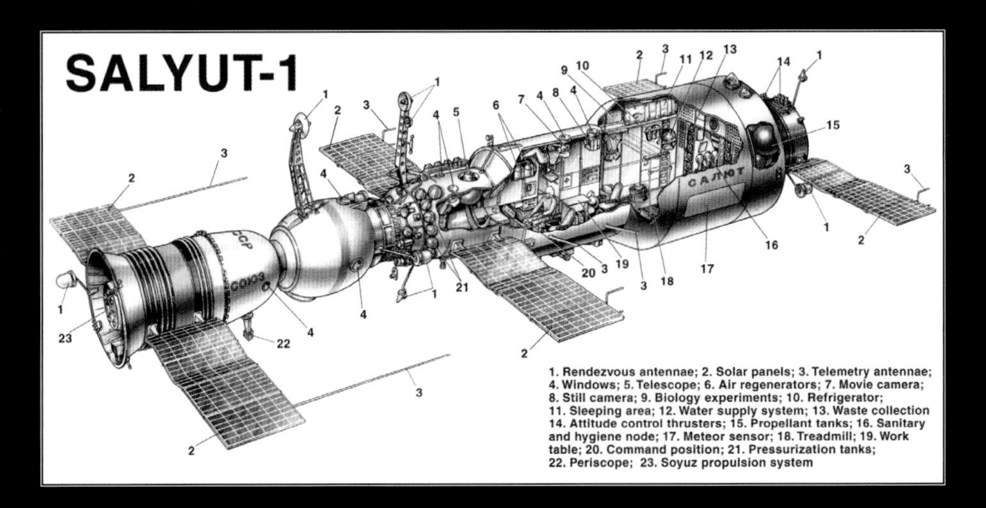

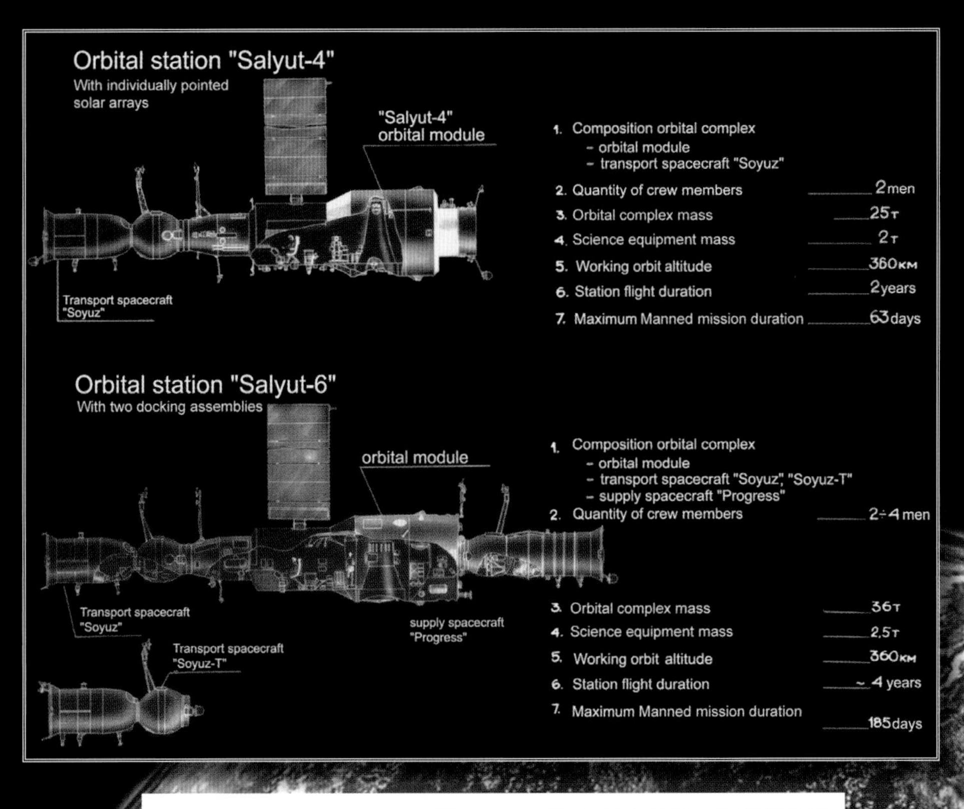

"We formed a team in our design bureau, headed by Konstantin Petrovich Feoktistov who was the main enthusiast of orbital flights. On December 26, Minister Ustinov gathered the leaders of the space program for a long talk. Having received this support from above, in literally just a few days, the activists of the project prepared the first official technical document "Osnovnyye polozheniya po orbital'noy stantsii" (Basic provisions for the orbital station), *which was signed December 31, 1969. The next decisive action was taken by Yuri Semenov, then the leading designer of the lunar spacecraft L1. Using even higher support, in January 1970 he managed to persuade Mishin to sign an order to organize work on the orbital station. Semenov was appointed the lead designer for DOS* (Dolgovremennaya Orbital'naya Stantsiya - Long Term Orbital Station) *and was assigned wide, unprecedented powers. Then, in March 1970, he managed to persuade, or rather to squeeze Chelomei (still with the highest support from above of course), who was then forced to give up the first hull of the orbital station. Looking back, I must say that Semenov very effectively used the opportunities he was provided. First of all, he made a huge contribution to the implementation of the project, to the solution of numerous problems that arose on a short but rich path, from the initial plan to the flight in space. In the end, this path led him to the leadership of our entire organization. From the outset the design work was in the hands of Feoktistov; his contribution to the project of the first and subsequent DOS was also huge. On the basis of this project at the direction of Ustinov, it was decided to develop and launch the first Soviet orbital station as soon as possible. In accordance with the decision of the Council of Ministers of February 9, 1970, our company was appointed the main organization for the project."* - Vladimir Syromyatnikov

Salyut 4 in the assembly stage.

When it became apparent that the funding for further lunar exploration would not be forthcoming it was decided that one of the unused Saturn V's could now be used to launch the Apollo Applications Program (AAP) "dry" orbital workshop. The Saturn IVB stage would be converted by McDonnell-Douglas into a fully equipped space station. In 1970 it had been named *Skylab*. In 1973 three crews would board and conduct hundreds of hours of experiments in low earth orbit on America's first space station, for a total of 167 days in space.

NASA/McDonnell-Douglas - "Skylab Orbital Workshop" - 1973

The NASA model seen at right is an early 1970s "wet" workshop from the AAP phase of development. The LM-A lab was deleted but a Solar Observatory Instrument was still mounted on the so-called *Apollo Telescope Mount* (center right).

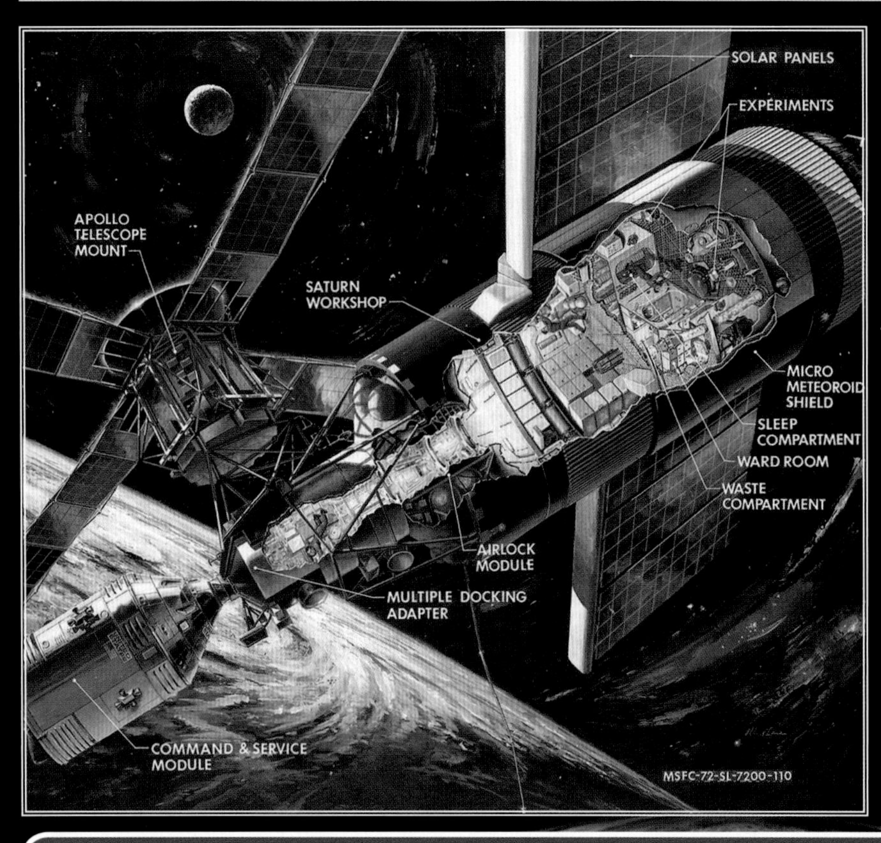

"The ATM, launched unmanned, has the upper portion of the Lunar Module (which is used, in Project Apollo, for landing on the moon) as an operator's cabin. It is equipped with special instruments for studying various solar phenomena including far ultraviolet, x-ray and gamma radiation, none of which can penetrate the earth's atmospheric shell. It also carries a white-light coronagraph for the study of the sun's corona. The telescope tube will be about 6½ feet in diameter, 12½ feet long, and will weigh more than a ton." - Wernher von Braun

"Skylab was fitted out with everything that we were going to have to have; all of the living quarters, plenty of food, all of the water we needed, and all of the telescopes that would be used to study the Sun. Skylab was launched ready for use. Then the crews went up on a smaller Saturn-IB, docked with it, and could conveniently go right to work." - Owen K. Garriott

After more than a decade of space station studies, involving a multitude of designs, Skylab like all of its predecessors was a product of expediency. The vast pressurized volume of the "dry" workshop was only possible because of the lifting power of the Saturn V lunar rocket. Skylab's layer-cake design was the most logical approach. Gone were any signs of centrifuges, inflatables or massive toroidal habitats. Instead Skylab was designed to house three people to conduct science for several months. Their supplies and scientific equipment would be spread across two main decks, a separate airlock and a multiple docking adapter.

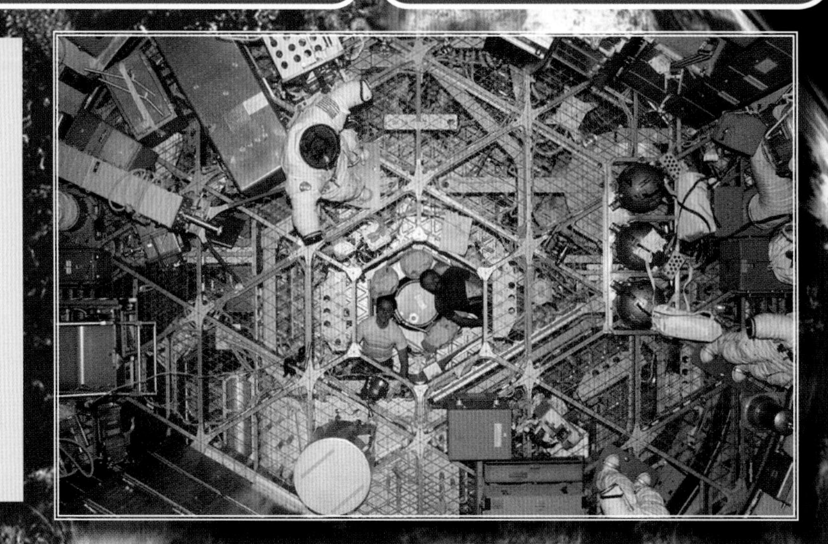

The LM-A/Apollo Telescope Mount could also be flown independently as a small pressurized laboratory/observatory. The Apollo command/service module could be docked directly to the lab as seen in this image. Although the LM-A was abandoned, the four large ATM solar panels would remain in the design that finally flew into space in 1973 (below). Despite years of meticulous planning *Skylab* arrived in orbit with one of the Orbital Workshop's massive solar panels permanently damaged and another with a problem which kept the station from being cool enough to be occupied. The first crew to arrive would need to conduct a difficult and dangerous space walk to untangle the remaining solar panel and deploy a gold mylar umbrella (seen below) to reflect the sun's heat.

"The MOL Program was going great. Then one morning I got a call, "Hey, it's cancelled."That was one of the lows in my life. We had fourteen crew members at the time. All of us came down to Houston and talked to NASA. In fact, we were at the Manned Spacecraft Center during the Apollo 11 Moon landing. "Deke" Slayton, NASA head of Flight Crew Operations, said, 'Hey, guys, I got more astronauts than I know what to do with.' But, Deke said, 'I'll split the group in half and take everybody that's thirty-five and younger.' When we came on board, Deke said, 'I've got lots of work for you to do, but there aren't going to be any flights until the Space Shuttle, and that's going to be like 1980.' The Shuttle wasn't even an approved program yet. But all of us came. Many of the things on MOL dealt with living in orbit for long periods and were directly associated with Skylab. Most of us were assigned to work on Skylab." - Robert Crippen

A robot arm had been considered for Skylab and dismissed, but the manual repair by space-walking astronauts opened the way forward for all future in-flight repair missions. Skylab went on to house three crews in 1973, the last crew staying aloft for three months. After being abandoned for several years plans were drawn up to use the upcoming Space Shuttle to push Skylab into a higher orbit, but in 1979 America's first space station burned up over the Pacific Ocean and Australia and the last of the Saturn Vs became museum exhibits.

"Skylab was a space station; a true space station. The initial goal was seven days, fourteen days, and twenty-eight days, each doubling; we did, twenty eight, fifty-six days and then eighty-three days. There were a lot of people that wanted to extend Skylab. They even tried to, and I think that was ludicrous. We designed the program to accomplish certain experiments; we did that, and it was over. We could have flown some more at great expense. We'd have just been repeating what we had already done." - Kenneth S. Kleinknecht

NASA - "O'Neill Sunflower" & "Stanford Torus" - 1975

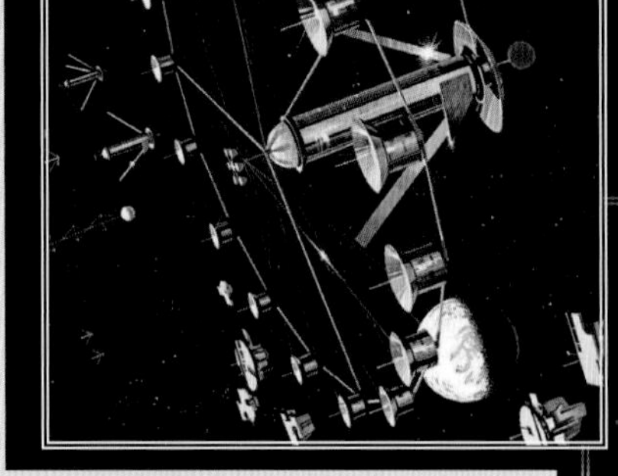

Despite the loss of NASA's heavy lift launch vehicle Princeton Professor Gerard K. O'Neill was not deterred from planning a permanent presence for humans in space. O'Neill was technical director for a series of studies conducted at the NASA Ames Research Center in 1975 which ended up being consolidated into an ambitious statement for future space settlement. The team assembled under O'Neill envisioned massive self-sustaining space colonies like the *O'Neill Sunflower* (above) and the *Stanford Torus* (right) which would be located at places such as the stable gravity LaGrange points of the Earth-Moon-Sun system. O'Neill had been influenced by Bernal and Cole when he devised his colony, which he said would rotate at 3rpm. However Larry Winkler, one of his team, insisted that 1 rpm was more feasible, which led to the enormous Stanford Torus.

NASA - MSFC/Douglas - "Manned Orbiting Facility" - 1975

The *Manned Orbital Systems Concepts* study done by Douglas for MSFC, was concluded in December of 1975. Making the most of the work which had been done by Convair on the RAM studies it proposed a *Manned Orbiting Facility* (right) housing four crew with a potential growth to twelve. It would have four modules - Subsystems, Habitability, Logistics and Payload, with a 90-day resupply cycle and a shirtsleeve environment. Two shuttle launches would make the station operational.

NASA - JSC/Grumman - "SSSAS" - 1976

In April of 1976 NASA issued two contracts for its 18-month *Space Station Systems Analysis Study* (SSSAS), one by McDonnell Douglas/JSC and one by Grumman/MSFC in Huntsville. The goal was to formulate plans for a permanent operational base and laboratory facility in Earth orbit, in addition to developing a space construction base design for implementing the program. An expended Space Shuttle external tank was to be the central core platform of the base, and additional pressurized modules could be added to provide laboratory facilities.

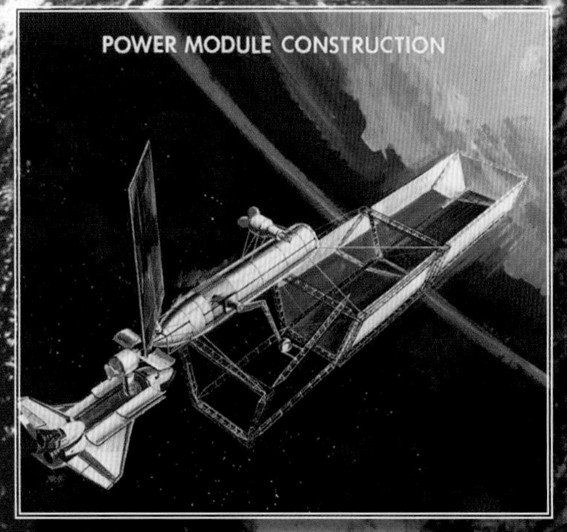

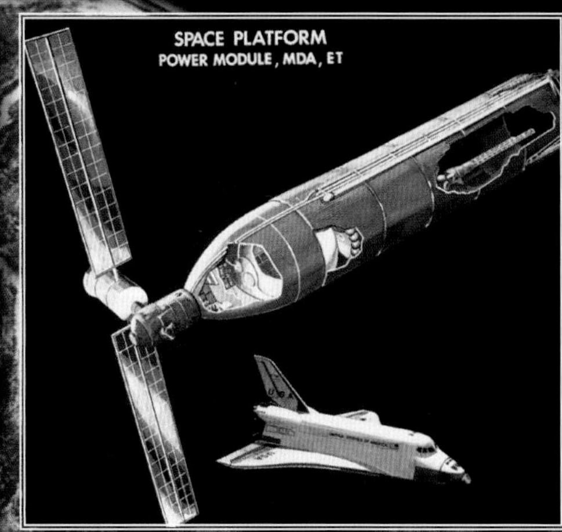

Considerable thought was given to the huge volume of the expended shuttle fuel tank as a building block for a space station. Studies involving the wet workshop back in the 1960s were still considered viable and if a station could be built using hardware which was being dragged into orbit already, this would help meet the budgetary constraints which had come into effect since the end of the Apollo program.

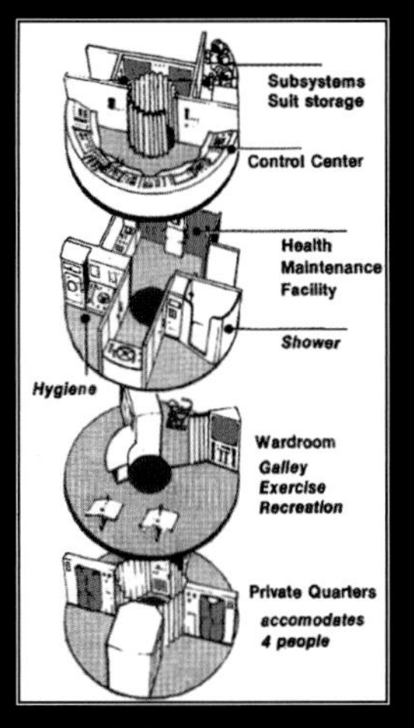

NASA JSC - "Space Operations Center" - 1979

The *Space Operations Center* would be a self-contained facility built up of several shuttle-launched modules, with resupply, and on-orbit refurbishment and maintenance. It would operate continuously for an indefinite period, usually with a crew, but it could be operated automatically. Safety of both the SOC and its crew would be a major design consideration. Through subsystem and module redundancy, SOC could sustain itself and its crew for 90 days without assistance in the event of a major failure, including the functional loss of an entire module. A rudimentary Space Operations Center would be able to sustain crew and operations for 90 days without assistance and without the presence of a shuttle.

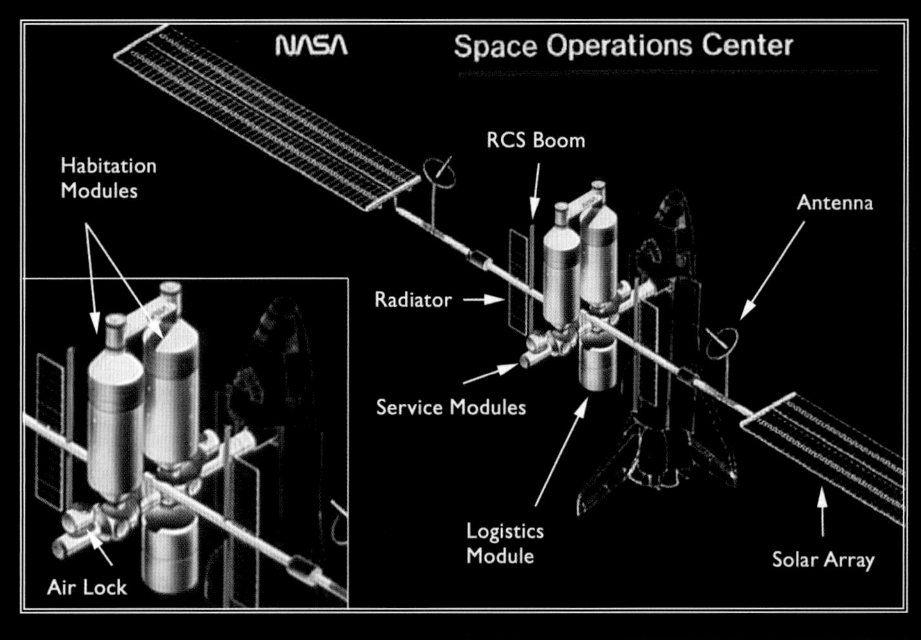

SOC's basic design was still evolving in 1982, after the shuttle had flown (below). *Key: 1) Utility module with power, comm, & thermal systems. 2) Habitat module, 3) Payload storage & repair bay, 4) command module & garage, 5) propellant storage & docking tunnel.*

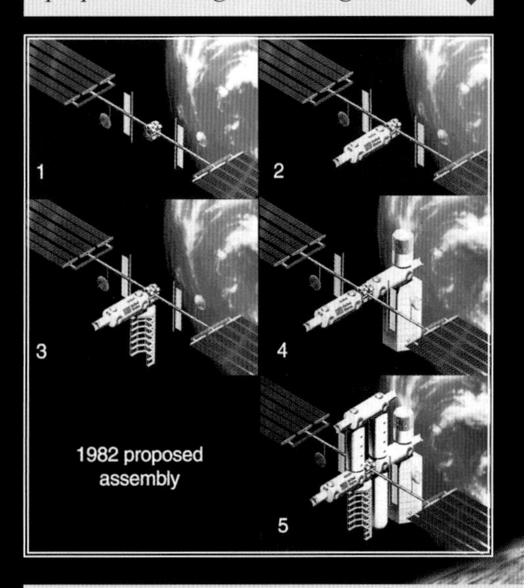

1982 proposed assembly

NASA JSC/Rockwell - "Modular Concept" - 1982

In the summer of 1982 NASA again sponsored industry studies to examine mission requirements for a possible U.S. space station. The studies analyzed future missions in space to determine what attributes a space station would have to fulfill these missions. While the studies focused on mission requirements and appropriate space station "architecture," several renderings were produced to enable a possible station configuration to be visualised. The picture at right shows a mature space station configuration which includes two solar panels to provide power; several modules for command, habitation, and experimental activity; a shuttle-sized unpressurized rack for storage of payloads; an advanced remote manipulator system for the assembly of large structures and the servicing/storing of satellites and instruments; and a docking/utility hub that might also serve as a safe haven in case of an emergency. The picture also shows an Orbital Transfer Vehicle (OTV) returning to the station after delivering a payload to higher orbit. Free-flying platforms were considered an integral part of the space station program. The one at bottom right may have operated in polar orbit and was part of the Phase B systems design work that took place as the program moved into the 1980s.

In 1984 US President Ronald Reagan committed his nation to building a permanent space station. Engineer John Hodge of NASA, who had been the head of the Space Station Working Group since 1982, began the process of putting his team together. One of the first things Hodge decided was that it was more important to determine what the space station was *for,* than what it would look like. As a consequence of this decision his group didn't specify a design in the early days, but the many contractors who had been paid to investigate the problem produced a deluge of conceptual art. It had been known since the space shuttle had been approved that whatever the space station eventually looked like, it would be constructed from components in orbit. Just before the President made his speech the NASA team concluded that the space station would be (1) An on-orbit lab; (2) an observatory; (3) a transportation node; (4) a satellite servicing facility; (5) a communications and data node; (6) a manufacturing facility; (7) an assembly facility and (8) a storage depot. Based on these requirements an assortment of different geometries were considered, some of which are pictured here.

NASA Space Station Working Group - 1982 to 1984

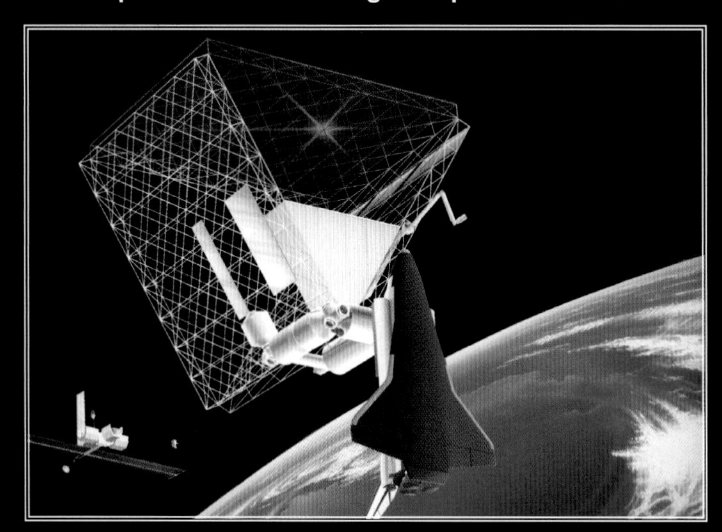

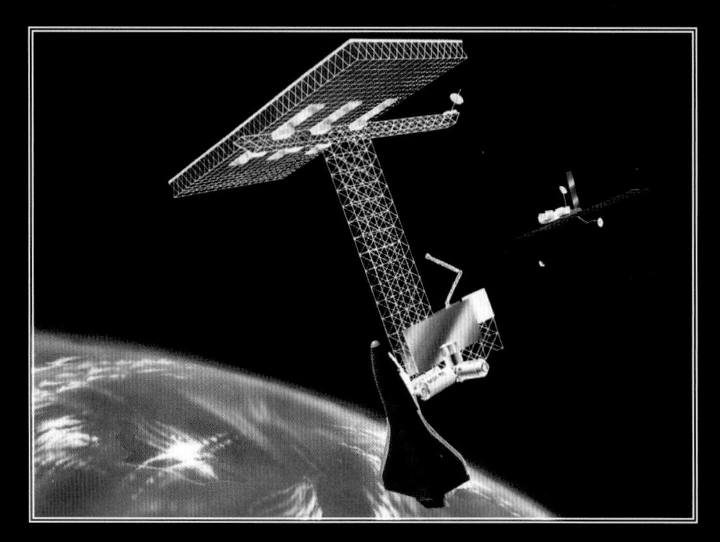

The Triangular or Delta (left) and the Big "T" (right), which was disqualified during the 1984 "Skunk Works" NASA study

"Once Jim Beggs became NASA administrator, he asked me if I would start a task force to set up the system to build a space station. The Space Station Task Force was probably the best thing I ever did in NASA—in my career. Jim Beggs gave me absolute freedom to put together the organization I wanted. I put together the authority structure, and I put together the idea of who's going to work for who and how. I based it on my experiences during the Project Mercury Space Task Group; there was so much work to do, that we didn't have to manage. We just had to say what the work was, and people would go do it. In the Station Task Force, we worked out all the jobs that had to be done, and then people would come in and they'd pick the jobs they wanted to do and go off and do them. It was the most efficient way to do it. So we didn't have a hierarchical structure at all. It was a very flat organization. We did that for a couple of years. We had about 100 people." - John Hodge

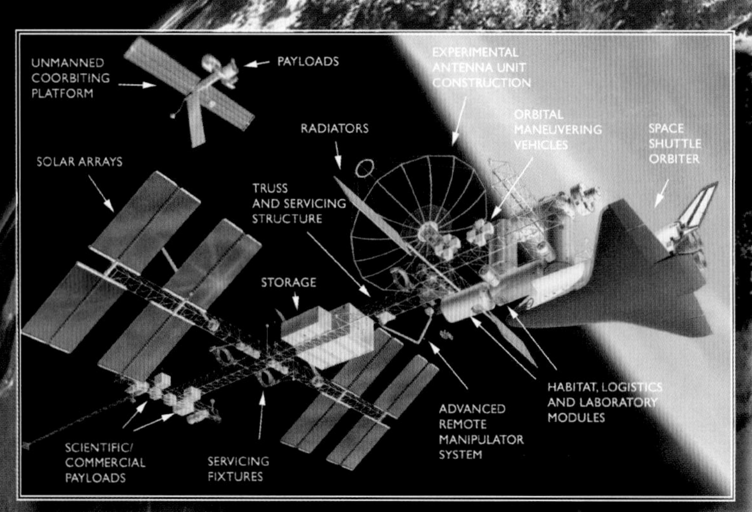

The Power Tower (left), the Concept Development Group (CDG) Raft or Planar (right)

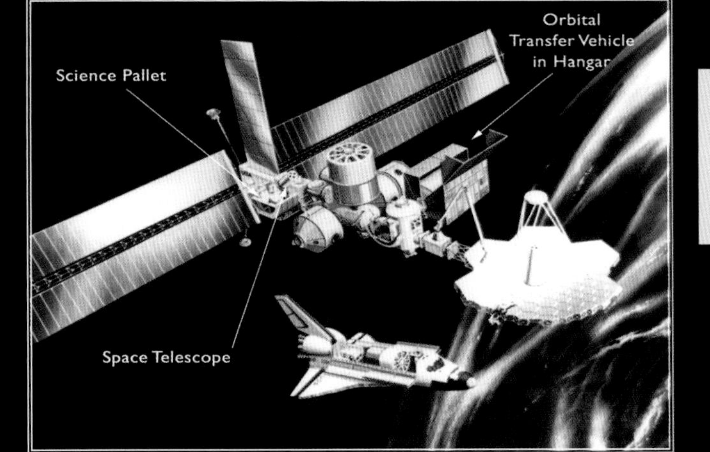

NASA - MSFC/Douglas - "Modular Concept" - 1982

This 1982 reference concept features a modular space platform of rotating pallets containing space science and applications payloads with airlocks joining three manned modules. The large dish was part of a study for a Large Deployable Reflector observatory with capability over a broad spectral range.

NASA - JSC - "Modular Concept" - 1982

The cylinders, with a connecting passageway, would serve as the command control center, living quarters and laboratory. The service modules would be set below the cylinders and would have the life support systems, power unit, communications apparatus, an airlock and docking berth for the Shuttle. This is an example (right) of the "raft" design.

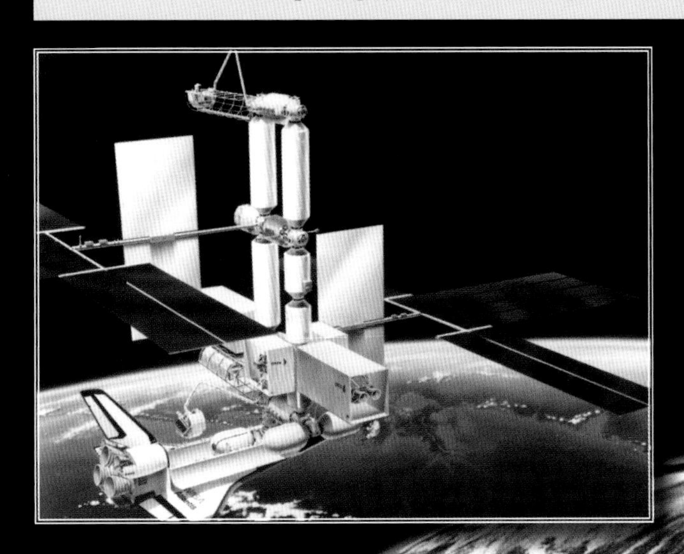

PATTERN / MODULE	RACETRACK	TWIN RAFT	FIGURE 8	RAFT

Three different module shapes and four configurations were considered (above). Shown at left is another space station by Johnson Space Center and TRW from 1982. This one shows a satellite fuelling depot and hangar in use in the foreground. This is an example of the "racetrack" design.

In 1983 engineers at the Langley Research Center, who had been working on stations since 1960, were able to construct a large model of an initial design. This model was used by NASA administrator James Beggs to convince President Ronald Reagan to make his announcement the following year. Reagan took the model to show Prime Minister Margaret Thatcher in England in June of 1984. Initially Thatcher suggested that she would commit Britain as a partner on the program but withdrew her support later.

NASA - Space Station Task Force - 1983

James Beggs with the Langley model, just after President Reagan's announcement that the United States was going to build a permanent space station.

"After Ronald Reagan's election, George Low served as the manager of the Reagan transition team for NASA. George thought 'the next step is how to work and live in space long duration, so the next logical thing we should do is the Space Station'; because we've been through the major expenditures on the Shuttle and the Shuttle was flying. I had a very, restricted objective for Reagan's presidency. My objective was to sell the space station." - James Beggs

"NASA began in earnest working on Space Station after Jim Beggs and Hans Mark were confirmed. That initiated the efforts at NASA to study more aggressively Space Station development. That was not really very visible from '81 to the end of '83, to us in Congress. It was mostly an internal thing. They let us know they had this Space Station Task Force. We knew that was going on; but we couldn't really follow it in any detail." - Jeff Bingham

Grumman - Initial & Evolved Space Station - 1983

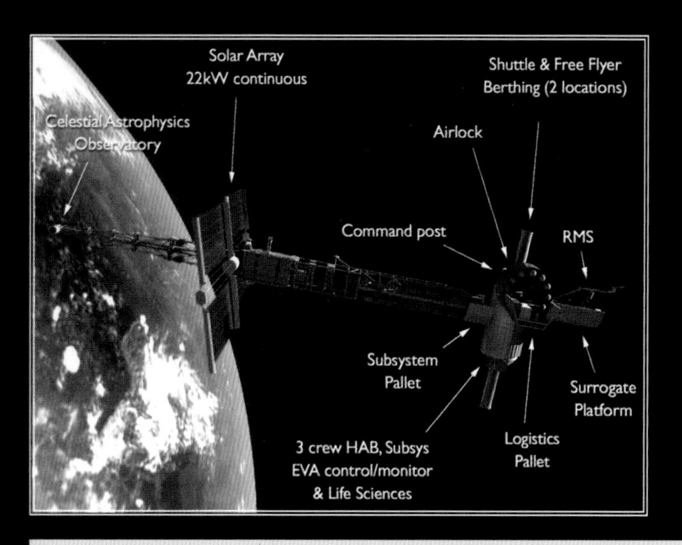

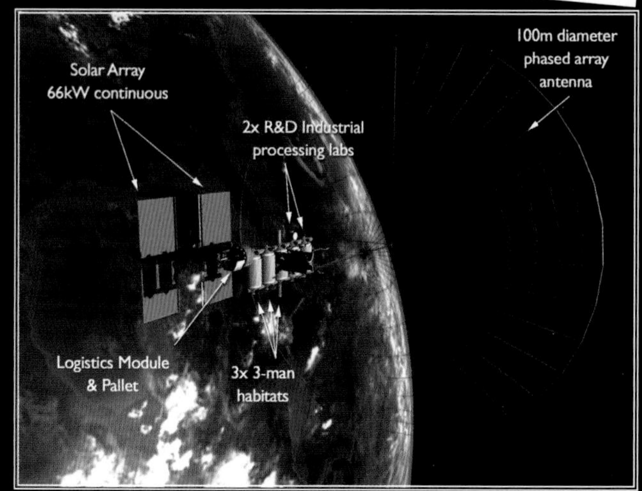

▲ Grumman's *Initial Space Station* could have been placed in orbit with two space shuttle launches. The station would have had four basic elements. The core, external subsystems, payload bays and an observation instrument tower. The payload bay would have a robotic arm and a cherry picker. The *Evolved* version of Grumman's station (right) would have used the "raft" design connecting five modules, each with two exits in case of emergency. By tripling the power available the crew could have been increased to nine. The most striking feature is the massive phased array radar, which was one of many things being studied at Grumman and elsewhere. ▲

NASA/TRW - "Expanded Space Operations Center" - 1984

By the end of 1983 serious consideration was again given to using the space station as a staging post for OLVs. This scaled-up version of the Space Operations Center (right) now included dual solar array sets, logistics modules, command and habitat modules, attached pallets for scientific inquiry, and hangars to prepare orbital transfer vehicles for launch. This concept was derived during mission analysis studies conducted for NASA by TRW Space and Technology Group. ▶

McDonnell Douglas - "Commercial Modules" - 1984

◀ Many companies began to realise the opportunity presented by a modular space station system. Unique modules could be built and flown for private purposes. This image from 1984 shows a purpose-built module (marked EOS) which was to be used by Johnson and Johnson for electrophoresis in space experiments. McDonnell Douglas had been proposing commercial space station modules since at least 1967.

By 1985 the European, Japanese and Canadian Space Agencies were all contributing concepts to the station. Canada's role would be to provide robotics to aid in the construction of the station and to minimize spacewalks, while Europe and Japan would both agree to build their own scientific laboratories. The result of these arrangements would ultimately begin to establish some inertia in the constantly changing design of the proposed space station, although the political winds would continue to blow designers off course for several more years. Eventually the end of the Cold War would be the single most important event in the final disposition of the space station.

CSA/ESA/JAXA - Robotics and Modules - 1985

How to dock a spacecraft without rocking the boat

(Read how de Havilland's system using the STEM technique avoids dangerous impact)

THE DE HAVILLAND DH CANADA AIRCRAFT OF CANADA LIMITED
SPECIAL PRODUCTS & APPLIED RESEARCH DIVISION
MALTON ONTARIO

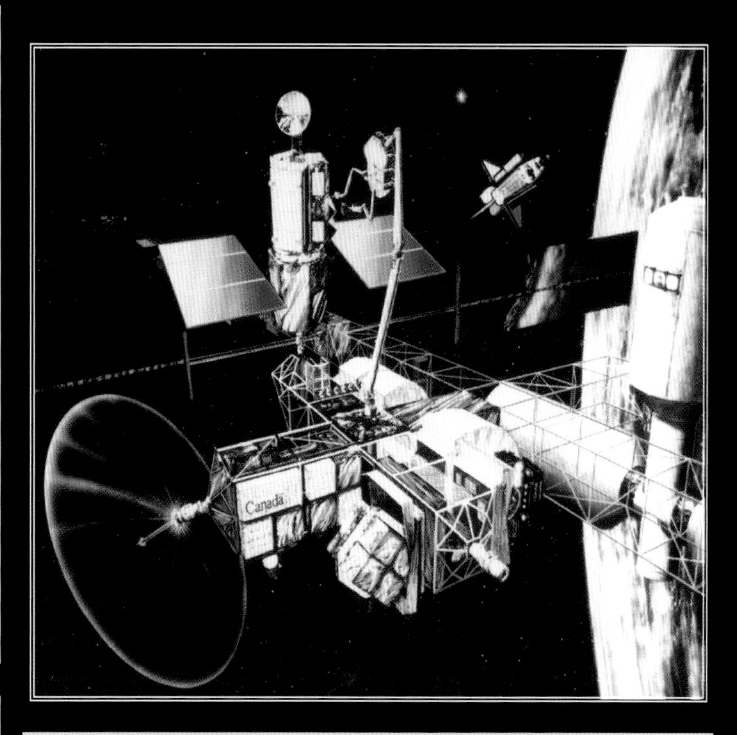

▲ Canada had been planning a robotic arm for a space station since at least 1962 as evidenced by this advertisement by De Havilland Canada from the mid 1960s which shows it attached to MORL. It was then declined again for Skylab. By 1987 De Havilland's Special Projects and Advanced Research division (SPAR) had evolved their concept into a sophisticated robotic mobile servicing system. ◥

Japan's chosen design for a space lab would include an exposed pallet, such as the one which had been suggested for the MOC. A robotic arm would also be available to take care of external experiments. The design for the JEM laboratory can already be seen taking shape in the early image and diagram below. ▼

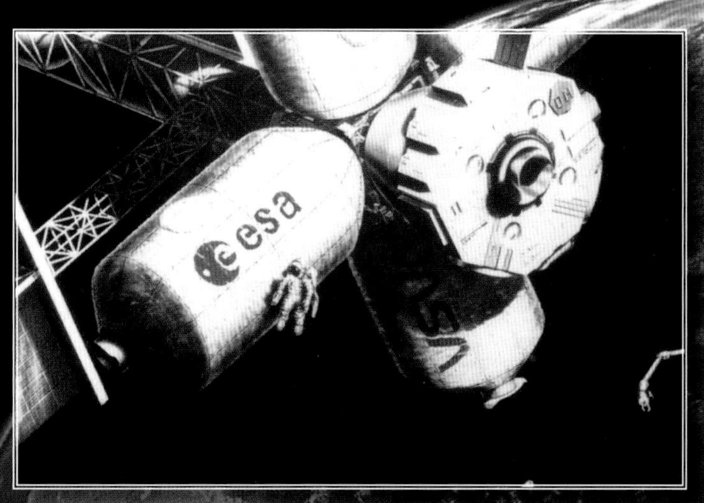

▲ Initially the Europeans were scheduled to build only one science module but as the negotiations progressed several of the planned modules for the space station would be built in Turin, Italy by satellite maker *Thales Alenia Space*.

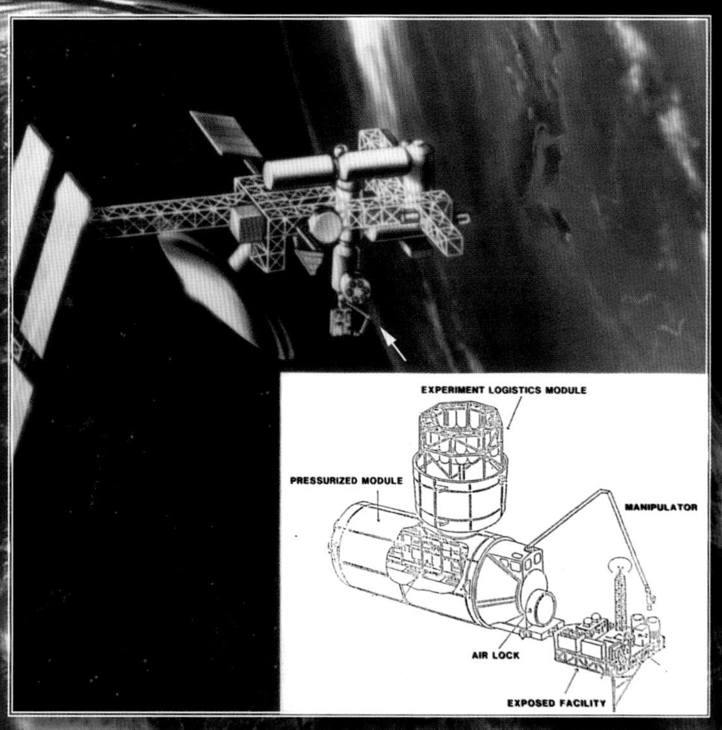

From 1982 to 1987 NASA's teams at five centers had worked to distill down a host of possible configurations. By 1985 the NASA design choices had been reduced to the *Power Tower*, the *Dual Keel* and a *Planar Truss* which had satellite storage and service bays and fuel storage as well as a hangar for an OMV. The next casualty of this slow attrition was the *Spinner* which would have used four spent shuttle tanks connected by a rim of solar panels and a zero-g hub.

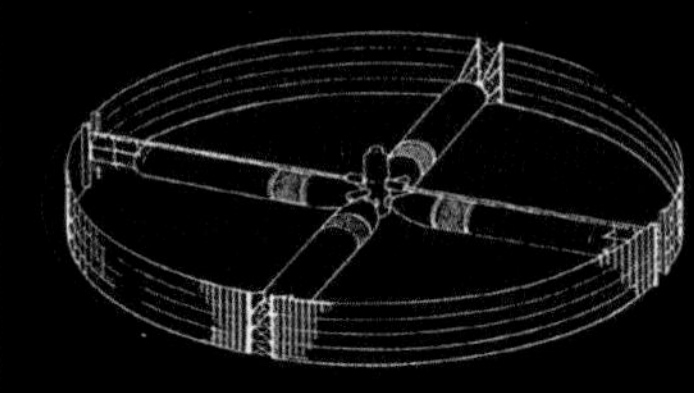

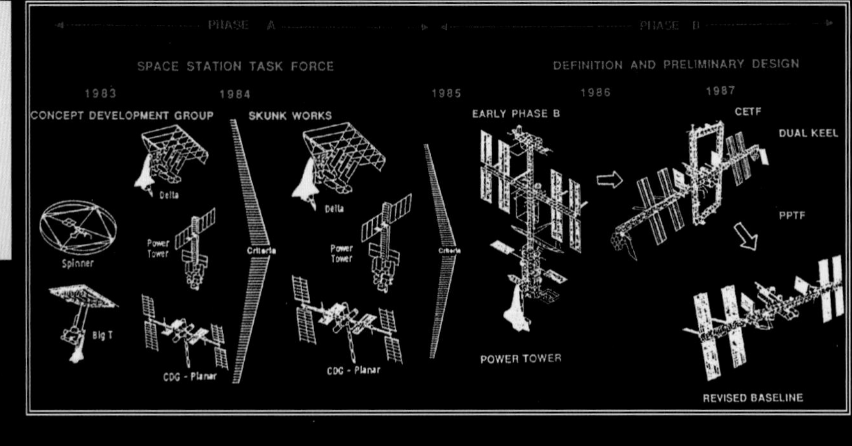

NASA/CSA/ESA/JAXA - "Dual Keel" - 1985 to 1987

By 1987 the choice was now between the *Dual Keel* and what became known as the *Revised Baseline*. This image shows the Dual Keel's appearance in January 1987 after the introduction in 1986 of two resource nodes. It also includes a single solar dynamic power supply, a system which had been absent from space stations for more than two decades. Researchers at NASA's Langley and Lewis centers, and at Rockwell, conducted in-depth studies of solar dynamics again and came up with many possible configurations (below right). Studies continued until at least 1989. The Dual Keel was dispensed with in 1987. It required too many launches. The April 1987 "Revised Baseline" Configuration (below) represented the template for the next several years until budgetary constraints began to take hold once again.

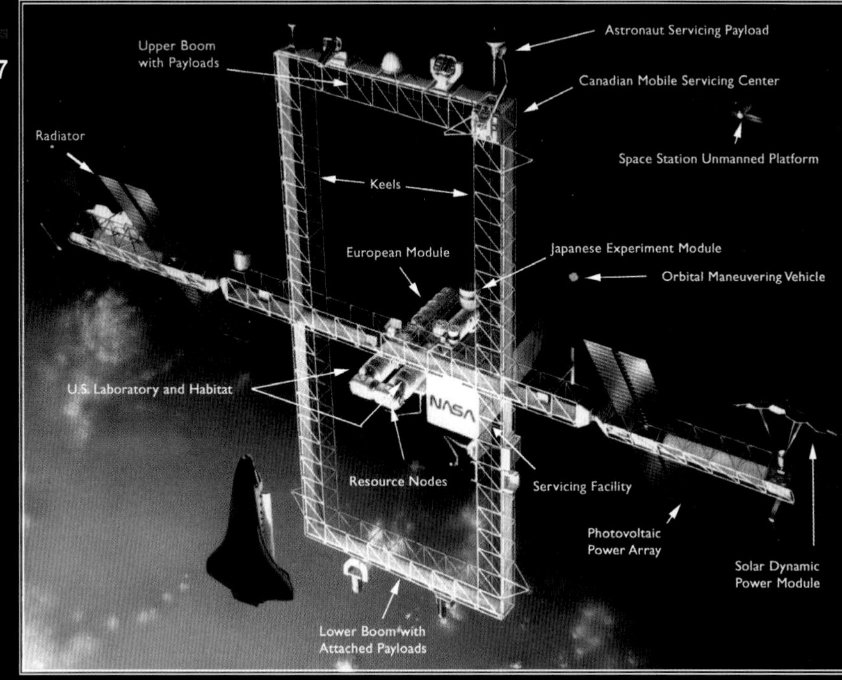

Solar Dynamics proposals (right, top to bottom) Lewis 275kw station 1989; Rockwell Dual Keel 1987; Langley two SD units and four SD units, both from 1989.

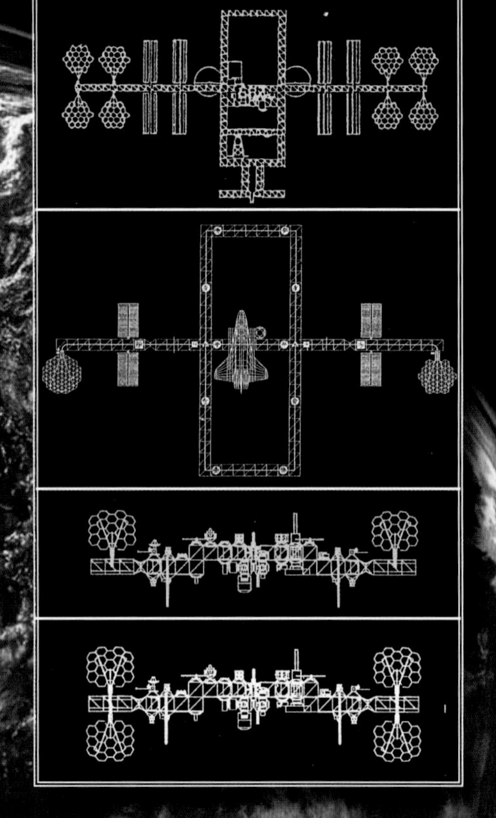

NASA/CSA/ESA/JAXA - "Baseline Configuration" - 1987

An aerospace alliance between East and West had begun in the 1970s reaching an early zenith with the Apollo-Soyuz Test Project flight in 1975. In 1977 further tentative agreements were made which would have involved a U.S. space shuttle docking to a Soviet Salyut space station but political differences didn't allow this to take place. The next major launch in the Soviet manned program after the Salyut series was the first component of a new, much larger, space station. It was to be called *Mir* and the first module entered orbit on February 20th 1986. The Mir consisted of six separate modules, several of which were based on the Chelomei *Almaz* system. When Mir was completed it contained a pressurized volume of over 90 m^3 and weighed 135 tons (see Mir chapter).

Energia - "Mir"/ "Nadezhda"/ "Mir-2" - 1986 to 2001

When the Soviet Union collapsed in the early 1990s Russia took over the remnants of the Soviet space program. *Mir* then represented the beginning of true international cooperation in space. An agreement was reached between the United States and Russia to conduct a series of joint missions to Mir. This involved installing a unique docking adapter which would allow the US Space Shuttle to dock with Mir. In late 1995, after Mir had already been operational for a decade, the Space Shuttle *Atlantis* became the first U.S. visitor. Over its fifteen year life-span Mir would be visited by thirty manned Soyuz spacecraft and nine American space shuttle crews.

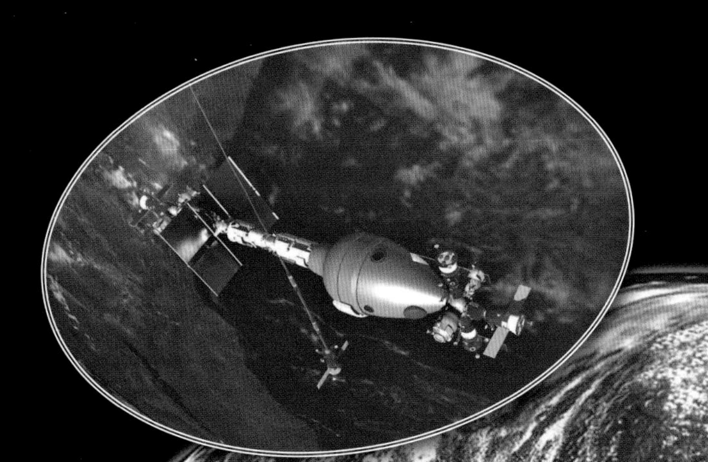

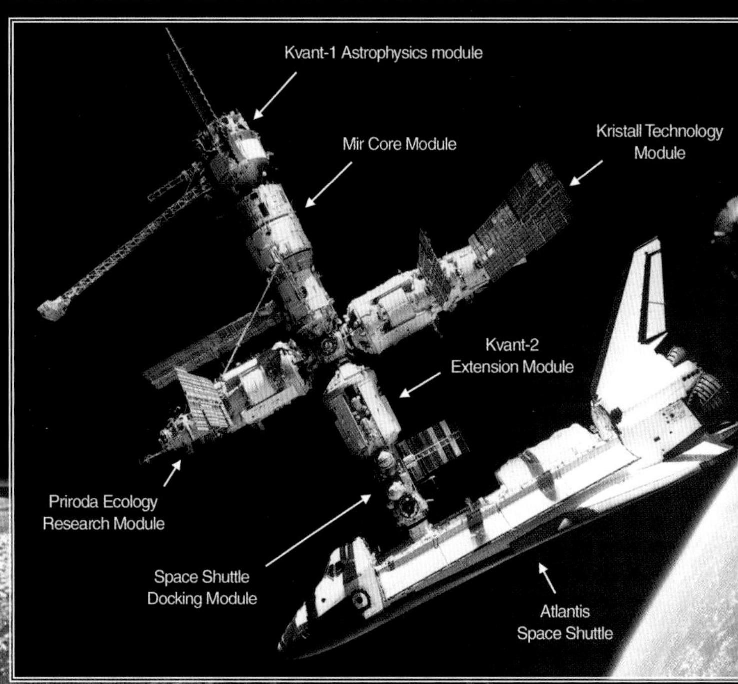

While a multitude of experiments were conducted by Russian and American crews aboard *Mir*, plans continued in the United States for a much larger space station. These plans would lead to a more robust partnership between the world's two pre-eminent space-faring nations, but not before people on both sides of the political divide tried to save *Mir*. A popular proposal was to deploy a long tether which would drag through the Earth's magnetic field and generate power for keeping the station in orbit without a constant need for refuelling. The Central Research Laboratory at the Moscow Aviation Institute designed the *Nadezhda* (Hope) station (above) which would have used many of Mir's core components and a tether. At both ends are docking adapters for Soyuz class vehicles. A basic habitat module weighing 20-23 tons with a diameter of 9m would be constructed in orbit. By the late 1980s Russian engineers were designing a much larger station to be named *Mir-2*. It went through at least a half dozen iterations including the version seen here (right) with eight solar dynamic arrays. The new Energia super booster would deliver a very large *Base Module* to orbit and would also be used to launch the Buran shuttle to the station for resupply and crew rotation. But just as funding was drying up in the USA the Soviet Union collapsed. The Buran and Energia booster programs ended and Mir-2 had to be downsized accordingly.

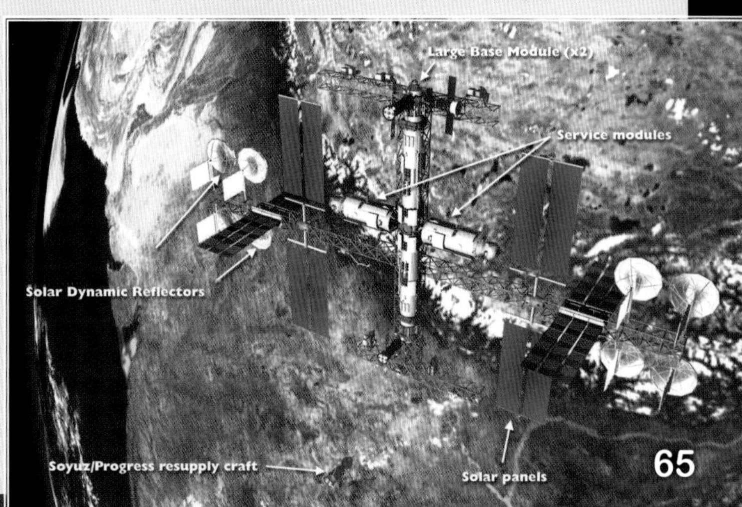

By 1990 plans for a much smaller *Mir-2* space station were on the drawing boards in Russia (right). Only two solar dynamic power sources were being considered for this less ambitious station as well as conventional solar cells delivering a total capacity by the on-board power system of 48 kW. Both Baikonur, and Plesetsk would be used for launches. Large elements would be launched by Proton and smaller elements by Zenit. *Mir-2* was to be in a 65° inclination allowing for a much more expansive coverage of the globe. Multiple lab and service modules would be available. Before cooperation with the United States was solidified Energia began conversations with the Europeans at Deutsche Aerospace to contribute a laboratory module called the Euro-Russian Technological Complex (ERTC). An agreement was signed in 1992 but not long after the ISS agreement put an end to Mir-2 and the ERTC.

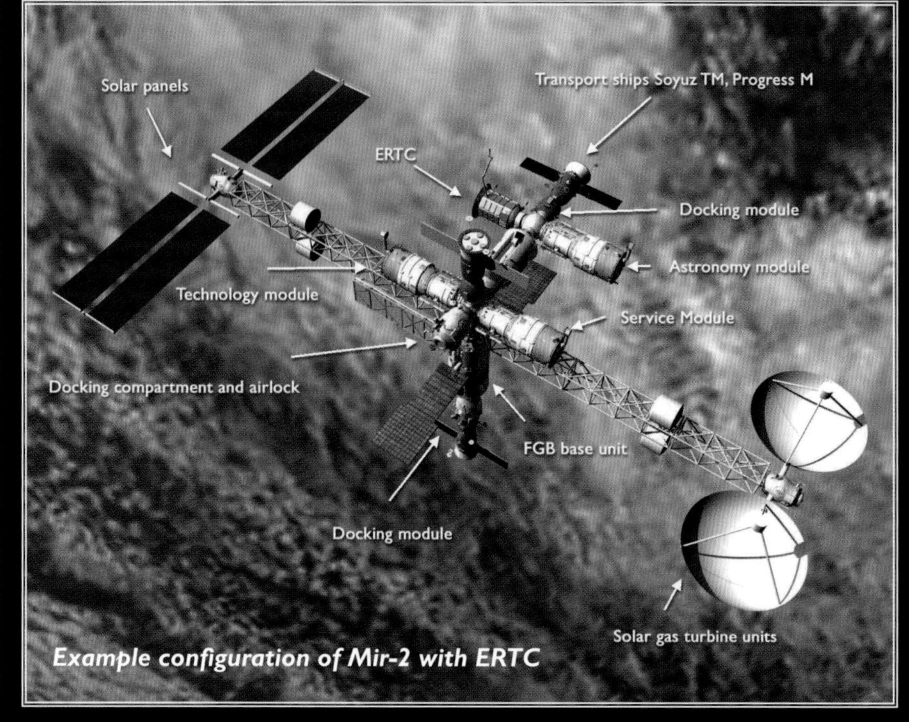

Example configuration of Mir-2 with ERTC

Space Industries Inc. - "Industrial Space Facility" - 1988

In the spring of 1988 a company named Space Industries Inc., run by spaceflight pioneer Max Faget, proposed this commercial module which was called the *Industrial Space Facility*. It would have 12 kilowatts of power coming from the solar voltaic system and four external docking ports. Inside, a shirt-sleeve environment of 71 cubic meters would allow crew to work in comfort.

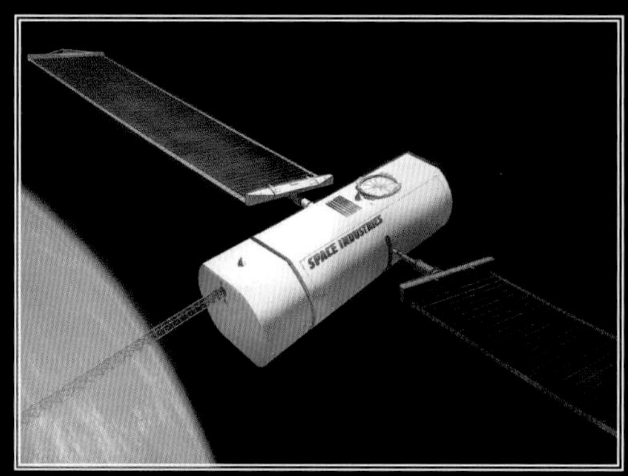

NASA/CSA/ESA/JAXA - "Space Station Freedom" - 1988 to 1993

In 1991 the U.S. Congress approved a modification which featured a shorter pre-integrated truss and modules. In July of 1988 President Reagan had officially named the project *Space Station Freedom*. But with the collapse of the Soviet Union the discussions which had been going on for more than a decade presented new possibilities. With budgets getting ever tighter a new hybrid of technologies whereby Mir-2 might be combined with the Baseline became not only possible, but attractive.

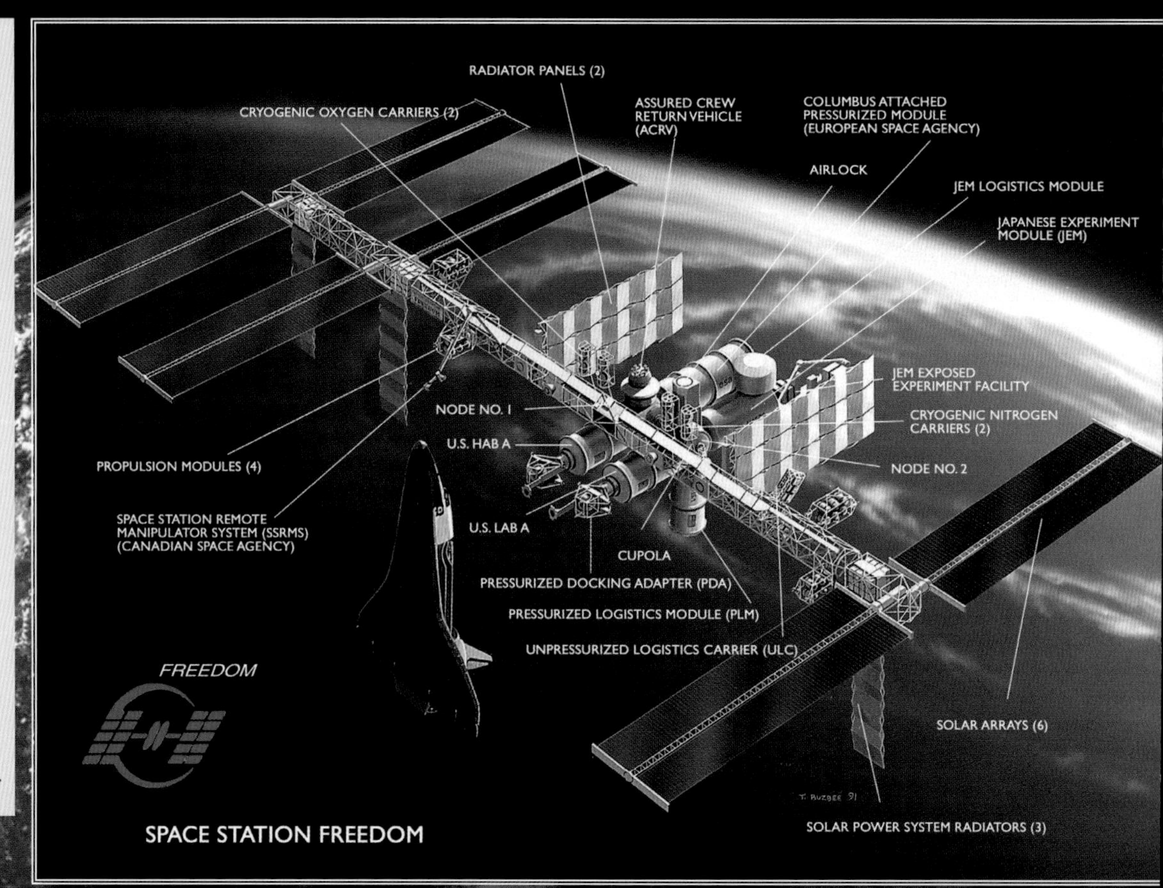

In 1992 US Vice President Al Gore travelled to Russia to investigate the Russian space program. His conclusion was that the Russians would be open to a bigger partnership in space activities. In the summer of 1993 the Clinton Administration ordered NASA to reduce the cost of building *Space Station Freedom*. Congress came within one vote of cancelling the project altogether. As a solution to this crisis, in June of 1993, the administration was presented with a choice of three less complex designs known as Options A, B and C. All of the options relied on the inclusion of Russian Soyuz escape vehicles.

NASA - "Space Station Freedom Redesign" - 1993

Option A (right) was a modular configuration which used a combination of Space Station *Freedom* hardware with a much shorter truss, a common core/lab rather than a node and lab, a simplified power system and a smaller airlock. ▶

Option B (right centre) was derived from mature *Space Station Freedom* designs. It made maximum use of existing systems and hardware to provide an incrementally increasing capability, emphasizing accommodations for users, adherence to international partner commitments, flexibility and growth potential, and recommended some changes to data, communications and tracking systems to save money. ▲

Option C (below) was to be launched on something similar to the proposed Shuttle-C cargo shuttle, which had been on the drawing boards for years. Internally it would revert to the old layer-cake format. It featured a 92-foot-long, 23-foot-diameter core module launched as part of a Space Shuttle using an external tank, solid rocket boosters and shuttle main engines. The module would provide 26,000 cubic feet of pressurized volume, separated into 7 decks connected by a centralized passageway. Seven berthing ports would be located on the module to accommodate the international labs and other elements. ▼

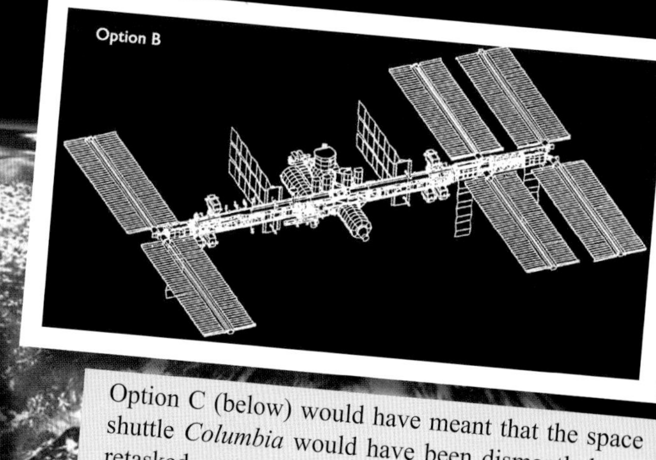

Option C (below) would have meant that the space shuttle *Columbia* would have been dismantled and retasked as a space station, harking back to Koelle and Hoeppner's ideas in 1951. ▼

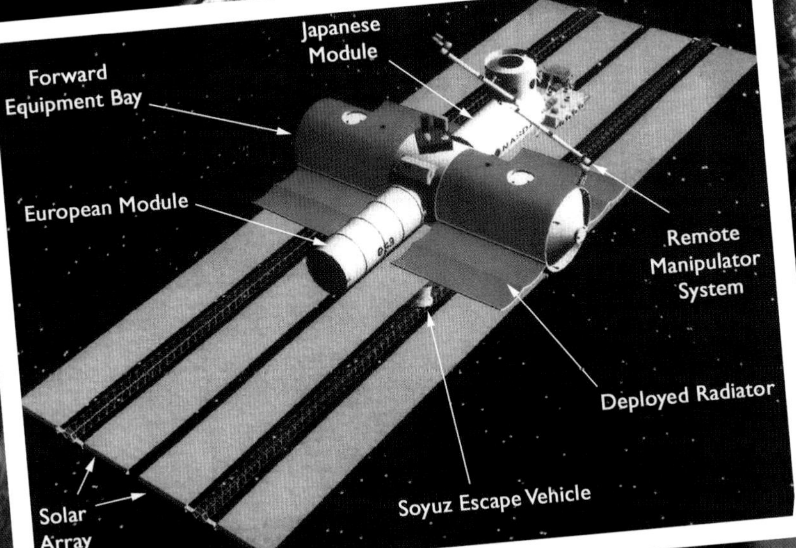

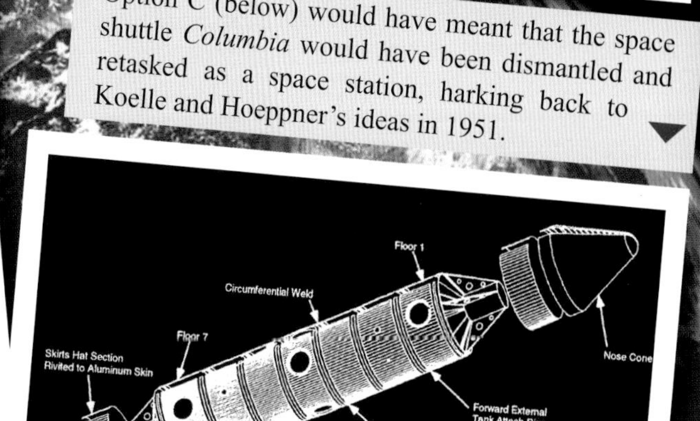

NASA/ENERGIA/CSA/ESA/JAXA - “Alpha/MIR” - 1993

In July of 1993 President Clinton gave NASA two months to develop a strategy to make the Option A design viable and cost-effective. NASA Administrator Dan Goldin presented what was called the *Alpha Station Program Implementation Plan* to the President’s office in September. *Space Station Alpha* was to use much of the material designed for *Space Station Freedom*, but it would use two Russian Soyuz crew return vehicles and a US Bus-1 or possibly a Salyut *Functional Cargo Block* for the main propulsion, navigation and guidance modules (right).

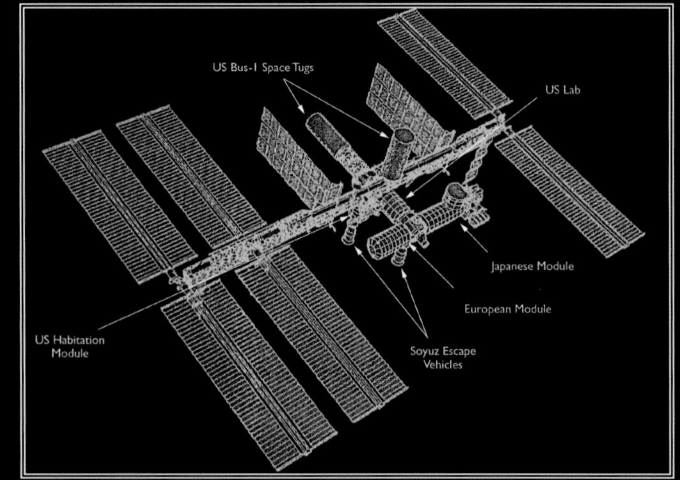

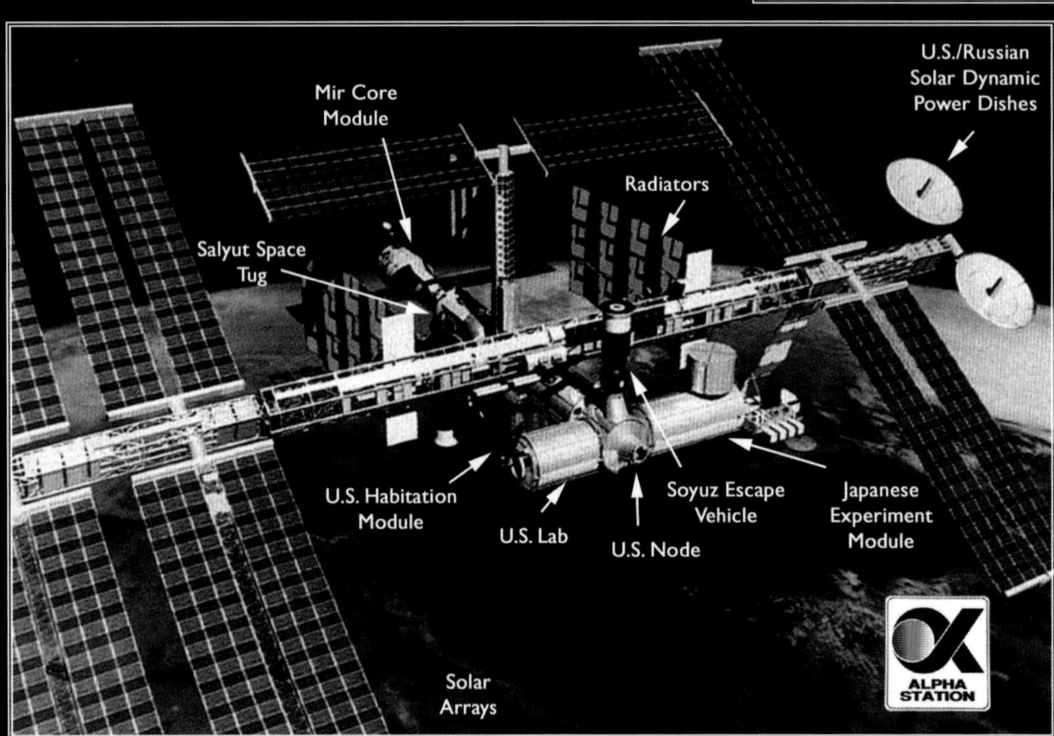

By November 1993 the USA/Russia partnership was ratified and the Salyut/Mir FGB was chosen due to its proven track record and the fact it could be serviced internally. The US/Russian docking module developed for Shuttle/Mir was also added to the design, along with a large power tower and several Russian science modules. This hybrid of an advanced *Option A* (Alpha), and what was to have been *Mir-2*, received support from the US Congress.

NASA/ENERGIA/CSA/ESA/JAXA - “International Space Station Alpha” - 1994

This image is from April of 1994 by which time the hybridization of *Mir-2* with *Alpha* began to take on the familiar shape of the *International Space Station*. The Russians would ultimately choose not to launch the large vertical power structure, or the science modules seen in the centre of this image. Also the solar dynamic power supply which had been briefly reintroduced to the Alpha Station was once again scrubbed from the platform.

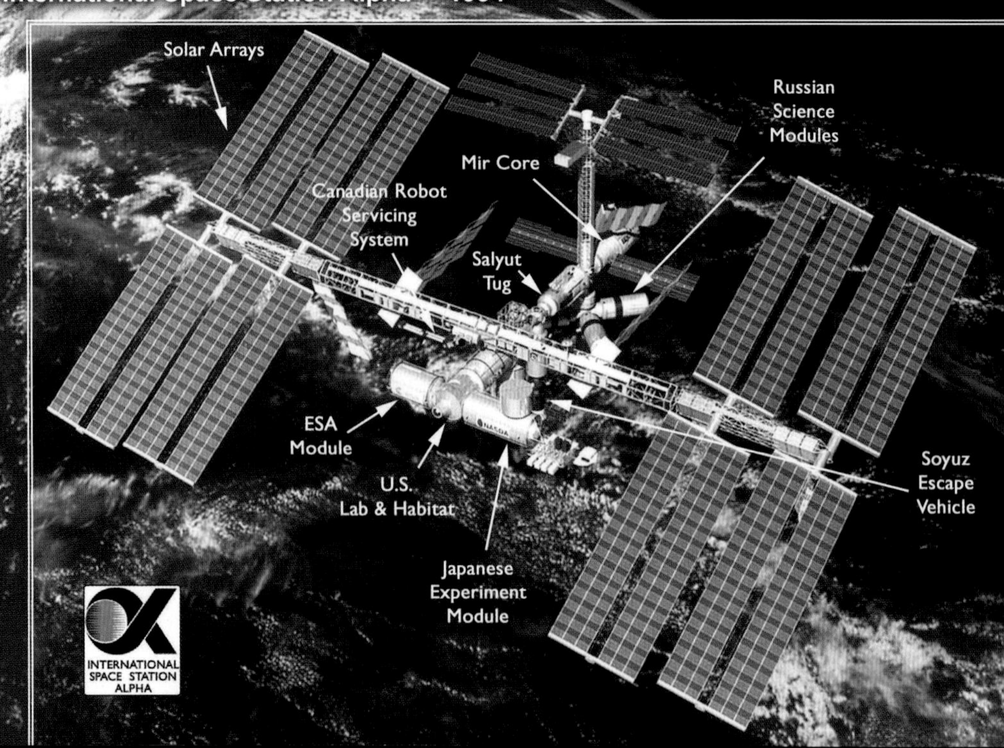

When the American Space Shuttle was proposed, it was to be one part of an infrastructure for human space transportation. This included the Shuttle ferrying supplies and crews to a large station in space, and an unmanned launch system based on the Shuttle, transport vehicles going to higher Earth orbits, a space station in low earth orbit, an Earth to Moon ferry system, outposts in lunar orbit and on the Moon's surface, and the first human expeditions to Mars.

But even before Apollo reached the moon, budgets were cut back and production of moon rockets had been terminated. The Space Shuttle remained in the plan, as the next goal after Skylab. It was needed to serve as a launch system for humans into orbit and a launch vehicle for payloads like space station modules. But there was no approval of a space station. There wouldn't be until after the Shuttle was already flying.

In 1984, President Ronald Reagan charged NASA with constructing a space station. It would be built by carrying individual components into orbit in the Shuttle.

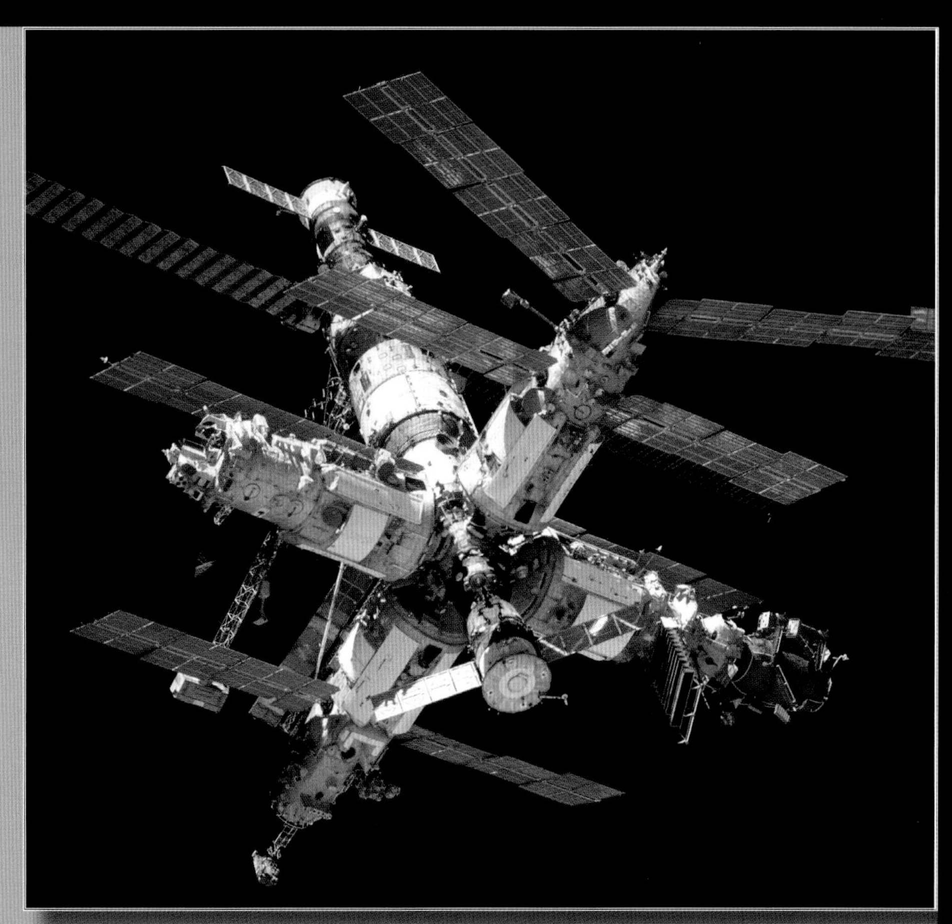

This 70mm frame affords a full view of Russia's Mir Space Station complex during approach for docking, back dropped against the blackness of space.

Beginning in 1971 with the first Salyut station, Russian flight controllers gathered extensive experience operating a series of space stations. The latest and most advanced, the Mir Orbital Station, was assembled in orbit beginning in 1986. Except for three unmanned periods, 28 expedition crews operated on the station for fifteen years from March 1986 until June 2000. For a ten year period, from 1990 to 1999, crews operated on board without interruption. Mission operations were supported by ground control on a continuous 24 hours a day, 7 days a week, 365 days a year schedule.

Before Mir, the Soviets had attempted nine earlier stations. Three were unsuccessful. Four first generation stations, Salyuts 1, 3, 4 and 5, operated with 7 crews that stayed in space for weeks at a time. Several of these were military in nature, used for reconnaissance and interception. Salyut 3 shot down an unmanned satellite with a cannon from a distance of several thousand meters. During Salyut 3 unmanned pods were returned to Earth, carrying canisters of exposed reconnaissance film. On Salyut 4, an unmanned, automated Soyuz showed the practicality of unmanned vehicles carrying logistics to and from the station.

Two second generation stations, Salyuts 6 and 7, provided docking ports on both ends which allowed crews, unmanned logistics vehicles called Progress, and large specialized modules, the size of the base station, to come and go. These showed the value of resupplying the station and of removing trash. Operations continued for months at a time with handovers between crews becoming routine. By the time Salyut 7 had completed its operations in 1986, the Soviets had been operating a series of space stations for fifteen years.

Beginning in 1978 with Salyut 6, and continuing on Salyut 7 and later on Mir, the Soviets hosted visiting cosmonauts from fourteen countries. This permitted the first citizens of countries other than the US or Russia to fly in space.

International missions to Salyut and Mir included visits from representatives of Czechoslovakia, Poland, the German Democratic Republic, Bulgaria, Hungary, Vietnam, Cuba, Mongolia, Romania, France, India, Syria and Afghanistan as part of the Interkosmos Program. Several additional cosmonauts flew commercially from Great Britain, Japan and the European Space

Agency. Towards the end of the Mir program, discussions had begun about flying a number of 'privately' sponsored missions, in which 'space tourists' would purchase tickets privately to fly on Russian Soyuz exchange missions. Delays caused by negotiations and the Mir's termination of operations, resulted in seven US, Canadian, and South African citizens flying to the International Space Station instead.

Following the ASTP mission in 1975, there were discussions about a US/Soviet, Shuttle/Salyut docking mission. But increased political tensions deferred any agreements between the two countries.

Mir was a third generation space station. Like Salyuts 6 and 7, docking ports were provided at either end of the Core Module. But unlike the Salyuts, on one end of the core module, a node provided four additional radial ports for docking large modules perpendicular to the core module's length. Smaller Soyuz and Progress logistics vehicles carrying crew and supplies would dock at the longitudinal ports on either end of the station.

In April 1987 the first large module, Kvant 1 (Квант or Quantum), intended originally as an astrophysics module, was launched. In December 1989, the second large module Kvant 2 joined the complex. In addition to scientific equipment, Kvant 2 carried an airlock to permit spacewalks by cosmonauts without depressurizing other areas of the station. Kvant 2 was followed by the Kristall (Кристалл or Crystal) material science module in June 1990. Kristall had docking ports on both ends. There were additional docking ports available on Mir, but at the time of the Soviet Union's collapse, it appeared that the Mir Orbital Station was complete with its four large modules. Funding shortfalls threatened to permanently defer the delivery of any further modules.

The first research laboratory to be launched, the Kvant 1 module, was permanently located at the aft port of the core module and included its own rear docking port and facilities to accept Soyuz and Progress vehicles. The later large modules all initially docked at the forward longitudinal port on the Mir core module and were then relocated to the radial ports using the 'Lyappa' manipulator arm. The first research laboratory to be launched, the Kvant 1 module, was permanently located at the aft port of the core module and included its own rear docking port and facilities to accept Soyuz and Progress vehicles. The Kristal module had docking ports on both ends and was to serve for docking the Soviet Buran Shuttle. The Buran Shuttle never visited the Mir but the US Shuttle did. Kristal did serve as an impromptu docking module for the first Space Shuttle to visit on STS-71. As the program continued, and more modules were added, the larger modules were moved around to maintain symmetry and stability.

Several additional modules awaited completion at the Krunechev State Research and Production Center in Moscow. They could have been used as back-ups to the Mir modules if any had failed early in their planned lives. As the Mir Orbital Station had now been operating well beyond its original planned five year life, the Russians tentatively planned to launch a second advanced orbital station.

In 1984, even before Mir was launched, Russian Energiya Company that had built many of the earlier stations was already planning a follow-on to Mir; a huge orbital complex was under consideration. This facility was identified by the Russians as OSET, which stood for *Orbitalny sborochno-ekspluatatsionnyy tsentr* (Orbital Assembly and Operations center). It would be used for orbital construction of large space facilities, as a possible refueling station and as a repair and servicing site for satellites. These were ideas similar to those NASA was considering for a Space Operations Center, prior to President Reagan's approval of the space station. The core module that had been the back-up to Mir was to serve as the central base. Other modules and facilities would be assembled and attached. Unlike the US station which required space shuttles with robotic arms and spacewalking astronauts to connect modules and assemble the station, the Russians used modules which flew autonomously, docked automatically, and were maneuvered into their final positions by remote operations from Earth.

As the Soviet Union collapsed and government funding of the space program dropped precipitously, the Russians set their sights on a smaller station which was similar to Mir. It would be composed of the same core module that had been Mir's back-up, with an atmospheric observation module Spektr (Спектр – Spectrum), and an Earth resources module, Priroda (Природа – Nature). The docking node on the new core module lacked lateral docking ports, so only the two research modules could be accommodated, unless additional modules were situated on the fore or aft ports of the core module.

The simultaneous planning for the US-led Freedom station and the Russian Mir follow on, led to consideration of a joint merged program. And so as Mir continued into its sixth year of operation, it was at this point that serious discussions began to take place concerning the Russians joining the already international US-led Freedom space station program.

There were several considerations to joining forces with the Russians. Freedom, while well along in design, was still several years away from the launch of its first elements. Mir had reached its planned lifespan, though continuously manned operations showed no signs of slowing down. In the aftermath of the Shuttle Challenger accident which occurred in 1986, most satellites were reassigned to unmanned launch vehicles and few large payloads were available for launch on the Shuttle. The Shuttle had tremendous capacity that would go unused until Freedom assembly began. The Soviet space program was desperate for cash as revenue from the Russian government dropped by more than 80%. Russian space companies had begun to sell their systems to countries not entirely friendly to the US and its allies. Meanwhile, cost overruns in the Freedom program demanded the elimination of some of the station's elements in order to save dollars.

In 1993, the new Clinton Democratic Administration took office. Both houses of the US Congress were under Democratic control. And while the space program had enjoyed bi-partisan support, the Republican-mandated Freedom space station program was threatened with termination in June, 1993 by a one-vote margin, owing to its incessant delays and cost over-runs. The President mandated that in order to rebuild support the program needed to serve his Administration's international interests.

NASA's new Administrator, Dan Goldin, saw this as a win-win opportunity. Bring the Russians into the Freedom Program. They could provide some of the more expensive elements that were needed early in the assembly phase, because their modules were capable of operating autonomously and provided critical functions like power, cooling guidance and propulsion and so could serve as a base from which to begin assembly of US elements of the station. An offer was made to the Russians to host one or more visiting Space Shuttle missions to Mir until space station Freedom assembly could begin. The US personnel would have an opportunity to learn about long duration space operations on the Mir. Contract negotiations between NASA and the Russians began.

Within a matter of weeks the space station program expanded. It was called Phase I of Space Station Freedom and later of the International Space Station Program. Although it started with only a single US Shuttle flight, eventually the decision was made that there would be ten Shuttle flights to Mir. Instead of a cap on the mass of US hardware to be placed on Mir, a joint working group would assess US and Russian hardware and supplies that could be brought to Mir and returned from Mir on the Shuttle. The two Russian modules that had been intended for use on a subsequent space station, Spektr and Priroda, would be outfitted as US laboratories

A Russian-built docking module, intended originally for the use by the Soviet Buran shuttle orbiter, was brought to Mir by the second visiting US Space Shuttle, STS-74. It was located on the end of the Kristal module to allow orbiters to dock with the station, maintaining safe clearance once additional Mir modules had been delivered.

The US would procure from the Russians the docking apparatus to be used for the Freedom space station. The Russians would provide to the Freedom space station the Mir back-up module that could provide power, propulsion and crew living quarters. They would provide Soyuz vehicles that would serves as emergency return vehicles to be based at the Freedom station and the Russians would provide Progress vehicles for carrying logistics supplies and for refueling the station. By 1 November 1993, NASA announced that the Russians would become a partner on the already ongoing international space station Freedom. Later, Boeing signed a separate contract with the Russians for one of their heavy modules outfitted to provide guidance, propulsion and power in advance of the Russian Mir back-up module.

The first Russian, Sergei Krikalev, flew on Shuttle STS-60 in 1994. In 1995 the series of Shuttle-Mir missions began. The first Mir rendezvous mission took place on STS-63.

American astronaut Norm Thagard launched on Soyuz TM-21 in March, 1995, and was recovered by the first Shuttle to dock at Mir, STS-71, in July. The series of nine Shuttle docking missions yielded experience in joint operations, Shuttle rendezvous and docking and proximity operations with a large space station, as well as long duration operations with US crewmembers. The Phase I joint program gave the Russians a valuable cash injection at a time when they were in desperate need and it allowed them to keep their spaceflight workforce intact to complete the delivery of modules to Mir and then continue with modules for delivery to the ISS.

The seven long duration missions of Americans on Mir provided a valuable database of biomedical data, ground control and tracking, scheduling launch and landing operations around each other's programs and in international cooperation over a prolonged period. All of this would be applied to the larger follow-on program (ISS).

The Americans found that flying long duration on Mir with the Russians was nothing like flying a short Shuttle mission, or even their own missions on Skylab twenty years before. The Americans found that their normal approach of detailed mission planning was impractical on Mir. The Russian approach was to plan a range of activities for the day and a list of priorities for the week or months, which might change in real-time as situations developed. If things were not done 'today or this week,' they could be delayed until later in the expedition or even moved to the next expedition.

Moving to Russia and living away from home for several months was also a challenge for the US astronauts, having to learn not only a new language, but also a new culture and approach to mission training and operations. There were also added difficulties with delays to the Russian and US flight schedules and Russian crew member changes, as well as the fact that direct communications with Mir for most of the orbit was difficult, as the Russian global communications network was not as extensive as the American network.

As with Salyuts 6 and 7, throughout the Mir Program, a series of Progress vehicles visited the Mir complex. The Soviet Shuttle called Buran, which was planned to dock with Mir, never did. It is ironic that while a shuttle docking with Mir was part of the original planning, it would be an American vehicle that achieved it.

The Russians recommended that they could detach the latest Spektr and Priroda modules from Mir, take them across to the new International Space Station and attach them to the Russian segment. This made some sense, as their full potential would never be realized on Mir having come so late in the station's operational life, but it was a task the Americans were not keen to attempt.

Mir EVA operations were extensive throughout its operational life. Activities included adding or replacing exterior hardware, systems and support equipment. There were also numerous EVAs supporting repairs and contingency operations. All this experience would be of benefit in the transition to ISS.

"When it became clear that the International Space Station was actually going to take place and US and Russians would join together, then there was a decision to have a transitional phase between Shuttle-Mir and ISS, called NASA-Mir. It was supposed to be a transition to ISS. We called ISS Phase 2, and we called NASA-Mir Phase 1. It was supposed to be a transition to gain as much experience as possible on the design and operation of various systems for ISS, Phase 2." - Victor D. Blagov

"In the first half of 1992, I was told "people want to look at a mission to fly the Shuttle one time to the Mir and dock with it". In July of 1992 we went to Moscow on the first trip with the task of working with the Russians to determine could we dock a Shuttle? What docking system would we use? What were the problems? There was a second trip to Russia in September, and a third trip in November. When we first started working with the Russians, it was one Shuttle flight. Some wanted to do ten flights. From 1992 to 1994, I was the pseudo program manager. We went from one flight to ten flights. Then they named me the program director. It was officially part of the ISS Program. It was highly integrated into the Shuttle Program." - Thomas Holloway

A Shuttle/Mir, Spektr/Priroda program patch includes the Shuttle docked at the Mir space station, in orbit over the earth. The patch is in the shape of a Soyuz capsule.

The external airlock assembly/Mir docking system is rotated into position for shipment to the Kennedy Space Center (KSC) in Florida. Jointly developed by Rockwell and the Russian company RSC Energia, the external airlock assembly and Mir docking system was mounted in the cargo bay of the Space Shuttle to enable the shuttle to link up to Russia's Mir space station.

"It was very frustrating for our international partners to have the Shuttle-Mir effort declared to be Phase I of the International Space Station, because by this time we had been working on Station with the other partners for 8 years. ISS Phase I was a practice run at negotiating with the Russians. The Russian Space Agency had just been established. We were establishing a relationship with this new entity. The NASA Administrator, Dan Goldin, decided that negotiating a contract with the Russians was a way of getting to know one another. NASA-Mir was good for the two sides to develop respect and trust and learn how to work together. We got to know one another as human beings and fellow engineers with a common goal. Initially there was to be a Russian cosmonaut who would fly on the US Shuttle and one US astronaut who launch on a Russian Soyuz and live on the Mir for a long duration mission. There was to be a single flight of a Space Shuttle to the Mir to recover the US astronaut. There was a limit, based on mass, of US hardware that would be placed on the Mir." - Lynn F.H. Cline

"I was the Mission Manager for the first Spacehab flights and had a chance to work with the Russian cosmonauts early on. By the time Titov flew on STS-63, I was involved in negotiating the Shuttle Mir contracts. Meanwhile I was encouraging Spacehab to look at doubling their Spacehab module to use it for flights to Mir and I presented a comparison study on the costs and efficiencies of Shuttle, Spacelab and Spacehab integration processes. Because of time and money, I told Tommy Holloway, Spacehab was the only option." - Gary Kitmacher

Sergei Krikalev removes a bioprocessing module from a stowage bag in the Spacehab commercial module prior to performing a BIOSERVE commercial experiment.

Vladimir G. Titov handles vials of samples for the Commercial Generic Bioprocessing Apparatus (CGBA) experiment in SpaceHab 3 Module onboard the Earth-orbiting Space Shuttle Discovery. Titov joined five NASA astronauts for eight days of research in Earth-orbit.

"The Gore-Chernomyrdin Vice Presidential Agreement, in September, 1993, established the Russian partnership on ISS. At that point it became clear there would be more than a single Shuttle flight to Mir. NASA Mir became Phase I of the ISS. The first cosmonaut flew on a Shuttle on STS-60 in February, 1994. The first rendezvous of a Shuttle with Mir occurred on STS-63, in February, 1995. Nine more Shuttle flights would dock with the Mir orbital station." - Robert E. Castle

To the right Gary Kitmacher, who negotiated the initial agreement to expand the NASA-Mir Program and then re-designed the Priroda Module to a US laboratory, signs drawings for the Priroda in Moscow. In the center Victor Volchkov of RSC Energia was the chief architect for Priroda. An interpreter sits at the far left.

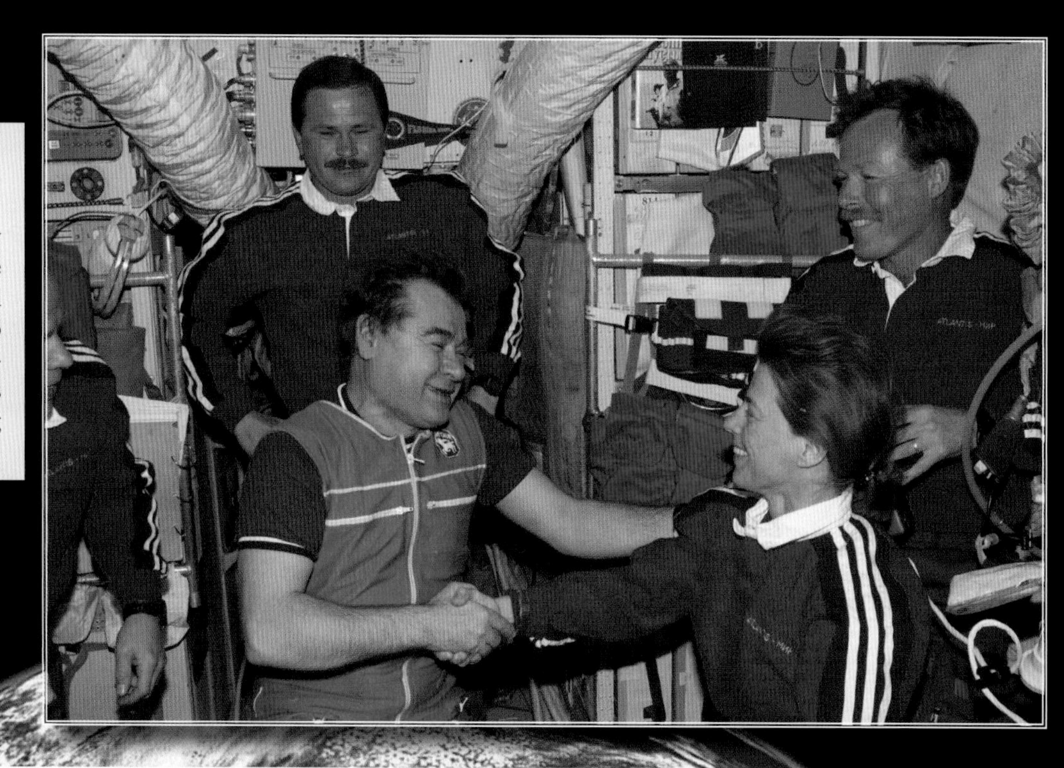

Aboard Russia's Mir Space Station, astronaut Bonnie J. Dunbar shakes hands with Gennadiy M. Strekalov. Looking on are Nikolai M. Budarin (left) and Robert L. Gibson. This photo was taken a short while after the Space Shuttle Atlantis performed its historic docking with Mir.

"During the Mir missions, we exchanged crew members, exchanged thousands of pounds of cargo, and tried dozens of science and technology experiments. Some were predecessors of what we eventually deployed on the ISS. We tested the Canadian Space Vision System which was eventually used for assembling the Space Station with the robotic arm. We tested a scale for measuring the mass of the astronauts. We had an opportunity to test hardware and protocols in the real environment before we deployed them and relied on them on ISS. That was risk mitigation. One example, we had the prototype of the Volatile Organics Assembly [VOA]. It looked for contaminants in the air. The prototype failed but then we had a chance to fix it. Another was the Japanese Real-Time Radiation Monitoring Device, RTMD. It measured radiation. It sent real-time data down to a team of Japanese researchers. International involvement should be recognized." - Bonnie Dunbar

"We were flying hardware and payloads on Mir as risk mitigation experiments; they were proof of concepts that could be tested on Mir years before they would have an opportunity to fly on ISS." - Jeffrey A. Cardenas

This photograph of the space shuttle Atlantis was taken from approximately 170 feet away by Shannon W. Lucid, winding up her duties as onboard Russia's Mir Space Station. Lucid was in Mir's Base Block Module. The Spacehab double module, a first time space flyer, is seen in the aft payload bay. Its tunnel can be seen connecting to both Atlantis crew cabin and the androgynous docking adapter. Also seen in the forward bay is the Ku-band antenna used for communications. Though not recognizable in this photo, several Atlantis crew members had their noses to the windows as NASA was about to make its first crew member exchange with Mir. John E. Blaha was onboard Atlantis as Lucid's replacement.

"We changed the whole integration approach for Shuttle flights to Mir. We depended on Spacehab modules for most of the Mir flights. Spacehab was a commercial venture. The Spacehab team did their own timelines and integrated everything they carried themselves. The Spacehabs looked the same on every flight. They flew with one module. They would get the module back at their facility, take the back end off of it, and reload it, and then roll it back to put back in the Shuttle. They didn't have as many people. They didn't spend a lot of time turning their flights around. It was a tight process that held together. Our NASA cargo organization had trouble with Rockwell over how to do this because on Shuttle there were drawings to change. The configuration management system integrated every piece of hardware on a flight by flight basis. Someone asked, "why are you doing that?" If all the flights look the same, then do the engineering analysis and drawings one time." - Denny Holt

"The NASA-Mir program would last about 4 years. Spacelab missions took about 4 years to prepare to fly. It normally took about 2 years to integrate payloads flying on the Shuttle. We could not afford a lengthy development and integration schedule for NASA-Mir. A good example of this was the Crew On Orbit Support (COSS) computer. This was to be a laptop used on Mir for computer based training, reference information, watching videos, and reading digital books. After his mission was over in July, 1995, the first astronaut on Mir, Norm Thagard, said he really had needed it. Shannon Lucid would be the next astronaut on Mir in early 1996. In 1994, we designed and built the US lockers, electrical system, computers and other equipment for Mir. In 1995, I was in Russia integrating the hardware to launch to Mir. I was busy and reluctant to get involved in the development of a new flight computer system. I went to the JSC office where they were already developing a COSS for the ISS. I asked the Project Manager to test her system on Mir. She laughed and said they'd been working on it for 5 years, and it would be another 5 years before it flew. They couldn't do anything in months. My own group of electrical and electronics engineers had recent experience developing the hardware for Mir. As a last resort, I asked my lead engineer, 26 year old Mariella Hartgerink, if she thought we could do the job. She spoke with the team. That afternoon she said "it would be fun; we're nearly done with all the other equipment, so we could keep the team together a little longer. So I let the Program Manager know that we would build COSS. We designed, tested and certified the hardware. We built 10 certification, training and flight units. We built a library of training and reference media. Thanksgiving week, 1995, I was at Star City training Shannon Lucid. The first COSS launched with Lucid on STS-76 in March, 1996. The process took 7 months." - Gary Kitmacher

Valeriy V. Polyakov, who boarded Russia's Mir Space Station on January 8, 1994, looks out Mir's window at the Space Shuttle Discovery. Onboard the Discovery were James D. Wetherbee; Eileen M. Collins; Bernard A. Harris, Jr.,; C. Michael Foale, Janice E. Voss, and Vladimir G. Titov.

Shannon Lucid is photographed looking at wheat growing in the Svet or greenhouse, which is located in the Mir's Kristall module.

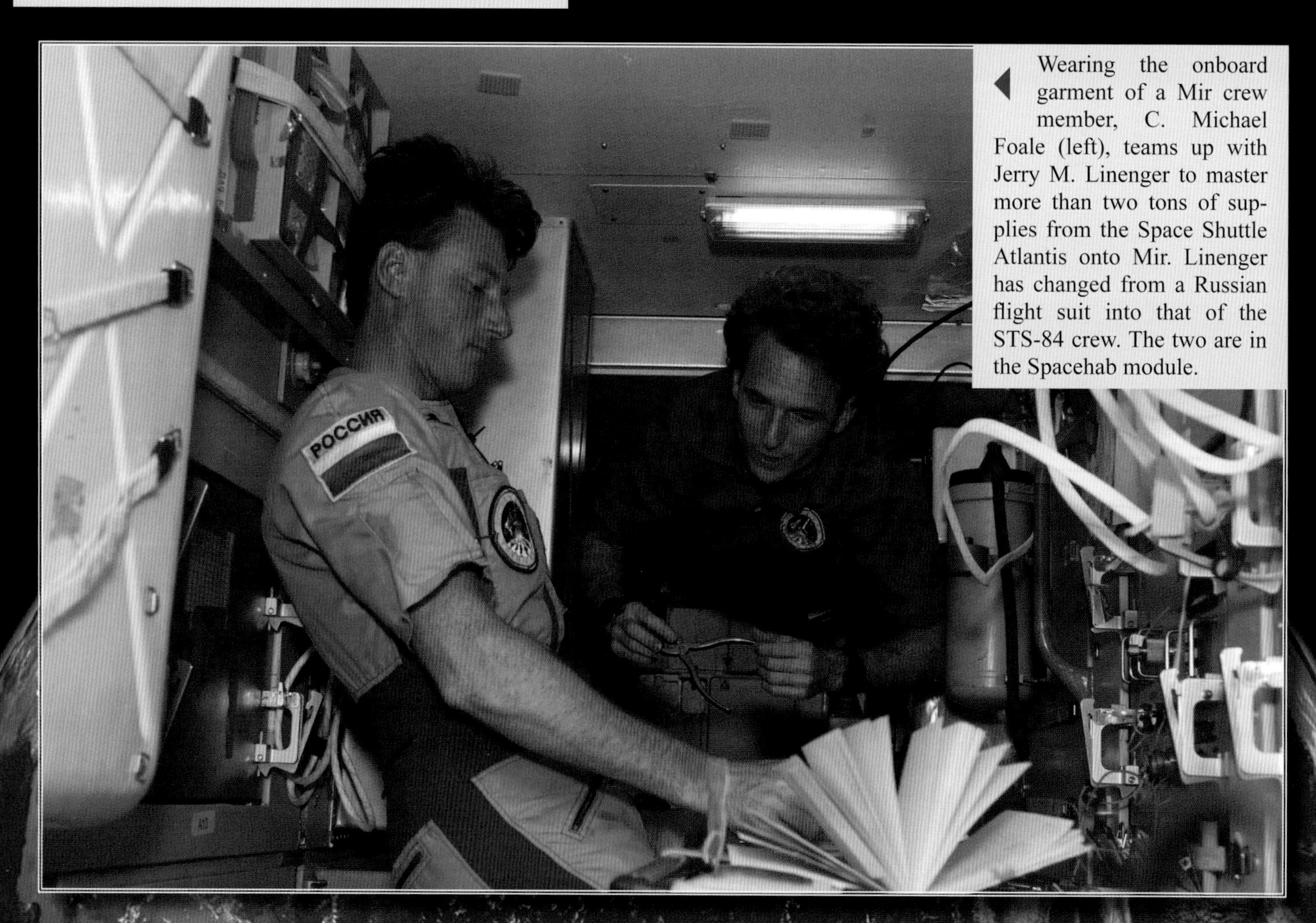

Wearing the onboard garment of a Mir crew member, C. Michael Foale (left), teams up with Jerry M. Linenger to master more than two tons of supplies from the Space Shuttle Atlantis onto Mir. Linenger has changed from a Russian flight suit into that of the STS-84 crew. The two are in the Spacehab module.

"On STS-63, I saw the Mir from afar as we approached it and flew around it. I couldn't relate to it. I knew I didn't have to live in there. It was like seeing the great wall of China or the pyramids from a distance. It was like being a tourist on a bus tour. We saw people on it. We saw Elena Kondakova and two kind of crazy guys waving to us, all excited. We didn't understand each other very well at all. I didn't know Russian well at the time. But we had Vladimir Titov on board the Shuttle, who could speak with them. We lingered there for about three hours. Then we went, "bye," and the Shuttle left. So I knew what it looked like. On STS-84, the Shuttle docked with Mir. Mir looked in better condition physically than I had imagined. When we first opened the hatch and traveled down the hatchway into the Mir's base block, I was expecting worse and saw something better. I saw less clutter. I saw brighter, more cheerful objects, more visible things, than the dull cellar-like impression I'd had in my mind. I was pleasantly surprised at the cheerfulness of the atmosphere there. It was kind of a warm, welcoming, cozy place, in spite of the masses of cables and equipment and wires that are on the walls. It looked like a home."- Mike Foale

"The NASA-Mir Program did several things for the ISS Program. The first thing was that it simply brought the Russians to the table. The Russians had a crew rescue vehicle; the Soyuz. It didn't require development. It was ready to go. The Russians had logistics and propulsion capabilities. They brought to the table several capabilities that NASA did not have. NASA-Mir enabled NASA to learn to work with the Russians. This was a major objective. I would characterize this in two or three categories. First of all, the Russians culturally, are different than we are. They're very capable in the work that they do. Their thought processes are much like ours, but there are a lot of things that are different about how the Russians do business and what they consider to be important and how they work and relate to one another. So I think NASA-Mir gave us an opportunity to begin to understand that. The Russian experience is based on long-duration flights; this is somewhat different than the US experience with predominantly shorter flights. An important thing we learned about was contamination. You have to worry much more about contamination in the atmosphere in a vehicle that you're going to more or less permanently maintain than one in which a crew resides for ten or twenty days. The Russian's overall approach to risk management is different. The US does not take any risk we do not need to. The Russians judge real risks and what are less important. NASA-Mir gave us an opportunity to learn to appreciate those things and learn to relate to those. Of course, there's some problems in all that, as you might expect. You know, they had a certain way of doing business, of setting a certain criteria, and we had a certain criteria. They'd want to do something one way and we'd want to do it the other way. We both thought we were right, and then that has to be resolved. And they also felt like they were in charge, to a certain degree. They've been doing this business for thirty-five years, and we're the new guys on the block relative to long-duration. We learned that personal working relationships are very important. Many of the Russians have been doing the same job for twenty or thirty years, and continuity and relationships are extremely important. The NASA-Mir Program provided a firm foundation for which to engage the Russians in the implementation of the ISS Program."- Thomas Holloway

John Blaha (foreground) and Shannon Lucid talk in the Spektr module of the Mir after docking with the shuttle Atlantis. Lucid is still wearing her Mir-22 crew flight suit.

"Mir was a place that had been lived in and worked in for twelve years by the time I went there. It was kind of like camping out. It's such a cluttered space. You're basically winding your way through, effectively, tunnels to go to one part of the station to another, so that equipment just isolates you from other parts of the station. There were times when I would suddenly pop out of my hole and Vasily would say, "Mike, have you seen Sasha? I haven't seen him all morning." I'd say, "No, I haven't seen him all morning." We didn't know where he was. I could easily spend a day without talking to the other crew members. We considered that to be not such a good thing. So we made an effort to try and tag up in the base block, especially for lunch and often just for a ten-minute tea break." - Mike Foale

"As the operations lead for Shannon Lucid's mission, I tried to set up a routine that would be repeated with the crew member every week. Sunday evening, it was time to start thinking about the work week. I first laid out a plan for the month. We would do microgravity science the first part of the month. In the middle of the month we would move on to combustion experiments. Towards the end of the month we would pick up some biology experiments. The astronaut had the big picture of where we were going. I would lay out the plan for the work week; what activities were scheduled that week. I would give Shannon an overview of what the week was going to be like. On Monday morning, I'd go over what was going on that day. By the end of the evening, I'd get from a data dump from orbit of what had been accomplished. Then I filled the astronaut in on the plan for the next day. We got into a routine that was the same all the way through the work week. Saturdays were typically relaxation days. Periodically on Saturdays, the Russians had a program in which they brought in entertainers for the cosmonauts. They were singers, or theatrical actors, or movie stars; they came in and would talk or sing to the crew." - William H. Gerstenmaier

"NASA Mir introduced us to what life on a space station would be like. The Russian elements of ISS, the Base Block and the FGB [Functional Cargo Block] were going to be very much like Mir. The Russians wanted to run the International Space Station exactly the way they ran Mir. We had our way of thinking how things should be run. The Russians didn't have to worry about any of that when it was their space station Mir, but with ISS they did. Those were tough negotiations. It is working, but from a U.S. perspective we probably didn't get everything we wanted, and from a Russian perspective they probably didn't get everything they wanted." - Kevin P. Chilton

"My first involvement in Shuttle-Mir was negotiating the contract to add several Shuttle missions to Mir, and negotiating for the resources to add US payloads on the Mir station. I'd been working Shuttle middeck types of payloads and had been finding and integrating NASA payloads for the commercial Spacehab. I knew we had no shortage of payloads we could fly on Mir. Plenty of investigators would jump at the opportunity. This was done before a Shuttle-Mir management organization had been established, so I could, and did very easily call scientists I knew, offered the opportunity, and placed them on-board. Later on it became more bureaucratic." - Gary Kitmacher

"Initially we were talking about putting a few experiments on Mir that would not be difficult to integrate. But then when the NASA Mir Program expanded, we almost completely outfitted their Priroda module with U.S. scientific hardware. It was an extensive integration effort putting significant experiment hardware in the Russian module. It took a significant effort to build all of that hardware on a very, very tight schedule, get it to Russia, get it integrated into the Russian vehicle, and get it through all their ground tests." - Rick Nygren

"The NASA Mir Program was instrumental in teaching people like me how to work with international partners. The reason I have the international science management job now for the Human Research Program is because I had the Russian experience in Phase One." - John B. Charles

"We established that we would have an American flight control team in the Russian control center and Russians in our Control Center. Our team was five people. We learned early on to get to know our Russian counterparts extremely well, and on a personal basis. In order to be successful with the Russians it all relied on personal relationships. Forget politics, forget protocol. You had to develop trust in their individuals and they had to trust you or you wouldn't get anything done. We established some really good relationships that still benefit the program today. It was a huge success." - William Reeves

This view of the shuttle Atlantis beginning its move away from Mir was photographed by the Mir-19 crew on July 4, 1995. Cosmonauts Anatoliy Y. Solovyev and Nikolai M. Budarin temporarily undocked the Soyuz spacecraft from the cluster of Mir elements to perform a brief fly-around. They took pictures while the STS-71 crew undocked. For a few minutes no one was left on Mir.

"Someone at the Energia Company had the idea of taking a picture of the Shuttle docking with the Mir, against the blackness of space. Energia's President, Mr. Semenov approved. The instructions were sent to the cosmonauts to get in the Soyuz and circle around and take the picture as the Shuttle undocked. NASA said, "No way. Absolutely not, you cannot do it." Semeyonov did it; he overrode NASA's concerns. The pictures were delivered to us at Energia USA. We sold the photo rights to a major New York image house. We wanted to control our brand. But the pictures started popping up on bookmarks, mugs in NASA stores, everywhere. NASA was releasing the pictures with no restrictions. It never occurred to NASA that Energia was a private company that owned the rights. Neither the American government, nor the Russian government owned the rights. Energia's space efforts were a business." - Jeffrey Manber

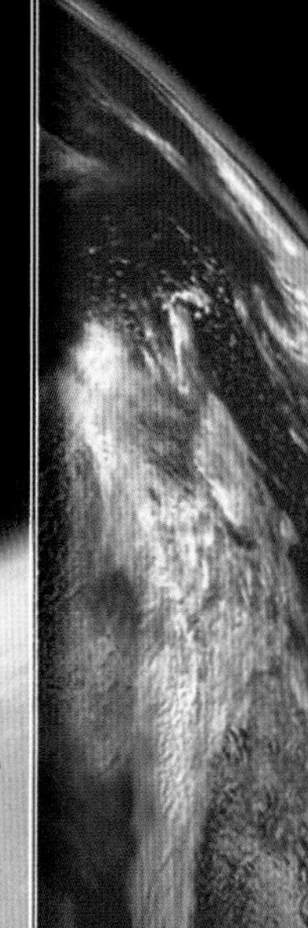

"Our goal for NASA Mir was to fly the Shuttle to the Mir to establish everything we needed to work with the Russians in space. The Russians did things totally different than the way we did things. The NASA Mir Phase I program really paid for itself. I would shudder to imagine how we could have ever pulled off building the ISS if we had not gone through the Phase I program. We started by talking to them about trying to set up simulations and training exercises to figure out how to dock the Shuttle to the Station. We wanted to get the flight control teams and the astronaut crews working and operating together. Simulation was a totally foreign concept to them. They didn't do things that way. What we did in Phase I went a long way toward making Space Station as successful as it became. We got closer to ISS." - William Reeves

John E. Blaha, a member of Mir-22 crew, takes notes in the Priroda Module on one of the many experiments stored there. Shortly after assuming his new duties, Blaha's attention was directed towards the bio-reactor experiment.

Russia's Mir space station is captured on film as it floats above the blue and white planet Earth during Shuttle-Mir final fly-around.

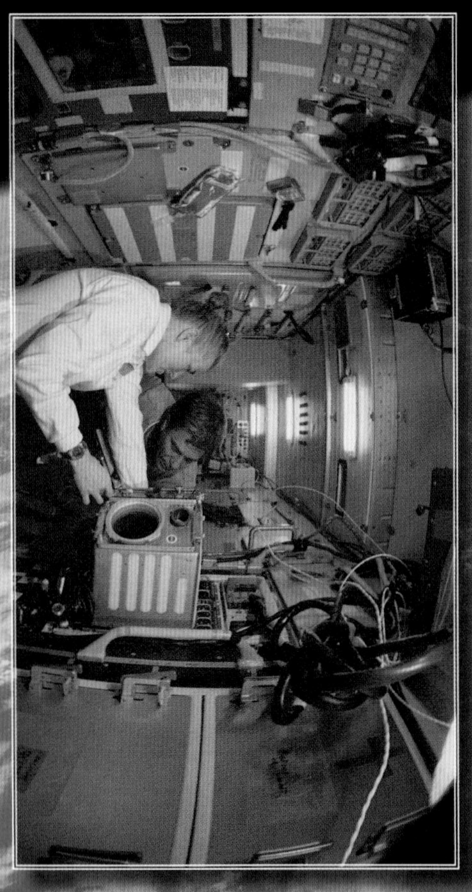

Shannon W. Lucid and John E. Blaha work at a microgravity glove box on the Priroda Module in Mir. Blaha, who flew into Earth-orbit with the STS-79 crew, and Lucid were the first participants in a series of ongoing exchanges of NASA astronauts serving time as long duration expedition crewmembers onboard Mir. Lucid went on to spend a total of 188 days in space before returning to Earth with the STS-79 crew.

"Mir was built without the ability to bring things back to Earth. It was designed not to have the American shuttle go to it, it was designed for the Buran to go to it, the Russian shuttle, but that program got cancelled with the fall of the Soviet Union, and so they had no return capability from Mir for any of the large hardware. So it was a little bit like a summer home where you constantly bring refrigerators and stuff but nobody ever thinks about taking stuff away. And so it was jammed full of things." - Chris Hadfield

On STS-86 Wendy Lawrence is photographed with Cargo Transfer Bags (CTBs) in the Priroda Module of Mir.

"On the Mir missions we had several setbacks; there was a fire on board; there was a collision of a Progress vehicle into the Spektr module. There is a natural tendency to stop everything and say, "let's shut everything down and turn it back on when we know what the resources are that we have access to." As the Chief Scientist my job was to say, "Turn off the minimum things you have to turn off, and don't shut down the science program, which is your justification for being here in the first place."" - John B. Charles

"In thinking about how to transfer cargo between the Shuttle and Mir, I showed one of my engineers a sketch of what I called a locker liner. It was a collapsible suitcase to hold the cargo and it would slide into the lockers on the Shuttle, the Spacehab and the Priroda, all of which were the same size. No one else had thought about the transfers and so we were working overtime right before the first docking, STS-71, to manufacture enough CTBs for the mission. They became the ubiquitous standard of measure for Mir and later for ISS. Eventually we built thousands of them in a range of sizes." - Gary Kitmacher

"ISS is a phenomenal platform. ISS tops the list in terms of a successful, mammoth, international construction project. The internal habitable volume of ISS is nice; which is important from the behavioral health and performance point of view. Volume affects people's well-being. The lessons learned on Mir were dark lessons. For our US astronauts, Mir was not all light and airy. They were a very difficult set of missions for our astronauts. Mir was in the final stages of its life. The Mir station was having a lot of integrity problems. Our intrepid astronauts did their time on Mir, and from those lessons we expected ISS to be more difficult than it turned out from a behavioral health and performance point of view. Only one US astronaut flew long duration on Mir and then long duration on ISS. Mike Foale said, "Mir and ISS are very, very different. ISS is better." Connectedness is a big deal on a space station. The Mir astronauts often had to do their family conferences over ham radio. It was not private. It was the mainstay of their connectivity with home. But, it was really isolating. It isolates someone not to be connected with their spouse, their social support, or their friends. ISS has multiple redundant methods of communication. There are redundant ways to communicate with the ground. The crew can e-mail, text, tweet [Twitter social media platform], get on the internet protocol (IP) phone and call any number on Earth at any time. There are multiple IP phones so multiple people can be on at the same time. There's the biweekly audio/video conference with the family. There's lots of ways to communicate. ISS has been at the top of its arc in terms of a life span. It has been a good time to be onboard a Space Station. It's not toward the end of its life. All the bugs were worked out on the front end." - Albert W. Holland

"The Shuttle-Mir science program was done jointly with investigators in Russia. We had investigators in the United States and in Russia working together for each investigation. I learned a lot about negotiating during that process. The first phase was all quid pro quo, so I only got what I could negotiate for." - Peggy Whitson

"NASA-Mir, Phase 1, was critical to the success of ISS. We funded Shuttle-Mir out of the ISS Program Office and it was integrated into the Space Station Program Office. We had a big investment; people spent years in Russia providing the infrastructure and the presence to make it successful. We learned all of those things that you had to do to be able to succeed in space." - Randy H. Brinkley

"Russia carried out grand projects like the Mir orbital station and which surpassed all expectations for the length of operation. We were proud of it, but without circumspect comprehension, we failed to answer the questions of: "why", "for what"? We failed to use it. Society failed to take a role in establishing the need for the project. We created a tremendous capability but Earthlings were not ready for it: there were no customers. We wanted to create digital maps of the topography and the forests at a scale suitable for conducting economic assessments. Our original analyses said we required resolutions of no greater than two meters. The "Resurs-4" satellites could only resolve to 30 meters. The Mir Priroda Module could resolve to 15 meters. We could have had the whole Earth in the palm of our hands. But society was not yet ready to utilize the capability. We have to do a better job of thinking cosmically." - Valentin Lebedev

"The Russian philosophy about science on the Mir station was expressed to Norm Thagard by one of the cosmonauts he flew with, when the cosmonaut asked, "Norm, why do they have you doing all this research? Don't they know Mir is not a research facility? It's an outpost." He was saying, we are showing the flag in orbit; the idea of doing all this research, that's not what it's about." - John B. Charles

This Image shows Andrew S. W. Thomas (on left) and David A. Wolf during hand-over operations inside the Priroda Module of the Russian Mir Orbital Station. Wolf is explaining the operations of this equipment to Thomas. ▼

David Wolf sorts samples for Microbial Investigations of Mir and Crew onboard the Priroda module of the Russian Mir Orbital Station. ▼

"We launched the last module, Priroda, to Mir April 26, 1996. I designed the module as a US laboratory with all of the hardware, systems and interfaces to directly accommodate payloads coming up on Space Shuttles or Spacehab modules. The last Shuttle flight to Mir was STS-91 in June, 1998. I felt that Priroda had just barely been used before we were terminating our operations after only 2 years. So I went to the ISS Program Manager and recommended they consider moving Priroda to the ISS. The orbital inclinations of Mir and ISS were going to be identical so if they launched into a coplanar orbit that was phased properly, it could have easily been accomplished. This was not well received." - Gary Kitmacher

◀ Andrew Thomas unpacks his bags in the Priroda module of the Mir Space Station.

A 70mm view of Russia's Mir Space Station's Spektr Module shows the backside of a solar array panel and damage incurred by the impact of a Russian unmanned Progress re-supply ship which collided with the space station on June 25, 1997, causing Spektr to depressurize. A radiator, which also was struck by the Progress, is out of view from this angle.

"NASA does not do a very good job at learning lessons and applying what they learn. As part of the NASA-Mir Program, we went to extensive lengths to record our lessons and what we had learned. We worked with the Russians to make sure we had it right and that we captured everything we could possibly capture. Then we worked with the ISS Program to try and transfer those lessons learned and that experience base. The ISS people wrote our experience off: "well, Phase I is not like Phase II, so it's not applicable". From a formal process of writing the lessons down and passing them on and incorporating them, NASA did not do as good a job as it could have." - Rick Nygren

United States astronaut Jerry M. Linenger works outside the Russian Mir Space Station during a joint United States-Russian space walk on April 29, 1997. He was joined by Mir-23 commander Vasili V. Tsibliyev (out of frame) for the five-hour Extravehicular Activity (EVA) designed to deploy scientific instruments and retrieve other science hardware. At the top of the frame is a Russian Progress re-supply capsule docked to the Mir's Kvant-1 module.

"The Russians said, early on, that "the Mir Space Station is a perfectly good space station. Why don't you just attach your new ISS onto our old station?" This was not well received on our side. I led the team that assessed this proposal and was told from the beginning it was not a good idea and to come up with the appropriate response. So I led the team that came up with many reasons. We didn't expend many billions of dollars on the world-class orbiting laboratory just to attach it to something that had a very limited remaining lifetime." - Charles Lundquist

Andrew Thomas signs his name on the Mir Space Station Base Block bulkhead.

"It was a rough transition when we finished Phase 1 and went to ISS. We had developed the procedures for certifying the safety of the cargo and experiments going to Mir. And it made sense to follow those same processes for ISS. The ISS people didn't want to have anything to do with what we'd learned with the Russians. It was very frustrating for the Russians too. NASA side for ISS had not developed processes for certifying crew supplies, cargo and experiments. The Russians said, "We have the Russian segment of ISS. We are part of the safety certification process; it is not just the NASA Safety Review Panel." We finally were able to convince the ISS safety people, "you will have to develop a process similar to what we had for Mir." Eventually they did. But the US ISS group was very much "you do it our way". It was unfortunate. NASA has a bad habit of not using the lessons from the past. The Russians had developed a process jointly with the US for Mir. And NASA didn't have a process to take its place. So eventually the process used for ISS was close to what we had used for Mir. Things would have gone much more smoothly if the US ISS Program had adopted what we had already been doing on Mir. It helped a lot that I was the one NASA point of contact from the start of the Shuttle-Mir that carried over to ISS." - Gary Johnson

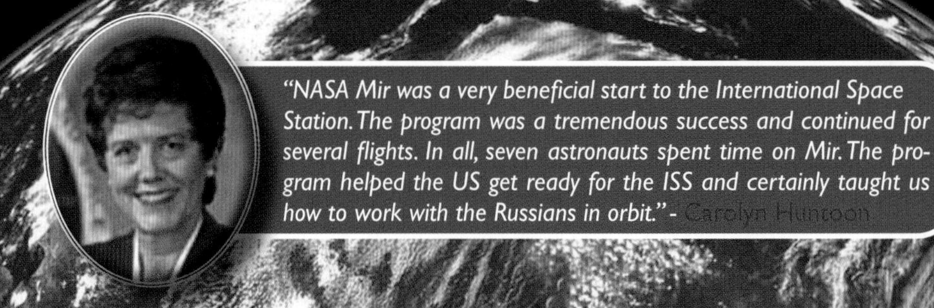

"NASA Mir was a very beneficial start to the International Space Station. The program was a tremendous success and continued for several flights. In all, seven astronauts spent time on Mir. The program helped the US get ready for the ISS and certainly taught us how to work with the Russians in orbit." - Carolyn Huntoon

In Russia our approach is to have people specialize in a specific technical area. In the US they rotate people frequently. In the 1970s, I was involved in Apollo-Soyuz. Many of the Russians involved in Apollo-Soyuz were still working during NASA-Mir and still doing the same work that they did in the seventies. The same people just transferred from Phase 1 to Phase 2. But only one or two Americans were still doing the same kind of work they had done in the 1970s. On the United States' side, it was not the same from Phase I to Phase II. A lot of new people came for Phase 2. It was like starting new, from scratch. I don't think that expertise is as important as learning to work as a single group. The Russian's task was to teach new people coming to Phase 2 how to deal with challenges, to try not to repeat the same mistakes." - Victor D. Blagov

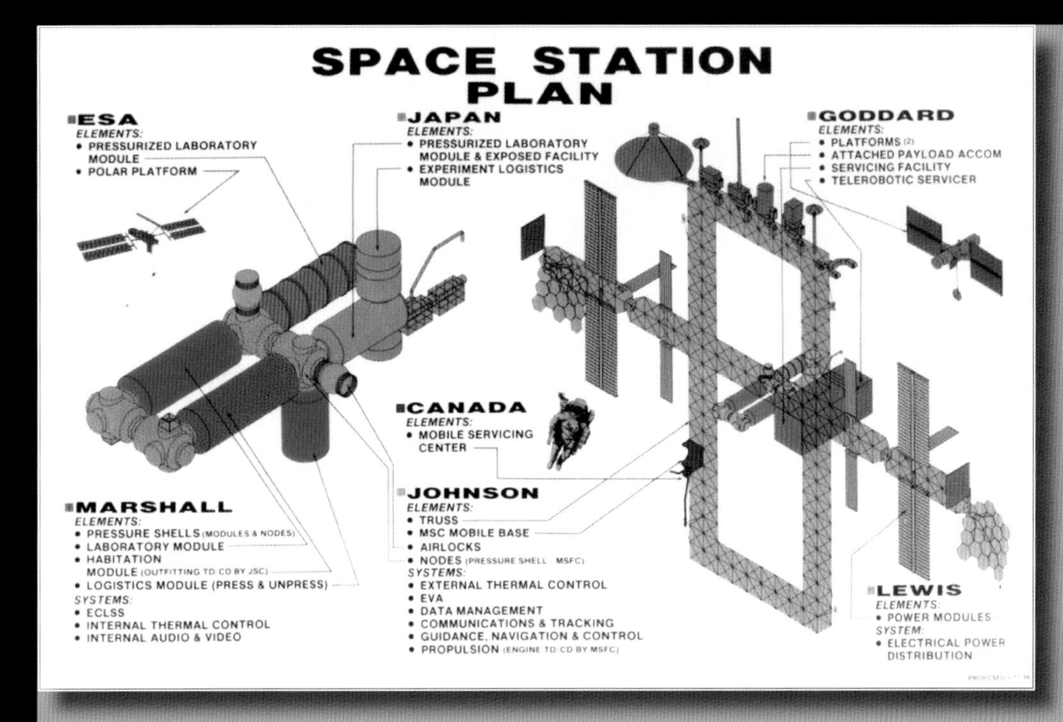

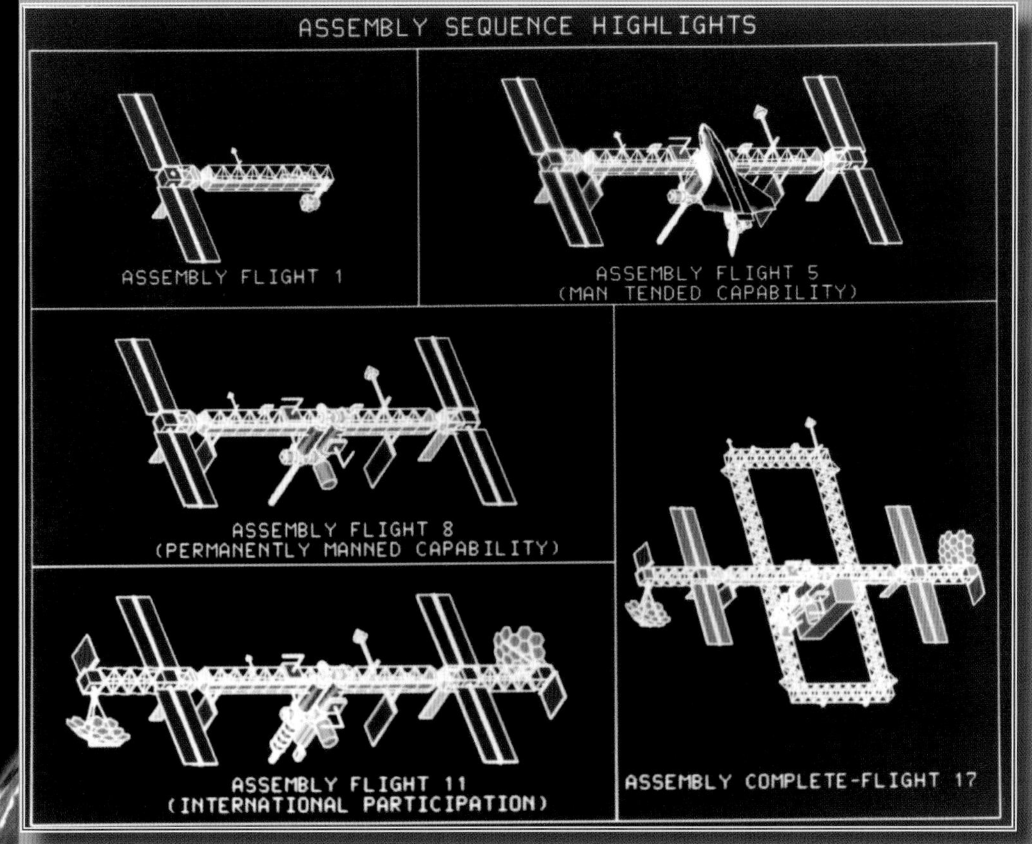

From the launch of the first element in late 1998 to the final Shuttle assembly mission, twelve years later, the ISS program encompassed a total of over 100 separate launches. This included 37 American Shuttle missions, 29 Russian Soyuz launches, 45 Progress resupply flights, two large Russian modules and the first unmanned European Ariane Transfer Vehicle (ATV) and Japanese H-II Transfer Vehicle (HTV) logistics flights. They supported the operations of 29 continuous expeditions. Continuous crew operations took place over an eleven year period beginning in 2000. Year by year the station grew.

But then, the Shuttle fleet was grounded in 2003 after the Columbia accident. The assembly went into a hiatus. Crew size was reduced from 3 to 2, but still a constant presence was maintained using Soyuz to carry the crewmembers up and back. The on board systems were maintained and the science program continued.

In 2006, Shuttle assembly missions resumed. The European and Japanese laboratories were added. The full length of the truss was completed and full complement of solar arrays enabled full power. The station transformed from a construction site to a full-fledged research facility.

Conceptual drawings of Space Station module configuration, final assembly configuration, ▲ and assembly sequence from 1986. At the time the power system was going to include two solar arrays, inboard and two solar dynamic power turbine generators, outboard. This image was made before the elimination of the two keels.

The first ISS element launched into Earth orbit was the Russian Functional Cargo Block, Zarya, on 20 November 1998. The following month, US Space Shuttle Endeavour delivered the first of three connecting nodes, Node 1, called Unity. The node carried on either end, two small elements, Pressurized Mating Adapters, PMA-1 and PMA-2. PMA-1 was used to link Node 1 to the FGB and Node 2 would serve as the primary Shuttle docking port.

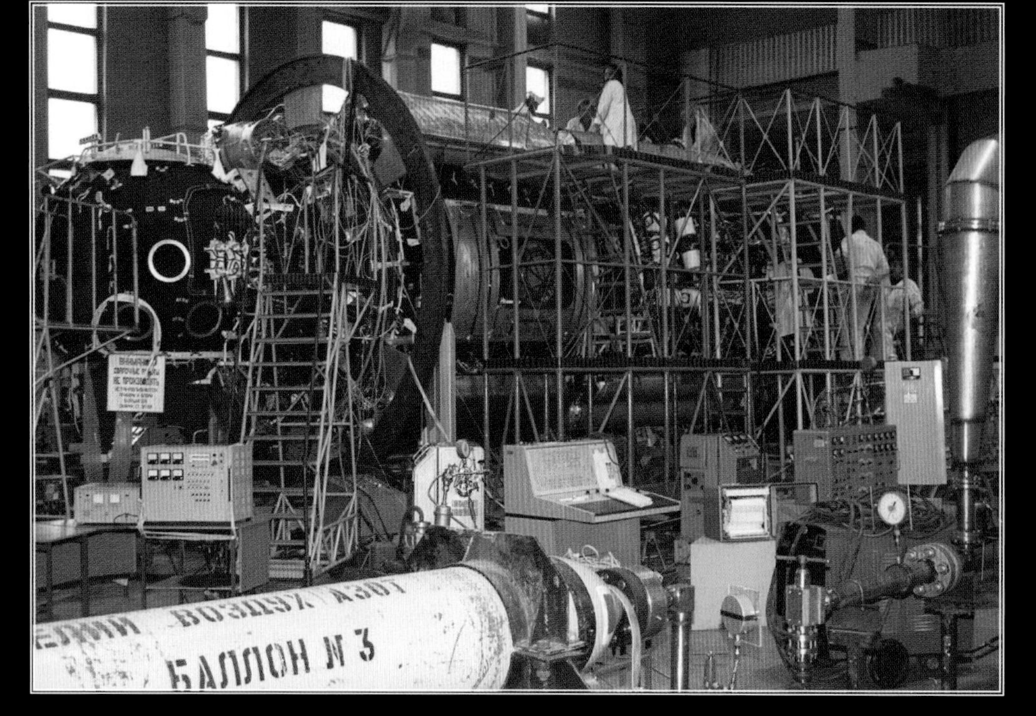

Khrunichev employees conduct a system test on the Russian built Functional Cargo Block (FGB). The FGB provided initial propulsion and attitude control for ISS. Khrunichev Space Industries was under contract to construct the FGB.

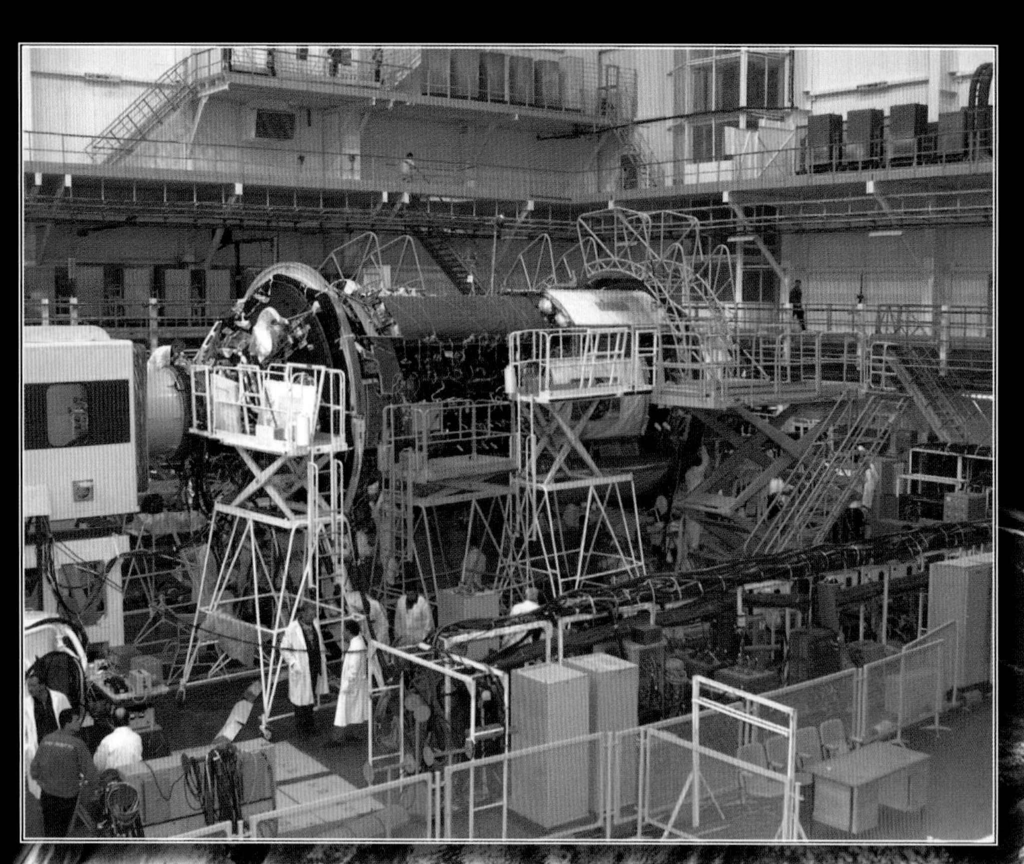

FGB, March 28, 1998 for integrated electrical testing (KIS), at Complex 254, Baikonur.

Launch of Functional Cargo Block (FGB), named Zarya (Dawn) from Baikonur Cosmodrome, Kazakhstan 1:40 a.m. (EST) on November 20, 1998.

The first element of the International Space Station to reach space, the Russian-built FGB, called Zarya, approaches the out-of-frame Space Shuttle Endeavour which carried the first U.S. ISS element to be launched, Node 1, called Unity.

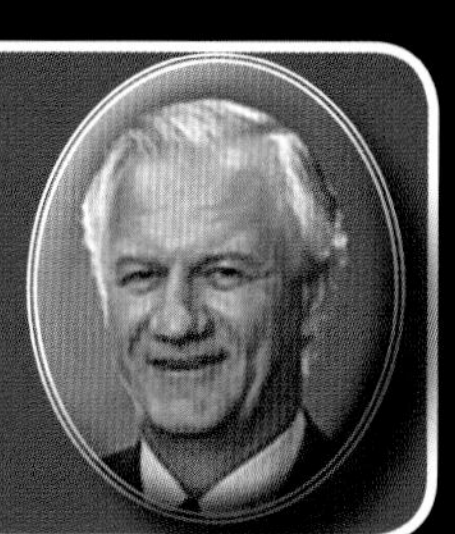

"When the Russians were pulled into the program, they provided some information on the FGB. In one weekend, we virtually re-designed the assembly sequence. That was what we wound up doing." - William Reeves

"We had an aggressive schedule for the Russians to build the Functional Cargo Block (FGB). FGB was a U.S. paid-for element that Khrunichev Production Center built. The Russian elements formed the backbone of the station." - Charles Lundquist

"The launch of the FGB was on a very, very cold November day. We had a perfect launch, and we waited until we had separation from the Proton launch vehicle. It was a great relief and a great sense of pride for all of us. Shortly followed by the launch of the Shuttle, STS-88 with Bob Cabana and his crew on the Endeavour in December, 1998 to deliver Node 1. Before the launch of the first Russian and US elements, the Russian hardware, the FGB, and the U.S. hardware, Node 1 had never been mated. The first time the interface was tested was in orbit. It worked; the power system, the software, the node physical interfaces. That was a big deal. There were many people who never believed we would ever develop the hardware, much less successfully launch it, and there it was, the first elements were in orbit. The successful launches changed the outlook. Both sides realized we could actually do this. It was a turning point for the Program. We still had 40 more impossible missions ahead. Just took it one at a time; we stayed focused." - Randolph H. Brinkley

"There was tremendous pressure to get something launched. We did not want to be in the middle of the program and run a risk of being canceled again. We got down to the first launch. The U.S. bought and paid for the FGB but the Russians paid to launch it. Then the next element that went up was the U.S. node, Node 1 (Unity), about two weeks later." - Melanie Saunders

▲ The Space Shuttle Endeavour lights up the night sky as it embarked on the first mission dedicated to the assembly of the International Space Station (ISS). Liftoff occurred at 3:35:34 a.m. (EST), December 4, 1998, from Launch Pad 39A at the Kennedy Space Center (KSC), Florida. Onboard were Robert D. Cabana, Frederick W. Sturckow, Nancy J. Currie, Jerry L. Ross, James H. Newman, and Sergei K. Krikalev.

"ISS mission 2A, was the first US element launched, Node 1. It was supposed to be earlier than 1998, but we had welding problems. There were all kinds of schedule problems. It was a very long time coming. It was wonderful to finally see it launch." - Lauri N. Hansen

◀ View from the IMAX camera of Node 1 and FGB, above the payload bay of STS-88. James H. Newman (left) and Jerry L. Ross work on the US built Node 1 (Unity). The Russian built FGB (Zarya), was launched weeks earlier by the Russian Space Agency (RSA). Newman is positioned in a foot restraint on the end of the Canadian-built Shuttle Remote Manipulator System (RMS) arm.

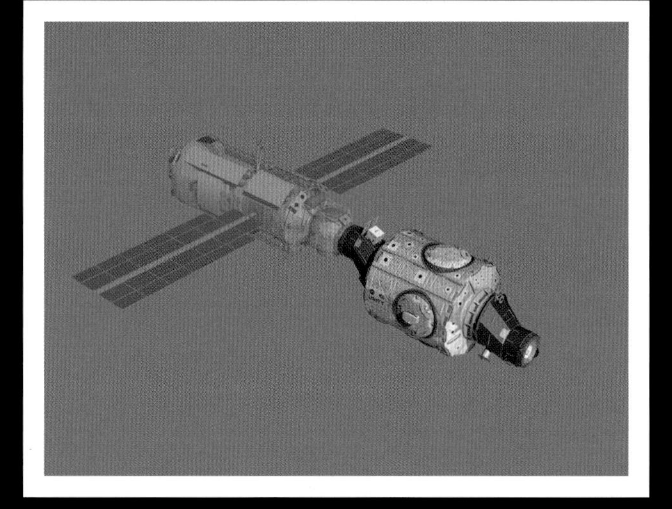

Artist's rendering of the International Space Station after flight STS-88. Space Shuttle Endeavour delivered the Unity node with two pressurized mating adapters. The STS-88 crew captured Zarya and mated it with the Unity node.

"We were in orbit, and we commanded to the Space Station, using IBM 760 XD laptops. We sent the commands that brought the Space Station to life and got everything going. I shared responsibility with Sergei Krikalev. We had different places in the procedures where I sent the commands, and he watched to make sure we were doing it right. I let him send some commands, and I watched. We sent the commands and it actually worked. Everything worked perfectly. It was just phenomenal. I attribute that to all the testing we did on the ground. The integration tests were just critical." - Robert D. Cabana

A view taken by STS-88 during fly-around of the nascent ISS, composed of Node 1/Unity and Functional Cargo Block (FGB)/Zarya. Pressurized Mating Adapters (PMA) 1 and 2 (PMA1, PMA2) are on either end of the Node.

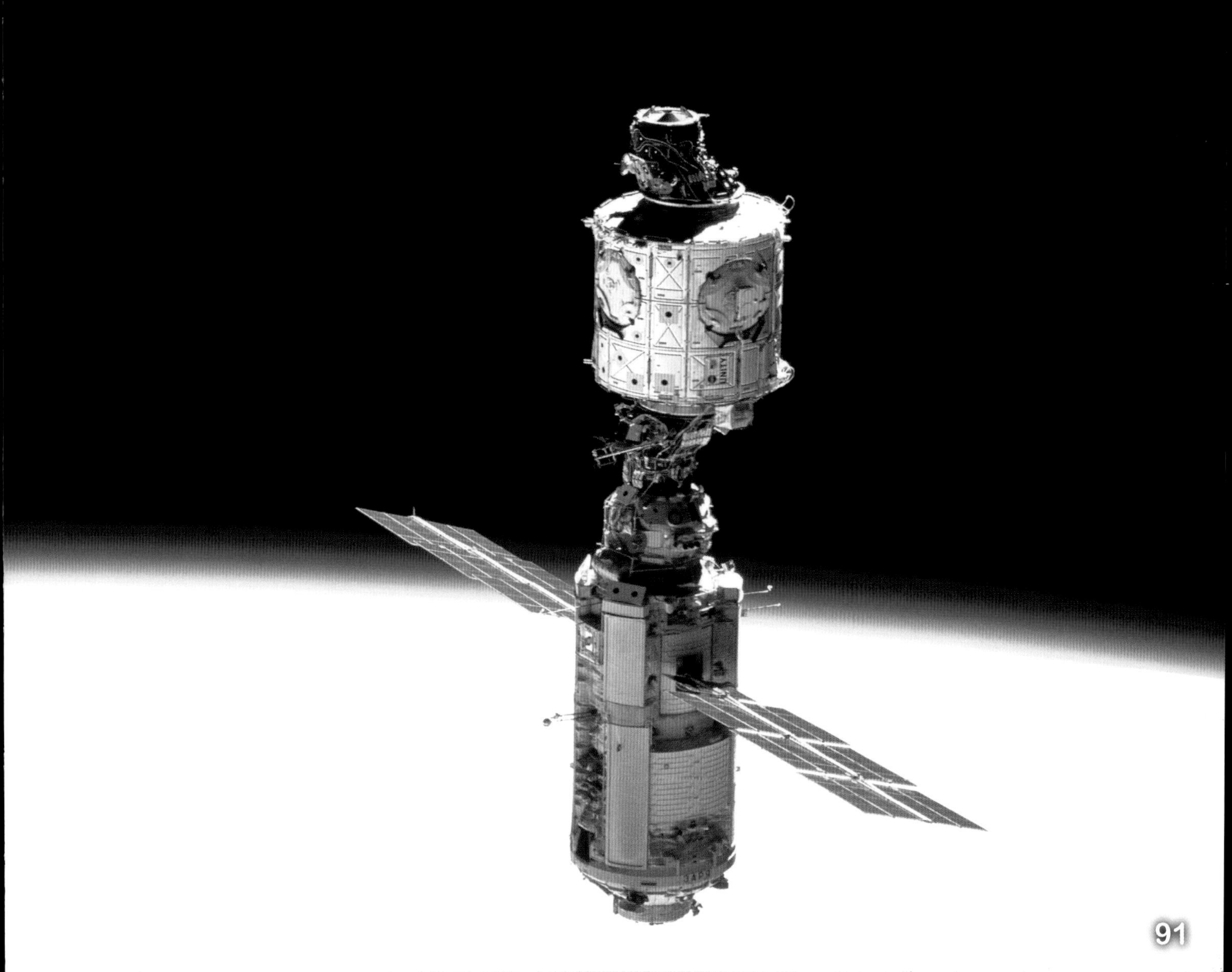

In 1999, the Shuttle STS-96 logistics delivery mission was the only flight to ISS. Because of funding shortfalls by the Russian government, the Russian Service Module, Zvezda was delayed. Technical issues with both the Space Shuttle and the Proton rockets, Zvezda's launch vehicle, delayed launches. Zvezda's launch was delayed repeatedly. Visiting Shuttle crews would continue assembly operations in 2000 but permanent residents on the station required the Zvezda Module and its life support system to sustain a crew.

The Zvezda Service Module was finally launched on July 12, 2000, docking autonomously with the Zarya/Unity combination a few days later. Shuttle logistics flights, STS-101 and 106, visited the outpost before and after the Zvezda launch. STS-92 delivered the Z1 truss sections from which the US solar arrays and supporting trusses would be attached. The first Progress unmanned resupply mission was also flown and adequate supplies were then in place to allow a permanent ISS crew.

Permanent occupation of the station began in October 2000 with the launch of Soyuz TM-31, carrying the first three long term residents of the station. They arrived on 2 November, almost two years after the assembly sequence started. The final mission of Shuttle mission of 2000, STS-97, attached the next element of the US truss system, Port 6 (P6) and the first US solar array. The new millennium saw the first expedition living on board and the first part of the main power system installed.

Continued large cost overruns forced NASA to rethink the Station design and assembly sequence and cancel several major elements. A US-made crew rescue vehicle was indefinitely deferred in favor of relying on the Russian Soyuz. Once the Russian FGB and Service Modules were in orbit and could provide propulsion, the US Propulsion Module was canceled. Then the US eliminated the US Habitation Module, the scientific Centrifuge Accommodations Module and significantly reduced the ISS scientific research budget to further reduce costs.

"Then there was a long gap because the Service Module [Zvezda] wasn't ready. Russians were having financial issues and their economy was tanking, and they were having a lot of challenges getting the Service Module finished. We have got a huge standing army, and every day that the Russians were late, we were suffering financially. In 2000 we went over to negotiate. We wanted to ensure that the Russians had the funding to finish the Service Module. We ended up buying their stowage and crew time. We couldn't buy hardware, because that would cost the Russians time and money. We had to buy something of benefit to the U.S. and wouldn't cost the Russians. We were trying to accelerate the Russian's completion of the Service Module, because that would save NASA money, but we had to do it in a way that was palatable to everybody. Once we got the Service Module launched, the first crew went up, and then things were going good." - Melanie Saunders

William Shepherd rehearses an extravehicular activity (EVA), June 7, 2000, with a full scale training model of the Zvezda Service Module at the Gagarin Cosmonaut Training Center Hydrolab near Moscow.

"The next big issue we had was the Russian schedule slip of the Service Module. Though we launched the FGB and Node 1, the Russians continued to slip the schedule for the Service Module. If they delayed too long, the gap could be so long that we could run out of on-orbit life or propellant before the Service Module was ready. The delays of the Service Module were the result of delays of funding by the Russian government. It caused a lot of angst. I ended up having to testify to Congress on several occasions. The program was burning $2.3 billion a year, so a year's delay of the Service Module cost the taxpayers $2.3 billion. It had a dramatic impact in terms of cost. It affected all the international partners because everybody was waiting around to fly." - Randolph H. Brinkley

Artist's rendering of the ISS as it appeared in September 2000 The STS-106 crew delivered supplies and performed maintenance on the station after the Service Module launch.

A view taken by STS-96, June 3, 1999

"In 1998, before the Space Station had flown and we were trying to find ways to make sure we bought services that would help the Russians facilitate the completion of their station contribution so we could get to flying, we negotiated the purchase of a dedicated Soyuz, all three seats going to the Space Station, for $65 million. We got shot down by somebody at the White House Office of Management and Budget (OMB) who thought that price was outrageous." - Melanie Saunders

Expedition Three crew Kenneth Bowersox, Vladimir Dezhurov and Mikhail Turin pose in front of the Service Module (SM) during the SM Crew Equipment Interface Test (CEIT) at Baikonur on May 22-25, 2000.

The Russian-built Zvezda module launch from Baikonur Cosmodrome in Kazakhstan aboard a 3-stage Proton rocket, July 12, 2000. The third ISS component, Zvezda served as the primary Station living quarters and provided propulsive attitude control and re-boost capabilities early during the life of the ISS. The SM was also the main docking port for Russian Soyuz and Progress spacecraft, as well as for the European Automated Transfer Vehicle.

Immediately after launch, July 12, 2000, the solar arrays of the Service Module (Zvezda) deploy in this artist's rendering.

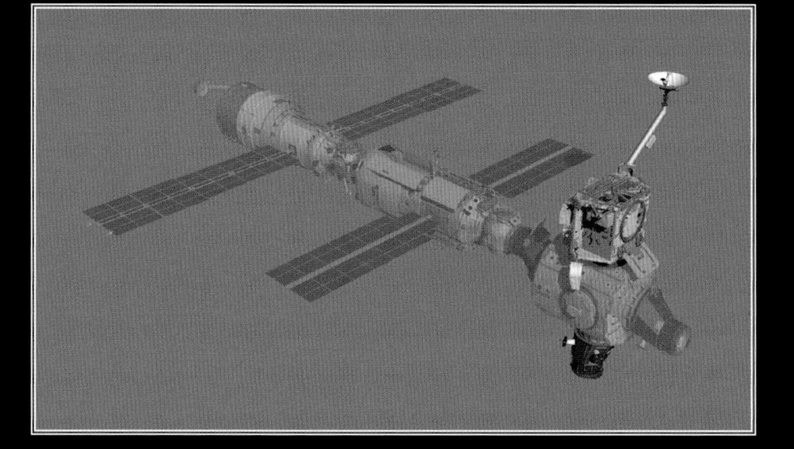

Artist's rendering of the International Space Station after flight STS-92. Arriving aboard Space Shuttle Discovery, the STS-92 crew installed the Z1 truss, a third pressurized mating adapter and a Ku-band antenna.

"The technical challenge on ISS is the assembly in orbit. The one reason we had to do that, was the Shuttle. It was decided many years ago to build it in orbit." - Kenneth A. Young

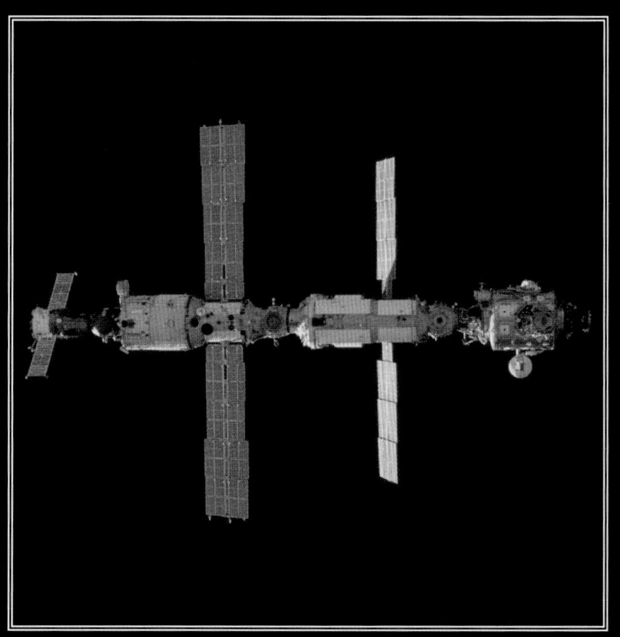

The International Space Station (ISS) as the Space Shuttle Endeavour approaches for docking during the first ISS expedition, December 2, 2000. The Soyuz spacecraft is docked at left to the Service Module Zvezda, which is linked to the Functional Cargo Block (FGB) Zarya. The Node 1 Unity is at right. The Z1 truss structure is largely obscured behind the Node. Three Pressurized Mating Adapters (PMA) are visible. PMA 1 is between Node 1 and the FGB. PMA-2 is on the right side. PMA-3 is on the Node 1 nadir side, facing the camera.

"Early on there was an office called the Space Station Assembly Office that was trying to define the operations required to build the Space Station. What sequence do you build things? How do you manifest the different pieces? What order do you have to put the Station together?" - William Reeves

"As ISS began, I led the Integrated Performance and Assembly Sequence area. Our assembly sequence at the time was 25 Shuttle flights. We spelled out the flights pretty specifically. As things evolved, they didn't work out like that. Each time we wanted to make a change, because the contract specifically defined what was to be on each assembly flight, every time things changed it was an enormous impact to the contract. This was very problematic. We thought, at the start, that we understood exactly what each assembly flight would look like. As things changed, we were wrong. It was a phenomenal integration exercise. Every time we looked at a change to the assembly flights, it became a big study impact and a big cost impact. One of the lessons I learned was that we need to be very careful about how specific we get on requirements. We thought that the more specific we got on requirements the better. But, requirements keep the provider in a box. If you understand that box really, really well, that's okay, but if you think it might shift a little bit or grow a little bit, you're hurting yourself by having it defined so specifically. If your requirements turn out to be wrong, you are going to pay for the changes later. ISS was a massive project. It was all these pieces. When we tried to pull it together, we didn't always get them 100 percent correct. It was a changing configuration." - Lauri N. Hansen

"The critical thing was to execute the assembly process. The assembly process effected the configuration and the design of the station. The process of the assembly made us very dependent on a very success-oriented set of things that had to happen. You certainly would like to not get hung up in the middle of mechanically assembling the various components that go up on each flight, and have to have it only partially completed when it was not yet a completed operational vehicle. As the station grew we could start having "internal" problems. At least if you have an operating vehicle and we have a crew there full time, we could work problems on an ongoing basis. The real critical facet of the early flights was to get it to the point where it was self-sustaining; where we had the time and the crew to continue to work any problems with the ground." - Fred Haise

The Soyuz spacecraft lifts off from the Baikonur Cosmodrome at 10:53 a.m., October 31, 2000, Kazakhstan time. Onboard were William M. (Bill) Shepherd, Yuri P. Gidzenko and Sergei K. Krikalev.

The Expedition One crew members pose for a final photo prior to their launch, October 31, 2000, aboard a Soyuz from the Baikonur Cosmodrome in Kazakhstan. William M. (Bill) Shepherd (center on steps) is flanked by Yuri P. Gidzenko (bottom) and Sergei K. Krikalev.

Against Earth's horizon, the International Space Station (ISS) is seen about a week after the first ISS crew docked to the ISS in their Soyuz in this photo taken by Shuttle STS-106. The Soyuz can be seen at the top, docked to the aft end of the Service Module.

The first Expedition crew inside the Zvezda Service module. From left to right are: William (Bill) Shepherd, Yuri Gidzenko and Sergei Krikalev. A model of the ISS in its present configuration floats in front of the trio.

STS-92, Shuttle Discovery leaves the ISS, October 23, 2000. The STS-92 added the first element of the truss, Zenith 1 (Z1) which was berthed to the top of Node 1. It also added a third Pressurized Mating Adapter (PMA), used to berth Shuttles and other vehicles at the US segment of ISS. In this view the Z1 is on the top of Node 1 and PMA 3 is at the bottom.

A view of the Space Shuttle Endeavour during STS-97, December 8, 2000. The photo was taken as the shuttle approached the ISS to deliver the Port 6 (P6) truss and the first set of U.S.-provided solar arrays.

Five STS-97 astronauts, in red, together with 3 Expedition 1 crewmembers, in blue, onboard the ISS in the Zvezda Service Module, 8 December 8, 2000. On the front row are (left to right) Brent W. Jett, Jr.; William M. Shepherd; and Joseph R. Tanner. On the second row are (from the left) Sergei K. Krikalev; Carlos I. Noriega, Yuri P. Gidzenko, and Michael J. Bloomfield. Behind them is Marc Garneau.

"I was on STS-101. With the delay in the Service Module they decided we needed to go to ISS to address some of the failures on the FGB and Node 1." - Jeffrey N Williams

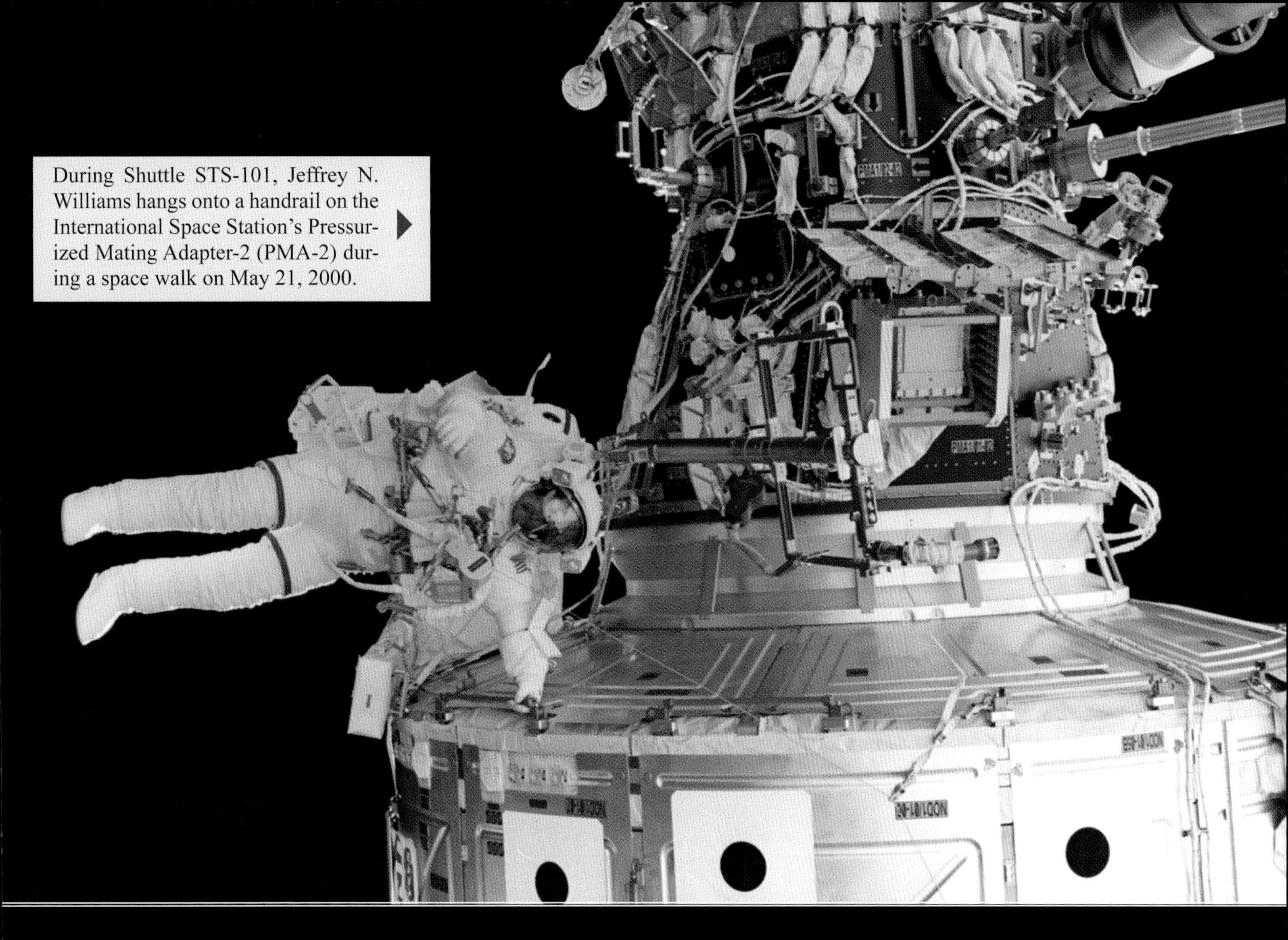

During Shuttle STS-101, Jeffrey N. Williams hangs onto a handrail on the International Space Station's Pressurized Mating Adapter-2 (PMA-2) during a space walk on May 21, 2000.

"Before retiring from NASA, I had always been on the operations side of things. After I left, I got into the hardware side of it. I found out there were people worrying about things that I never knew people worried about. I'd run into materials and processes (M&P) people. They are experts on materials: metals, paints, solvents, glues, everything that goes into building a space vehicle. They are just unbelievably smart in their narrow world. They know what kind of glue to use for what and what its limitations are and what kind of metal to use for this screw and for this bracket. It's just amazing. Then I found out there are people that worry about nuts, bolts and washers. Every little teeny part of this monstrous vehicle. Just millions of parts. The specifications and the requirements and the test requirements and everything that goes along with all of that. There are the integrated systems experts. There are specialists on each system: mechanical systems, data systems, software, computers, electrical systems, crew equipment and stowage systems, maintenance, habitability systems, environmental control, spacesuit systems. There is testing constantly going on all over the world on every one of these and all of the components that each system is made of. These people manage and keep track of each of these. There is program integration. It is a totally different world. It is the analytical world as opposed to the hardware world. One of their biggest responsibilities is systems engineering and integration (SE&I). They define the environments the vehicle fly through and the astronauts operate in. They certify that the vehicle stays within all of its limits. They talk about the acceleration environment, the thermal environment, the electromagnetic field environment, the structural environment. When vehicles dock to the Space Station, they analyze the structural loads between the docking vehicles. SE&I is where a lot of the hard-core engineering skills are. The cargo engineering office is responsible for the design and fabrication of all of the interface hardware for payloads. There are different kinds of interface cables. They define the characteristics of each payload, the required connectors. They design those connectors and get them built. They make sure the payloads fit where they are supposed to. In program integration, there is configuration management. These are the people who take care of all the documentation for the program. They support all of the change boards and keep the agendas, get the presentations, write the minutes of the discussions, and track the actions that come out of the meetings, and make sure every action is followed up. If any requirements or documents need to be updated, they make sure they are updated. The management integration office is in charge of all of the program schedules and the flight manifests. There is a group that is responsible for packing everything that goes inside the crew cabin, and everything that goes inside the logistics carriers that go back and forth, to and from the Space Station. They track everything that goes in the Russian, European, Japanese or commercial vehicles, make sure every item is properly certified and ship and track each item to the proper place. Everything that comes back from ISS, they track it, de-manifesting it and make sure it gets to where it's supposed to go. The Information Technology (IT) office is responsible for the software and the digital equipment to keep those systems up and running." - William Reeves

The initial launch of 2001 was STS-98 carrying the first dedicated scientific research module, the US Laboratory Module, Destiny. The next mission, STS-102, was a logistics and outfitting mission. It carried the first European-built Multi-Purpose Logistics Module (MPLM). That mission also delivered the Expedition 2 crew and brought home the first expedition crew after they had spent 188 days in space.

Four additional Shuttle missions launched in 2001. STS-100 carried the Canadian robotic arm, Canadarm2. The arm would prove instrumental in all subsequent assembly operations. STS-104 added the American EVA airlock, called Quest, enabling astronauts to spacewalk from the orbiting complex. An automated Russian assembly flight launched the Pirs (Pier) airlock on an unmanned Soyuz booster carried by a Progress 'bus'. Two additional Space Shuttles and four unmanned Progress freighters carried logistics supplies and equipment to continue outfitting the growing station.

The second ISS expedition crew, ISS-2, became the first crew to conduct a spacewalk. It was inside of the depressurized Zvezda module to relocate docking hardware. The second expedition crew was replaced by the third, ISS-3, expedition launched on STS-105. The third crew became the first to conduct EVAs from the station without a Shuttle docked.

While the expedition crews flew to and from the station on the US Shuttle, the Russian Soyuz were always present and continuously supported emergency crew evacuation. Each Soyuz has a lifetime of about 6 months and each was exchanged for a fresh vehicle prior to the lifetime limit being reached. These replacements were called ferry missions or taxi flights. Each would carry a crew of three. The fresh Soyuz would be left for the resident crew, and they would return home after about a week in the older Soyuz. Each Soyuz required a crew of two Russian cosmonauts, so the third Soyuz seat was available. Sometimes ISS partner countries or agencies (Canada, Japan or ESA) would purchase the seat or sometimes they were sold to fare paying 'passengers' or space-tourists. The seat sold for tens of millions of dollars, escalating over the course of the program. The funds provided a valuable injection of resources to the cash-starved Russians.

In 2001, the first of these fare paying 'space tourists', Dennis Tito, flew on the taxi mission to exchange the Soyuz TM-31 craft with TM-32. A few months later, French ESA 'career' cosmonaut Claudie Haigneré flew a similar short mission on TM-33, returning home on TM-32.

As the station entered the second full year of operation in 2002, a regular cadence had been established. Nine flights reached the station: four Shuttle flights, two Soyuz missions and three unmanned Progress flights. Three of the Shuttle flights (STS-110, 112 and 113) added the Starboard-0 (S-0), S-1 and Port-1 (P1) truss and solar array assemblies to the structure, while STS-111 was a logistics supply mission which also delivered the ISS-5 crew and brought home the ISS-4 crew. STS-113 exchanged the ISS-5 crew with the ISS-6 crew later in the year. The second Soyuz mission of the year used a new, more advanced TMA version of the Soyuz. It had upgraded systems, longer orbital lifetime and the capacity to fly taller astronauts to and from the station.

Prospects for 2003 looked good. Three people served on each ISS expedition crew typically for a six month stay. Science operations had begun; with the addition of the ESA and Japanese laboratories and the expansion of the crew size to six in the years ahead, considerably more progress at a more rapid rate was anticipated. It was expected that Station assembly could be completed in the next few years.

◀ Sergei K. Krikalev is positioned at a porthole on the Zvezda service module of the ISS as the Space Shuttle STS-98, Atlantis approaches for a link-up on February 9, 2001. The crew cabin and forward section of the Space Shuttle Atlantis can be seen in the window. The aft part of the cargo bay, where the Destiny laboratory was stowed, is not visible in the scene.

In the grasp of the Shuttle's Remote Manipulator System (RMS) robot arm, the US Destiny Laboratory Module is moved from its stowed position in the cargo bay of the Space Shuttle Atlantis. The photo was taken by Thomas D. Jones, who was participating in one of three STS-98 spacewalks.

A series of photos of the International Space Station (ISS) taken during a fly-around by the STS-98 crew after the orbiter Atlantis had successfully added the US Laboratory Module Destiny to the ISS complex. The US Lab is the largest US module and the core of the US segment, housing critical life support, data management and other systems provided by the US.

A view of the ISS taken from STS-113, December 2, 2002. This was the configuration the ISS would be left in during the 2½ years following the STS-107 Columbia accident. This view shows the Port One (P1) truss segment, S-zero (SO) and Starboard 1 (S1) truss segments mounted across the top of the US Laboratory Module, and the Port 6 (P6) Truss and its photovoltaic solar arrays, mounted vertically at the top of Node 1. The Space Station Remote Manipulator System (SSRMS, or Canadarm2), Pressurized Mating Adapter (PMA) 2, the Quest Airlock (A/L) and Soyuz Spacecraft are also visible.

"At the start of Station assembly we had the node, the power system, and the laboratory flights all right together, because we wanted to get them all done as fast as we could and try not to expose ourselves to too much risk. We wanted to get people on board. I really wanted to do Space Station assembly, and I knew from what I'd been through on Space Station how hard it could be to assemble and how much trouble each flight could encounter. I thought I was in a position to help, and it was something I wanted to do. I had pretty much decided that once the Space Station Remote Manipulator System (SSRMS) robotic arm got into orbit, then that would be the last flight I would work. I wanted the Station arm, as much as anything because I'd worked so much with the Canadians during the Freedom Program, and I had a good rapport with them. I loved working with the Canadians, because they were absolutely the most matter-of-fact, pragmatic bunch of people in the world. If it was something that was right to do, then that was what they did." - Denny Holt

"The Russian government was not funding the Russian Space Agency to the level that it needed to be in order for them to be able to execute the program on the schedule we had agreed to. The Russians struggled with minimal funding. The Russians flew billionaires to the Space Station. They did so with NASA's passive blessings. They got $10 million or more to fly each passenger to the Space Station, because they needed the money. There was great angst about who would approve it. But initially NASA concurred. NASA's initial understanding was that the passengers would be astronauts from the European Space Agency, Japanese Space Agency, and that that was where the money would come from. But the Russians were more entrepreneurial than that. They went after and sold those launches to billionaires and we were stuck with our concurrence. NASA paid for a lot of things that you could argue that the Russians should have contributed to; but if we hadn't paid, we would have been further delayed so it would have cost the US even more. I think our Russian counterparts did the best they could with what they had." - Randolph H. Brinkley

Unexpectedly the tragic events of STS-107 on February 1, 2003, threw all of the ISS Program plans into disarray. 2003 became a struggle merely to keep ISS flying. Station assembly and the scientific research program had to take a back seat.

When Columbia was lost with her seven crew members, the ISS expedition 6 crew of Russian Nikolai Budarin, and Americans Ken Bowersox and Don Pettit were on the station. The Shuttle fleet was grounded pending the investigation into the Columbia tragedy, and it was anticipated Shuttle would not return to flight for years. Replacing the station expedition crews would have to depend on the only human rated spacecraft then available, the Soyuz.

The loss of the logistical capacity of the Shuttles meant that consumables and supplies would also have to be ferried to the station on the much smaller Soyuz and Progress vessels. This restriction in the amount of supplies available at the station made the support of three-person crews marginal at best. The decision was made to reduce the ISS expedition crew size to only two. Each Soyuz would ferry the two person ISS crew. The third seat on Soyuz could be offered to partner countries or to paying tourists. The flight schedule was severely reduced. The ISS expedition 6 crew returned to Earth aboard Soyuz TMA-1. The expedition 7 and 8 crews were launched aboard Soyuz TMA-2 and 3. In 2003, three Progress missions carried much needed supplies to the station.

In 2004, Soyuz TMA-4 and TMA 5 delivered the ISS expedition 9 and 10 crews respectively. Four Progress missions delivered supplies. On 14 January 2004, President George W. Bush announced that the Shuttle would return to flight and fly until assembly of the ISS was complete. It was expected assembly could be completed in 2010.

"When Space Shuttle Columbia, STS-107 fell, it was devastating. It was a very traumatic time from the point of view of everybody understanding what we were going to do. There was chaos; there were debates whether to abandon Space Station or not. We were determined that abandoning station wasn't the right answer. I concentrated on figuring out how we could keep flying with just the Soyuz and Progress. We spun up the whole team to figure out how to keep the ISS alive; what we would have to do, how we could keep going without having Shuttle to bring stuff up. There was a huge amount of energy associated with coming up with all the plans to make that happen." - Leonard Nicholson

"After Columbia, STS-107, I was a part of a team assessing whether we should continue building Station, or whether we should stop where we were, because we didn't want to use the Shuttle. We didn't know how safe the Shuttle was going to be. I think our decision to continue was because we hadn't met our international partner commitments to the European Space Agency and the Japanese Space Agency. I think it was important that we did continue. Continuing was important for us and for our future. For future international collaborations, we needed to not lose that trust with our partners and to continue to propagate that trust so that we will have future endeavors, whether Moon, Mars, wherever it is that we end up going, it will be advantageous to all of us if we do it as an international community." - Peggy Whitson

"We were in the middle of building the Space Station and then the Columbia accident happened. We had a crew in orbit. The chances of losing the Space Station would go up exponentially if you didn't have a crew onboard. The Space Station Processing Facility at KSC was chock-full of Space Station elements in their final testing phases. We had to figure out how we could sustain the crew. There were many restrictions and prohibitions on buying goods and services from the Russians. So suddenly we had a need for the Russians to do more, but we couldn't buy it from them. We had a crew of three onboard, but we went down to a crew of two. In the beginning of 2004, President Bush rolled out his Vision for Space Exploration that included retiring the Shuttle. Our plan for building and using the Space Station was dependent on the Shuttle. Station had been going since 1984, so everyone was under a lot of pressure. The Space Station was partially built, but the European Laboratory, Columbus and the Japanese Lab, Kibo were not in orbit yet. Part of the Canadian Canadarm II, the hand, Dextre the Special Purpose Dextrous Manipulator, was not up there yet. Before Columbia, we had said "we don't want to use Soyuz". But now the Russians were thinking "now you need me." They knew we were desperate. The Russians could have really gouged us on cost, but they didn't. They were good partners about that. They made money on it but they didn't do what they could have done." - Melanie Saunders

A view of the ISS taken after STS-114 left on August 6, 2005. Visible are the U.S. Laboratory / Destiny, S0 and S1 Trusses, P1 truss and P6 Truss, Pressurized Mating Adapter 2 (PMA2) Service module Zvezda, FGB/ Zarya module, Progress spacecraft and the Space Station Remote Manipulator System (SSRMS) / Canadarm2.

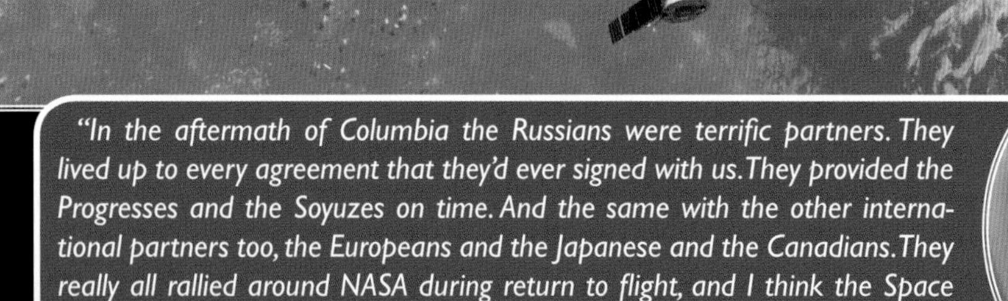

"In the aftermath of Columbia the Russians were terrific partners. They lived up to every agreement that they'd ever signed with us. They provided the Progresses and the Soyuzes on time. And the same with the other international partners too, the Europeans and the Japanese and the Canadians. They really all rallied around NASA during return to flight, and I think the Space Station partnership got a lot stronger for it." - William Readdy

"My first mission was on Expedition 6 in 2002. From 2002 to 2003, when I first flew to Station, I likened it to living in a house while we were still building the house. The facilities weren't built, we had one-fourth the electrical power, we had about half the pressurized volume, and we had half the crew size. Resources were limited in terms of power, volume, and crew time. We were learning how to operate the equipment. It was a different era than now that Station is complete. Station operation now is a well-oiled machine, training is a well-oiled machine, and most of the crew time now goes to utilization and research. When we started, Expedition 6 was planned to be one of the shortest. We were going to launch in October, and come back in January after 3 months. We did launch in mid-November. The Shuttle that was going to pick us up, STS-114, was pushed to February, but of course that changed with Columbia. The unexpected extension to our expedition allowed us to get a whole lot more work done. "Sox" [Bowersox], Nikolai Budarin, and I, worked really well together. I was the rookie, and both of them were very experienced people. We ended up doing two EVAs. We had quite a few science experiments onboard. There was no time to do them during our planned mission, but during the extended mission now there was time to do them. There was lots of construction and repair work. Things that were broken that we now had time to fix. We fixed things in orbit that were never intended to be fixed. On Expedition 6, we had Nikon F3 cameras and Hasselblad cameras, and they were film-based. When Columbia happened we didn't fly Shuttles for two and a half years. If film stays on orbit more than a few months, it deteriorates. We could barely get enough food and water to keep the crew going. There was no room for film going either up or down. We buried the film from the first half of Expedition 6 behind a wall of water. Water does an amazing job of protecting things from the galactic cosmic rays, and it stayed there for two and a half years. It came down on STS-114, two and a half years later. After the first half of the expedition we switched completely to digital cameras." - Donald R. Pettit

In the first post Columbia, Shuttle mission STS-114, on July 28, 2005. Space Shuttle Discovery was about 600 feet from the ISS when Sergei K. Krikalev and John L. Phillips photographed the spacecraft as it approached the station and performed a backflip to allow photography to evaluate the condition of Discovery's heat shield. Eileen M. Collins guided the Shuttle. The Italian-built Raffaello Multi-Purpose Logistics Module (MPLM) is visible in the cargo bay.

In-flight portrait of the Expedition 6 and Expedition 7 (Expedition 6 in light blue, and Expedition 7 in dark blue) crews in the Zvezda Service module. Back row: (l.-r.) Kenneth D. Bowersox and Nikolai M. Budarin. Front row (l.-r.), Ed Lu, Donald R. Pettit, and Yuri I. Malenchenko.

In 2005, the first post-Columbia Shuttle mission, STS-114, was launched carrying supplies to the ISS in a European-built Multi-Purpose Logistics Module (MPLM). The ISS expedition 11 and 12 crews were delivered on the Soyuz TMA-6 and 7 missions, and four more Progress flights delivered more supplies.

The ISS expedition 13 and 14 crews were launched on Soyuz TMA-8 and 9. Three Progress missions were flown to supply logistics and dispose of some of the accumulated trash. STS-121 carried the first ESA expedition crew member, German astronaut Thomas Reiter, to re-establish three-person crews for the first time since the accident. In March, 2006 the ISS partners agreed on a new assembly sequence of 16 Shuttle flights, leading to the Shuttle's retirement. After addition of the European and Japanese modules and with the completion of solar array and truss, the station would be ready to support six-person expedition crews.

During 2006, the Shuttle returned to a more regular and accelerating flight schedule. Three Shuttle flights went to ISS in 2006, another three in 2007, then four in 2008 and five in 2009. After the first two post Columbia missions carried ISS supplies, the next eleven Shuttle flights in a row, completed in mid-2008, were assembly missions which completed the ISS truss and full complement of solar arrays, added an additional US node, and then the European and Japanese laboratories.

▲ The ISS shortly after STS-115, Space Shuttle Atlantis left the orbital outpost September 17, 2006. Atlantis left the station with a new, second pair of 240-foot solar wings, attached to a new 17.5-ton section of truss with batteries, electronics and a giant rotating joint.

"I launched with Pavel Vinogradov, and with us was Marcos Pontes, who was a Brazilian astronaut. Brazil had arranged for his flight directly with Russia, so it was an agreement between those two countries. He launched with us. We had about a week handover period with Bill McArthur and Valery Tokarev onboard. Then Marcos returned to Earth with them, and Pavel and I began our expedition. Halfway through Expedition 13, STS-121 launched... they brought along with them Thomas Reiter from Germany. So now we had a Russian, a German, and an American onboard; later on Expeditions 21 and 22, I flew with a Canadian crewmate, and later a Japanese crewmate, and many other Russian cosmonauts. It really highlights the international flavor." - Jeffrey N. Williams

"Expedition 14 taught me how cool it was to really live in space and adapt to space and feel like I would be at home in space. Part of that was three spacewalks with Mike López-Alegría. One of our solar arrays didn't quite retract all the way, so we had to go up and shake it, to get it to move and do a couple of different on-the-fly types of things to get the thing to work. My three spacewalks were critical parts of changing the Space Station from a temporary living quarters to a more permanent living quarters. We changed out the heating, cooling and electrical power systems from temporary to permanent. The tasks were not simple and proved difficult. They had never been done before. They didn't all go as planned and went out of order. Expedition 14 and 15 were all about getting station to its more permanent configuration. That included spacewalking, but it was inside, too. We did a lot of rewiring for the computer systems on the inside as well as oxygen generation and other things to make sure the inside of the Space Station was ready to go. STS-117 was supposed to come up and add on another solar array and take me home. However, they were delayed because their external tank got hit with hail out on the launch pad. The ground was like, "this next Shuttle is not going to come for a while." That gap that followed was a time when I was left on ISS with two Russian colleagues, after Mike L.A left. I was the only American up there. It was a little surprising how significant that felt. The delay allowed more time to do more science. My good buddy Clay Anderson came up on STS-117. They added Clay and they brought me home after 195 days, which was a little bit longer than anybody had expected. As a relatively young guy, that was important to me because it was eating up my radiation allocation. It was good for me, and it allowed Clay to get up there and jump into his Expedition, do a couple of great spacewalks with his Russian counterparts, and then get ready for 118, which was the expedition he had training with." - Sunita L. Williams

Group portrait (from left to right, top to bottom): Piers J. Sellers, Michael E. Fossum, Mark E. Kelly, Stephanie D. Wilson, Steven W. Lindsey, Lisa M. Nowak and Thomas Reiter, Pavel V. Vinogradov and Jeffrey N. Williams. Photo was taken in the U.S. Laboratory during Expedition 13 and STS-121 joint effort.

Space Shuttle Atlantis, STS-117, leaves the ISS on June 19, 2007 after 8 days of work with the Expedition 15 crew. The mission gave the station a new configuration with balancing trusses and solar arrays on port and starboard sides. The P6 truss remains mounted vertically off Node 1, but its solar arrays have been collapsed in preparation for a subsequent move to a position outboard on the port side.

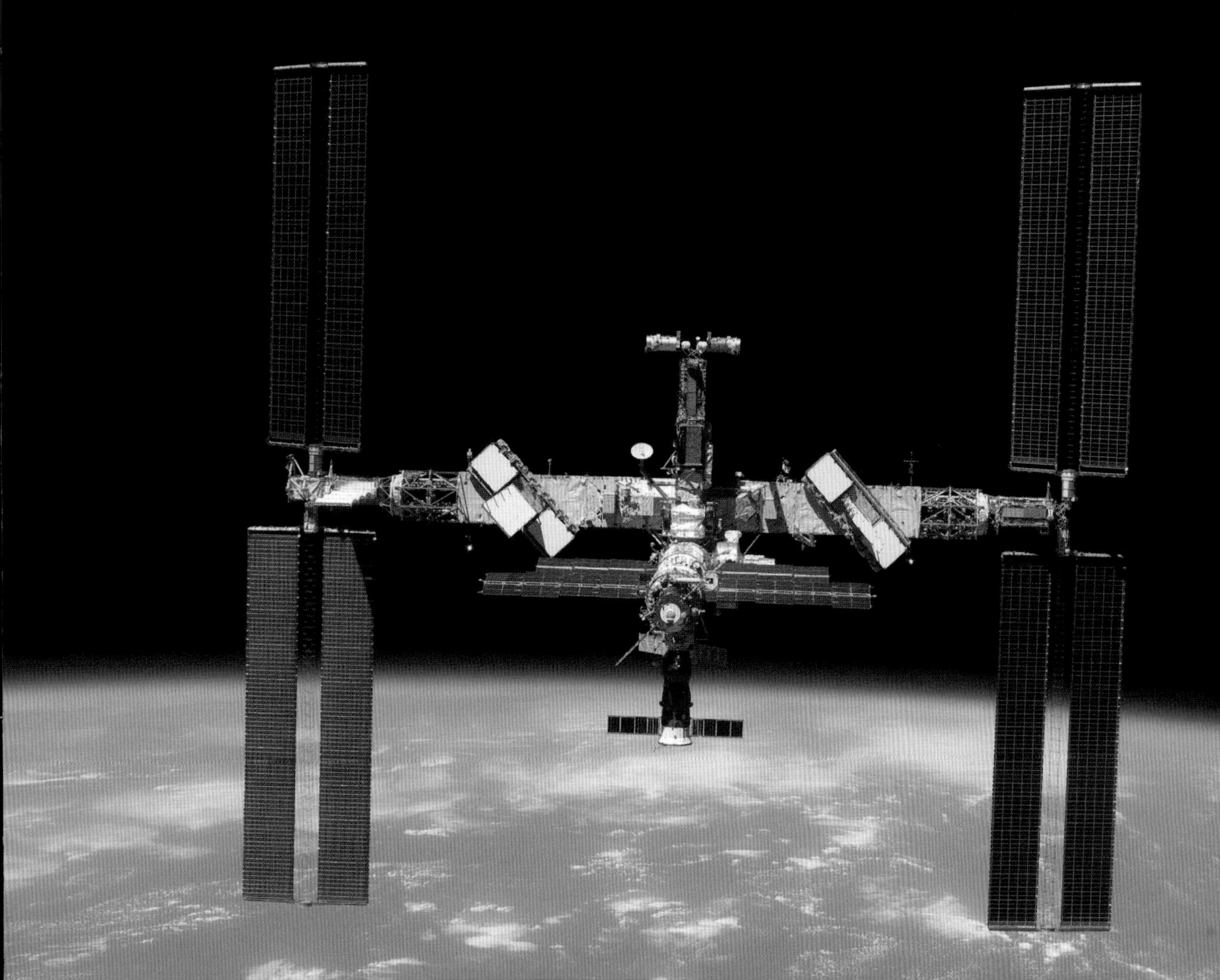

2007 – Resuming the construction

Nine missions were launched to the ISS during 2007. Four Progress missions brought more supplies, and Soyuz TMA-10 and TMA-11 brought the next two expedition crews, ISS-15 and 16. Two Shuttle missions, STS-117 and 118, delivered the S3/S4 and S5 trusses and solar arrays. A third Shuttle, STS-120, delivered Node 2, Harmony. Node 2 was the module to which the European and Japanese research modules would attach. The three Shuttle missions also carried members of the expedition crews, maintaining the three-person crew complement on the station.

2008 – International facilities

Eleven missions were flown during 2008 as the launch pace accelerated. Four Progress resupply missions (including the first of a new variant Progress M-01M) were supplemented by the first flight of the European Automated Transfer Vehicle (ATV), designated 'Jules Verne'. The Expedition 17 and 18 crews were delivered by Soyuz TMA-12 and 13; four Shuttle flights completed the year's program. STS-122 delivered the ESA Columbus Module, while STS-123 and STS-124 delivered elements of the Japanese Kibo facility and the Canadian remote manipulator system called 'Dextre'. Shuttle STS-126 delivered supplies. Each of the Shuttle flights exchanged one member of the expedition crew.

During the spring, the Russians announced that in order to complete the various modules planned for their segment, Russian federal government funding would have to be doubled. Now that the European and most of the Japanese elements had been launched, the Russian segment of ISS was the last area requiring completion.

Sunita L. Williams and Michael A. López-Alegría, don their extravehicular mobility unit (EMU) space suits on February 8, 2007 onboard the International Space Station prior to one of several Expedition 14 spacewalks. This marked their third spacewalk in nine days.

While anchored to a foot restraint on the end of the Station robotic arm joined to the Orbiter boom sensor for extra length, Scott Parazynski conducts a spacewalk to install "cuff links" to strengthen a damaged solar array on November 3, 2007. Once the repair was complete the array was successfully deployed.

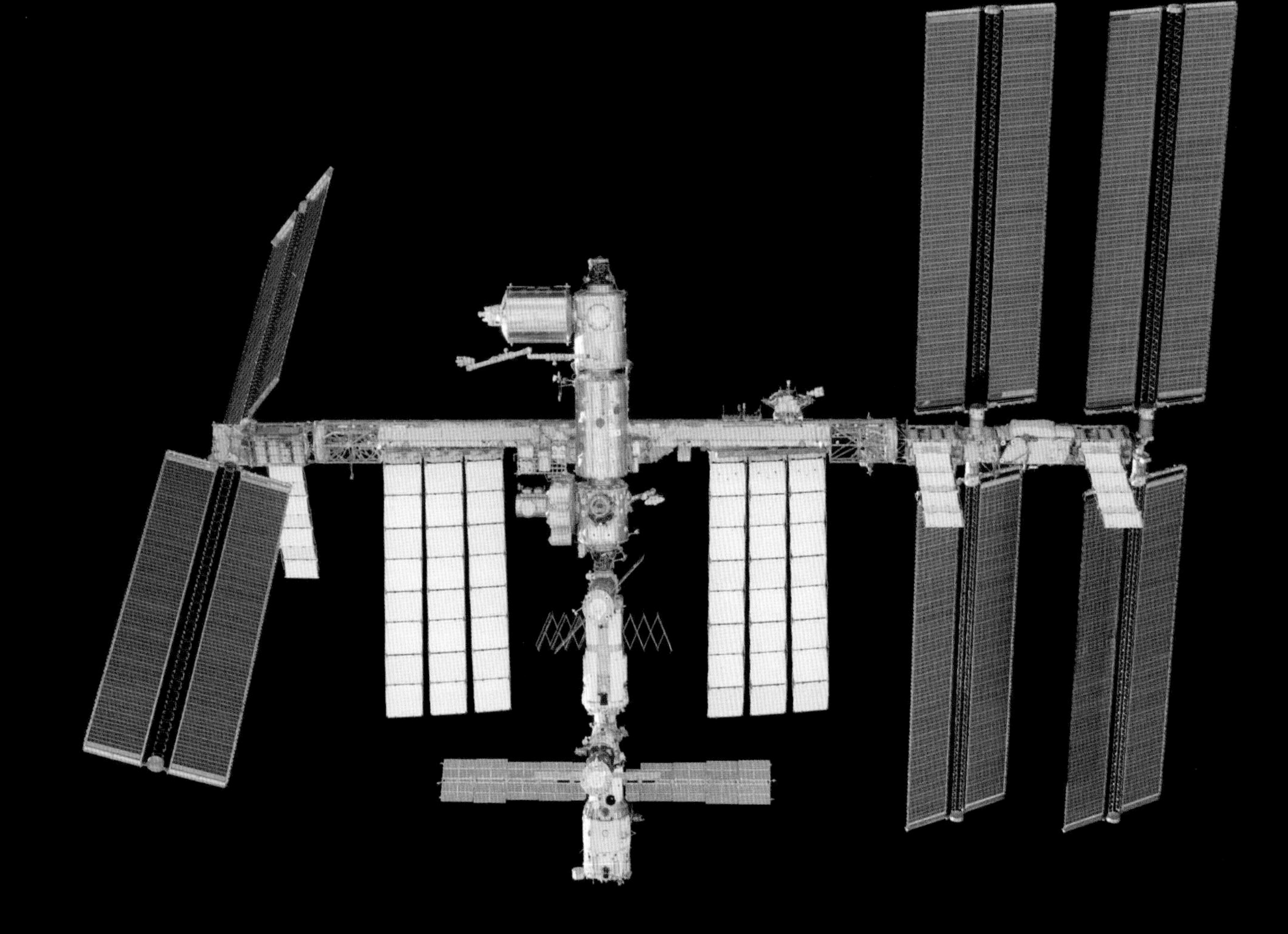

The ISS is seen from Space Shuttle Atlantis as the two separate on February 18, 2008, after 9 days of joint work between the STS-122 and Expedition 16 crews. In this picture, seen from below, the complete port truss with solar arrays is in place while the starboard truss awaits the addition of the final piece, S6. The European Space Agency Columbia laboratory is in place mounted on Node 2 but the Japanese Kibo module is not yet in place. The Canadian Space Station Remote Manipulator System (SSRMS, or Canadarm 2) can be seen mounted on Node 2, parallel to the Columbus module.

"My most significant accomplishment during Expedition 16; we had to install Node 2 into its final position. Node 2 enabled the addition of the ESA Columbus and the Japanese Kibo modules. Normally all the spacewalks and the robotics involved for placing and connecting the Node would have been done by a Shuttle crew. But in this case the Shuttle crew could not do it, because the Shuttle was docked to the port where we had to attach the Node. So the Shuttle had to leave. Our crew of three, did all of the disconnections, spacewalks, robotics, and reconnections. First we disconnected the Pressurized Mating Adaptor, and robotically moved it and reconnected it to Node 2. Then we robotically moved the Node to the forward end of the Lab. Then we installed the umbilicals to connect all the power, data, and thermal connections between the modules and Node 2. Then we connected those systems to the European and Japanese modules. That was my most significant contribution." - Peggy Whitson

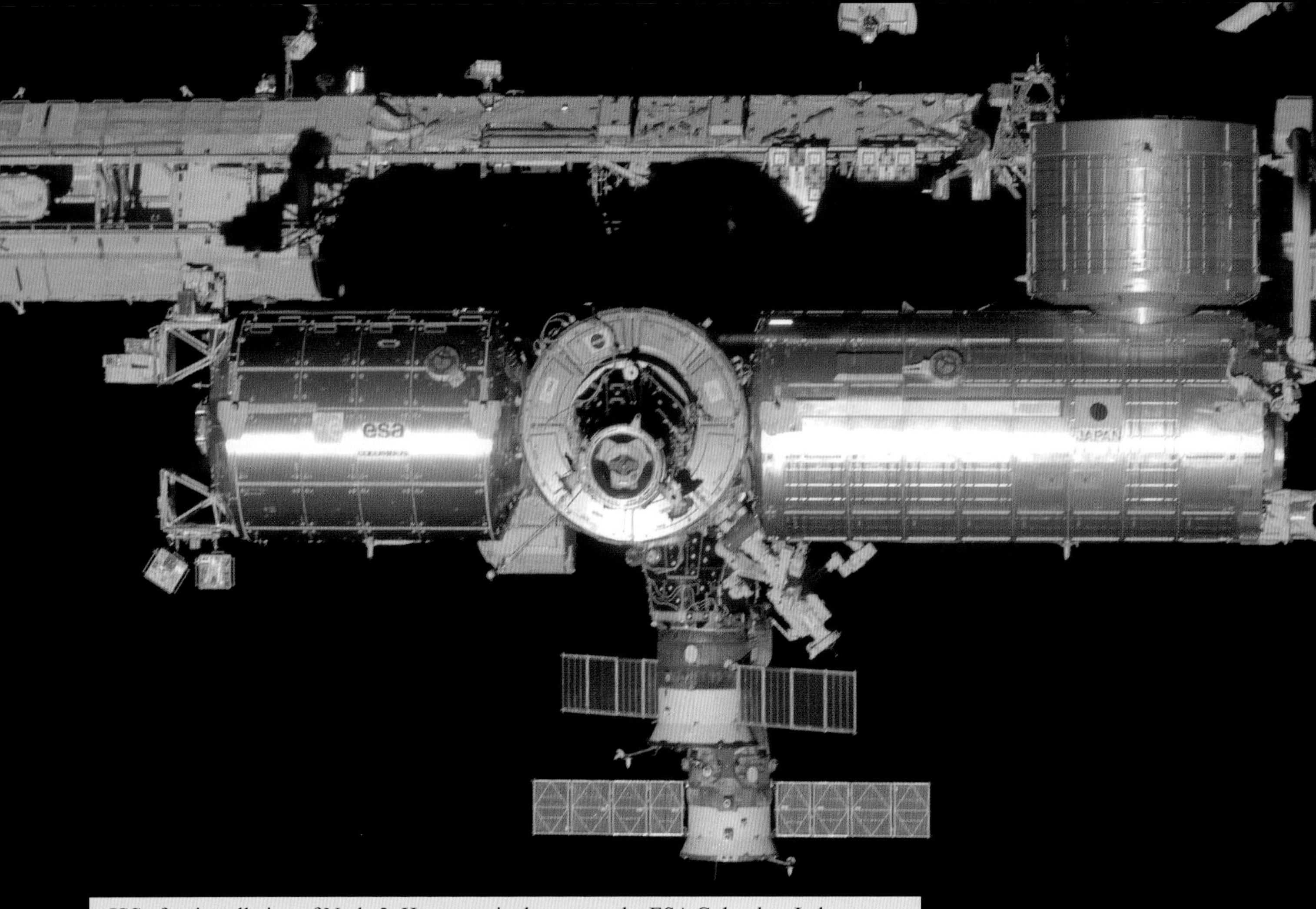

ISS after installation of Node 2, Harmony, in the center, the ESA Columbus Laboratory, on the left, and the Japanese Kibo laboratory, to the right, on 11 June 2008. The Canadian-built Dextre manipulator is just under and between the Node 2 and Kibo modules. Two Russian spacecraft are docked below the station. This photo was taken at the end of the Shuttle STS-124 mission and marked the completion of installation of the major Japanese, European and Canadian contributions to ISS.

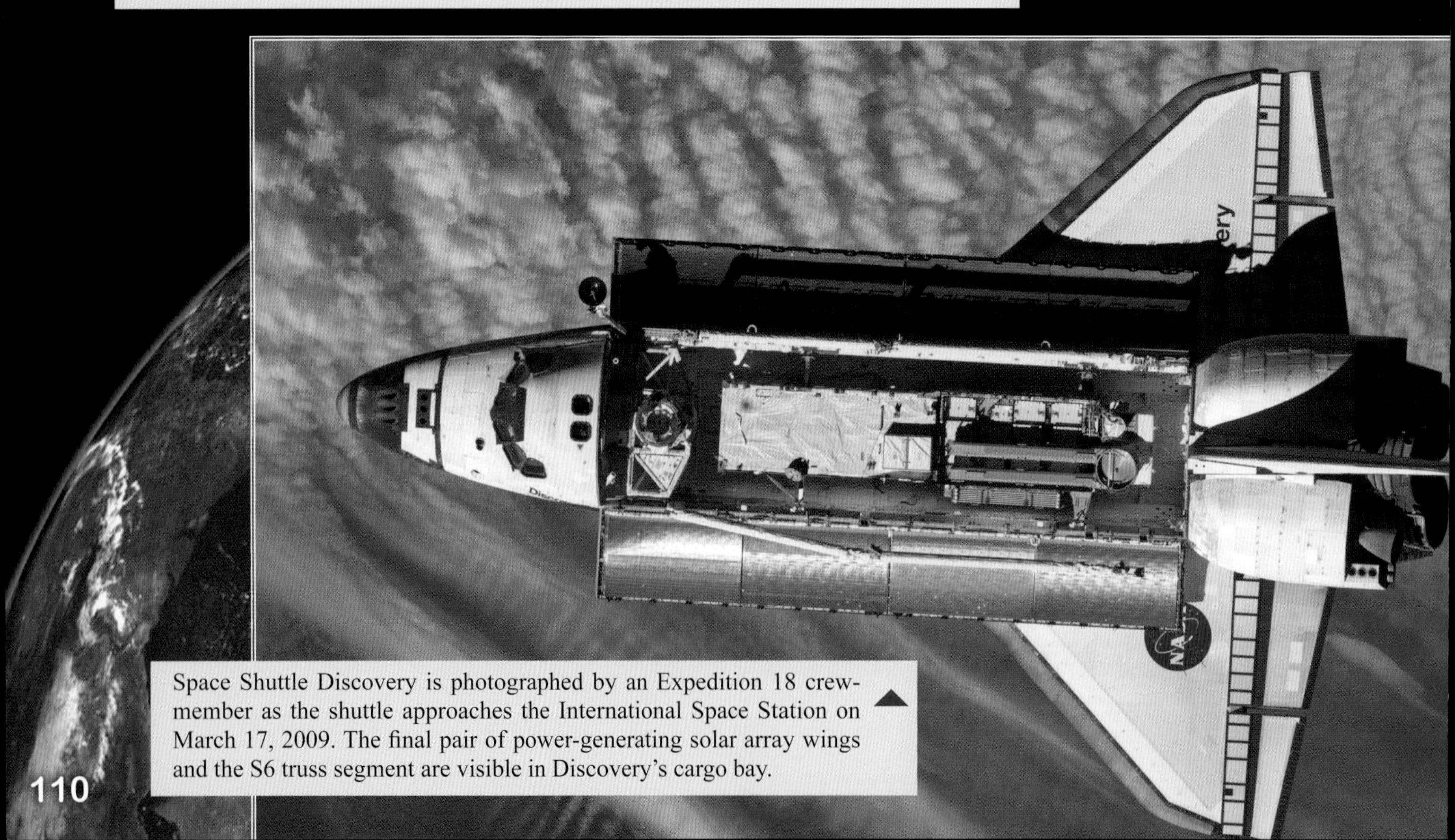

Space Shuttle Discovery is photographed by an Expedition 18 crewmember as the shuttle approaches the International Space Station on March 17, 2009. The final pair of power-generating solar array wings and the S6 truss segment are visible in Discovery's cargo bay.

2009 – Six person capability

In 2009, Shuttle STS-119 delivered the final segment of the truss (S6), and STS-127 launched and assembled the last portion of the Japanese Kibo. Shuttles STS-128 and 129 delivered tons of supplies and facilities to the station. More supplies were delivered on four Progress missions. The Japanese H-II transfer vehicle (HTV) completed its first mission to the station. An automated Progress 'bus' delivered a second Russian airlock/research module, Poisk (search) to the Russian segment. ISS expedition crew size was increased to six. The ISS expedition 19 crew arrived on Soyuz TMA-14. Then three ISS expedition 20/21 crewmembers arrived; the six crewmembers forming the first six-person expedition crew, ISS expedition 20. For a short time, there were nine expedition astronauts aboard the station; by the time the ISS expedition 20 crew departed, the expedition 21/22 crew had already arrived aboard Soyuz TMA-16. The ISS expedition 21 crew returned a few days before the ISS expedition 22/23 crew arrived, then reducing the expedition crew complement to three for a while. The final expedition crew member to be transported via the Shuttle, Nicole Stott, returned on STS-129 in November.

In January 2009, a new US President was inaugurated and the Democratic political party took the leadership from the Republicans in the Congress. A year later, the Obama administration cancelled the Constellation Moon return program. It now appeared that the ISS would be the only program to ensure American astronauts could remain in space for some time. ISS was formally extended from 2015 to 2020 and work began to certify on-orbit elements to 2028. By 2028, the first ISS elements to be launched would have been in space for 30 years.

"The construction of the ISS was very complex. Every part was very much required for the final complex. Getting the final life support systems up there in 2009 was key, because that allowed us to go to six person crew operations. That transitioned us from a construction and assembly focus to a science focus, which is what Space Station was about. That was a key turning point for us." - Peggy Whitson

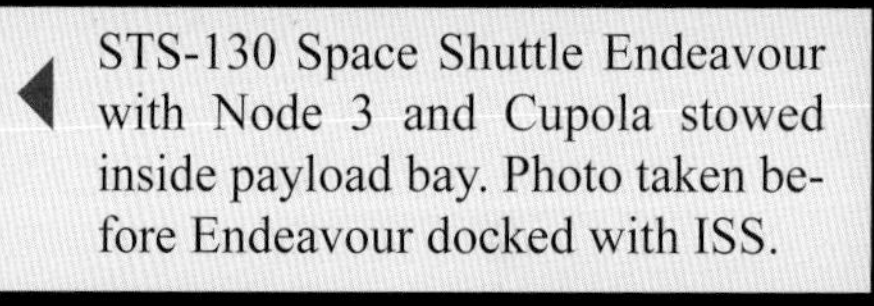
STS-130 Space Shuttle Endeavour with Node 3 and Cupola stowed inside payload bay. Photo taken before Endeavour docked with ISS.

Node 3, Tranquility and the Cupola being maneuvered into place on the ISS by Shuttle and the Space Station Remote Manipulator Systems (SSRMS) during Expedition 22 / STS-130 operations.

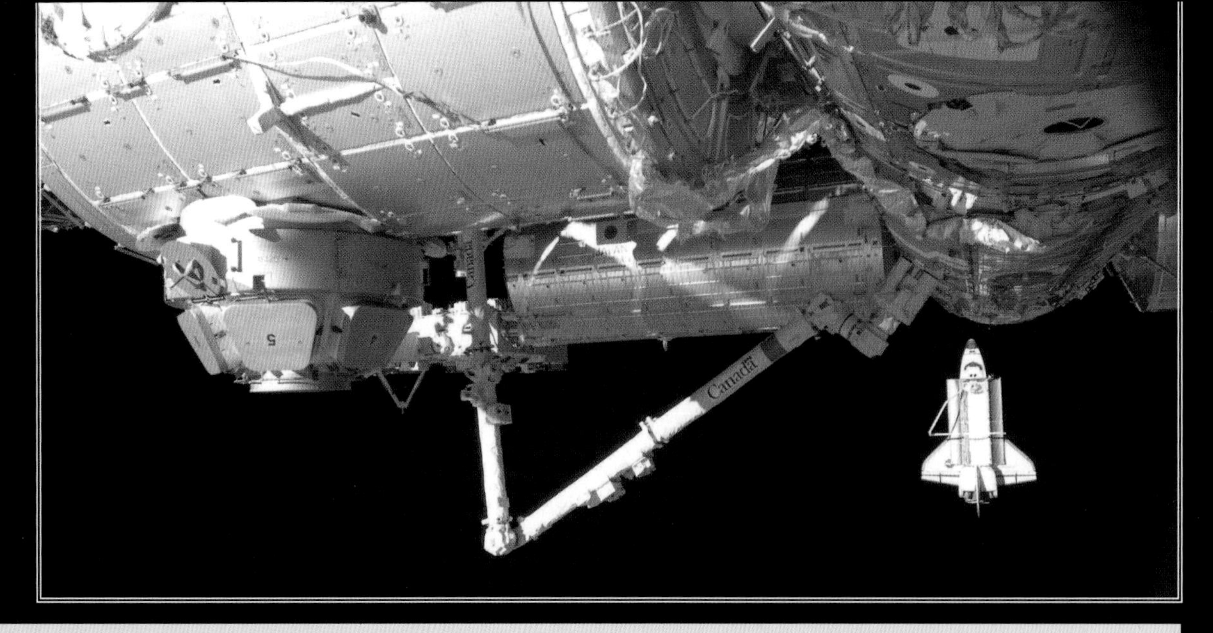

Space Shuttle STS-130, Endeavour as it departs the ISS during Expedition 22, Feb. 19, 2010. Node 3, Tranquility and the Cupola Module are visible just in front of the Japanese Experiment Module (JEM) Kibo. 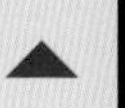

"I can recount every flight to assemble the Space Station. Our class of astronauts got here right in the beginning. As we were getting trained for it, I never thought that all these things were actually going to happen. I was thinking, "we'll see when that happens." But it went so smoothly, I became a believer. If you watch the assembly sequence or look at how the Space Station morphed over course of the assembly flights, we had to put in temporary power and temporary heating and cooling in the beginning and then switch it around as we were building one piece after another onto the Space Station. We had to learn and operate many different configurations of the Space Station. We lived in it throughout all of these changes, and for all those expeditions, right up to assembly complete. It was an incredible thing to experience. One of the amazing things were the people who wrote the procedures. I was part of the procedure validation team in the beginning and part of the group that did a lot of the procedures through the assembly sequence. For those people who understood how the technical parts of the Space Station were changing and when and how those changes were important to the procedures; how in contingencies the crew was going to handle malfunctions. They were working night and day to write those procedures, get them onboard, and make sure we understood what was changing. That was a mountain of work and incredible. The teams that planned that assembly process were pretty awesome." - Sunita L. Williams

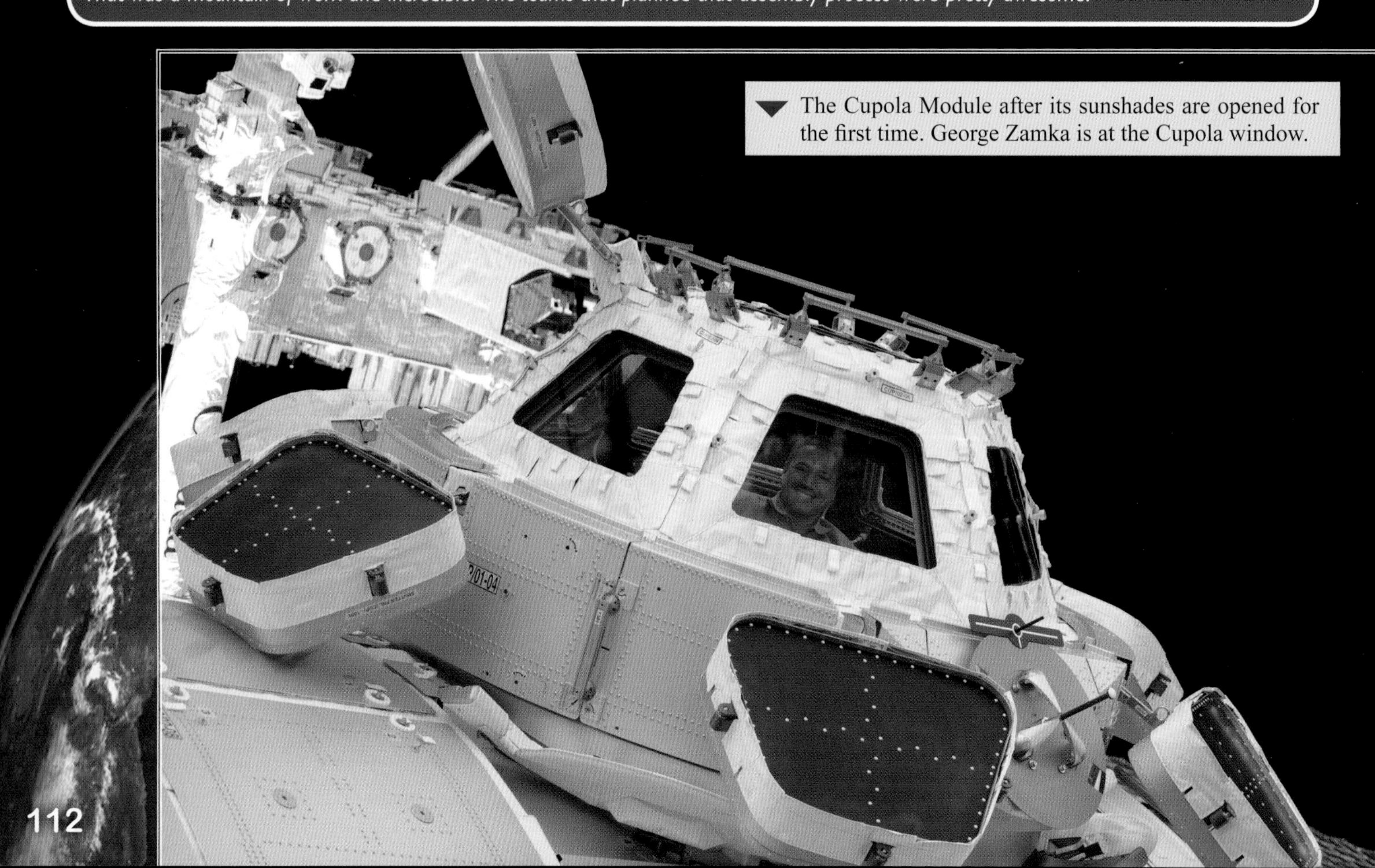

The Cupola Module after its sunshades are opened for the first time. George Zamka is at the Cupola window.

2010-2011 - Completing the assembly

The Node 3, Tranquility Module and the Cupola observation modules were launched on STS-130, the first flight of 2010. Three Shuttle launches in 2010 and 2011 carried multipurpose logistics modules (MPLMs), prepositioning tons of spare components in preparation for Shuttle's retirement, when the Shuttle's launch capabilities would no longer be available. Shuttle STS-132 carried the Russian Mini-Research Module Rassvet or Dawn. Rassvet could carry cargo, be used for scientific experiments, and served as a fourth Russian docking port. The flight of one more Shuttle logistics mission was authorized and STS-135 became the final Shuttle mission to fly, in July 2011. One of the three MPLMs was modified to become a permanent ISS module and the converted Leonardo Permanent Logistics Module provided additional stowage volume.

Five Progress missions visited ISS in 2010. In 2011, four more Progress vehicles, the second Japanese HTV and the second ESA ATV all carried cargo to ISS. The delicate balance of continued presence on the station without the Shuttle became clear with the launch failure of Progress M-12M in August 2011. The launch failure also delayed Soyuz launches since the two vehicles use the same launch vehicles. The failure highlighted the possibility of abandoning the station in the event of delays caused by interruption of the resupply fleet. But the Progress problem was soon identified and resolved, and Progress resupply missions resumed later in 2011. New crew members arrived after only a few weeks' delay.

▲ Extravehicular Activity 3 during STS-130 to prepare Cupola. Nicholas and Robert Behnken remove insulation covers from Cupola Module windows. The Space Shuttle Remote Manipulator System (RMS) is visible.

"We could not have done what we did on Station without the Shuttle. Another vehicle, unless it was something extremely different from what I've heard of, it would be impossible to construct something like the International Space Station we've put together. We needed the Shuttle to complete that construction." - Robert Crippen

"The number of launches it took to put the Space Station up, the complexity of the technology in each piece that went up, then design teams that had to integrate each of those pieces preflight.... Sometimes you did it by attaching the hardware together on the ground and doing testing, but in many cases you couldn't do that because the Station components were launched from different places in the world, and we didn't have the luxury to do that on-the-ground preflight integration." - Jeffrey N. Williams

The nearly complete International Space Station and the docked space shuttle Endeavour was photographed by the returning Expedition 27 crew from the Soyuz TMA-20 following undocking on May 23, 2011. Onboard the Soyuz were Dmitry Kondratyev; Paolo Nespoli, and Cady Coleman returning after 159 days in space.

A port side view of the International Space Station and the docked space shuttle Endeavour during Expedition 27/28 as seen from the Soyuz TMA-20 following its undocking on May 23, 2011.

Final departure, 19 July 2011, of a space shuttle from the ISS. Shuttle STS-135, Atlantis carrying an MPLM. Onboard the station were Andrey Borisenko, Sergei Volkov, Alexander Samokutyaev, Satoshi Furukawa, Mike Fossum and Ron Garan. Onboard the shuttle were Chris Ferguson, Doug Hurley, Sandy Magnus and Rex Walheim.

Russian Segment
Functional Cargo Block (FGB) 'Zarya'

The first ISS element launched was the Functional Cargo Block 'Zarya' (Sunrise). It was purchased by the US through a contract between the ISS prime contractor, Boeing with the Russians. The module was manufactured by the Russian Khrunechev factory with final integration, test and verification performed by the RSC Energia Company. Though US owned and financed, FGB would become a part of the Russian segment of ISS because of its commonality with other Russians systems and elements. The FGB was based on the Soviet/Russian Almaz military module design and similar modules had been used on earlier Salyuts and the Mir Orbital Station. Mir had used four similar modules: Kvant II, Kristal, Spektr and Priroda. Mission: The first element of ISS, FGB was critical because it was automated and fully self-supporting, providing the nascent ISS' primary source of propulsion for orientation and to maintain its orbit, and the initial source of electrical power for the complex; FGB also provided rendezvous and docking capability for the Service Module (Zvezda); FGB could serve as a fuel depot and provide fuel storage capability for fuel supplied by unmanned Progress freighter spacecraft.

▲ Backdropped against the darkness of space, the Russian-built, US owned FGB, 'Zarya', approaches the Space Shuttle Endeavour and the U.S.-built Node 1, 'Unity' during the first ISS assembly mission in late 1988.

Functional Cargo Block (FGB) 'Zarya' Configuration

The FGB module was designed for an operational life time of 15 years. This was surpassed in November 2013. After 20 years, the module continued to perform as originally designed. The FGB design includes an equipment and cargo compartment (PGO) that is functionally divided into three smaller compartments. A pressurized adapter (GA) contains the systems for docking to other spacecraft. The pressurized adapter is separated from the PGO by a pressurized spherical bulkhead, incorporating an 800 mm (31.2 inch) hatch.

Of the 71.5m^3 (280.56 ft.^{3}) total pressurized volume, 64.5m^3 (253.09 ft.^{3}) is made up of the PGO and 7.0m^3 (27.468 ft.^{3}) is the GA. There are two zones within the habitable space: the equipment zone houses the on-board systems, while the habitation zone is the crew work area. This also houses on board systems-monitoring controls and displays, as well as the caution and warning system. To gain access to the equipment zone from the habitation zone, crew members must open or remove panels. The three compartments in the PGO are designated PGO-2 (the conical FGB section), PGO-3 (the cylindrical section adjacent to GA) and PGO-1 (a cylindrical section between PGO-2 and PGO-3).

▲ End-on view showing the docking target of the Functional Cargo Block (FGB)/Zarya module as seen from the STS-88 orbiter Endeavour prior to rendezvous and grapple. FGB was the first element of the ISS to be launched.

There are three docking interfaces on the FGB. On the forward end frame of the PGO is the active hybrid docking system, to which the Zvezda Service Module is mated. This end is forward during launch but faces aft in orbit. The rear end frame of GA is the passive androgynous peripheral docking system (APAS), which is mated with the US Pressurized Mating Adapter attached to Unity Node-1. The third passive mechanism is located on the nadir side of the GA and provided a cone for Soyuz and Progress spacecraft docking. Since 2010 the Rassvet docking module has been docked to the nadir port. During the Shuttle mission STS-88, the FGB was stabilized and then captured by the Shuttle which then linked FGB to the US Pressurized Mating Adapter 1 and Node 1 using the Shuttle Remote Manipulator System (RMS) 'Canadarm'.

PROPULSION AND ATTITUDE CONTROL

For orbital maneuvers, the FGB has two regeneratively cooled turbo-pump engines, each developing 416.13 kg (917.4 lbs.) of thrust. They do not gimbal. The cant angle of 36 degrees (plus or minus 10) was set prior to launch to provide thrust through the initial module center of gravity.

FGB is oriented for all maneuvers using 40 attitude control thrusters. 24 of the thrusters are rated at 39.91 kg (88 lbs.) thrust, and 16 at 1.31 kg. (2.9 lbs.) thrust. The thrusters can be used both for orientation and for small translational maneuvers. FGB's two main engines are fueled from 16 pressure-fed bellow tanks that hold approximately 5715.36 kg (12,600 lbs.) of usable propellant. FGB was launched with 4717.44 kg (10,400 lbs.) in the tanks. The propellant is hypergolic nitrogen tetroxide and unsymmetrical dimethyl hydrazine in a ratio of 1:85. To allow refueling from Progress freighters, three compressors restore nitrogen from propellant tanks to nitrogen storage bottles.

Power: 2 x 28m² (100.46 ft.²) solar arrays can generate a maximum of 13.0kW, supported by six nickel-cadmium batteries.
Life Support: Primary source of oxygen supply for Zarya is the Elektron water electrolysis system, based upon providing 20-25 liters (35.2- 44 pints) per person per hour. On board systems also include a non-condensing heat exchanger, circulation fans, fire detection and suppression equipment, and pressure regulators.

Functional Cargo Block (FGB) 'Zarya' Specifications

Launch:	20 November 1998
Launch Vehicle:	Proton
Launch Site:	Baikonur Cosmodrome, Kazakhstan
Mass:	19,325 kg (42,600 lb.) unfueled 24,568 kg (55,045 lb.) fueled
Length:	12,990 m (42.6 ft.)
Max Width:	4.2 m (13.5 ft.)
Pressurized Volume:	71.5m³ (2525 ft.³)

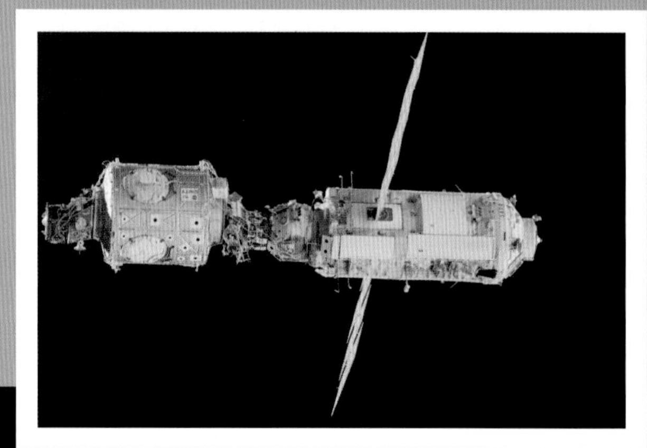

The U.S.-built Unity Connecting Module and the Russian-built FGB 'Zarya', (with solar panels deployed) are backdropped against the blackness of space.

"I remember going over to Russia and training. It was so cool. You've got the docking system for the Space Station, same one we used on Shuttle-Mir, the APAS [Androgynous Peripheral Attach System]. We're getting trained on it by the guy who designed it, Vladimir Sergeevich Syromyatnikov. Here's the guy that designed the system; it's essentially the same one that was used on Apollo-Soyuz [Test Project]. To be trained on something by the designer that actually designed the system was pretty cool. Getting to work with Khrunichev on the FGB. Of course, the FGB, although it was built by Khrunichev in Russia, it was a module paid for by the United States, and Boeing was the contractor that we worked with. To be able to see the FGB as it was being built, learn the systems—it was absolutely great." - Robert D. Cabana

"The Russian elements of ISS, which included the Base Block and the FGB [Functional Cargo Block] were very much like modules on Mir." - Kevin P. Chilton

This aft zenith view of the International Space Station shows the Zarya FGB module and the Zvezda Service Module (foreground).

"We had an aggressive schedule to build the FGB [functional cargo block], which was a U.S. paid-for element that Khrunichev [Production Center] built. The Russian elements formed the backbone of the station." - Charles Lundquist

Views of James Newman (left) and Sergei Krikalev (right) taken in the FGB/Zarya module.

Susan J. Helms changes out some hardware in the FGB Zarya as the seven-member STS-101 crew prepare the interior of the ISS for the arrival of the first expedition crew later in the year, 2000. Cargo Transfer Bags (CTB) and stowed hardware surround her. At the time the FGB was the primary stowage location and CTBs took up about the half of the volume inside the FGB.

"Probably one of the most critical events for the International Space Station was the launch of the first element, the Functional Cargo Block (FGB). That was in November of 1998. The FGB had the environmental control and life support system, the orientation system, the propulsion system and the electrical power system for the first stage of the program. The FGB was a Russian vehicle, although the United States paid for its development through Boeing to Khrunichev. If the FGB launch had failed, or if that element had failed, it was a single path failure. We didn't have any additional backup modules to launch. It was all or nothing. five months earlier, there had been a failure of the launch vehicle, the Proton. That was the launch vehicle from Khrunichev that the FGB was going to be launched on. We spent five agonizing months going through root cause analysis, to make sure that we understood what caused the previous launch failure of the Proton and that it wouldn't happen again, because if it did, the Space Station Program would have never survived. There were no backups. It would have been a catastrophic failure for the Program and for NASA. That one was probably the most significant because there weren't any alternatives." - Randolph H. Brinkley

"I can remember when I was chief of the Astronaut Office, getting that first flight ready to go. the media kept asking, "Who's going to be the first one in?" I wouldn't tell anybody, and I didn't even tell the crew. As we opened the hatch, I said, "Sergei (Krikalev), come here," and I grabbed him. Every hatch, from the PMA [Pressurized Mating Adaptor] into the node into the PMA-2, the end of the FGB—every hatch that we opened, we opened the hatch, and he and I entered side by side. Because I said, "It's going to be an International Space Station, it needs to be an international crew entering the modules together." So I was the first American, and he was the first Russian in the International Space Station. There was no first person in. We went in side by side, to every module." - Robert D. Cabana

"There's a lot of legacy designs that are really good in the Russian section of the space station, where they originally thought of them for Almaz and Salyut stations and then they were developed and improved for Mir and then developed and improved for ISS. But they have a definite familiar feel. They're from the same company." - Chris Hadfield

"FGB's two main functions – it is our foothold in orbit. It is where we start. It is what we build off of. It has to maintain orbit and hold attitude close enough so that the shuttle can dock to it, allow us to put the node on and then it docks with the service module. FGB has a guidance navigation system that allows it to hold attitude. Its propulsion system is nitrogen tetroxide and unsymmetrical dimethylhydrazine (UDMH). On both ends, FGB has docking systems that allow the attachment of other elements to it. On the forward end FGB has an APAS androgynous docking system. The same kind of mechanism on the Pressurized Mating Adapter allows the PMA to dock to the FGB. At the nadir end FGB has a probe and cone docking mechanism used for smaller vehicles like Soyuz and Progress, and a larger mechanism similar to the APAS. This is where the service module will dock to the FGB. The FGB is provided by Krunichev, a major aerospace manufacturer. They've been a major part of the Russian Space Program. They've built vehicles like the FGB before; the Mir modules, other than the Base Block, look a lot like the FGB. NPO Energia, another Russian aerospace contractor, is the integrator for the whole space station. The FGB avionics only lasts, is only certified for four hundred days. So you're taking a risk. The longer you expose that system by itself, without having the Service Module up there to take over, you're risking the overall station. So you don't want that stuff to be up there any longer than it needs to be before the Service Module comes up. I came on the beginning of 1994. We were struggling with who was going to pay for the FGB. The Russians, after our initial discussions, believed that the FGB was really not required, that we could have started building off the service module, which is what happened on Mir. The RSA [Russian Space Agency] then basically said, "Well, if you really want to do this thing, then you can pay for it." In the end, we agreed that the United States paid for development, design, development, and manufacturing of the FGB. RSA then pays for operations, sustaining engineering, and for launch. When we first started, we were working hard on agreements that were needed at the beginning of a program. Energia did the majority of that work. We did not see much hardware in those early phases. Some of the Russian hardware for ISS has been upgraded because of the Mir experience. But a lot of it is very, very similar." - Mark Geyer

"After, it must have been a couple of years, I remember being in Moscow and finally, after continually nagging the Russians, "we want to see your drawings of the service module," they walked us down these labyrinthine hallways and finally to this room, it was hardly bigger than a broom closet, and we walk in there and there's this wall of drawings rolled up, like old parchment, papyrus paper. They rolled it out, and it was all well-worn, kind of like your favorite book. It was covered in pen-and-ink changes, where in contrast, our drawings were all in an electronic system. You'd never have this broom closet with old worn-out drawings; it's all computer CAD operated and everything. So they rolled it out, and there it is. There were like two of us in the room and we got to see it. They rolled it up and put it back in the wall, and we left. That was how they finally shared their data." - Charles Lundquist

"After the Service Module launch, the Service Module becomes the guidance navigation and control system for the station. The Service Module is the key guidance and navigation system for station. After the Service Module, FGB's main function is really as a gas tank. It has the main tanks, holds upward of six tons of propellant; it can be refueled and reused. It has some pressurized volume that we use. We can store items." - Mark Geyer

Multipurpose Laboratory Module (MLM) 'Nauka'
Functional Cargo Block-2 (FGB-2)

Expected to be the next ISS element launched, the MLM 'Nauka' (science), has been repeatedly delayed for many years. MLM is a similar design to the FGB and served as the FGB back-up.

Mission: Should it be launched the MLM will be fully automated and fully self-supporting. It will serve as the active element during rendezvous and docking to the ISS at the Service Module (Zvezda) Nadir port; which will require the undocking and disposal of the Pirs docking module; MLM is intended for use for science experiments and will provide a scientific airlock for deployment of experiments outside the pressurized module. MLM will also launch with an ESA provided robotic arm that can assist in deployments. MLM will provide new Russian crew sleeping quarters, galley and waste management compartments. This will be the first time that an Almaz design module will be used for these habitation functions.

Russian Segment
Service Module 'Zvezda'

The Service Module (Zvezda, which means 'Star') was the first and one of Russian's primary ISS contributions. The module remains a critical element of ISS as it provides the primary docking port for Progress resupply and logistics, and provides propulsion for orientation control and re-boost. Early in the ISS Program, the module also served as a primary source of electrical power for the station and as the early crew living quarters. The Service Module is similar in design to the Mir core module and earlier Salyut space stations. The Service Module was planned initially as a back-up core module for Mir, and later was planned as the core module of Mir-2 which would have followed the original Mir orbital station. The Service Module provided the living quarters for crews living on the ISS for about a decade until additional life support systems were orbited in US-segment modules. The Service Module can support a crew of three. It provides a complete life support system, recycling waste water and generating oxygen for breathing; the module also provides an electrical power generation and distribution system, and data processing systems. The data processing system was upgraded for ISS from earlier systems by the Russians working in collaboration with the European Space Agency. The Service Module was designed to be autonomous or remotely operated from ground-based mission control. It has a complete flight control system providing orientation, propulsion and communications capabilities. As ISS developed, many of the on-board systems were replaced and many new systems were launched to augment the original capabilities, but the Service Module continues to be the functional core of the Russian ISS segment.

The Service Module is comprised of three compartments. The spherical 'Transfer Compartment', or node, is at the forward end. This has three docking ports. The forward port is attached to the FGB (Zarya) module. Two additional ports are located on the nadir and zenith sides of the Transfer Compartment. Originally they were intended for the addition of the Russian Science and Power Platform on the zenith port and a Universal Docking Module on the nadir port. These elements were subsequently replaced by smaller multiple use airlock, docking and laboratory modules. A fourth port, at the rear of the Service Module, has been in nearly continuous use throughout the program, accommodating Progress, Soyuz and European Automated Transfer Vehicle (ATV) dockings. A Kurs ('Course') automated rendezvous and docking system enables rendezvous and docking capabilities for Russian and ESA logistics vehicles. Aft of the forward Transfer Compartment is the cylindrical main 'Working Compartment'. This includes two personal sleep stations, toilet and hygiene facilities, a galley with a refrigerator-freezer, water dispenser, and a table for securing meals or work equipment. The module is also fitted with a treadmill and exercise bicycle-ergometer. There is provision for data, voice and TV links with mission control sites in Moscow and Houston. The final compartment is the 2.0m (6.564 ft.) diameter cylindrical aft Transfer Compartment, which is a tunnel leading from the Work Compartment to the rear docking hatch. Around this is an unpressurized Transfer Chamber which houses external equipment, including the propellant tanks, maneuvering thrusters, docking lights and antenna.

The Service Module provides 14 windows from which the astronauts can view the exterior. The 14 windows in the module include three 22.86 cm (9 in.) diameter windows in the forward Transfer Compartment for viewing docking activities; one large 40.64 cm (16-in.) diameter window in the Work Compartment; an individual window in each of the two crew sleep compartments; and a selection of Earth observation and inter-module observation windows around the periphery. The Service Module can support EVA by crew members wearing Orlan M EVA suits, although none have been performed directly from Zvezda. Following orbital insertion, the Service Module followed a pre-programmed sequence of on-board commands to activate systems and deploy solar arrays and antenna. The Service Module served as the passive target, while the combined Zarya/Unity ISS performed the rendezvous and docking via control from the ground and Kurs automatic system. After successful docking, the Service Module guidance and propulsion systems took over from those on the FGB. FGB was then relegated for use as a propellant storage and transfer compartment between the Service Module and Node 1.

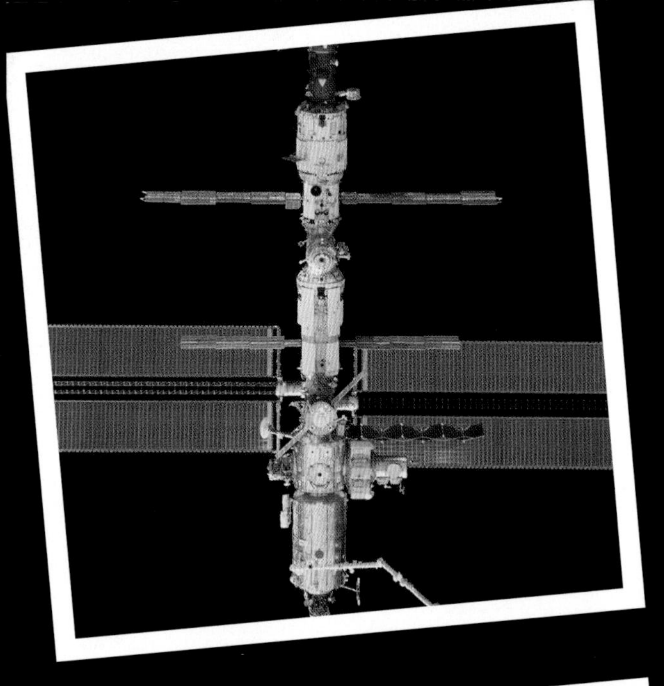

The International Space Station in December 2001. This view of the nadir side during the undocking and final flyaround procedures for the STS-108 mission. Visible are the Soyuz spacecraft, Service Module (SM) 'Zvezda', Pirs docking compartment and airlock, Functional Cargo Block (FGB) 'Zarya', Progress vehicle, solar arrays extended from the P6 Truss, mounted on the zenith side of Node 1, Node 1 'Unity', Joint Airlock 'Quest', Pressurized Mating Adapter 3 (PMA3), and US Laboratory Module 'Destiny', Space Station Remote Manipulator Systems 'Canadarm2' and Pressurized Mating Adapter 2 (PMA2).

"We hadn't appreciated the work that was required to get the [Zvezda] Service Module ready for the arrival of Expedition 1. ....Originally it was focused on the Russian Service Module...but with the delay in the Service Module they decided we needed to go there to address some of the failures... on the FGB and the Node ...that were going on prior to the Service Module." - Jeffrey N. Williams

The partially retracted aft radiator of the P6 truss is in the near field of view in this image photographed 4 Feb. 2007 during a spacewalk by Michael E. Lopez-Alegria and Sunita L. Williams. The Service Module 'Zvezda' and the FGB 'Zarya' module are visible at left.

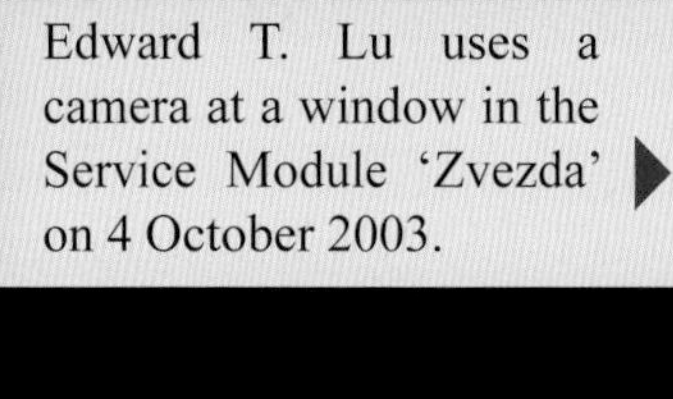

Edward T. Lu uses a camera at a window in the Service Module 'Zvezda' on 4 October 2003.

View of engine thruster firing during sunset. Service Module (SM)/Zvezda is visible. Photo was taken during Expedition Six on the International Space Station (ISS).

"We had similar schedule problems with the Destiny lab. It was not as late as the Service Module and it wasn't as critical; because the Service Module provided the environmental control and life support system for the whole Space Station, the only way we could have humans working and living on the Station was with the Service Module. We did have a contingency capability in work just in case the Service Module was further delayed. We were funding FGB2 that we could launch and attach to FGB1. This would extend Station's time in orbit until the Service Module got there. But these weren't the days of Apollo where we had all the money we needed. A lot of the contingency plans weren't funded. We did fund some of the work for FGB2; it was the test vehicle for the first FGB, and we looked at how to modify it from a test bed to an actual FGB2 flight vehicle." - Randolph H. Brinkley

"We did not meet—the Russians didn't meet their initial promised date; they were actually two years late on the Service Module. But of course we put that time to really good use; we had used the time to do a lot more ground testing of the U.S. segment and were able to really give us a lot higher confidence in the pedigree of the hardware that the U.S. was sending to Space Station. So that's what I think was the ultimate driver of the progress that we made." - Charles Lundquist

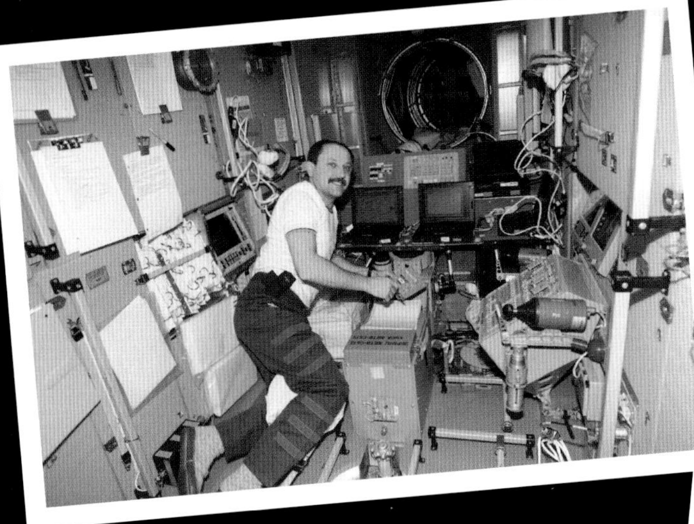

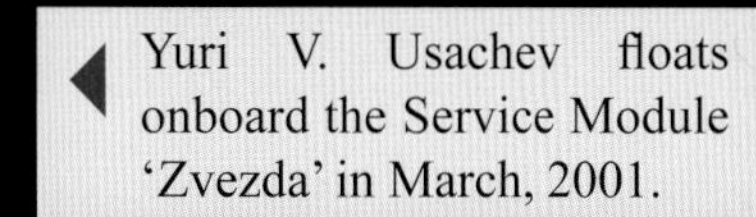

Yuri V. Usachev floats onboard the Service Module 'Zvezda' in March, 2001.

The Expedition Three crew members assemble for a crew photo in the Service Module 'Zvezda' in September 2001. Frank L. Culbertson, Jr. (center) is flanked by Mikhail Tyurin (left) and Vladimir Dezhurov.

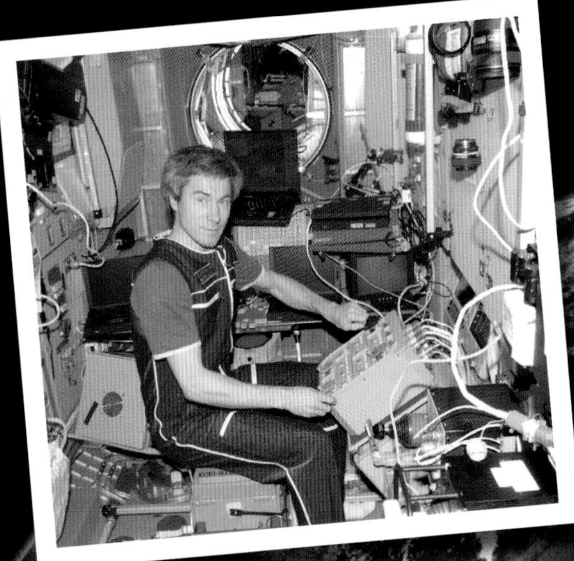

Sergei K. Krikalev tests the newly installed Proximity Communications Equipment (PCE) hardware of the ASN-M satellite navigation system for the European Automated Transfer Vehicle (ATV) "Jules Verne" in the Service Module 'Zvezda' 28 June 2005.

Yury V. Usachev holds a small-scale model of the International Space Station that crewmembers use for planning purposes. Usachev is in the aft end of the Service Module near the galley. To the left is the main control station and TORU control station for control of incoming unmanned vehicles.

Service Module 'Zvezda'
Propulsion and Attitude Control

Spinning electrically powered gyrodynes inside the Service Module can maintain orientation without using rocket thrusters or fuel. The Service Module has two 313 kg (3070 N, 690 lb) force thrusters and two 1645 kg (16,130 N, 3626 lb) force rocket motors that are used to maintain orbit altitude and attitude/orientation control. Fuel is pressure fed from four tanks. A total of 860 kg (1896.3 lbs.) of nitrogen tetroxide and unsymmetrical dimethyl hydrazine hypergolic reactants can be stored.

SERVICE MODULE (ZVEZDA) Specifications

Launch date:	12 July 2000
Launch Vehicle:	Proton
Launch Site:	Baikonur Cosmodrome, Kazakhstan
Mass:	24,604 kg (20.6 tons, 54,242 lbs.)
Length:	13.112 m (43 ft.)
Diameter:	4.2 m (13.5 ft.)
Wingspan:	29.7 m (97.5 ft.) across its extended solar arrays
Pressurized Volume:	89.0 m^3
Windows:	14

Russian Segment
Russian Docking Compartment/Airlock 'Pirs'

In September 2001, Russia launched the Progress M-SO-1 carrying the Russian docking compartment/airlock to the ISS. It is also known as Docking Compartment 1 (DC1). The Russians gave it the name Pirs (Pier). The compartment serves two primary functions. It is a docking port for the Soyuz crew transport and Progress cargo vessels and it is used as an airlock for spacewalks (EVA) by two space station crew members when they use Russian Orlan suits.

At the front of Russian Airlock is an active androgynous docking system with hybrid probe/drogue, while at the rear is a passive probe/drogue docking system. This allows either the Progress or Soyuz to dock with the rear port of the facility. The Russian Airlock also has the capability to transport fuel from the tanks of a Progress freighter to either the Zvezda Service Module Integrated Propulsion System or the Zarya Functional Cargo Block and can also transfer propellant from either Zvezda or Zarya to the propulsion systems of any Soyuz or Progress docked with it. Air and water can also be pumped from a Progress to the Russian Airlock Module.

RKK Energiya constructed the airlock between 1998 and 2000 and conducted verification tests in 2001. In addition to the flight unit, a mock up for underwater EVA simulations and training at GCTC Gagarin Cosmonaut Training Center, and a 1-g training unit for crew familiarization were built. Originally intended for an orbital lifetime of five years, Pirs exceeded this duration in September 2006 and continues in use more than twelve years later. Pirs was due to be replaced by the much larger Universal Docking Module (UDM), but in 2002 these plans were changed and Pirs continues in use in 2018

Russian Airlock 'Pirs' Specifications

Launch:	15 September 2001, Progress M-S01
Launch Vehicle:	Soyuz-U (R7)
Launch Site:	Baikonur Cosmodrome, Kazakhstan
Mass:	3,838 kg (8.461 lb.)
Length:	4.9m (16 ft.)
Max diameter:	2.55 m (8.4 ft.)
Pressurized Volume:	13 m3 (460 cu ft.)
Windows:	2

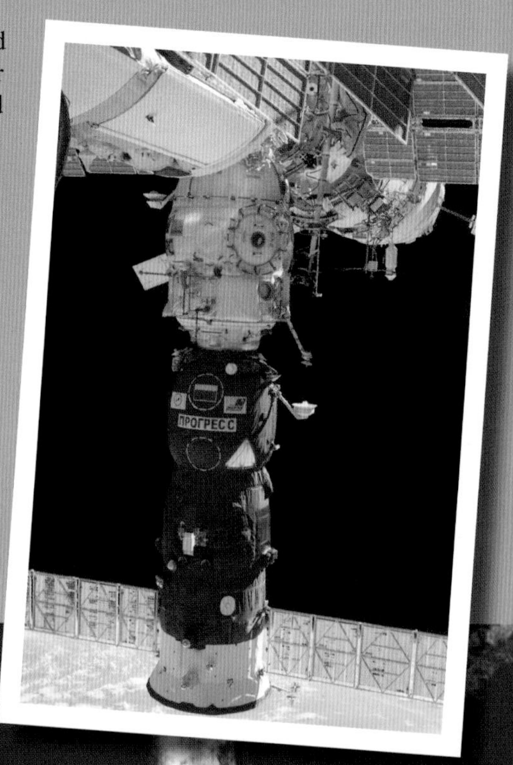

Alexander Misurkin on Expedition 36, attired in a Russian Orlan spacesuit, participates in a spacewalk outside of the Service Module 16 Aug. 2013. The Pirs docking compartment and airlock is docked to the Service Module nadir port, and a Progress vehicle is docked to the nadir end of the Pirs.

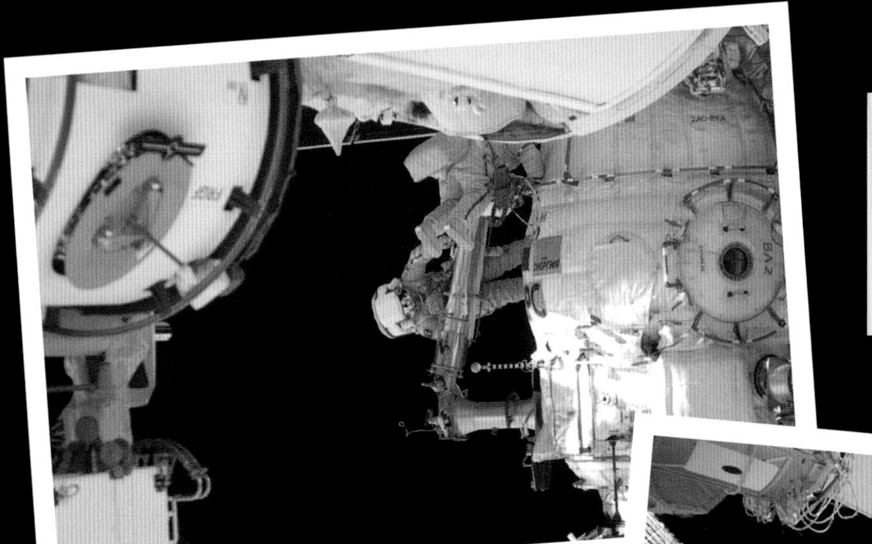

View taken 15 Nov. 2010 during a Russian Extravehicular Activity during Expedition 25. Cosmonauts are visible working outside the Pirs Docking Compartment (DC1). Part of the Service Module and FGB are also visible. Oleg Skripochka (top) and Fyodor Yurchikhin, wear Russian Orlan spacesuits..

Two Russian spacecraft, docked with the International Space Station, are featured in this image photographed 11 May 2010 by an Expedition 23 crew member on the station. View of the docked Soyuz TMA-17/21S is in the foreground and Progress 37P spacecraft is docked to the Pirs Docking Compartment (DC1).

Russian Segment
Mini Research Module (MRM)-2 'Poisk'

Essentially identical to the Pirs airlock module, the Mini Research Module 2 (MRM 2) was originally designated as Docking Module 2. Named Poisk (search, seek or explore), it was designed to provide a fourth docking facility for Soyuz and Progress spacecraft, with the added capability of delivering new fuel supplies to the station, and it can support EVAs, providing an alternative point of exit and entry for the Russian segment. Several improvements and upgrades were made to the new module over Pirs, in that it could support applied scientific and research experiments both inside the pressurized vessel and outside the module. Utilizing the same design features and production process as Progress and Pirs, Poisk was constructed of aluminum alloy, stainless steel and titanium alloys. It features two docking ports, one active the other passive, on the longitudinal axis of the vessel. There are two EVA hatches on each side of the pressurized module, adorned with internal and external handrails to assist with exiting and entering the module during EVAs wearing Orlan pressure garments. Though designated Mini Research Module-2, it was actually launched first, because MRM-1 was not ready. Launched from the Baikonur Cosmodrome on 10 November 2009 by Soyuz U it docked to the zenith port of Zvezda two days later. Cosmonauts Oleg Kotov and Maxim Surayev of Expedition 21/22, completed an EVA in January 2010 to prepare for future Soyuz and Progress dockings. Surayev and astronaut Jeff Williams relocated their Soyuz TMA-16 spacecraft from the aft end of the Zvezda module to the zenith facing port of Poisk, making it the first operational use of the new docking module. The new port gave far greater flexibility for the docking and relocation of support spacecraft, and it also afforded both better views of Earth during approach and a new perspective of the station which helps in the on-going program of exterior examination of the station as it ages.

Scientific program: MRM-2 Poisk was also intended for scientific research.

Delivered Cargo: When it was launched, Progress M-CO2/Poisk carried 751 kg of supplies, including:

- 88 kg of water
- 3 kg of atmosphere measurement devices
- 2 kg of maintenance and repair equipment
- 160 kg of sanitation and hygiene equipment
- 236 kg containers of food
- 83 kg of medical equipment, clothing, personal hygiene items
- 74 kg of protective clothing
- 16 kg of inter module ventilation equipment, and 11 kg of on board documentation and personal parcels for the crew members.

Mini Research Module (MRM)-2 'Poisk' Specifications

Mass:	3670 kg (8090 lb.)
Max diameter:	2.55 m (8.4 ft.)
Length:	4.049 m (13.3 ft.)
Pressurized volume:	14.8 m³ (525 cu ft.)
Habitable volume:	10.7 m³ (380 cu ft.)
Number of hatches:	2 x 1 m (3.3 ft.) diameter
Windows:	1

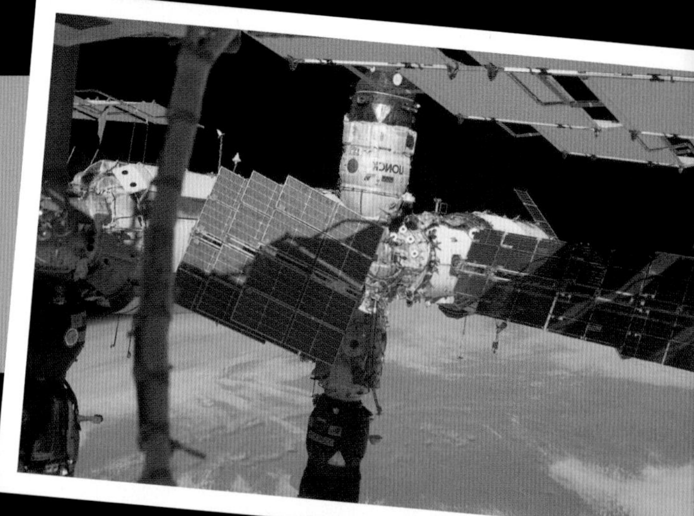

The Poisk Mini Research Module 2 (MRM2) is docked to the zenith port of the Service Module 'Zvezda' at top center; and a Progress resupply vehicle is docked to the Pirs Docking Compartment at bottom center. The FGB 'Zarya' (partially obscured by solar panels) is at right center.

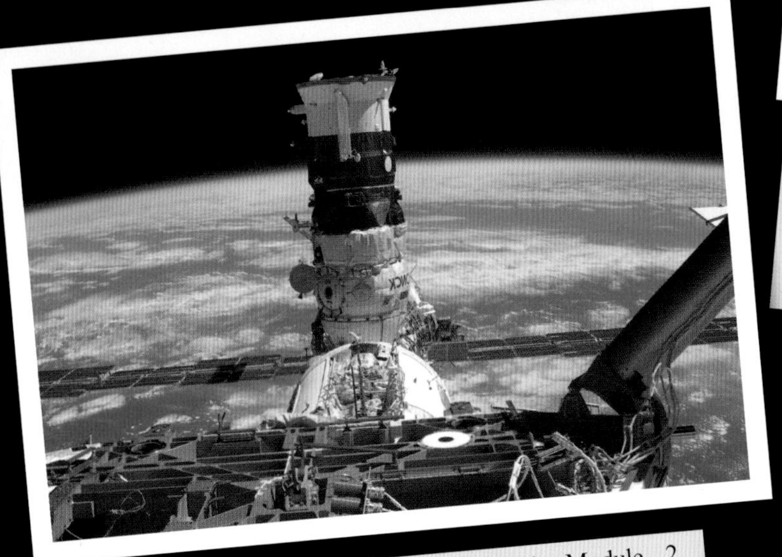

Exterior view of the Mini Research Module 2 (MRM2)/Poisk docked at the zenith port of the Service Module. Photo taken by an STS-129 crew member during a spacewalk 21 Nov. 2009. The zenith side of the Z1 truss segment and the aft end of the FGB are in the foreground.

The unpiloted Russian Mini-Research Module 2 (MRM2), 'Poisk', approaches the International Space Station two days after launch

Russian Segment
Mini Research Module (MRM)-1 'Rassvet'

Initially called the Docking Cargo Module (SGM), and later re-designated MRM-1 and named 'Rassvet' (Dawn). The module was constructed from the already built hull of the Science Power Module (NEM). It provides two docking ports. The first was for docking to the nadir port of Zarya and the opposite side for docking future Soyuz spacecraft. After Node 3 (Tranquility) was berthed in early 2010, MRM-1 was attached in order to move the Soyuz docking facility further away to provide a greater clearance from Node 3. Mini Research Module MRM-1 was launched on the Shuttle STS-132 aboard Atlantis in May 2010. The module was installed on the nadir port of the Zarya FGB. The module also carried a total of 1400 kg (3090 lb.) of US supplies as well as outfitting hardware for the Russian segment.

Mini Research Module (MRM)-1 'Rassvet' Specifications

Length:	5.5 m (18 ft.)
Diameter:	2.2 m (7.2 ft.)
Mass:	4.7 tons (10,360 lb.)
Internal Volume:	14.5 m³ (512 cu ft.); 5.5 m³ (194 cu ft.) for cargo
Internal habitable volume:	6.0 m³ (212 cu ft.)

Russia's Soyuz MS-07 crew ship (foreground) and Progress 68 cargo craft are seen docked to the Earth-facing ports of the International Space Station's Russian segment on Feb. 11, 2018. The Soyuz is docked to the Rassvet module and the Progress is attached to the Pirs docking compartment.

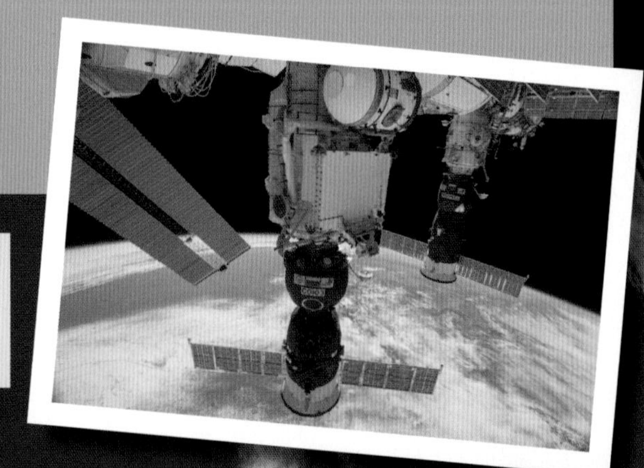

Alexander Misurkin (right) and Chris Cassidy work in Mini Research Module 1 'Rassvet' April 3, 2013.

As Space Shuttle Atlantis leaves the ISS, the Mini-Research Module 1 (MRM1) 'Rassvet' is in the foreground during Expedition 23/23 in May 2010. The SSRMS was used to position MRM1 on the nadir docking port of the FGB 'Zarya'.

US Segment Connecting Nodes

With the first module of ISS, the FGB 'Zarya', in orbit, the first US developed element was launched. This was Node 1, called 'Unity'. Node 1 was the first of three connecting modules, with common berthing mechanisms that would permit the attachment of subsequent new elements as they arrived. Node 1 'Unity' provides permanent attachment to the Russian segment through the Pressurized Mating Adapter (PMA)-1, which berths to the FGB Zarya. Node 1's six ports provide berthing locations and connections to the Z1 Truss (zenith port), the US Laboratory Module 'Destiny' (forward port), the US Airlock 'Quest' (starboard port), Node 3 'Tranquility' port), PMA-1 (aft port) which is permanently attached to Node 1 and to the FGB, and PMA-2 and 3 which were moved repeatedly to different locations on Node 1 and other elements, and providing docking facilities for the Shuttle and other vehicles. Node-1 also provided berthing facilities for the Multi-Purpose Logistics Modules (MPLMs) when they were delivered during Shuttle logistics missions. Node-1 'Unity' was named for the ISS spirit of multi-national co-operation; the 'unity' between the US and the international partners. Boeing manufactured the Node 1 between January 1995 and June 1997.

Node 1 is constructed of aluminium. Its primary structure is a pressurizable cylinder made of 1.25 cm (0.49 inch) thick aluminum. The module was constructed in sections and then welded together to form a cylinder with two end-cones and six berthing ports. To allow passage of electrical and fluid lines, feed-throughs were welded into the structure between the inner pressure vessel and an outer unpressurized secondary structure, also made of aluminum. Titanium was used to connect the two structures together, its high strength and low thermal conducting qualities preventing transfer of solar heat during the sunlight portions of each orbit. After assembly, welds between the structural assemblies were x-rayed to locate any cracks, and the structure was proof pressure-tested to an internal pressure of 24 psi, which is 1.5 times the normal on-orbit pressure. This was followed by a leak test, using helium injected inside the module and detection devices outside to 'sniff' for any leaking gas.

Final assembly of Node 1 included the installation of the secondary structure, checking for interference or unacceptable gaps, drilling bolt holes, and bonding brackets to the outer structure. The secondary structure of the Node supports wire harness clamps, meteoroid/orbital debris shields, multi-layer insulation (MLI) blankets, electrical equipment, fluid clamp lines, interface brackets, PMA-2 Grapple Fixtures, EVA and IVA handrails, and interior Standard Interface Racks. Node 1 can accommodate 4 racks. Nodes 2 and 3 are slightly longer than Node 1 and each can accommodate 8 racks. Node-1 'Unity' was launched with a single International Standard Interface Stowage Rack in place. Three more racks would later be added. Node 1 features active Common Berthing

Mechanisms on five of its six hatches. Node 1 contains 50,000 mechanical components, 216 fluid and gas transfer lines in 9 different fluid systems and 121 internal and external electrical cables, using six miles of wire.

Node 1 was launched with two Pressurized Mating Adapters (PMA 1 & PMA 2) berthed to its two end-cones. PMA 1 permanently connects Node 1 to the FGB. PMA 2 was used initially on the forward end of Node-1 for Shuttle dockings, and later was moved to the forward end of the US Laboratory Module after the US Lab was connected to The Node 1's forward berthing port. Node 1 was fabricated by Boeing at a facility located at Marshall Spaceflight Center in Huntsville, Alabama.

View of the Pressurized Mating Adapter 1 (PMA-1), Node 1 'Unity' module and the US Laboratory Module 'Destiny' 28 July 2005. Also visible is the Quest airlock, and Pressurized Mating Adapters (PMA) 2 and 3 as well as the Space Station Remote Manipulator System (SSRMS) 'Canadarm2'.

Each Node has two halves. One half carries a single longitudinal berthing port on an end cone. This half houses standard-sized racks that carry the systems and equipment that supports operations. The other half of the node carries the other five berthing ports; one in the other end cone on the longitudinal axis and four arranged around the circumference of the cylindrical body of the node. Longer than Node 1, Node 2 'Harmony' and Node 3 'Tranquility', can each accommodate eight International Standard Racks, as compared with the four that can be accommodated in Node 1. The later Nodes 2, Harmony, and Node 3, Tranquility, are similar in overall configuration to Node 1 but were designed by ESA and its contractor Thales Alenia Spazio at their facility in Turin, Italy. Nodes 2 and 3 were based on tooling originally designed for the ESA laboratory module Columbus.

The Boeing Company provided several major components for Nodes 2 and 3, including the Active Common Berthing Mechanisms (ACBM), hatches, racks, and internal outfitting lights, fans, power switches and power converters, air diffusers and smoke detectors. Nodes 2 and 3 were constructed with pressurized cylindrical hulls, approximately 4.5 m (14.7 ft) in diameter, and with a conical section at each end-cone. The pressurized shell for the Nodes is made from aluminum, covered with multi-layered insulation blankets for thermal stability in orbit. On each Node, there are also approximately 75 aluminum panel sections on the outside of the cylinder, which serve as a protective shield against impacts from space debris or micro-meteorites. Between the internal pressure cylinder and the external aluminium debris shields are layers of Kevlar and Nextel. Low and moderate temperature water loops carry waste heat from the module interiors to the station's exterior ammonia lines, via two heat exchangers located externally on the end cones of each Node's structure. The ammonia lines run through exterior radiators where the heat is radiated into space.

Nodes 2 and 3 were fabricated for ESA in a barter arrangement with NASA. The agreement was signed 8 October 1997 between ESA and NASA as partial payment for the launch of the ESA Columbus science laboratory on the Space Shuttle. The Columbus Launch Offset Agreement is also known as the 'Nodes barter'. In what was described as a 'win-win agreement', the budget intended to pay for the launch of the ISS European laboratory would instead be spent in Europe to enhance European industry human space flight engineering. In return, NASA would pay for the Shuttle launch of the Columbus. As part of the ESA agreement with NASA, Italian ESA astronaut Paolo Nespoli accompanied the Node 2 module as a Mission Specialist aboard the STS-120 mission, to oversee the node's installation on the station.

ISS nadir side view includes, from bottom to top, the Pressurized Mating Adapter 2 (PMA-2), US Laboratory Module 'Destiny', Node-1 'Unity'. On the port side of Node-1, the Joint Airlock 'Quest' is berthed. On the opposite side, Pressurized Mating Adapter 3 (PMA-3) is visible in front of the S0 truss segment. Pressurized Mating Adapter 1 (PMA-1) is hidden behind a Soyuz spacecraft. The FGB 'Zvezda' is in front of the P-6 truss with its extended solar arrays. The Pirs docking compartment and airlock is berthed to the Service Module 'Zarya' nadir port. The Service Module 'Zarya' is trailed by a Progress spacecraft.

Don Pettit (left) and Andre Kuipers eat a snack in the Node-1 'Unity'. Visible in this view are three of the Nodes' six hatches. Pressurized Mating Adapter 1 (PMA-1) is behind them, lined with Cargo Transfer Bags (CTBs). The Joint Airlock 'Quest' is to the left.

"Another memorable experience was getting ready for the first U.S. element to fly, the node. There were a lot of processes we were learning at the time, and so we had mountains and mountains of open paper before we fly. We had like 4,000 pieces of open paper, and I got the lucky job of closing out all these individual issues with people responsible for closing them that you had to reach out and touch. Everybody was an owner on this or that, had to sign off on it, so there was tremendous push, meetings all day long. I think back on it, and it was great. I got to talk to hundreds of different people to try to get this mountain of paper closed. The lessons learned out of that is there is a lot of paper to close before we fly. Getting the hardware built and tested is important, but there is also a lot of certification, verification that has to occur as well. There a chance that that mountain of paper would prevent a launch. You know, during the COFR [Certification of Flight Readiness] process or flight readiness review process, you basically have to sign up all the verification is complete, all the open discrepancies have been dispositioned acceptably. The list goes on and on of the different things you have to sign off, that all the "i's are dotted, t's are crossed." You don't want to go into a flight readiness review and say, "I've got 3,000 pieces of open work that address the verification, the certification that this thing meets its requirements." And so yeah, it needs to get done. Of course there is a prioritization, there are some key things and some not so key things, and so that obviously plays into it as well." - Charles Lundquist

"At the start of Station assembly we had the node, the power system, and the laboratory all right there; the systems that power ISS. The 3-A, 4-A, and 5-A flights were all just right together, because you wanted to get them all done just as fast as you could and try not to expose yourself to too much risk." - Denny Holt

"There was tremendous pressure to get something launched and be in the middle of the program and not want to run a risk of being canceled again.... we get down to the first [Space Station] element launch in November 1998. It was an element built in Russia that the U.S. bought and paid for.... Russians paid to launch it.... the next element that went up was the U.S. node, Node 1 [Unity].... about two weeks later, December 4." - Melanie Saunders

Node 2 was named 'Harmony' in March 2007. The name was chosen as the winner in a US school competition that engaged students across the United States. The competition was intended to educate students about the ISS, and inspire them to careers in engineering. 2200 students in 32 states, from kindergarten through high school, participated in the program. They learned about the space station, built a scale model, and wrote an essay explaining their proposed name for the module. The name 'Harmony', submitted by six schools, reflected a desire for harmony not only in the ISS program, and in space and on Earth. Harmony was the first US element of ISS to be named by people outside of NASA.

Node 3 was named in an on-line public poll beginning in March 2008. Four names were offered by the agency: Earthrise, Legacy, Serenity, and Venture. Users could also submit their own recommended names. Serenity led the poll, but in March 2009, American TV personality Steven Colbert suggested that his viewers submit the name 'Colbert' in the poll. This prompted several other groups to nominate their respective names. NASA announced that 'Colbert' had won the poll, with NASA's name 'Serenity' coming in second. But the rules governing the poll gave NASA the final say. In April, Steven Colbert threatened legal action if Node 3 was not named after him and one Congressman planned to use Congressional power to force NASA to honor the winning name. Later that month astronaut Suni Williams appeared on The Colbert Report and revealed the name to be 'Tranquility' to honor the 40th anniversary of the Apollo 11 landing on the Moon at the Sea of Tranquility. However, in a compromise, NASA renamed the ISS on-board treadmill to be "COLBERT", an acronym standing for Combined Operational Load Bearing External Resistance Treadmill. Steven Colbert accepted the compromise. The COLBERT treadmill launched to ISS on STS-128 in August 2009 and was installed in the Node-3 'Tranquility' in 2010.

Once Node 2 was attached, it enabled the attachment of the Japanese and European science laboratories. It increased the habitable volume of the station from a three-bedroom to a typical five-bedroom house. The attachment of these modules marked the 'US segment core complete' stage of ISS assembly. Node 2, berthed to the forward end-cone of the US Laboratory Module, Destiny, serves as a primary utility hub for the US segment of ISS, connecting the US Laboratory 'Destiny', ESA Laboratory 'Columbus' and Japanese Laboratory 'Kibo' with air, electricity, water, heating, communications, data management, and other essential support. Node 2 also serves as a primary berthing location for visiting logistics vehicles. Node 2 provides crew quarters for 4 crew members. Node 3 was launched on STS-130. During the launch, the Cupola was mounted on the end-cone that would eventually face away from the station. Five of eight Standard Interface Racks (SIR) were in place during launch. These were two avionics racks and three stowage racks containing pallets for storing cargo.

Following launch, Node 3 was berthed to the port side of Node 1. The Cupola was relocated to the nadir (Earth-facing) port of Node 3. The three cargo pallet racks were removed and returned to Earth on STS-131 in March 2010. Node 3 provides the station with additional habitation volume as well as additional environmental control and life support systems. After Node 3 was added, much of the closed cycle Environmental Control and Life Support System (ECLSS), including the racks for air revitalization, air composition monitoring, and carbon dioxide removal, oxygen generation, and water recovery were moved into Node 3. The ECLSS processes urine and waste water to reclaim drinking water. Node 3 also contains the Waste Management Compartment or 'bathroom' for crew hygiene. Crew Health Care System exercise equipment, including the COLBERT Treadmill and Vibration Isolation System (TVIS) and the Resistive Exercise weight-lifting Device (RED) are also accommodated. Once these systems were in place and functioning, it enabled the increase of the ISS crew size from 3 to 6 crew members.

"When I became Deputy Program Manager of Operations for International Space Station program I found out the program was behind schedule, over budget, and had a serious technical problem with Node 1. Node 1 had failed its first pressurized structural test. The program had been run with a large number of IPTs, integrated product teams. I can't remember the number, but a very large number of integrated product teams. And, there weren't any formal control boards in the program. We decided that since Apollo worked out pretty well, perhaps we should adopt some of those organizational constructs from that program. Jay Green set up the Configuration Control Board, which was an engineering board." - Kevin P Chilton

"The second launch was with [Robert D.] Bob Cabana and his [STS-88] crew on the Shuttle Endeavour in December [1998] to deliver Node 1. I really believed that Node 1, Unity, was going to do well once it was there. We decided that since we couldn't name the Space Station we would name all the elements. We decided on our own that Node 1 was Unity, because it represented the bridge between the Russian segment and the U.S. and the other segments and international partners. That was how Unity came about. We took that name. That was how Unity became Unity." - Randolph H. Brinkley

Node Specifications

Node 1 'Unity' Specifications

Launch:	US Space Shuttle Endeavour, STS-88, 4 December 1998
Mass:	11,895 kg (26,225 lb.)
Length:	5.5m (18 ft.)
Max Width:	4.5 m (14.7 ft.)

Node 2 'Harmony' Specifications

Launch:	US Space Shuttle Discovery, STS-120, 23 October 2007
Mass:	14,787 kg (32,599 lb.)
Length:	6.7 m (22 ft.)
Max Width:	4.5 m (14.7 ft.)

Node 2 'Tranquility' Specifications

Launch:	US Space Shuttle Endeavour, STS-130, 8 February 2010
Mass:	17,992 kg (39,655 lb.)
Length:	6.7 m (22 ft.)
Max Width:	4.5 m (14.7 ft.)

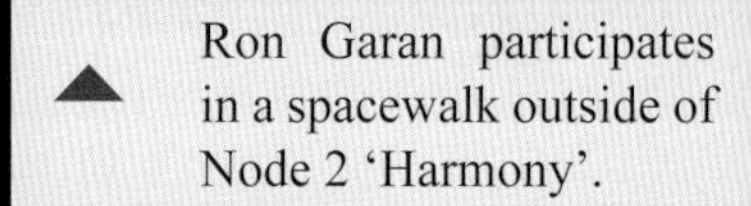

▲ Ron Garan participates in a spacewalk outside of Node 2 'Harmony'.

◀ In the grasp of the International Space Station's Canadarm2, the Node-2 'Harmony' with the Pressurized Mating Adapter (PMA-2) berthed to the endcone on the right is moved to the front of the US Laboratory 'Destiny', 14 Nov. 2007 in this view from NASA TV.

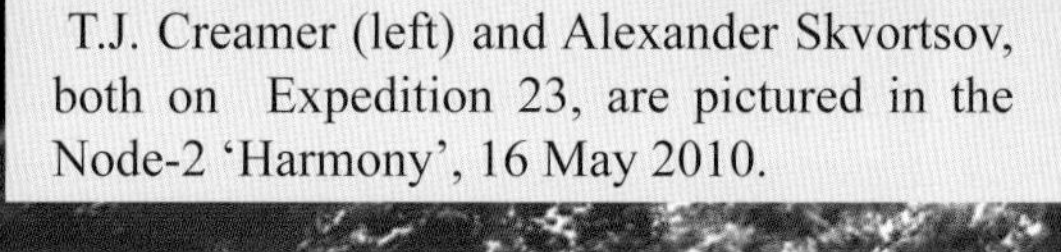

T.J. Creamer (left) and Alexander Skvortsov, both on Expedition 23, are pictured in the Node-2 'Harmony', 16 May 2010. ▶

"The first US element, the first Node, launched in 1998. Originally it was supposed to be considerably earlier than '98. It kept dragging out. We had welding problems—it's amazing to me. We always seem to be surprised when we have welding problems. There were all kinds of schedule problems. It seemed like it was a very long time coming. It was wonderful to finally see it launch." - Lauri N. Hansen

"The two nodes were initially supposed to be supplied by Boeing. It happened that Boeing, while developing Node 1, had some design problems with the structure in the area of the berthing ports. Italy had a proposal in which we had a different design; the design corrected the structure in the area of the ports and it provided longer modules. So there was strong support for our design. ESA bartered in exchange for the Columbus launch, we bartered the development and the supply to NASA of the two nodes plus the cupola, based on our ideas of how to develop them." - Alan Thirkettle

"During the development of Node 1, Boeing got themselves into a mess and NASA was annoyed with them. The problem with unity Node 1 was a structural problem. The nodes have got lots of docking ports on them and it's quite complicated structurally to interface two cylinders at right angles to one another if you want to do it in such a way that you never get any leakage from the joint. We had a sea level atmosphere inside and vacuum outside, and if air starts to escape, either you have to replace that air, or people die. Leakages are a big problem. Boeing had to change their design to reinforce the Unity node. When NASA was looking at the requirements for Node 2 and Node 3, they realized that maybe they should be bigger. That introduced more problems." - Alan Thirkettle

"Three of us were the core of the team that designed the space station for the Man-Systems Division. We felt that anything the human, the astronaut, came in touch with was our responsibility. I would write all of the requirements for all of the hardware in the Man-Systems Architectural Control Document, the ACD. The other two of us were Malcolm Johnson, who was an Apollo veteran, and who guided much of our effort, and Jay Cory, who was an industrial designer; really could dream up designs to answer the question of what things should look like. The three of us were constant companions for several years. One day we were eating lunch at McDonalds, just out the gate of the space center. Up until this time our idea of the station modules were all the same: a 15 foot diameter 45 foot long cylinder, about the maximum the Shuttle could carry. There were hatches at both ends and 4 hatches around the cylinder towards one end. It was called the common module. Well at McDonalds we were eating kid's meals. And in the kid's meals they gave us a kid's toy called pop-ems. They were little plastic cylinders and some were 'nodes' that would link cylinders together. The plain cylinders were longer and only had connection points on the ends. The nodes were short and had six connection points. We started playing with these toys and discussed, 'you know it would be a good idea to study the efficiency of mass and usable volume for long modules with only two hatches on either end, versus short modules with six connection points. I was the scribe and analyst of the group, and did some calculations in the next day or two. Then showed Malcolm and Jay this was a good idea. We put together a set of charts, briefed it to our Man-Systems group and then to the Program Manager. We sold the idea within a matter of days. Good ideas are accepted quickly. That was the origin of the Nodes and the long modules; we were supposed to have four: a habitation module and three laboratories. NASA was building the nodes, and there were four of them too." - Gary Kitmacher

In the grasp of the station's Canadarm2, Node 3 'Harmony' module is moved from its stowage position in the cargo bay of the Space Shuttle Discovery (STS-120) to the International Space Station, 26 Oct. 2007.

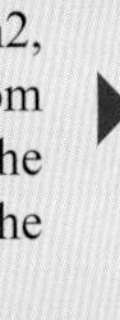

The nadir/forward side of Node-3 'Tranquility' and the Cupola berthed on its nadir port. The Japanese Experiment Module 'Kibo' is visible behind Node-3. Thermal control system (TCS) radiators are visible beyond Node-3 to the right and the starboard truss is visible to the far right. In this view the SSRMS 'Canadarm2' is in the process of berthing Pressurized Mating Adapter 3 (PMA-3) to the starboard end port of Node-3,

"Node 3 was launched when I was the program manager. The cupola was launched at that same time. ESA providing Node 3 was a way to pay for the Columbus launch. The Columbus launch was, if I remember, was 23,000 pounds on the shuttle. Between 10 and 12 tons; that doesn't come for free. We could pay for the launch with cash, which, obviously, we didn't want to have European taxpayer money going directly to NASA. Or we could barter via goods and services. And for this reason, we built the two nodes to pay our debt for the service that NASA was delivering to us. The second option was of course a win-win, because we could spend the money in Europe, and we could also, develop further engineering knowledge by building those relatively complex elements that are now in the station and we can claim that we have built 50% or more of the pressurized volume of the U.S. segments." - Bernardo Patti

"The design of the primary structure of Node 2 and Node 3 is exactly the same. We had some improvements in the technology for meteorite debris protection. Node 2 is an interconnecting element that provides services to the lab, or to the JEM, to the Japanese Kibo, and to Columbus. So it manages the power, thermal, water flow, it is analyzing, it is sampling the air. It provides the crew quarters in which the crew rests. Internally, Node 3 is completely different and is considered one of the most complex elements of the ISS because of the functions it implements. Node 3 has to guarantee a comfortable temperature and comfortable environment for the crew and this is maintained through ventilation. But Node 3 is much more compressed than Node 2. It provides the same functionality as Node 2 in term of being the interconnecting element. It distributes power, fluids, and air. But it also has to accommodate the water processor, the galley, and waste management. It makes it really crowded. The air treatment for the labs reside mostly in Node 3. So the capability to regenerate and clean the air of the us segment is mainly in Node 3. Node 3 needed to regenerate the air and control the humidity. Without the systems in the U.S. lab and in Node 3 you cannot keep six or seven crew on board. The system has to be fully maintainable. One of the primary challenges for the systems which are manned, especially in low earth orbit is the maintenance. The system has to provide the functionality while providing easy access, easy replacement, and maintenance. Everything has to be planned before it enters the design phase. Every element has to be designed and checked on the ground to be maintainable and to be replaceable." - Walter Cugno

US Segment
Pressurized Mating Adapters (PMA)

The funnel-shaped Pressurized Mating Adapters (PMA) provide internal pressurized passageways for crew, equipment and supplies connecting berthing ports at their wide ends to androgynous docking ports at their narrow ends. The berthing ports are used for permanent or long term location of ISS modules. The docking ports are used in the Russian segment and for some logistics vehicles, such as the Space Shuttle.

PMA Configuration

Conical in outward appearance, the PMAs are pressurized and heated, providing an interior shirtsleeve environment. Each is fitted internally with hand rails to ease movement from one station element to another in microgravity. Airtight seals are created with adjoining modules at berthing and docking ports. Three PMAs were built by Boeing. Each of the three units are slightly different but they function in a similar way. The PMAs appear as truncated cones, shaped like a funnel, from the exterior but they are actually formed of multiple cylinders with a 24-inch axial offset in the diameters between each end. The structure are fabricated from 2219 aluminum alloy roll ring forgings that are welded together, creating a ring-stiffened shell structure. The Passive Common Berthing Mechanisms (PCBM) were built by Boeing and an Androgynous Peripheral Attach System (APAS) docking system was supplied by the Russians. EVA provisions include a Portable Foot Restraint (PFR), a top-mounted Worksite Inter Face (WIF) fixture, two Flight-Released Grapple Fixtures (FRGF), camera and laser targets, a number of Space Vision Systems (SVS) targets, and a number of handholds and hand rails.

PMA-1

PMA-1 was permanently integrated with Node 1 before launch. PMA-1 is the connection between the US and Russian segments of the station. It connects to the FGB 'Zarya' using an APAS and internal hatch arrangement. The other end of the PMA is attached to Node-1 (Unity) by a PCBM. PMA-1 was linked to the two elements on December 7, 1998 and has not been moved since.

PMA-2

PMA-2 was attached to Unity before launch and provided the initial docking interface for Shuttle missions with ISS. The end attached to Node-1 'Unity' was connected by a PCBM, while the other end used an APAS to connect with the Shuttle's docking system mounted on its external airlock. PMA-2 has a hatch with an 8-inch view port. Attached to the exterior of PMA-2 are MDM computers, which provided command and control of the Unity node. The MDMs allowed the Shuttle crew to control the Shuttle-end PMA-2 docking mechanism from the orbiter's aft flight deck. On 9 February, 2001, PMA-2 was relocated to the forward Common Berthing Mechanism (CBM) of the Z1 truss to allow the US Laboratory Module Destiny to be berthed to Node 1. Once the US Laboratory Module 'Destiny' was berthed, PMA-2 was moved to the US Laboratory Module's forward longitudinal docking port. On 12 Nov 2007, STS-120 delivered Node 2 'Harmony' to the ISS. PMA-2 was relocated to the forward berthing port of Node 2 using the Space Station Remote Manipulator System (SSRMS) 'Canadarm 2'. Two days later, the combined Node 2 and PMA-2 was berthed to the forward longitudinal berthing port of the US Laboratory Module. PMA-2 is planned to remain on the forward port of Node-2 until the completion of ISS operations.

PMA-3

Almost two years after delivery of the first two PMAs, STS-92 carried PMA-3 to the station. The third PMA was intended to provide a back-up docking port for Space Shuttles. On 13 Oct 2000 PMA-3 was berthed to Node-1's 'Unity' nadir port during STS-92. The first use of PMA-3 for a Shuttle docking was during mission STS-97 in December 2000. The following mission, STS-98, in February 2001 was the final Shuttle mission to dock to PMA-3. On 10 Mar 2001, PMA-3 was relocated to Node-1's port-side berthing port by the STS-102 crew, to make room for berthing of a Multi-Purpose Logistics Module (MPLM). On 30 Aug 2007, PMA-3 was moved from Node-1's 'Unity' port-side to Node-1's nadir port to allow the temporary berthing of Node 2 'Harmony' during mission STS-120 in October. On 25 Jan 2010, PMA-3 was moved from Node-1 'Unity' to the zenith port of Node-2 (Harmony) to allow berthing of Node 3 'Tranquility' during STS-130. On 16 Feb 2010, following the movement of the Cupola from Node-3's 'Tranquility' outer longitudinal port-side berthing port to the nadir port, PMA-3 was relocated to the outer port-side berthing port of Node 3. In March 2017 PMA-3 was relocated from the Tranquility module and attached to the zenith port on the Harmony module.

Closeup view of the forward docking port, Pressurized Mating Adapter-2 (PMA 2). PMA-2 served as the primary docking port for most Shuttle missions as well as for later commercial vehicles.

A view through the hatch porthole of Node-2, into Pressurized Mating Adapter (PMA-2). With his arm extended to left foreground is Kent V. Rominger. Clockwise from Rominger's position, are Umberto Guidoni, Scott E. Parazynski, Chris A. Hadfield, Jeffrey S. Ashby and John L. Phillips. Yuri V. Lonchakov's head emerges at bottom center.

The exterior of PMA-2. Much of the exterior is a black carbon composite designed to absorb solar heat. Yellow EVA handrails are visible around the perimeter.

US Segment
Joint Airlock 'Quest'

The attachment of the US-built Joint Airlock 'Quest' in July 2001, on the starboard side of Node 1 'Unity' enabled spacewalks or extravehicular activity (EVAs) to be performed from the ISS without the loss of environmental consumables, like breathing air, and without the need to have a Shuttle docked. Between December 1998 and the first use of the Joint Airlock 'Quest' in July 2001, twenty-three station-related spacewalks were conducted using airlocks of the docked Space Shuttles Endeavour, Atlantis or Discovery; another "internal EVA" was conducted in the node of the Service Module 'Zvezda' in June 2001 by the ISS Expedition 2 crew. In July, 2001, the Joint Airlock 'Quest' was delivered by Shuttle STS-104 and then two spacewalks, working in conjunction with the Space Station's Robotic Manipulator System (SSRMS), installed the Joint Airlock and high pressure nitrogen and oxygen tanks on the airlock's exterior. The first EVA through the airlock was by STS-104 astronauts on 20 July 2001 and verified installation and systems performance. The first station-based spacewalk using the Joint Airlock facility without a Shuttle being docked occurred on 20 February 2002, by ISS Expedition 4 crew members Carl Walz and Dan Bursch. The Joint Airlock is called "Joint" because it can support EVAs by astronauts wearing either the US Extravehicular Mobility Unit (EMU) suits or the Russian Orlan suits. However to date the Joint Airlock has primarily been used for spacewalks on the US segment of ISS using the US spacesuits.

The Joint Airlock 'Quest' consists of two cylindrical chambers attached by a connecting bulkhead and hatch. The two cylinders are of different dimensions. The innermost cylinder is berthed to the US Node 1 'Unity' and is a large diameter, similar to other modules of the US ISS segment, and long enough to accommodate a standard rack. Two racks house air and avionics equipment, while batteries, power tools and EVA supplies can be stowed in specially designed storage areas. This large diameter portion of the module is called the 'Equipment Lock' and was designed to service US Extravehicular Mobility Unit (EMU) spacesuits in orbit. It acts as a storage area for EMU hardware and as a staging area during the preparation for EVA. The Equipment Lock can be used as a 'camp out' facility, where EVA astronauts sleep the night prior to their EVA. While they sleep, the atmospheric pressure in the compartment is reduced from 14.7 psi to 10.2 psi. This purges nitrogen from their bloodstream and prevents decompression sickness (the bends) when the crew adjusts to the 4.3 psi pressure inside their suit. High pressure nitrogen and oxygen tanks are used for storage of atmospheric gases and are mounted on the exterior of the wide cylinder. The depress pump and pressure equalization valves located inside the hatches give the airlock its depressurization and re-pressurization capabilities. The 'Crew Lock' serves as the Joint Airlock's passage to and from space. The Crew Lock's cylinder is much narrower in diameter than the Equipment Lock, and is just wide enough to accommodate two or three space suited astronauts. In its side is an EVA hatch that provides access to open space. The Crew Lock is essentially the same design as the Space Shuttle's airlock, which is why the hatch to space is mounted on the Crew Lock's side rather than on the end. The Crew Lock provides lighting, handrails and an Umbilical Interface Assembly (UIA) which allows the space suits to obtain electrical power or breathing air from the spacecraft. The UIA supplies water, waste water return, oxygen, communications and spacesuit power for two suits. Prior to exiting the facility, the crew depressurizes the Crew Lock to 3 psi, then to zero psi, and then open the outer hatch for exit into space. The Joint Airlock was constructed from aluminium by Boeing at the Marshall Space Flight Center (MSFC) in Huntsville Alabama.

Node-1 'Unity' with the Joint Airlock 'Quest' berthed to its starboard side. The SSRMS 'Canadarm2' is mounted on Node-1. A Soyuz spacecraft, and beyond it the Russian Pirs airlock and docking compartment are on the nadir side of the ISS.

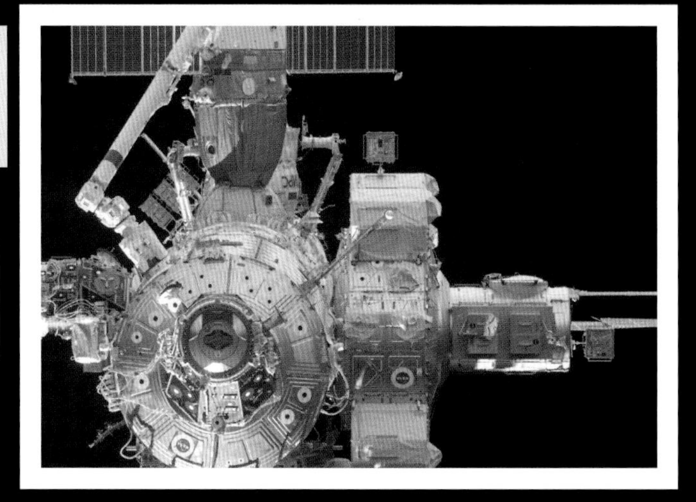

"Just a couple of months after the flight to emplace the Station arm, the next flight the U.S. [Joint] Airlock was going to be flown. Only the Canadarm2 could install the airlock where it needed to go." - Tim Braithwaite

Nadir side of the ISS shows Pressurized Mating Adapter-2, to the top, Node-1 'Unity' and the US Laboratory Module 'Destiny'. Berthed to Node-1 on the port side is Node-3 'Tranquility' and on the starboard side, the Joint Airlock 'Quest'. The Cupola is visible berthed to Node-3's nadir port. Pressurized Mating Adapter-3 is berthed to the port end of Node-3. The central portion of the truss is visible behind the pressurized modules.

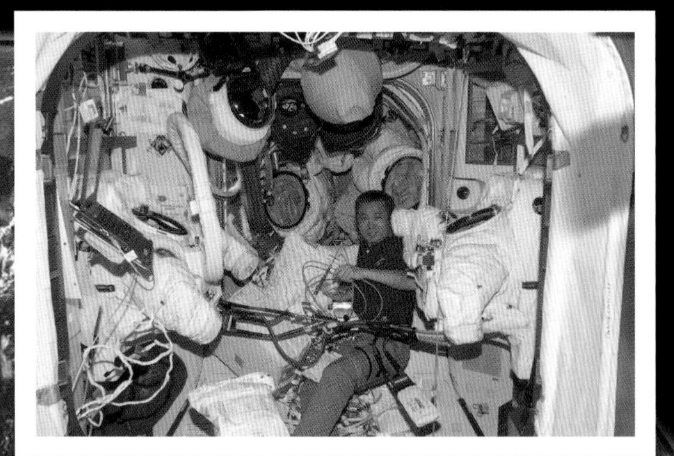

Koichi Wakata works in the Joint Airlock 'Quest'. Wakata was conducting maintenance in the wider diameter Equipment Lock with the hard upper torsos of two spacesuits to either side. Behind Wakata is a hatch into the narrower diameter crew lock in which is situated two of the lower portions of the spacesuits.

Rick Mastracchio (left) and Clayton Anderson, in their Extravehicular Mobility Unit (EMU) spacesuits, prepare to exit the Joint Airlock 'Quest' Crew Lock to enter space and begin a spacewalk.

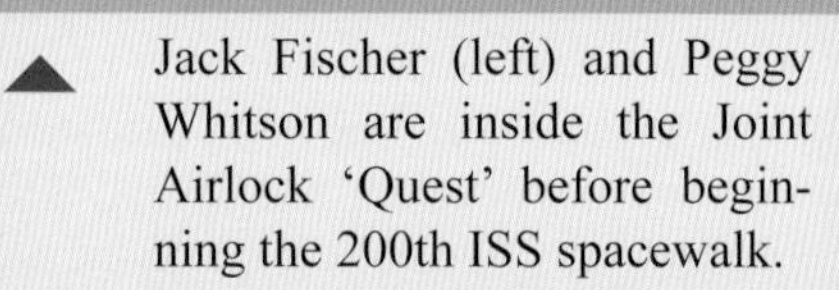

Jack Fischer (left) and Peggy Whitson are inside the Joint Airlock 'Quest' before beginning the 200th ISS spacewalk.

US/Joint Airlock 'Quest' Specifications

Launch:	12 July 2001
Launch Vehicle:	Space Shuttle Atlantis, mission STS 104
Mass:	9,923 kg (21,877 lb.)
Length:	5.5m (18 ft.)
Max diameter:	4.0 m (13.1 ft.
Pressurized Volume:	34 m3 (1200 cu ft.)

US Segment

Cupola

The Cupola serves as an observation deck and flight deck for the ISS crew, affording the visibility to support control of the Space Station RMS (SSRMS) 'Canadarm 2', as well as general external views of Earth, celestial objects, visiting vehicles and the space station itself. Because the cupola permits '3 dimensional immersion viewing' it provides a level of situational awareness that can be gained in no other way, except for perhaps by being outside during a spacewalk. Requirements and designs for the Cupola were initiated by NASA in the mid-1980s. The small module was initially called the 'Windowed Workstation' to contrast it with other computer workstations in which the crew would depend upon television views; later it was renamed 'Cupola', after the raised observation deck of a railroad caboose. Cupola is an architectural term and is Italian for 'Dome'. Initially the Cupola was expected to serve as the primary control and observation point for astronauts during assembly of the ISS while using the Space Station Remote Manipulator System (SSRMS) or conducting spacewalks. However the Cupola was repeatedly delayed, cancelled, then reintroduced to the program. This had the effect of delaying its availability until after assembly was all but complete. In a barter agreement, signed in 1999, ESA agreed to do the final design and assembly of the Cupola observation facility, previously planned and designed by NASA, in exchange for the launch and return of five ESA Laboratory 'Columbus' payloads on the Space Shuttle. While components of the Cupola, including windows, were manufactured by Boeing under contract to NASA, the final design and assembly of the Cupola was performed for ESA by their contractor, Alenia Spazio (now Thales Alenia Space) in Turin, Italy. The launch of the unit aboard STS-130 in early 2010 saw the final main element of the US segment of the station launched to the ISS. While the launch of the Cupola so late in the assembly phase of the ISS meant it would not serve its original operational purpose, the Cupola has proven of considerable benefit for the well-being of crew members, offering them views of the Earth and space that are normally only gained during EVAs. The quality and range of Earth observation photography, targets of opportunity, public outreach and education have all been enhanced by the addition of the Cupola and many astronauts have reported it is their favorite place on the ISS. The primary unit of the Cupola is constructed from a single forged and machined aluminum ingot. This provides superior structural integrity. It resembles a dome welded to a skirt. Its machining was completed in October 2002 at the Ratier-Figeac facility in Figeac, France.

A Passive Common Berthing Mechanism (PCBM) is bolted to the skirt and connects the Cupola to the port of Node 3 'Tranquiility'. A Micro-Meteoroid and Orbital Debris Protection System (MDPS) covers the outer surfaces of the Cupola. The windows include four glass pane layers and window heaters with thermostats. The circular top window is 80 cm (31.5 in) in diameter, making it the largest window ever flown in space. The six trapezoidal side windows are arranged around the dome structure. All the windows are composed of fused silica and borosilicate glass, using advanced technologies to ensure that they are protected from years of exposure to the space environment, solar radiation and orbital debris impacts. Each window features four panes. An inner scratch pane protects the two pressure panes from accidental damage by the crew from inside the Cupola. The two 25 mm (1 in) thick pressure panes help maintain both pressure and the environment inside the Cupola, with the outer pane as a back up to the inner one. Finally, a debris pane on the outside protects the pressure panes from space debris when the shutters are open. The shutters can be opened by the crew from inside the Cupola by a simple turn of the shutter control. When operations in the Cupola are completed, the shutters are closed to protect the windows and maintain a suitable temperature inside the facility. The Cupola cannot be returned to Earth for repair, but multiple redundant pane layers enable the capability to replace a window while it is in orbit. The scratch pane, debris pane or whole window can be replaced by the crew, although replacing a complete window could require an EVA to fit an external pressure cover over the window to be removed.

Several secondary structures in the Cupola include internal closure panels, equipment and harness support brackets and a crew system kit (seat tracks, handrails, handholds and tethers). Two, flight replaceable grapple fixture interfaces were used by the SSRMS for relocating the Cupola. The manually operated shutters for each of the windows also serve as elements of the MDPS. Two window change out covers are stowed in orbit to support any future on-orbit window assembly replacement. Operations in orbit include periodic work at the robotic work station, updating computer systems, portable lighting systems and foot restraints to support crew operations. Removable panels give access to some of the subsystems to support maintenance, repair or inspection, while retaining internal pressure and life support to allow shirtsleeve operations by the crew.

Koichi Wakata peers out of one of the windows of the Cupola on the Earth-orbiting International Space Station. A crewmate inside the Pirs docking module took the photo. 

Samantha Cristoforetti in the Cupola after capturing the SpaceX Dragon cargo craft using the Canadarm2 robotic arm. Dragon carried more than 2 tons of equipment, experiments and supplies for the Expedition 43 crew aboard the station.

The Cupola with its open window shades, mounted on the nadir side of Node-3 'Tranquility'. Node-3 is berthed to the port side of Node-1 'Unity'. To the upper right, the Permanent Multipurpose Module (PMM) 'Leonardo' is berthed to Node-1 nadir. Beyond the PMM, also on the nadir side of ISS, are the Rassvet with a docked Soyuz and then the Pirs with a docked Progress spacecraft.

"The Cupola project has certainly been a successful one. The favorite place for the ISS crew to be. Being able to work there, controlling the robotic arm to capture visiting vehicles provides the crew with great oversight and visual feedback during critical operations. It is also a great recreational place and it offers the possibility to view the Earth in all its greatness and fragility, and all what happens in cities, atmospheric/physical events (hurricanes, lighting, auroras, etc). Most of the greatest pictures of the Earth are taken from the Cupola. I would personally like to be there and spend a few hours." - Daniel Laurini

"We were working on the idea of computer workstations being used for controlling many of the functions on board the ISS. This was in the mid-1980s, 1985, 86, 87. I maintained the Man-Systems Architectural Control Document (ACD). Today it would be politically incorrect to call it "Man" but then we felt it was critical to highlight the significance of the human's role in the space station. We had already come up with the concept of standard racks for packaging all of the equipment inside of the modules and so some of these racks I called "command and control computer workstations". These were not laptop computers. Laptops did not yet exist as we know them today. They were full racks. A couple of our human-computer interface (HCI) specialists, we mainly had 2, and I got into a discussion with a couple of them about how would the astronauts control the robotic arm? How would they control any vehicles coming in to dock and bring supplies? How would they ensure astronauts during spacewalks were being adequately monitored? Our HCI specialists argued they could do this on computers and supplement that with TV screens. The idea of putting TV views on a computer was something that had not yet happened. I said I felt that the astronauts would really want to see these kinds of mission events with their own eyes. A lot of people in our Division, the Man-Systems Division, felt likewise and joined my campaign. A lot of astronauts supported this position too. We got letters of support from the Chief of the astronaut office and the astronaut assigned to Station. So I wrote the requirements. I really modeled it after the aft flight deck of the Shuttle. I got the Program Manager's approval. Initially I called it the "Windowed Workstation". But a couple of us thought that was too cumbersome. A friend of mine and I were avid model railroaders. I asked what those windows were on the top of a caboose. It turned out it was called the Cupola, so I rewrote the ACD and I renamed it the Cupola. That was probably in 1987. Our big concern was that when the Phase C/D bids were received from the contractors bidding for Johnson Space Center and Marshall Space Center work, that they included the Cupola. We were afraid they might try to drop it. They didn't." - Gary Kitmacher

The Cupola supports two astronauts operating side-by-side. The Cupola's internal layout is dominated by upper and lower handrails inside the cabin and the facility has a limited volume for the crew, who must orchestrate their movement into and inside of the Cupola. Larger versions of the Cupola had been considered early in the design process but the smaller hexagonal layout version which permits only the astronaut's upper torso inside was selected because of mass and cost considerations. Many of the Cupola's systems are parasitic and dependent upon the adjacent Node.

Orbital Activities:

8 Feb 2010: Launched on STS-130

11 Feb 2010: Cupola attached to Tranquility (Node 3) and installed on ISS on port side of Unity

Feb14 2010: Cupola relocated from Tranquility forward port to Earth facing (nadir) port.

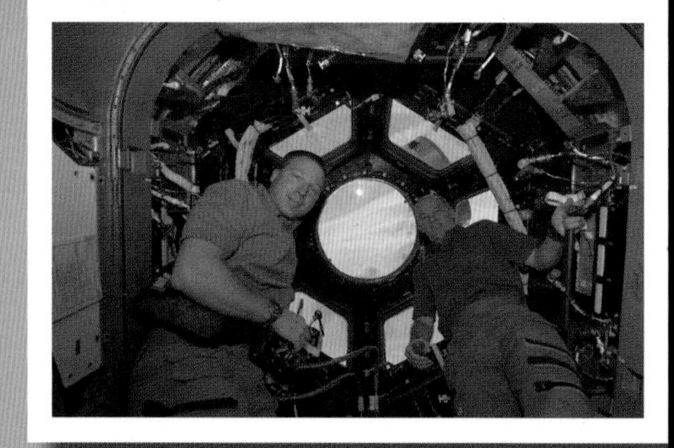

Terry Virts (left) and Jeffrey Williams inside the Cupola shortly after its installation. Many of the installed control computers and fixtures for the mechanisms for opening and closing the window shades can also be seen.

Cupola Specifications

Launch:	8 February 2010
Launch Vehicle:	Space Shuttle Endeavour, mission STS-130
Launch Site:	Pad 39A, Kennedy Space Center, Florida, USA
Mass:	1880 kg (4,136 lb.)
Length:	3m (9.8 ft.)
Height:	1.5 m (4.7 ft.)
Max diameter:	2.95 m (9.8 ft.) inclusive of the Micro-meteoroid and orbital Debris Protection System (MDPS) with window shutters closed and the flight reusable grapple fixture
Capacity:	2 crewmembers with portable laptops
Windows:	6 sides, 1 top

"Nothing like the cupola had ever been done before. The key was the technology, because cupola is made from a single casting with bolted on parts. There is no welding; it is completely machined. We had to have the right design and the right technology to then do the machining. We learned how to do seals on the machined cylinder and on the berthing port. The port is also machined from a plate which starts from 3.5 tons, and then ends as a 50 kilogram end product. In the future we might improve on the technology and develop an alternative solution for the window, because this is technology underwritten and owned completely by one U.S. company. So we might develop an alternative solution or technology for the window, and also perhaps look for technology to develop an inflatable cupola. The capability of cupola helps the astronauts psychologically. It also is important in operations to see the real view of what's happening outside as an incoming vehicle approaches." - Walter Cugno

"NASA was developing the cupolas and had already procured the glass. NASA got as far as the critical design review. When the agreement was made for ESA to provide the cupolas, Alenia in Torino would be the prime contractor and they would start from square one they had to be responsible for the design. All the interfaces had to be identical. The design or configuration of the cupola with the seven windows was a given to ESA. The glass is flat, about three inches thick, three panes of glass. It cannot be curved and still have optical qualities, so the glass is flat, but the overall shape of the cupola is as close as possible to being spherical, which is what you want from a structural point of view. Its optimal for a pressurized vehicle. So it was a hexagon. In a square or a rectangle the corners would be too highly stressed. 25 flat windows would be better, but the windows would be too small. So this was the compromise. Two cupolas would have been better than one. But it was part of the descoping that they were only going to do one. The cupola in its position underneath the ISS, has limited viewing of the ISS elements. The second cupola would have overcome that. NASA knew that the cupola would be the most popular place in the station." - Alan Thirkettle

"The Cupola has been an incredible success story. At one point there were 2 Cupolas in the program. Then the Cupolas were cancelled and then we brought them back, but then cancelled one of them. Then when the Shuttle flights were reduced, when the Shuttle was going to be retired, the Cupola was strongly considered for cancellation again. People did not see the benefit of it. The main operational benefit that was recognized was to be able to help berthing the HTV. Some did not think we could do without it. But almost nobody anticipated the impact the Cupola would have once it was up there. By having that view of the world, even though operationally maybe it was not required, to my mind it changed the whole station for the whole world. Think about it. Every science fiction movie you've ever seen has always had a place with these big windows to look out into deep space. I can't imagine a deep space vehicle not having some means for the crew to look outside. Inside of a can for 9 months would be very tough for them. For the astronaut's psych support, it's been an incredible boost to their morale to be able to look out." - Dan Jacobs

US Segment
Permanent Multipurpose Module (PMM) 'Leonardo'

Formerly known as the Multipurpose Logistics Module (MPLM) 'Leonardo', the re-fitted Permanent Multipurpose Module (PMM) 'Leonardo' was delivered to the station by the final mission of the Discovery (STS-133) in March 2011 and permanently attached to Node 1 'Unity'. In space pressurized volume is always at a premium. This is never more so than on a growing and developing space station. Prior to the retirement of the Space Shuttle, there was an opportunity to attach one of the three MPLMs to the station permanently in order to create an additional 70 m^3 (2472 cu ft.) of stowage volume. The primary contractor of the three MPLM, Thales Alenia Space, under contract to the Italian Space Agency – ASI, was responsible for the modifications made to the MPLM to allow for permanent attachment to the space station. As MPLM Leonardo had flown seven round trips to the station, this was the unit chosen to be the PMM. Its eighth and final trip would be one way. The module resembles the European Columbus laboratory in dimensions. The MPLMs had been built first and Columbus used the MPLM design as its basis. For permanent use on ISS, the MPLM was modified to allow for an expected ten year life attached to the station. The improvements included:

1. Enhancement to shielding with the addition of special micrometeorite debris protection to meet more stringent penetration requirements.
2. Provision of on-orbit maintenance capability, featuring re-routing of the internal electrical harnesses and brackets to provide easier access by crew members.
3. Installing interfaces to make it easier to use the retained resources in the module
4. Provide a life extension certification for all the equipment and subsystems in the PMM
5. Updating software to eliminate erroneous alarms.

The PMM was launched to the ISS on the last launch of the Shuttle. Besides providing the additional internal stowage volume, another reason was to provide as much maintenance spares, logistics and supplies as possible on ISS. So, on the STS-133 mission the module was filled with 12,850 kg (28,353 lb.) of payload, including a variety of racks and experiments.

The Space Station Remote Manipulator System (SSRMS) Canadarm2 grasps the Italian-built Multi-Purpose Logistics Module Leonardo on 14 July 2006.

Permanent Multipurpose Module (PMM) 'Leonardo' Specifications

Length:	6.6m (21.7 ft.)
Diameter:	4.5 m (15 ft.)
Mass (structure):	9900 kg (21,817 lb.)
Pressurized Habitable Volume:	70 m3 (2,472 cu ft.)
Windows:	None

Cady Coleman poses with three Extravehicular Mobility Unit (EMU) spacesuits in the Permanent Multipurpose Module (PMM) of the International Space Station.

Daniel W. Bursch floats inside Leonardo Multi-Purpose Logistics Module built to serve as pressurized, reusable cargo carriers and was later modified for permanent installation on ISS.

US Segment
US Laboratory Module 'Destiny'

The US Laboratory Module 'Destiny' is one of the primary work and habitation locations for astronauts in the US segment of ISS. The module has been described as a world class, state-of-the-art, research facility in a microgravity environment. It provides the ISS crew with a shirtsleeve environment for research in life sciences, microgravity sciences, Earth sciences and space science research from a program of high quality experiments. The US Laboratory Module is fabricated of aluminium. It consists of three cylindrical sections and two end cones, with Common Berthing Mechanisms (CBMs) and hatches in the end cones. The module's aluminum cylindrical structure exterior is machined with a 'waffle' isogrid pattern that strengthens the hull while minimizing weight. Insulation blankets and Kevlar layers are installed over the aluminium shell to protect the module from the temperature extremes and space debris or micro-meteoroids. Outer aluminum debris shield panels provide an additional layer of protection and reduces the load on the air conditioning subsystem by reflecting sunlight. A single, 50.9 cm (20 in) diameter round window is located on the floor or nadir side of the US Laboratory module. The window is made of high quality optical glass. The window is designed to be used for Earth observation experiments. A rack, called the Window Observation Research Facility (WORF) is often located in front of the window and can be used for remotely operating cameras or other observation equipment looking through the window. Outside of the module pressure shell, a shutter protects and shields the window from orbital debris damage during the life of the station. The shutter must be manually opened by the crew. Common Berthing Mechanisms (CBMs) are located in each end cone of the US Laboratory Module. The aft Passive CBM (PCBM) connects the US Laboratory directly to the Node-1 Unity module and was planned as a permanent connection. The hatches between the US Laboratory and Node 1 normally remain open unless there is a situation such as a pressure leak or contamination that requires isolation of either of the modules. Pressurized Mating Adapter-2 (PMA-2) was attached to the laboratory's Active CBM (ACBM) on the forward port and served as the primary Space Shuttle docking interface until the arrival of Node 2 (Harmony). After the arrival of Node-2, PMA-2 was moved to the forward end cone of Node 2. Both hatches at either end of the US Laboratory Module have viewing ports. The windows can be covered or uncovered from either side.

US Laboratory Module 'Destiny' Systems:
STAND-OFFS AND STANDARD RACKS

Within the cylindrical structure of the US Laboratory and other US segment modules, four 'stand-off' structures at upper and lower 'corners' support utility runs for the electrical power, data management systems, vacuum systems, cooling, water and air conditioning systems. The stand-offs support the location and attachment of Standard Equipment Racks (SIR). The racks themselves are typically 185.4 cm (73 in) tall and 106.7 cm (42 in) wide. Racks can have a mass of up to 700 kg (1550 lb.). Each rack is about the size of a household refrigerator, and most are fabricated from a graphite composite shell. Power, cooling, video and data interfaces are standardized at most rack locations. The racks are integrated for their respective purposes. Some are designed for passive stowage; others hold experiment facilities. Some serve as the structure for mounting functioning systems hardware. The racks are designed to rapidly swing away from the module pressure shell in the event of a penetration by orbital debris or micrometeoroid penetration. In this case the crew would be able to apply a patch to seal any leak. Many rack locations have standard interfaces to systems and the racks can easily be disconnected in one location, moved to another location in the same or a different module, and reconnected in the new location. All of the modules in the US segment have a similar design from the perspective of being able to host standard racks. The racks are modular by design and so can frequently be moved from one module to another in order to optimize layout and use of a module. Racks are oriented to form a floor, ceiling and two walls within each module. Further, the floor of most modules is usually oriented towards the Earth. The two upper corner stand-offs contain light fixtures for general lighting, and airflow registers for blowing fresh air into the module. Lower stand-offs have air intakes. Launched with five systems racks providing life-sustaining support, including electrical power, cooling water, air revitalization and temperature and humidity control. Six more racks were supplied by MPLM on STS-102 and a further 12 (also delivered by MPLM) are used for scientific experiments. In total, the US Laboratory 'Destiny' can house 23 racks on all four internal faces, six each on the port, starboard and overhead and five on the deck of the module. As other modules of the US segment were subsequently placed in orbit, racks were moved from the US Laboratory Module to the other modules.

US Laboratory Module 'Destiny' Specifications

Launch date:	7 February 2001
Launch Vehicle:	US Space Shuttle Atlantis, mission STS-98
Launch Site:	Pad 39A. Kennedy Space Center, Florida, USA
Launch Mass:	14,515 kg (32,000 lbs.)
Orbital Mass:	with racks and outfitting, the mass increases to 24,023 kg (52,962 lbs.)
Length:	8.5 m (28 ft.)
Width:	4.3 m (14 ft.) diameter
Pressurized Volume:	106 m3 (3750 cu ft.)
Windows:	1, 50.9 cm (20 in) diameter

"I was asked to be the deputy launch package manager for the U.S. Laboratory Module. I was very excited to work on the U.S. Laboratory Module, it is the heart of Space Station from both a systems perspective as well as from the science perspective of being a laboratory. After about a year I was selected to manage the launch package for the U.S. Laboratory until launch. It was an integration role across the program, but also across [NASA] Centers. I came on there in '97, and it launched in 2001. The requirements had been defined, it was in the final build integration test phase, the interfaces to the other modules as well as the interfaces of all the systems and hardware within it. My role with Destiny ended when it was launched and integrated. We worked real-time ops [operations] just during the actual flight until it was integrated onto Space Station, then my role moved to later flights." - Suzan C. Voss

A view of the nadir and forward sides of the US Laboratory 'Destiny'. The SSRMS 'Canadarm2' is mounted on the US Laboratory. Pressurized Mating Adapter-2 is berthed to the forward US Laboratory port and Node-1 'Unity' to the US Lab berthing port. The cover over the US Lab's large scientific window is also visible.

Stephanie Wilson works the controls of the Space Station Remote Manipulator System (SSRMS) 'Canadarm2' in the US Laboratory Module 'Destiny'.

Greg Chamitoff on Expedition 17 looks over a checklist in the US Laboratory 'Destiny' of the International Space Station.

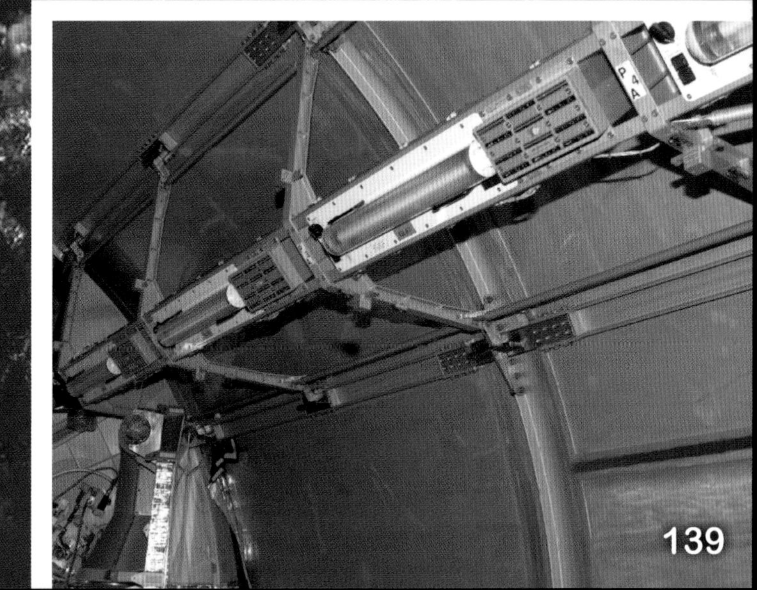

Photographic documentation of the module standoffs showing secondary structure attachment to primary structure cylinder, utility runs, air vents and luminieres. Stand Off seat track overall on the aft port side.

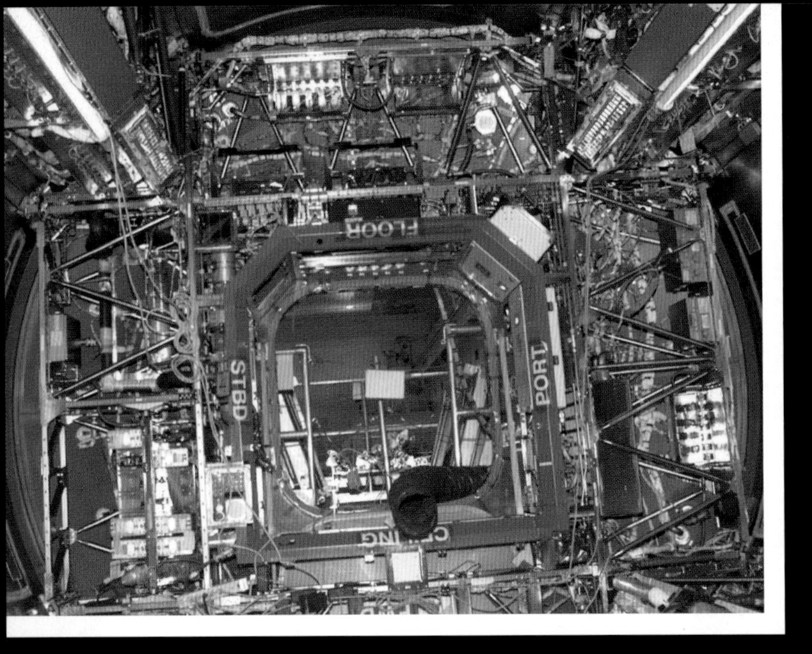

Looking forward inside of the US Laboratory Module showing the installation of utility runs and system components prior to the placement of Standard Racks.

James S. Voss works with a series of cables in front of Express Rack #1 in the U.S. Laboratory. EXPRESS (Expedite the Processing of Experiments for Space Station) were designed to mount Space Shuttle sized stowage lockers. After 20 years of Shuttle missions, many payloads and facilities had been designed to fit inside the standard 2 cu ft (.6 cu m) Shuttle lockers.

Robert Thirsk (right) and Frank De Winne of Expedition 20 relocate the Crew Health Care System (CHeCS) rack from the US Laboratory 'Destiny' to the Japanese Laboratory 'Kibo'.

Mikhail Tyurin (foreground) and Thomas Reiter work with the Direct Current-to-Current Converter Unit (DDCU-2) rack.

European Space Agency (ESA) Laboratory 'Columbus'

The European Space Agency (ESA) Laboratory Module takes its name from the famous Genoa navigator, Christopher Columbus, who conducted notable voyages from Europe to the Americas between 1492 and 1504. The ESA Laboratory Module supports experiment facilities both inside and outside the pressurized structure; it offers the ESA sponsored researchers the opportunity to conduct research in the fields of life sciences, fluid physics and other disciplines. Columbus extends and enhances science operations beyond those conducted during the Spacelab Shuttle missions and on the Russian Salyut and Mir orbital stations in the 1980s and 1990s. Although originally intended to be longer, in order to reduce expenses, the ESA Laboratory Module was reduced in length and its structure based on the Italian built Multi-Purpose Logistics Modules (MPLM). In order to reduce launch costs, the shortened module was able to accommodate 2500 kg (5510 lb.) of experiment facilities and other hardware during launch, rather than requiring a later logistics launch on another Space Shuttle mission. Columbus has a cylindrical hull with end cones. The primary and secondary structures were constructed from aluminium. The pressure shell was then covered with a multi-layer insulation blanket for thermal protection. Layers of Kevlar and Nextel protect the pressure shell from orbital debris or meteoroid penetration. On top of these layers, aluminium alloy panels provide additional protection.

Although the ESA Laboratory Module is the station's smallest laboratory, it offers similar payload volume, power and data services as in the other ISS laboratory modules by placing most subsystem equipment within the end cones. Located in the central area of the starboard cone are the subsystems for video monitors and cameras, switching panels, audio terminals and fire extinguishers. When fully operational and outfitted, Columbus has sufficient volume to support three station crew members during their research studies. The same rack designs are used throughout the US segment of ISS for standardization, allowing interchangeability throughout the US segment. The racks are arranged in a floor, ceiling and wall orientation, similar to most other modules of the US segment. The Columbus laboratory can accommodate 16 standard racks. Three racks inside the ESA Laboratory module are dedicated to laboratory subsystems, including water pumps, heat exchangers and avionics. The remaining three racks are available for general storage purposes. Ten of the sixteen racks in Columbus are the International Standard Payloads Racks (ISPR). Experiment racks designed for Columbus are intended as multi-user facilities incorporating a high degree of autonomy, maximizing the limited research time available to the astronauts on orbit. Some 13.5 kW of the 20 kW of electricity is available for experimental research facilities. In addition to the internal facilities, Columbus features four external mounting points for experiment payloads. These can be related to applications in space science, Earth observation, and technology and innovation. The crew can control the temperature and humidity inside the module (16-27° C). For a comfortable environment inside the module, continuous air flow ventilation is provided by sucking in air from the Node 2, to which Columbus is attached, then returning 'stale' air back into Node 2 for refreshing and removal of carbon dioxide. Contamination levels inside the module are monitored by various subsystems. A water loop system, which is connected to the ISS heat removal systems, supports the removal of heat from all experimental facilities to prevent overheating. Cabin air condensation is removed by an air/water heat exchanger and there is a system of electrical heaters available for those times when the station's attitude induces extremely cold temperatures. Management of ESA research on ISS is the responsibility of the ESA Columbus Control Center (Col-CC), located in Oberpfaffenhofen, Germany, in close cooperation with both MCC-Houston and Moscow and with the Payload Operations and Integration Center (POIC) at NASA's Marshall Space Flight Center in Huntsville, Alabama. The latter retains overall responsibility for ISS experiment payloads.

The ESA Module 'Columbus' is berthed to Node 2 'Harmony' at bottom left. The newly installed Japanese Logistics Module - Pressurized Section (JLP) is visible at top center on the zenith berthing port of Node 2. ▶

Hans Schlegel installs cables connecting the newly installed ESA Laboratory Columbus to Node 2 Harmony, during STS-122.

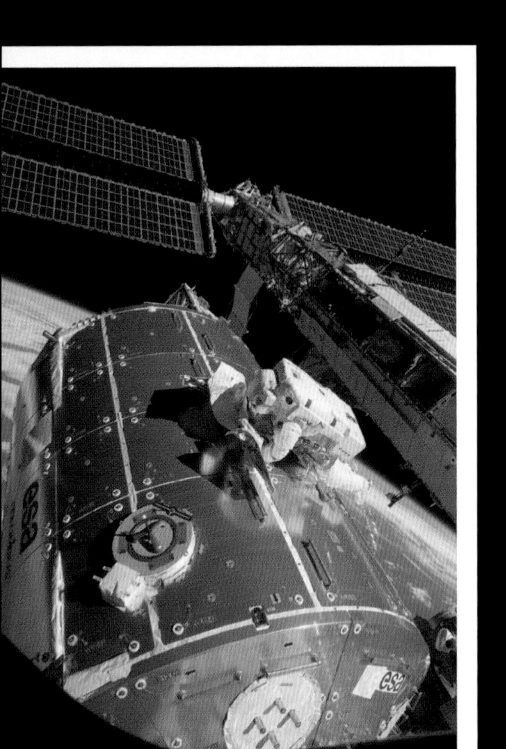

Hans Schlegel from Shuttle STS-122, works on the ESA Laboratory Columbus during a spacewalk on 13 Feb. 2008.

The ESA Laboratory is berthed to the starboard port of Node 2 'Harmony'. Attached to Node 2 and below the station is the SS-RMS 'Canadarm2'.

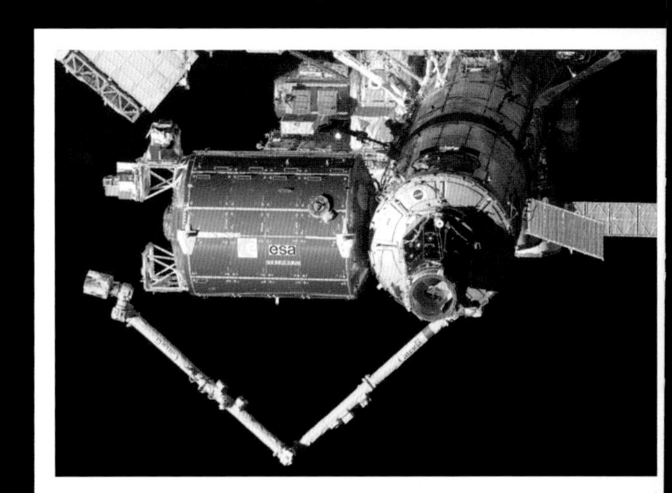

Catherine (Cady) Coleman uses a still camera to photograph Paolo Nespoli in the ESA Laboratory 'Columbus'.

European Space Agency Laboratory 'Columbus' Specifications

Launch:	7 February 2008; attached 11 February
Launch Vehicle:	Space Shuttle Atlantis, mission STS-112
Launch Site:	Pad 39A, Kennedy Space Center, Florida, USA
Mass:	Total weight of Element 21,000 kg (46,300 lb.) max
Mass without payload:	10,300 kg (22,700 lb.)
Launch mass:	12,800 kg (28,220 lb.), of which 2500 kg (5510 lb.) was payload.
Length:	6.8 m (22.6 ft.)
Max diameter:	4.5 m (14.7 ft.)
Pressurized Volume:	75 m³ (2650 cu ft.);
Volume of payload racks;	25 m³ (885 cu ft.)
Cabin temperature:	16-27°C
Air pressure:	959-1013 hPa
Total power:	20 kW provided by the station.
Payload power:	13.5 kW

"Maintenance is an integral part of the design philosophy for the modules of the space station. On Spacelab, maintenance was not a consideration. Maintaining avionics, air conditioning and general life support things like color schemes, ceilings and floors, acoustics attenuation, making sure that the environment inside the module is soft; we learned a lot about all of these things, these technologies evolve, and so you learn about it the first time you do it, and the next time you do it, you've got to start again because the technology has moved on and it is different. Columbus was designed to be maintained -- now that you haven't got the capability of even ATV-- there's no more ATV flights-- if a rack fails or there's a serious problem on board Columbus, the rack can be repaired or replaced or upgraded. There is a difficulty in replacing the rack because the ATV cannot carry a rack for replacement. It has racks inside but they are not transportable through the Russian docking mechanism because the docking mechanism is not large enough to get a rack through. It is also much more crowded in the Russian segment so that a rack cannot be carried through. But the Japanese HTV, is similar to ATV, but it can carry racks. The racks go from HTV through the American segment to the port to Columbus. The SpaceX dragon can also carry racks. And the Orbital Sciences' Cygnus can carry racks as well. So the logistics modules that service the station can all carry racks. The ATV was the barter offset for paying common operations costs for the ISS. ESA would pay for power, and operations, by providing an ATV. One ATV flight corresponded to 18 months of common operations costs. But after we had flown five ATV'S, industry said, "We want design and development work, not reproduction work. We don't learn anything by building more of the same. We want to do new things." The Americans were developing their new launcher, the SLS with the Orion moon and Mars capsule. The US said, "The resource module part of the ATV could be used as the resource module for the Orion spacecraft. So if you'd like to do that, we can count that for common operations costs." So ESA'S industry is now doing the development of the Orion resource module." - Alan Thirkettle

"Columbus was finally ready. I was selected to integrate Columbus onto the ISS. All of a sudden I had a flight. I felt myself lucky and I was happily training with my crew. I was selected and trained to do spacewalks, EVAs, to integrate the Columbus module. That got me a lot of training responsibilities. We did a lot of reviews of hardware at the Cape during integration. We were looking at space flight hardware which would be integrated onto the ISS later. That was a completely new kind of responsibility, integrating hardware onto ISS after launch. As we approached the Columbus launch, I began to think about how I would do. My first flight was 15 years earlier. How much of the experience from 15 years ago could I use now?' Going EVA, I wondered 'how will I perform?' These questions and concerns helped me, or bothered me, and then I was in space. I realized then that it was like driving, or like riding a bicycle, or like swimming. These come back to you. It is very natural. So, in the first day or two I got out of that launching rocket and into a Space Station mindset. The moment the hatch in the airlock was open and I looked outside, I said 'wow, now this is just beautiful; it is wonderful'. I realized that for me this was the most rewarding part of space flight. Training in the Neutral Buoyancy Lab (NBL), and having seen the views in virtual reality, I really felt at home. We completed our activities in the Shuttle's payload bay, and then it was time to move over to the Station. Everything was going pretty well. Alan Poindexter, our pilot on the Shuttle said, "Hans in five minutes we are flying over Germany, so prepare to take some time." I hadn't thought about it before; the day was fairly stable, meteorologically, which only happens if there are easterly winds. The whole middle of Europe was free of clouds; so I looked out. I saw shades of green and brownish fields in spring not yet showing vegetation. I saw, there was a river; this was exactly the winding of the River Rhine at Cologne. There's the bridge I walked over all my childhood to school. There is the railway station, it's a huge railway station. And next to it that must be the dome of the cathedral; over the east, and that must be Arnhem, where I studied; that must be Hamburg. Between Arnhem and Cologne was about an hour drive by car. It's about 75, 80 kilometers; but with the Space Shuttle the trip took 10 seconds. You just barely look, you focus your eyes, another three seconds for orientation and then you recognize something and then the next and the next. Ten seconds have passed and I realize where I am. My hairs stood up on my neck. As I flew over Germany; I could imagine my family; maybe they were thinking about me. Maybe they were looking up. It was the most wonderful personal moment of my flight." - Hans Schlegel

US Segment

Japanese Experiment Module (JEM) 'Kibo'

The Japanese Experiment Module (JEM) is known as 'Kibo' (Hope). It is comprised of several major elements:

- JEM Pressurized Module
- JEM Experiment Logistics Module- Pressurized Section (ELM-PS)
- JEM Exposed Facility
- JEM Exposed Facility Logistics Element
- JEM Remote Manipulator System (RMS)

JEM Pressurized Module

The internal working environment of the JEM is maintained in a shirtsleeve condition. As in other modules of the US segment, much of the internal hardware of the JEM is located in standard racks. The racks in the JEM are arranged along the floor, ceiling and two walls. The JEM accommodates 23 racks. Six racks line three walls with five racks on the fourth wall. The racks are modular with similar interfaces and can be moved from module-to-module within the US segment, installed, detached and replaced on orbit. Thirteen of the racks are normally dedicated to JEM storage and systems. Eleven of the JEM racks support systems for power distribution, communications, air conditioning and thermal control.

The Japanese Experiment Module (JEM) Laboratory 'Kibo' berthed to the port side of the Harmony node of the International Space Station is photographed from Space Shuttle Discovery as the two spacecraft begin their relative separation. The Japanese logistics module is visible on the top berthing port of the JEM. The ESA Laboratory 'Columbus' is berthed to the port side of Node-2. The Canadian-built Special Purpose Dexterous Manipulator (SPDM) 'Dextre' is just below the JEM. Two Russian spacecraft are docked with the station, below Node 2. A portion of the truss is above and beyond the assembly of pressurized modules. This picture was taken June 11, 2008.

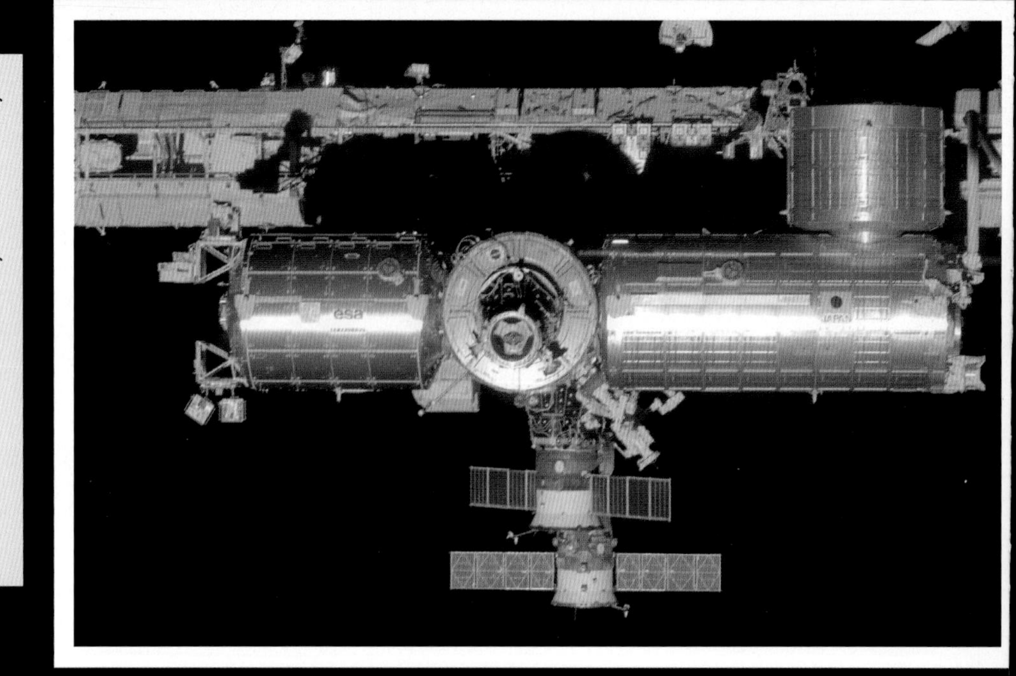

The remaining ten racks are International Standard Payload Racks (ISPR), used for experiments. Primary use of the JEM science racks is for life and material sciences. Two of the ISPR Experiment Racks are used for systems that support scientific experiments. One is an ambient stowage rack and the other is a refrigerated stowage rack. As with all elements launched on the Shuttle, the structural body of the PM had to be designed to withstand the launch and ascent loads. To protect the pressurized environment inside the aluminum structure, orbital debris shields cover the pressurized cylinder. There are two windows on the side of the EF docking port for external views.

JEM Airlock

An equipment airlock is situated at the end of the JEM opposite the berthing port linking the JEM to Node 2. The airlock allows experiments to be transferred from inside the pressurized module to be placed on the JEM Exposed Facility. Experiments being placed on the Exposed Facility can be manipulated by the JEM RMS. The JEM airlock also has a system for launching small satellites. The use of the equipment airlock in conjunction with the JEM RMS alleviates the need for spacewalks to support external payloads. This airlock was designed for transferring equipment, supplies or experiments between the pressurized module and the exposed facility, without the need for astronauts to perform EVA. Astronauts cannot use the JEM Airlock as it is too small for the spacesuits with the attached backpacks. Cylindrical in construction, the airlock measures 2.0 m (6.5 ft.) long, with the diameter varying between 1.7 m (5.6 ft.) on the outside and 1.4 m (4.6 ft.) on the inner side within the pressurized module. There is an inner hatch on the PM side, with an observation window included. Equipment and experiments to be exposed to space can be fixed to a sliding table, which is sealed inside the unit before opening the outer hatch. The table can then be extended out into the space environment. The table measures 0.64 x 0.83 x 0.80 m (2.1 x 2.7 x 2.6 ft.), with capacity to carry a maximum of 300 kg (660 lb.) mass to transfer.

JEM Experiment Logistics Module – Pressurized Section (ELM-PS):

Located on top of the PM, this element was launched by the Space Shuttle carrying payload racks. The ELM-PS now acts as a stowage facility and is used to stow maintenance tools, payloads and samples, and spare items. It has a smaller internal volume than the PM, with a capacity for up to eight racks. The ELM-PS distributes 3kW 120V DC power. Its internal environment is maintained between 18.3 to 29.4° C and a humidity of 25-70%.

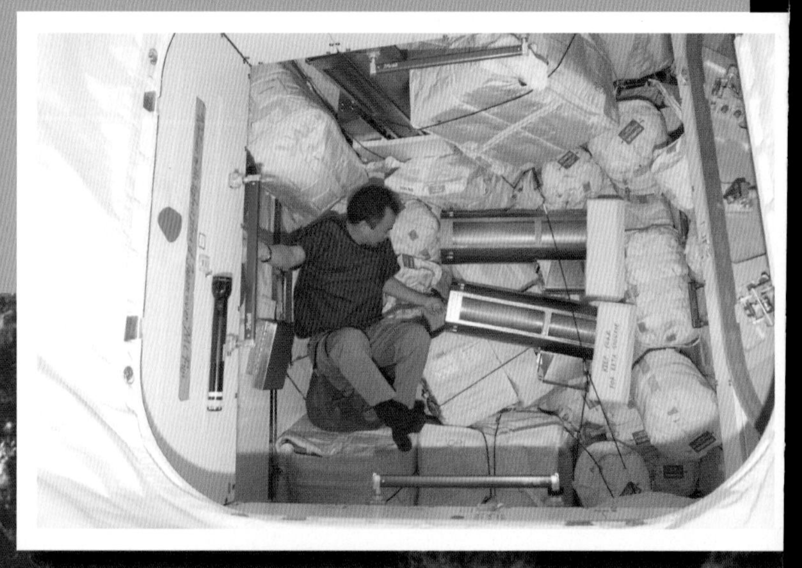

Soichi Noguchi works among cargo transfer bags (CTBs) in the Japanese Logistics Module - Pressurized Section (JLP).

JEM Exposed Facility – (EF):

The JEM-EF is an external experiment platform attached to the PM. The EF provides electrical, thermal control, and data system connections. Modular payloads designed for mounting on the EF can easily be exchanged. The EF provides a maximum of 11 kW of power (of which 1 kW is for systems and 10 kW for payloads) and 120V DC. The data management system features a 16-bit computer system and a 100 Mbps maximum high speed data link. There are no requirements for environmental control. Equipment Exchange Units (EF-EEU) are used to attach the payloads to the EF. A maximum of 12 payloads can be accommodated. Two locations are reserved for supporting systems and temporary storage. Each standard payload envelope measures 1.85 m x 1.0 m x 0.8 m (6.1 x 3.3 x 2.6 ft.), and can accommodate a mass of 500 kg (1100 lb.). Individual payloads provide a Payload Interface Unit (PIU) that plugs into power and data interfaces and a grapple fixture by which the JEM-RMS handles the payload.

▲ The Japanese Experiment Module - Exposed Facility photographed by an Expedition 32 crew member through a window of the Japanese Experiment Module (JEM) 'Kibo' Laboratory.

Experiment Logistic Module- Exposed Section (ELM-ES):

The Experiment Logistic Module – Exposed Section (ELM-ES) assists in transferring system and payload replacement units to the Exposed Facility. Payloads can also be temporarily stored during the exchange after an investigation is completed. Payloads can be removed, transferred and retrieved for storage using the JEM-RMS. Electrical power can be supplied to each payload on the exposed facility at a maximum of 1.0 kW, 120V DC. A variety of scientific experiments that require microgravity and vacuum exposure have been developed for use on the EF. Research has focused on Earth observations, communications, engineering, materials science and other disciplines.

Japanese Experiment Module Remote Manipulator System (JEM-RMS):

The JEM-RMS is a robotic arm system, developed in Japan to support operations on the JEM. It has been used in conjunction with the Space Station Remote Manipulator Systems (SSRMS) Canadarm 2 system and the Dexterous Manipulator Dextre unit. In August 1997, a Manipulator Flight Demonstrator (MFD) prototype test model was tested on Shuttle mission STS-85 and verified some of the systems and functions intended for the JEM-RMS. A second prototype unit was launched aboard Engineering Test Satellite VII (ETS-VII) in November 1997. It further evaluated the system and helped the Japanese to develop basic robotics techniques.

Kibo element delivery at ISS

Unlike the ESA and US Laboratory Modules which both grew smaller after their initial design in the late 1980s, the Japanese Experiment Module Kibo elements remained reasonably constant both in terms of the quantity, size and mass. Three missions were required to lift the JEM as it was too large and heavy to launch on a single Shuttle mission. STS-123 in March 2008 delivered the Kibo Experiment Logistics Module-Pressurized Section (ELM-PS) and initially berthed it to the zenith port of Node 2 'Harmony'. STS-124, in June 2008, delivered the main JEM Pressurized Module (PM) and the JEM Remote Manipulator System (RMS) robotic arm; once the JEM-PM was berthed in place, the JEM-ELM-PS was moved to the zenith port of the JEM PS. STS-127 in July 2009, delivered the Exposed Facility and Experiment Logistics Module – Exposed Section.

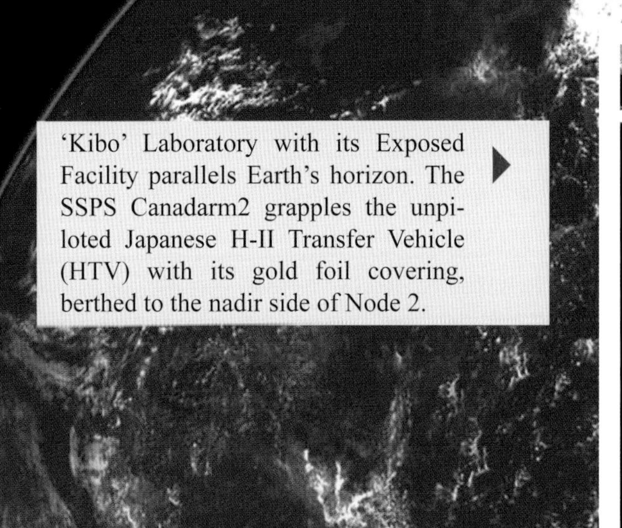

'Kibo' Laboratory with its Exposed Facility parallels Earth's horizon. The SSPS Canadarm2 grapples the unpiloted Japanese H-II Transfer Vehicle (HTV) with its gold foil covering, berthed to the nadir side of Node 2. ▶

Japanese Laboratory Module 'Kibo' Specifications

Launches:	
11 Mar 2008 STS-123; Space Shuttle Atlantis (STS-123)	
1 Jun 2008 STS-124; Discovery (STS-124)	
15 Jul 2009 on STS-127; Endeavour (STS-127	
Mass:	
JEM:	15.9 tons (35,050 lb.)
ELM-PS:	4.2 tons (9260 lb.)
EF:	4.1 tons (9040 lb.)
ELM-ES:	1.2 tons (2650 lb.)
RMS:	1.6 tons (3530 lb.) (including RMS console)
	RMS main arm maximum lift capacity:
	6350 kg (7 tons, 14,000 lb.)
Length:	11.2 meters (36.7 ft.) (PM)
	4.2 m (14 ft.) (ELM-PS)
	5.6 m (18.5 ft.) (EF)
	4.2 m (14 ft.) (ELM-ES)
	12.2 m (40 ft.) (RMS – main arm 10 m, small fine arm 2.2 m)
Max Diameter:	
(PM)	Outer diameter 4.4 m (14.4 ft.); inner diameter 4.2 m
(ELM-PS)	Outer diameter 4.4 m; inner diameter 4.2 m
(EF)	5.0 m (16.5 ft.)
(ELM-ES)	4.9 m (16 ft.)
Windows:	2

▲ Laboratory 'Kibo' is berthed to the port side of the Harmony node and the ESA Laboratory 'Columbus' is berthed to the starboard side. A portion of the truss is above and beyond the assembly of pressurized modules.

Integrated Truss Assembly (ITA)

The 'backbone of the ISS', the Integrated Truss Assembly (ITA) is a lattice-work structure which resembles a girder. Although originally envisioned as hundreds of individual rods and balls which would be assembled into a truss in orbit, in 1989-1990 it was redesigned to become the "Pre-Integrated Truss" which would have essentially all major components assembled on the ground and then launched in 9 segments, minimizing spacewalk assembly requirements. Each truss segment is numbered. "P" or "S" designates port or starboard side of the ISS, with lower numbers towards the center and higher numbers further out. One segment is designated by the letter "Z" for zenith, and does not connect with the remainder of the truss assembly. On completion, the assembled truss segments stretch to 110 m (360 ft.) in length; as long as a football field. The ITA houses batteries, computers, communications, navigation and motion control equipment. Mounted on the truss are several communications and four GPS antennae.

The truss houses much of the Station's external electrical power system. Eight solar array wings convert sunlight to electrical power. Two alpha gimbals, midway on port and starboard sides turn the outer ends of the truss. Eight beta gimbals turn each solar array wing. The arrays can track the Sun as the ISS orbits Earth in order to produce the maximum amount of electrical energy. Electrical cables route electrical power from the arrays through electrical components that convert or condition power for use in different parts of the station: the Main Bus Switching Units, DC-to-DC Converter Units and Secondary Power Distribution Assemblies. The truss serves as a conduit for plumbing that carries the external active coolant, ammonia, used in the Thermal Control System. The plumbing is channelled through six 6.4 m (21 ft.) wide deployable radiator panels, used to radiate heat transported from the truss' electronics boxes. On the forward-face of the truss, two rails extend from truss segments P-3 to S-3, providing a track along which the Mobile Transporter moves the station robotic arm. This allows the arm to translate along the Integrated Truss Structure, carrying equipment or spacewalking astronauts from one end to another. The ITA is attached to the US Segment Pressurized modules in two locations. The Z1 truss segment berths to the zenith of Node 1 'Unity'. The integrated truss assembly is mounted on the zenith of the US Laboratory Module 'Destiny'.

Zenith (Z)-1

The first truss segment launched was Zenith (Z)-1. It is mounted to the zenith Common Berthing Mechanism (CBM) on Node 1. The truss segment Port (P)-6 was subsequently mounted on top of Z-1 and for more than seven years provided the first major power source for the US segment. Once this power source was available, it marked a significant increase in operations on ISS. Z1 houses DC power converters that provide power for four massive Control Moment Gyros; the rapidly spinning gyros can be turned on gimbals; as the gyros are turned, the ISS is reoriented around them. Z-1 also houses part of the S-band communications system used for the Tracking and Data Relay Satellite Systems (TDRSS); a Ku-band communications system; the primary video and payload digital data return link; thermal control hardware; mechanical interfaces for connection to the pressurized modules and to the P6 truss segment; to the robotics systems; and to support EVA operations. Z1 has remained separate from the Integrated Truss Assembly, although it formed part of the first permanent lattice structure on the station and set the stage for the latter expansion of the main truss assembly.

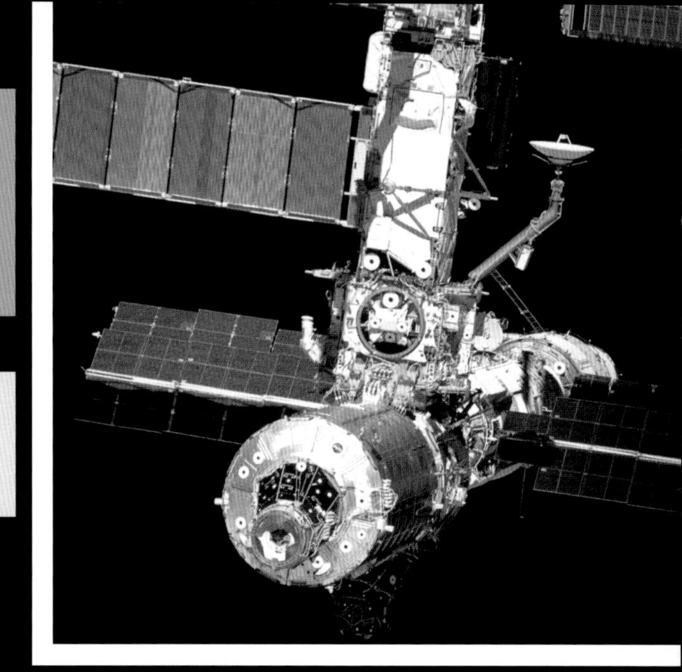

Forward nadir view of the Destiny laboratory module and the Z1 truss mounted atop the Node 1 Unity, as seen by the STS-98 crew during fly-around. The P6 Truss with deployed radiators and solar arrays is mounted on top of the Z1.

Port (P)-6 and Starboard (S)-6 Truss Segments

P-6 provided the ISS electrical power generation system for the ISS US segment during the first phase of the program, throughout much of the assembly phase of the station. P-6 was delivered to the ISS in November, 2000, by Shuttle STS-96 and installed on the Z1 Truss. The P-6 remained attached to Z1 until the STS-120 mission in 2007 when it was relocated to the end of the P5 truss. P-6 and S-6 are essentially identical and are the last truss segments on either end of the Integrated Truss Assembly (ITA). Each consists of three main sections:

A spacer provides adequate clearance from the next truss segment so that the solar array wings can be rotated on their beta gimbals. An Integrated Equipment Assembly (IEA) houses power system electronics for storing, regulating and distributing electrical energy, including the storage batteries, computers used to control functions of the P6 or S6 truss segments; and the thermal control system fluid plumbing and radiators that dissipate excess heat generated into space.

Linda M. Godwin and Daniel M. Tani during a spacewalk outside of the P6 solar array during its initial deployment in December, 2001. The Photovoltaic (Solar) Array Assembly (PVAA) is comprised of upper and lower solar array wings. On each wing, solar cells are affixed to blankets. During launch the solar array blankets were folded, accordion style, and lay flat inside of blanket boxes. In space an extendible mast pulled the blankets until the arrays were relatively flat.

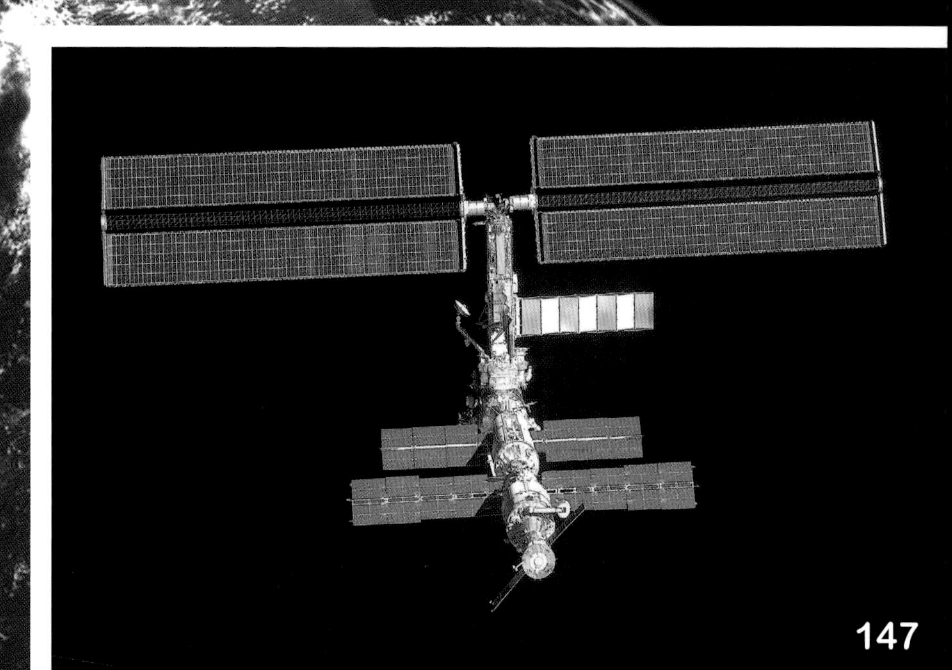

In 2001, the P6 truss segment with its solar arrays were the first to be deployed and provide electrical power to the US segment of the station. For the first phase of the station's life this was the US segment power source. It was mounted on top of the Z1 truss segment, which in turn was berthed to the zenith port of Node 1. This photo from the STS-100 mission was taken 29 April 2001. Visible are the Zvezda / Service Module (SM), Functional Cargo Block (FGB) / Zarya, Node 1 / Unity, Z1 Truss and P6 Truss. On each side of the Station, the S3 and P3 truss segments house the Solar Array Alpha Joints (SARJ). Each SARJ rotates all of the arrays on their respective side of the ISS to track the Sun as the ISS orbits around the Earth.

Starboard (S)-0 Truss Segment

In 2002, the centrally-located Starboard-0 (S-0) element of the Integrated Truss Assembly (ITA) was mounted on the zenith side of the US Laboratory Module 'Destiny'. S-0 was delivered by Shuttle STS-110) in April 2002. It is the central segment of the integrated truss assembly. The S-0 connects the pressurized modules to the huge US solar-Electrical Power System (EPS) arrays and to the Thermal Control System (TCS) radiators. S-0 is attached to the top of the US Laboratory Destiny and acts as the central junction from which all the external utilities are routed to the pressurized modules. These utilities include power, data, video, and ammonia for the external Active Thermal Control System.

Starboard-1 (S-1) and Port-1 (P-1) Truss Segments

S-1 and P-1 are essentially identical and were the first truss segments added on either end of the Integrated Truss Assembly (ITA).
S-1 was delivered by Shuttle mission STS-112 in October 2002 and attached to the starboard end of the S0-Truss. P-1 was delivered by Shuttle mission 11A (STS-113) in November 2002 and was the first port side truss element added.

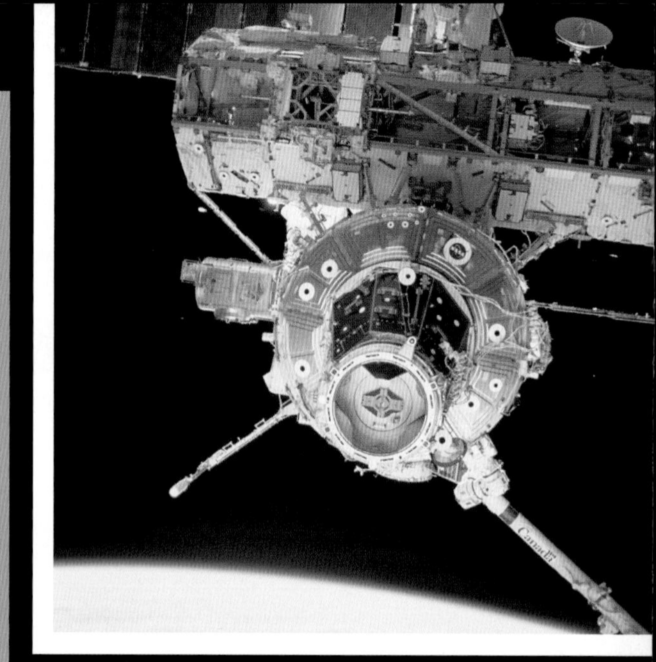

▲ Newly equipped with the 27,000 pound S0 (S-zero) truss; S0 was the first segment of the integrated truss structure mounted in a permanent location, on the zenith side of the US Laboratory Module 'Destiny', seen below the truss. Pressurized Mating Adapter 2 (PMA-2) is berthed to the forward end of the US Lab.

The International Space Station (ISS) in November 2002, as the Space Shuttle Endeavour approached the orbital outpost for a November 25 docking. At the time the S1 truss segment had been attached but not the P1, which meant there was a significant imbalance. ▶

S-1 and P-1 provide structural support for the primary Active Thermal Control System (TCS) radiators and are known as the Port and Starboard Side Thermal Radiator Trusses. S-1 and P-1 provide mounting Space Station Remote Manipulator System (SSRMS) rails, for cameras and lights, and for S-band communications antenna support equipment. The S1 truss includes support for the Canadarm2 and astronaut support carts. It transports 290 kg (640 lb.) of anhydrous ammonia through three heat rejection radiators, and is known, together with the identical P1 Truss, as the Port and Starboard Side Thermal Radiator Trusses. S-1 and P-1 were each attached to opposite ends of S-0. Two months later the Shuttle Columbia accident temporarily halted ISS assembly. It was fortuitous that the interruption came when the truss was relatively evenly balanced with P6 mounted on Z-1 on the zenith side, and with S-0, S-1 and P-1 mounted evenly across the top of the US Laboratory module. If the truss had not been evenly balanced this could have introduced orientation control problems, in turn requiring excess fuel usage.

Port-2 (P-2) and Starboard-2 (S-2) Truss Segments

In the original design of the Space Station, these were intended to support the rocket thrusters to provide re-boost capability to the station. Once the Russian elements were incorporated into the redesigned International Space Station, after 1993, the development of the P2 and S2 truss segments was cancelled, and the segments were removed from the ISS design.

Completing the Integrated Truss

Once Shuttle flights resumed, and after a 4 year hiatus, additional truss segments with their solar arrays and radiators were added to ISS in 2006 and 2007.

Port-3/4 (P3/P4) and S-3/4 (S3/S4) Truss Segments

Shuttle assembly mission STS-115 delivered the second port truss segment, which was attached to the P1-Truss. This second starboard truss segment was delivered by Shuttle assembly mission STS-117 in June 2007. It was attached to the first starboard truss segment (S1). P-3/P-4 and S-3/S-4 each provided the first set of solar arrays to be mounted to either end of the integrated truss. In addition to the solar arrays, each truss also carried a truss radiator and beta gimbals for pointing the arrays. When each of these arrays was first installed, power from their arrays was not yet connected to the ISS electrical power system. During the December 2006 flight of Shuttle STS-116, the astronauts completed a major electrical rewiring of the station, routing the power from the P3/P4 arrays to the station grid. The S-3/S-4 second truss segment was delivered by Shuttle mission STS-117 in June 2007. The truss segment was attached to S-1.

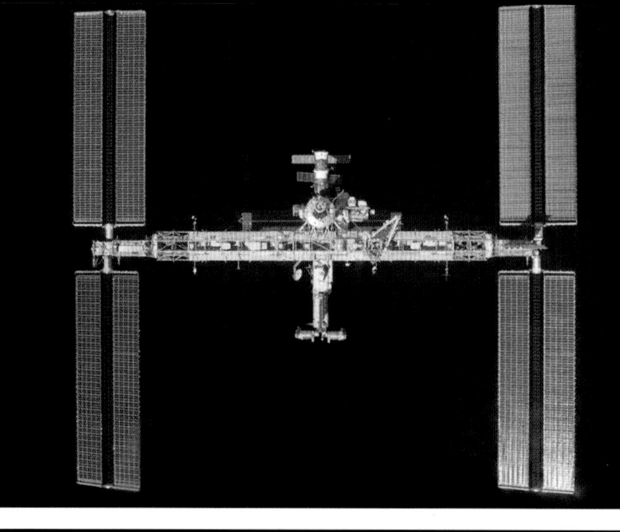

Backdropped by the blackness of space, the International Space Station moves away from the Space Shuttle Atlantis. Earlier the STS-117 and Expedition 15 crews concluded about eight days of cooperative work onboard the shuttle and station.

Port-5 (P-5) and Starboard-5 (S-5) Truss Segments

Shuttle mission STS-116 delivered P-5 in December, 2006, and attached it to the P3/P4 truss segment. During the mission, the astronauts completed a major electrical rewiring of the station, routing the power from the P3/P4 arrays to the station grid. The P6 truss solar array which had been functioning as the primary source of electrical power for the ISS, in place atop the Z-1, since November, 2000 was deactivated and the solar arrays retracted. In August 2007, Shuttle mission STS-118, attached S-5 to the S3/S4-Truss. During STS-118, the P-6 truss was relocated from the top of the Z-1 Truss to its final location attached to the P-5 Truss. The P-6 solar arrays were retracted; P-6 was removed from Z-1 and relocated to the port end of the integrated truss and the P-6 solar arrays were re-deployed. During retraction of one of the solar array panels, a guide wire repeatedly snagged on grommets. Then during the re-deployment, one section of the folded array blanket began to separate and then tear. The ground devised a fix, in which a spacewalking astronaut cut away the frayed guide wire, and then using a series of sutures and "cuff-links", tied the separating solar array blankets together. The improvised braces held as the solar array was fully deployed, placed about 70 pounds of continuous force to stretch the array blanket. The truss structure was completed during STS-119 in March, 2009.

The zenith side of the ISS at the end of STS-113, shows the balanced S1 and P1 portions of the truss just above and behind the P6 solar array. The station would remain in this configuration for the next several years as the Columbia shuttle accident occurred in February, 2003.

"As we got the Space Station going. We went through several design iterations. it is just amazing how much the real vehicle turned out to be like the one we have today. It had a backbone truss structure, solar panels and heat rejection devices on it, and all the basic big design parameters of a Space Station." - Neil Hutchinson

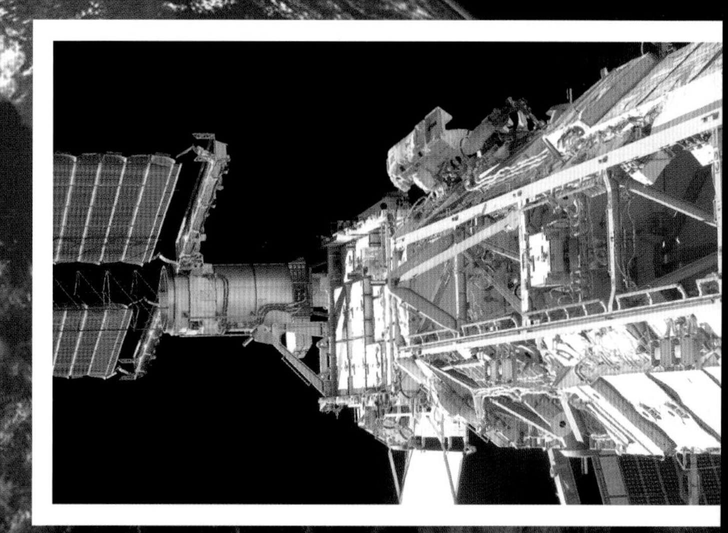

Patrick Forrester and Steven Swanson (out of frame) working at the Solar Array Rotary Joint (SARJ) on S3/S4 Trusses. Among other tasks, Forrester and Swanson removed all of the launch locks holding the 10-foot-wide solar alpha rotary joint in place and began the solar array retraction.

"The structures community fell into two camps, those that thought that you could do some really neat thing, that had an elegant solution. Trusses had the sticks and balls and all, the tinker-toy-type arrangement. If you're structures people, then that has the elegant simplicity of being lightweight and high strength, so you can get a very large station for very light weight; "I can get more Station for the same mass than if I just were to build up chunks of it and stick them together." At CDR [Critical Design Review], McDonnell Douglas said, "Here is how it's all put together." But when McDonnell Douglas showed their performance chart, we were 100,000 pounds overweight at the permanently manned configuration, and about 60,000 pounds overweight at the man-tended configuration, which occurred on assembly flight 5. So, you needed about another 60,000 pounds of Space Station in the first five flights, just to have something that would function. The project's response was, "we need a weight scrub." We packaged a briefing for management. We looked at all the alternatives. It became obvious that a pre-integrated truss was more efficient; hang everything on it, put it where you wanted it, weigh it, tailor it to fit in the Shuttle cargo bay, and then carry it to orbit and screw it together." - Denny Holt

"After I became program manager, there was a technical challenge on Assembly Flight 10A, where we were trying to move P6 out to the outboard truss. After some struggle, we were finally able to get the solar arrays retracted. We put P6 out at the end of the truss and we were ready to deploy those arrays, and at the same time, on the starboard side, we had installed the starboard truss, and we were rotating it. We were noticing some problems with the alpha joint. We then were trying to deploy the array, when the array ripped. We had no spare array and that wasn't a problem we'd ever planned for. That was a major moment. I can remember thinking that my job is to be cool, calm. "This is just a problem. We'll sort it out, we'll move on." Meanwhile, we had this limp array out there that we couldn't maneuver. We couldn't do anything because it was in the state it was in. We couldn't stiffen it because of the rip. We did an EVA [Extravehicular Activity] to go look at this alpha joint. This is the big alpha joint. We also found on the EVA that the whole array had just been basically chewed up. I decided almost immediately that we couldn't work two structures problems. We could work one structures problem at a time, and the Shuttle was docked. I got challenged by the press, "Why don't you split up the team? You go work on this, you go work on that." I said, "No. As a team, we're going to focus on the solar array. That's the most important thing. I can live with the alpha joint for now, but the array has got to get worked now." I had to focus everybody. Because all of these systems were new, it wasn't like we had a lot of experience with them. We were learning how to operate them. Folks came up with a design to fix the array. My job was to say, "Yes, I think that's a good design. You guys have figured it out. Go do it." That's what the team did." - Michael T. Suffredini

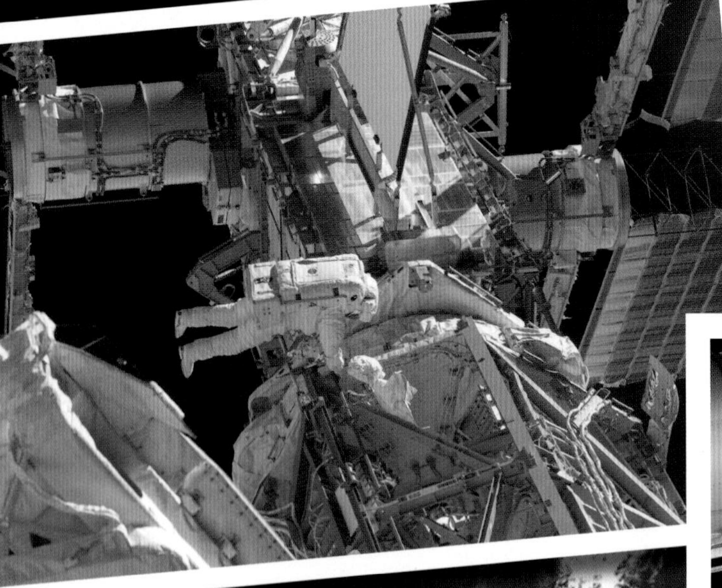

Steve Bowen during (EVA) as construction and maintenance. During the six-hour, 52-minute spacewalk, Bowen and astronaut Heidemarie Stefanyshyn-Piper (out of frame), worked to clean and lubricate part of the station's starboard Solar Alpha Rotary Joints (SARJ) and to remove two of SARJ's 12 trundle bearing assemblies.

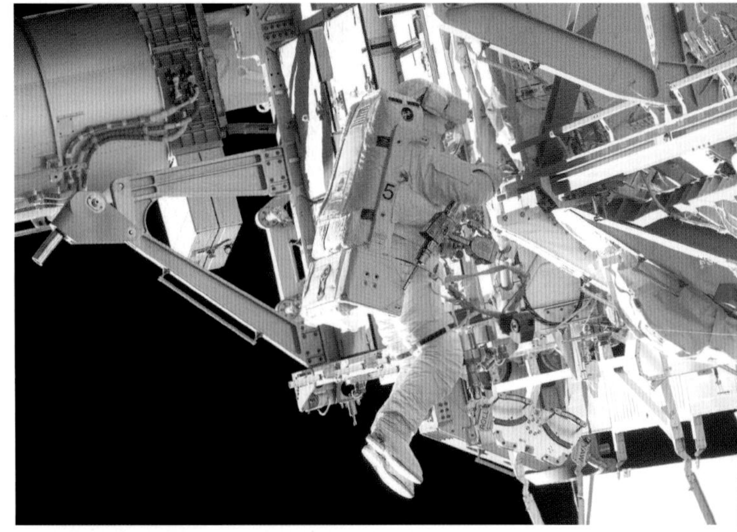

Patrick Forrester and Steven Swanson (out of frame), working at the Solar Array Rotary Joint (SARJ) on S3/S4 Trusses during Extravehicular Activity 2 (EVA 2).

Truss Specifications

Element	Shuttle	Launch date	Major systems components
Z1 truss	STS-92	2000 Oct 11	Orientation gyros, comm system, berthing mechanism
Integrated Truss Structure (ITS)			
P6 truss solar array	STS-97	2000 Nov 30	Beta gimbals, batteries, electric power converters
S0 truss	STS-110	2002 Apr 08	US Lab mount, umbilicals, power converters
S1 truss	STS-112	2002 Oct 07	Primary radiators, MDM computers, Comm system
P1 truss	STS-113	2002 Nov 23	Primary radiators, MDM computers, Comm system
P3/P4 truss solar array	STS-115	2006 Sep 09	Alpha gimbal, MDM computers, solar arrays
P5 truss spacer	STS-116	2006 Dec 09	
S3/S4 truss solar array	STS-117	2007 Jun 08	Alpha gimbal, MDM computers, solar arrays
S5 truss spacer	STS-118	2007 Aug 08	
P6 truss array relocation	STS-120	2007 Oct 23	(relocation; previously launched in 2000)
S6 truss solar array	STS-119	2009 Mar 15	Beta gimbals, batteries, electric power converters

	Length (m)	Diameter (m)	Mass (kg)
Z1	4.9	4.2	8755
Integrated Truss Structure (ITS)			
P6	73.2*	10.7	15824
S0	13.4**	4.6	13971
S1	13.7	4.6	14124
P1	13.7	4.6	14003
P3/P4	13.8	4.8	15824
P5	3.37	4.55	1864
S3/S4	73.2	10.7	15824
S5	3.37	4.55	1818
S6	73.2	10.7	15824
		107.5m	117831

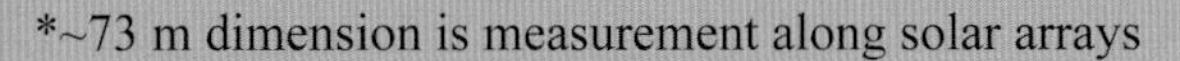

*~73 m dimension is measurement along solar arrays
**~13m dimension is measurement along truss

Z: zenith, S: starboard, P: port

During STS-133 in 2011, Steve Bowen and Alvin Drew, installed several pieces of hardware and exposed the Japanese "Message in a Bottle" experiment. ▲

MOBILE SERVICING SYSTEM

The Space Station Robotic Manipulator System (SSRMS) and Mobile Servicing System (MSS), 'Canadarm2', is Canada's primary contribution to the ISS Program. This second generation mobile robotic manipulator system has been described as being a 'bigger, better and smarter' version of the first generation Robotic Manipulator System (RMS) used on the Space Shuttle, which was also developed by Canada. SSRMS has had a crucial role in the assembly and maintenance of the station. The MSS is capable of moving equipment and astronauts around the station. It can support astronauts during EVA, or service instruments and other payloads attached to the exterior of the station. The robotic 'arm' was fabricated by SPAR Aerospace of Toronto, Canada, which was also the prime contractor for the Shuttle RMS.

MSS is comprised of three main sections: The Space Station Remote Manipulator System (SSRMS) or "Canadarm 2"; The Mobile Base System (MBS); and the Special Purpose Dexterous Manipulator (SPDM) "Dextre". The MBS is installed on the Mobile Transporter, which rolls across the station on a rail system mounted on the Truss. SSRMS has the capacity to move end-over-end from the U-Lab fixture to the MBS and then across the truss to 10 different work sites on the station. The twin-armed 'SPDM' has advanced stabilization and handling capabilities, making it capable of performing delicate tasks that have traditionally required the human touch. SPDM thus offers increased crew safety, reducing the amount of time the astronauts are required to spend outside performing routine maintenance and allowing them to concentrate on scientific activities.

Space Station Remote Manipulator System (SSRMS) or Canadarm2:

Delivered on STS-100 in April 2001, this is an improved version of the Shuttle RMS. Originally it was controlled from a robotics workstation in the US Laboratory 'Destiny'. When fully extended it stretches 17.6 meters. SSRMS has seven motorized joints and has been used to handle large payloads and assist the Shuttle RMS operator in relocating major elements from the payload bay of the visiting Shuttles to the required locations on the ISS. The SSRMS features a Latching End Effector, allowing it to be attached to ports incorporated across the station's external surfaces. This allows the arm to be 'walked' across the station to work in areas it would not be able to extend to if fixed at one location (like the Shuttle RMS). Both ends of the arm can thus act as either the attachment point to the station or the working end effector. Command and control of the robotics systems can take place from computer workstations in the Cupola, other locations on ISS, or from control centers in the US and Canada.

Shuttle Canadarm RMS vs. Station Canadarm2 RMS

- Mission Profile: The Shuttle's RMS returned to Earth with each mission; the station's Canadarm2 remains permanently in space attached to the station.
- Range of motion: The reach of the Shuttle RMS was limited to the length of the arm. The Station RMS is much more adaptable, capable of moving end over end so that it can work at many locations across the exterior of the station. The limiting factor of its relocation is the availability of the Power Data Grapple Fixtures (PDGF) on the exterior of the station to which the arm attaches. These fixtures feature power, data and video links through the Latching End Effectors (LEE). Canadarm2 is also able to travel the length of the Mobile Base System, riding on the Mobile Transporter.
- Fixed Joint: The Shuttle RMS was attached on the forward longeron bulkhead port side. The Station arm has no permanent fixed end. Each end of the arm has LEE, with support subsystems to allow movement across the station as required.
- Degrees of Freedom: Similar to a human arm, the Shuttle RMS featured 6 degrees of freedom, with a shoulder (2 joints), elbow (1 joint), and wrist (3 joints). In comparison, the Station RMS incorporates 7 degrees of freedom: shoulder (3 joints) elbow (1 joint) and wrist (3 joints) and has the added capability of changing its configuration without having to relocate its 'hands' (end effectors).
- Joint Rotation: The Shuttle RMS elbow rotation was limited to 180°, while the Station RMS has all 7 joints capable of 540° rotation, giving a larger range of motion than the human arm.
- Senses: The Shuttle RMS had no sensation of 'touch'. However, incorporated in the Station RMS are force motion sensors which do imitate such a sensation. There is also an automatic vision feature to aid capture of objects, as well as an automatic collision avoidance system.
- Dimensions: The Shuttle RMS was 15 m (49.2 ft.) long, while the station RMS measures 17.6 m (57.75 ft.). The mass of the Shuttle RMS was 410.5 kg (905 lb.) compared to 1800 kg (3970 lb.) for the SSRMS. The exterior diameter of the composite boom of the Shuttle arm was 33 cm (13 in); that of the station arm is 36cm (14 in).
- Mass handling capacity: The original RMS used on Shuttle was capable of deploying or retrieving payloads with a mass of up to 32.5 tons (65,000 lb.) in the vacuum of space. By the 1990s the arm control system had been redesigned to increase the payload capacity up to 293 tons (586,000 lb.) in order to support space station assembly operations. The SSRMS can handle a mass of 116 tons (255,740 lb.).
- Operating speed: Unladen, the Shuttle RMS could move at 60 cm/sec, reducing to 6 cm/sec when fully laden. In comparison, the SSRMS moves unladen at 37cm/sec, or 2 cm/sec when laden during station assembly, 16 cm/sec during EVA operations and 1.2 cm/sec during Shuttle Orbiter operations.

◀ The SSRMS 'Canadarm2' appears to be waving good bye to the Space Shuttle as the orbital outpost moves away from the Shuttle Discovery in this picture taken 6 August 2005.

- Composition: The Shuttle RMS comprised 16 plies of high modulus carbon fiber-epoxy. The SSRMS arm is constructed from 19 plies of high strength carbon fiber-thermoplastic.
- Repairs: The Shuttle RMS returned to Earth for any repairs and modifications at the end of each mission. The SSRMS is intended to be repaired in space and incorporates Orbital Replacement Units in its design, along with built in redundancy.
- Control: The Shuttle RMS had autonomous operation or astronaut control modes from the aft flight deck station on the Shuttle flight deck. The SSRMS features autonomous operation or astronaut control, from inside the station. This was initially from a computer workstation within the US Laboratory Module 'Destiny', but moved to the Cupola after its installation in 2010.
- Cameras: The Shuttle RMS featured 2 cameras; one on the elbow and the other on the wrist. The SSRMS incorporates 4 color cameras; one at each side of the elbow with the other two on the Latching End Effectors.

Mobile Base System (MBS):

MBS is a work platform that travels along rails along the length of the main truss structure. This provides lateral mobility for the SSRMS 'Canadarm 2' as it moves across the station's truss. The MBS was installed during the STS-111 (UF-2) mission in June 2002. Resembling a railroad flat car, the MBS moves in conjunction with two Crew Equipment and Translation Assembly (or CETA) EVA astronaut support cart. These save the astronauts from expending energy translating by hand across the long truss structure.

Size:	5.7 x 4.5 x 2.9 m (18.7 x 14.7 x 9.5 ft.)
Mass:	1500 kg (3307 lb.)
Mass handling transport capability:	20,900 kg (46,076 lb.)
Degrees of freedom:	Fixed
Peak power (operational):	825 W
Average power (keep alive)	365 W

Subsystems:

Payloads/Orbital Replacement Unit Accommodation (POA): This is used for carrying large structural elements along the length of the station, using a support structure similar to the Canadarm's latching end effectors but without the 'sense of touch'.

MBS Orbital Replacement Unit Base: The MBS is constructed, like all elements of the station, with a number of separate and interchangeable units called Orbital Replacement Units (ORU). These can be replaced during EVAs or by Dextre. The ORU include the two computer units, the video distribution units, and the Remote Power Control Module.

Power Data Grapple Fixture (PDGF): The Mobile Base System (MBS) has four Power Data Grapple Fixtures (PDGF), or anchor points, which can serve as a base for the SSRMS 'Canadarm 2' and Special Purpose Dexterous Manipulator (SPDM) 'Dextre'. They provide power to both the robotic arms and any payloads they may be supporting. Transfer of computer commands and video signals initiated from the Robotic Work Station inside the station are also included in the fixtures. The power and data exchanges to payloads are via an Umbilical Mating Assembly (UMA).

The Mobile Base, with the SSRMS 'Canadarm2' is at center. The view also shows portions of the S1, S3, and S4 truss segments, the Solar Array Wing (SAW), radiator panels

Command Attachment Systems (CAS): The CAS consists of three payload alignments guides with ready-to-latch indicators. The system incorporates a powered claw to grip a special capture bar on payloads, a camera target for payload berthing and the UMA.

Camera/Light/Pan and Tilt Assembly: A color camera was used to monitor the mating of the MBS to the MT. It was then placed on the POA mast to provide overall views of the MBS during operations.

Mobile Transporter: This US-built movable base is set atop the Integrated Truss, with the MBS mated to it. It allows the SSRMS 'Canadarm 2', payloads and SPDM 'Dextre' to travel across the Station Truss and allows the MBS to relocate to 10 pre-designated Space Station work sites. Controlled by complex software, there are 20 different motors which are energized in sequence to run the transport from one location to the next. They then latch down the transporters and supply the required power. The transporter was delivered and installed on mission STS-110 (8A)

Specifications:

Size:	27.4 x 26.2 x 9.6 m (90 x 86 x 31.5 ft.)
Speed (maximum rate):	90 m (295 ft.) per hour or 2.5 cm (1 in) per second
Mass:	886 kg (1953 lb.)
Mass handling transport capacity:	20,954 kg (46,196 lb.)

Special Purpose Dexterous Manipulator (SPDM) or 'Dextre':

SPDM 'Dextre' is a small, two-armed robot capable of handing the delicate assembly tasks currently handled by astronauts during EVA. It was delivered to ISS aboard STS-123 in March 2008. SPDM features manipulator systems with a more refined and precise control capability, reducing the need for routine maintenance, repair and servicing directly by crew members during EVAs and thus allowing the crew to focus on other internal activities. This has the cumulative effect of increasing both crew safety and effective work time by reducing the time spent outside the station and the time spent preparing for (and maintaining equipment after) EVAs.

SPDM 'Dextre' Facts:

Height:	3.67 m (12 ft.)
Width:	2.37 m (7.75 ft.) across shoulders
Arm length:	3.35 m (11 ft.) each, Arm 1 and Arm 2
Mass:	1560 kg (3439 lb.) approximately
Performance:	
Handling capacity:	600 kg (1323 lb.)
Positioning Accuracy:	2 mm (incremental)
Positioning accuracy:	6 mm (relative to target)
Force Accuracy:	2.2 newtons
Average operating power:	1400 watts

SPDM 'Dextre' can be placed on the end of the SSRMS to perform sophisticated operations. It can handle tools (such as specialized wrenches and socket extensions) for delicate maintenance and servicing. The unit is also fitted with lights, video equipment, a tool platform and four tool holders.

Station Mobile Remote Service Base System Specifications:

	SSRMS	SPDM	MBS
Length (meters)	17.6	3.5	5.7 x 4.5 x 2.9
Mass (approx.)	1641 kg	1662 kg	1500kg
Mass handling capacity	116000kg	600kg	20900 kg
Degrees of freedom	7	15	Fixed
Peak power (operational)	2000 W	2000 W	825 W
Average power (keep alive)	1200 W	600 W	365 W
Stopping distance (under maximum load conditions)	0.6m	0.05m	N/A

In the grasp of the station's robotic Special Purpose Dextrous Manipulator (SPDM) 'Dextre'. A section of a solar array wing and a cloud-covered part of Earth provide the backdrop for the scene

"The first Shuttle flight to launch part of the SSRMS delivered the robotics workstation; it was the console that the crew used to operate the arm. It had the hand controllers, monitors, and switches. We flew the robotics workstations before we flew the arm. Our objective for that flight was to have the crew assemble and test one robotics workstation. We had to do one thing, and that was to prove that with the Shuttle docked we could route video from the Shuttle camera looking at the payload bay through the Shuttle, into the station, to the station video system, and display it on a Robotic Workstation monitor. The ISS crew had these monitors, little windows on the world, because there were no actual windows. They relied completely on the cameras; they needed that view. The night finally came-- March 12, 2001. The Expedition 2 crew, Susan Helms, and Jim Voss, assembled the first Robotics Work Station (RWS). Then it came time to power up....we were not able to power the workstation.... we had an anomaly. We started working the problem. We thought it through. Leakage through the grounding circuit We realized, 'okay, if we just unplug two of the three monitors maybe there would be less leakage through the grounding circuit and maybe that would allow the circuit to hold.' The next flight was in April, 2001. This was the big moment when the big arm, Canadarm2, the Space Station Remote Manipulator System, the SSRMS, arrived on the Shuttle with Chris Hadfield, the Canadian astronaut. And there were all these steps to do the spacewalk assembly of the arm. Then the arm had to be powered up and walked off the Shuttle and onto the Space Station. Then it would be home." - Tim Braithwaite

ORBITER BOOM SENSOR SYSTEM (OBSS)

Length:	15 m (49 ft.) The combined length of the Shuttle RMS and boom was 30 m (98.5 ft.).
Mass:	211 kg (465 lb.)
Primary Contractor:	MD Robotics, Brampton, Ontario, Canada

The loss of the Columbia orbiter during the final stages of its mission in February 2003 highlighted the need to inspect the exterior of the Shuttles prior to committal to the entry phase. Inspecting the underside of the orbiter required an inspection boom to provide the needed additional reach. As a result, the OBSS was created as an extension based on the existing Shuttle RMS. The boom could be grappled either by the RMS or station SSRMS.

Operations: Delivered to KSC in December 2004, the boom assembly was flown for the first time on STS-114. During STS-120 the boom was used as an extension to the SSRMS 'Canadarm2'. On that mission, the P6 solar array had become damaged during redeployment. The Canadarm2 grappled the center grapple fixture and astronaut Scott Parazynski rode the end of the boom to make repairs.

Enhanced ISS Boom Assembly (EIBA):

The ability to use the boom during EVAs at station led to the decision to leave the OBSS at the station during Shuttle mission STS-134 so that it could be used in conjunction with the SSRMS. Modifications were made to the assembly, which was then renamed the Enhanced ISS Boom Assembly (EIBA). The modifications included the addition of a Power Data and Grapple Fixture. This allows the boom to be mated to the Canadarm 2 grapple fixture during station operations. On 27 May 2011, the boom was relocated to the S1 truss during the fourth EVA of STS-134. As the OBSS sensors used to examine the exterior of the Shuttle were no longer able to withstand the thermal conditions outside the station without a power supply to keep them warm, they were turned off during the EVA. At some point in the future, modifications to the grapple fixture could allow this or similar monitoring and examination equipment to be utilized on the boom.

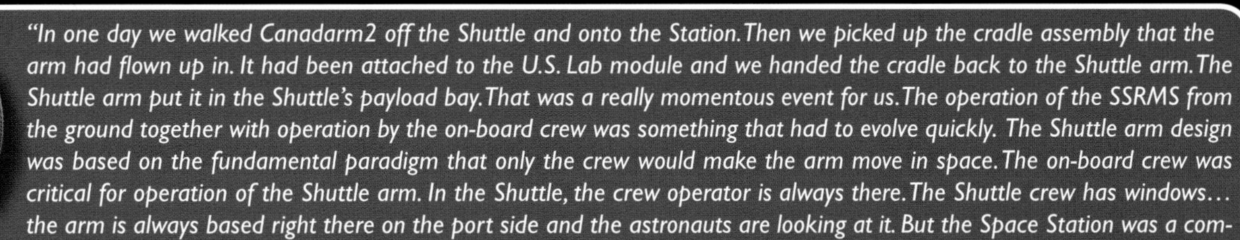

"In one day we walked Canadarm2 off the Shuttle and onto the Station. Then we picked up the cradle assembly that the arm had flown up in. It had been attached to the U.S. Lab module and we handed the cradle back to the Shuttle arm. The Shuttle arm put it in the Shuttle's payload bay. That was a really momentous event for us. The operation of the SSRMS from the ground together with operation by the on-board crew was something that had to evolve quickly. The Shuttle arm design was based on the fundamental paradigm that only the crew would make the arm move in space. The on-board crew was critical for operation of the Shuttle arm. In the Shuttle, the crew operator is always there. The Shuttle crew has windows… the arm is always based right there on the port side and the astronauts are looking at it. But the Space Station was a completely different environment. On those early flights, there really were not any windows on the Station in the same way as on the Shuttle, so it was a different concept of operation. We had to rely on the camera views. Eventually, years later we got the Cupola on the Station. The Cupola is an amazing window that allows the astronauts to stick their head into space to be much more aware of the environment around the Station. Then they could define by their eyes the coordinate frame where the arm was working; which way was up and which way was down. This was a fundamental thing that needed to be intuitive for the crew if they were driving the arm. Now, most SSRMS operations involving the crew take place in the Cupola, though the Cupola was not there throughout most of the assembly phase. We simply didn't have it, and there were no windows that offered any useful views. But even now, the crew cannot always see where the arm is from the Cupola. For free flyer captures and releases, those are designed to happen right there in front of them. It is a great thing from a situational awareness and a safety point of view. You can see the giant, multi-ton object you're about to capture with the arm. But, if we're doing a spacewalk somewhere along the truss, the Cupola may or may not offer a good view of that. Then we have to rely on camera views. The robotics workstations each have three monitors and extra laptop computers can supply extra screens set up to give them as many views and provide the situational awareness aids the crew needs. Windows are helpful, but they often don't look in the right direction." - Tim Braithwaite

"One event that totally stands out was STS-120 when we had the solar array tear and we ended up having to repair it. So I worked on that mission with another couple of Canadians, actually we were on the prime team, but I remember we were near the end of the flight there's only a few more days and they're rolling out the array and it started ripping and then stopped. Initially we're like, oh well, that's PHALCON's (Power, Heating, Articulation, Lighting Control Officer MCC) *problem you know we can't do anything about it. Our arm can't reach it, so well, good luck to them, they will figure out something, but it's pretty serious, and then we went home from our shift and then we came back and it was like, nope, we're going to do it. We're going to go try to repair it by going and grabbing the OBFF which was used for surveying the underneath of the shuttle and we are going to grab it with the Canadarm and stick an EVA spacewalker out at the very end which it was never intended to do. We're going to go and try to repair it because we think we can just reach it. Dextre, originally it was for fine repair maintenance but the different ways that it's been used could have not been thought of when it was first established. If it's there it will get used in some way."* - Charadean Smith

"Dextre is really the part of the system that looks like a real robot. It has a head, and has arms, and hands. Dextre is operated as a part of the Mobile Servicing System (MSS), which also includes Canadarm2, and the mobile base, and the robotics workstations. All of the MSS is operated by the same flight control team led by the ROBO in the Mission Control Center (MCC). It makes sense because the architecture is so integral between the elements. Although Dextre doesn't look very much like Canadarm2, the architecture of Dextre, the way it communicates, the way the computers are designed, is actually very similar. It is one system that operates as an integral unit. So it makes sense for the entire system to be operated by the one team. We call Dextre, Canada's two-armed handyman. Dextre is somewhat anthropomorphic, and it's got these two giant arms. What looks like the head is the grapple fixture to which the arm attaches. Dextre provides remarkable capability. It has tremendous precision. Dextre has two little arms; cameras are mounted on them. When you look through the cameras at the end of those arms you are working at millimeter level precision. Canadarm2 operates at centimeter level precision, which is still pretty good. For example, once we capture a Dragon spacecraft, our Canadarm2 team's part is largely done. Then other parts of ground control take over and move the Dragon, and attach it to a berthing interface, a Common Berthing Mechanism (CBM) on the Space Station. That still requires pretty tight tolerances, and that is all possible fed by visual feedback from cameras. Dextre takes handling to a new level. We can control precision to a couple of millimeters." - Tim Braithwaite

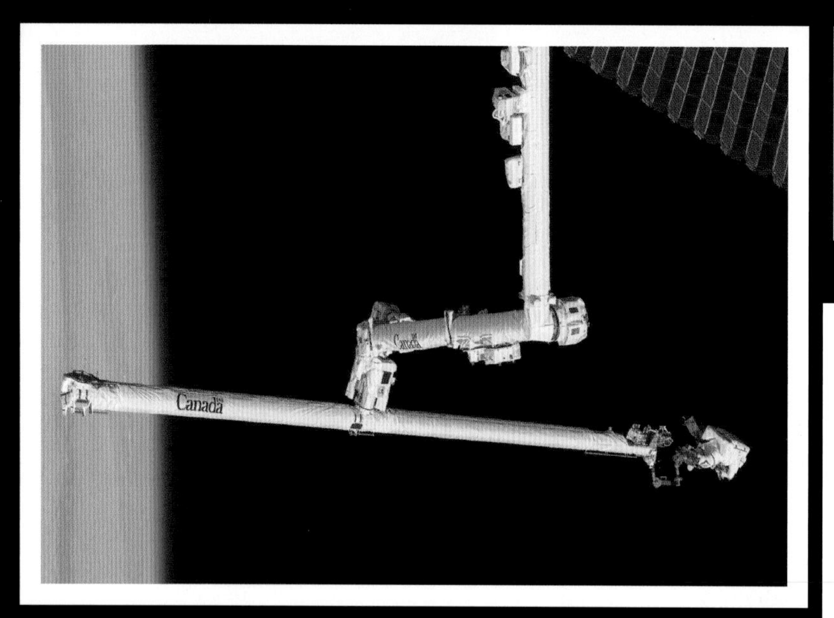

While anchored to a foot restraint on the end of the Orbiter Boom Sensor System (OBSS), Scott Parazynski during a spacewalk where Parazynski reached the limit of the SSRMS reach in order to install homemade stabilizers designed to strengthen a damaged solar array.

Peggy Whitson controls the SSRMS at the Robotic Workstation in the U.S. Laboratory 'Destiny'.

"For Canadarm2 everything that we do is simulated on computers. So we have full robotic planning simulators that allow us to go through and look at the trajectories we have to go through and to plan those more carefully. And once that is simulated there it goes through more of the dynamics analysis as well to make sure that the proper parameter files and loads and whatnot are being used appropriately in that assessment. It goes through several iterations of running it in sim. The astronauts have all come back and said the sim is great in terms of training and it's quite realistic. Nothing quite beats the cupola views you have on the ISS but we try and get as close as possible. It's not quite automated. We're still doing ground commanding of all the robotic systems. In terms of repairs we've actually done quite a few recently, both robotically and using astronauts. It's a bunch of these orbital replacement units that can be taken out and then fully replaced. We are able to use DEXTRE or SPDM which is the smaller robot with the two arms and the ability to rotate its body, and we could pick that up with Canadarm2 and then robotically change out these things. So we're able to do that completely from the ground, but it's not automated, it's planned on the ground then the sequence is commanded by the flight controllers here. The systems are becoming more modularized, like plug and play." - Kristen Facciol

"We can build world-class hardware for the least understood of environments. Canadarm, the one that flew on the shuttles, invented space robotics. It wasn't like we could ask anyone else. We started with a clean slate and came up with a whole bunch of very original ideas at a very low power draw that have been copied and modelled by everyone who has ever built space robots since - including ourselves when we built Canadarm2." - Chris Hadfield

If the various elements and structure of the ISS are its visible limbs and body, then the numerous subsystems are its circulation and central nervous system. The crew normally monitors the systems, awaiting warning alerts but mostly focused on performing science, housekeeping and maintenance chores. While the crew works or rests, teams of ground-based controllers and support staff in control centers around the world plan and program the ISS system operations.

Electrical Power System

Electrical power for the ISS is derived from solar arrays. Batteries store energy for use during the night-side passage of each orbit. Some Russian modules have their own self-supporting electrical power systems so that they can operate independently. The US segment has a distributed, integrated Electrical Power System composed of components that feed, distribute and regulate power throughout the ISS.

Russian Segment

The first power source for the ISS came from the Russian FGB and Service Modules. Each was fully autonomous.

FGB Module 'Zarya' and FGB 2 Module 'Nauka'

The FGB has twin solar arrays that span 21.94 m (72 ft) when deployed. The area of arrays totals 300 m² (3230 ft.²). They provided a daily average of about 3 kW with the arrays fully oriented towards the sun, up to 2kW of which was supplied to the US elements. FGB has six nickel-cadmium batteries. Battery output varied between 24 and 34 volts. Maximum power available was 13 kW in sunlight and 6 kW in the shade. On-board guidance and navigation sensors kept the solar panels oriented towards the sun during daytime. The solar arrays remained motionless during the shaded night time passes. As the US segment truss elements were launched and assembled and Thermal Control System radiators were deployed, the FGB solar arrays were partially retracted in September 2007 in order to allow the US radiators to rotate. This decreased the energy the arrays could provide.

Service Module 'Zvezda'

The Service Module (SM) provided power for early ISS operations and still supports the Russian segment of ISS. The SM arrays' span is 29.6 m (97 ft). The two arrays have a surface area of 76 m² (818 ft.²). The solar arrays generate 9.8 kW, supported by 8 nickel-cadmium batteries. The SM arrays are oriented to face the sun. Orientation of the arrays is by automatic command from the station or ground, or by manual commands via laptops used by the crew. In the event of electrical drive failure, manual reorientation is possible via EVA.

US Segment

Solar Arrays

There are 16 solar array blankets on the US segment. Each blanket is 4.6 m (15 ft) wide and 35m (112 ft) long when extended. Each blanket has 16,400 solar array elements, each measuring 8 cm (3.15 in) square, grouped into 82 active panels, each consisting of 200 cells with 2,050 diodes. A single solar array blanket covers an area of 156 m² (1680 ft²) and is capable of generating between 5.25 to 7.5 kW of DC power. A pair of solar array blankets attaches to a single extendible/retractable mast. Each mast extends to 35m (115 ft).

Solar Array Wings (SAW) and Photovoltaic Module

A pair of opposed solar array masts forms a Solar Array Wing (SAW). The masts that extend each pair of arrays extend from cannisters on either side of the ISS truss. The canister stores the retracted mast. A storage canister measures 1.8 m (5.9 ft) in diameter and 2 m (6.6 ft) tall. A Beta Gimbal Assembly (BGA) deploys and retracts the arrays and rotates them around the longitudinal (Beta) axis to track the sun. The BGA includes the mast canister, motors that rotate the arrays, and electronics that control SAW power. A single SAW is 73m (240 ft) by 12 m (39 ft) spanning across four solar array blankets, two opposed masts, and the truss segment in between and has a mass of 1088 kg (2400 lb.). Each solar array wing provides an array surface area of 626 m² (6720 ft²) and is capable of generating 33 kW DC. Each SAW or Photovoltaic Module resides on a separate truss segment: P-4, P-6, S-4 and S-6.

Solar Array Power Supply

Once the integrated truss was assembled in 2007, there were four SAWs, comprised of sixteen solar array blankets. Together they cover a total surface area of about 2500 m² (26,900 ft²) and provide up to 120 kW of electricity. Some energy is routed to batteries for storage. On Earth's night side, station batteries can supply about 78 kW. Nickel-hydrogen batteries were used early on, but were replaced with lithium ion batteries. The new batteries store more energy and a single lithium ion battery can replace two nickel-hydrogen batteries.

Solar Alpha Rotary Joint (SARJ), Alpha Joint or Alpha Gimbal

Two Solar Array Rotary Joints (SARJ) also called Alpha Joints or Alpha Gimbals, are located midway on either side of the integrated truss, between the P4 and P5 and the S3 and S4 truss segments. The 'Alpha' joint rotates 360° each orbit to track the sun. Each SARJ measures 3 m (10 ft.) in diameter and has a mass of approximately 1135 kg (2500 lb.). During 2007 an anomaly was detected in the starboard SARJ. An inspection revealed excessive premature wear. Cleaning and lubrication during a spacewalk resolved the issue.

Power Conditioning and Storage:

The ISS system generates power at up to 240V DC. The electricity is routed to transformers that convert the power to 124V DC for 'consumer' use in the laboratories, living quarters, robotic and other systems.

EPS Cooling

Each solar array photo-voltaic electrical circuit has its own radiator panel to radiate away waste heat from electrical system components. These radiator panels are mounted on the same truss segments and at right angles to the solar arrays.

"One of the most striking differences between MIR and ISS is available power. Mir was always underpowered and use was carefully husbanded. We had to turn things off in order to turn something on somewhere else. We couldn't run as many dehumidifiers and fans so parts of Mir collected condensation. The ISS has so much more electrical power that we can do so many more things on it." - Chris Hadfield

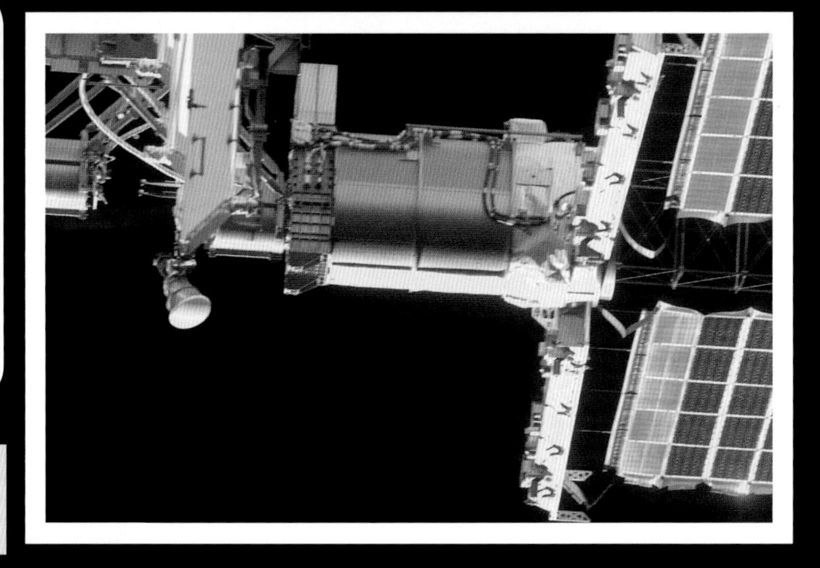

A mast canister, Solar Array Blanket Boxes (SABB) and Solar Array Wing (SAW) of the P6 Truss.

Deployment of the P4 Solar Array blanket in 2006.

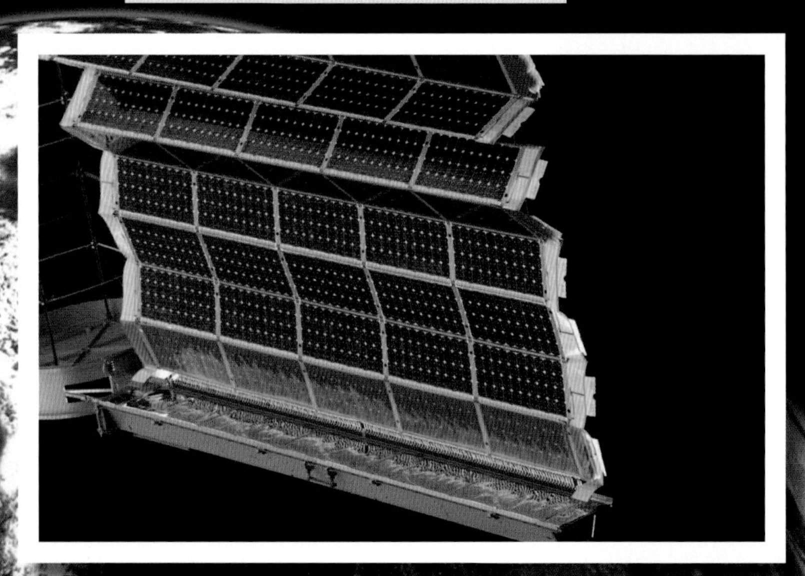

Robert L. Curbeam, Jr., (red stripes) and Sunita L. Williams, highlight their relative sizes as compared with the canisters and masts of the P6 Solar Array Wing in 2006.

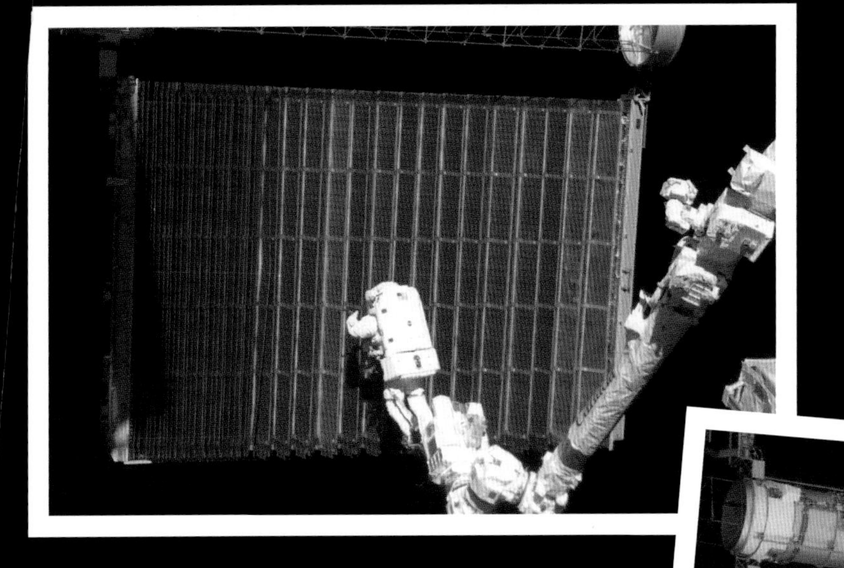

Anchored to the SSRMS Canadarm2, Robert L. Curbeam Jr., works near the P6 SAW in 2006.

A close-up view of two opposed solar array canisters and with a solar array radiator panel in the foreground.

The SSRMS Canadarm2 mounted on the Mobile Base, and the solar arrays on one side of the ISS in 2007. The Solar Array Rotary Alpha Joint is located between the large width of the truss closer to the camera and the narrow truss segment holding the solar arrays, further from the camera.

Thermal Control System

ISS orbits the Earth once every 90 minutes and as it does so it spends roughly half of each orbit in daylight and half in darkness. Exposure to direct solar lighting and radiation can cause severe heating, up to temperatures of 150°C (302°F). Shaded from sunlight, the temperature can plunge to as low as -130°C (-202°F). The entire ISS undergoes swings of 280°C (-504°F) twice each orbit causing thermal stresses on the structure and systems. Equipment inside and outside of the ISS must be maintained at proper working temperatures. In space where there is no up or down, there is no convection so hot air does not rise. Therefore cooling relies on conduction, where heat is transferred through direct contact. The excess heat has to be channelled away from hot equipment to radiate the heat into space. Using a combination of passive and active subsystems, the Thermal Control System (TCS) maintains the temperatures of the structure, fuel and liquid lines, and internal and external equipment, within allowable limits.

Passive Thermal Control Systems

Passive thermal control systems are those that do not use circulating cooling fluids or pumps. The passive thermal control system uses a combination of coloration, multi-layer insulation (MLI) blankets, heat pipe radiators (HPR), shell heaters and temperature sensors. Temperatures can be controlled passively using coloration. Panels with surface coatings or specific coloration are used outside of modules. White surface panels reflect heat while black panels absorb heat. The panels radiate away waste energy. While surfaces exposed to the extreme temperatures of space can themselves have extremely high or low temperatures, insulation can help to isolate internal surfaces from the temperatures of the space environment. Multi-layer insulation (MLI) Blankets consist of layers of metal, metallic foils, and fabrics. Heat pipes are hollow tubes containing ammonia. Several tubes can be aligned together such that one end of the pipes is in contact with the warm electronics, and the other end of the pipes is mounted a short distance away from the heat source. When the heat from the electronics is transferred to the ammonia in the tubes, the ammonia turns to vapour. When the ammonia vapour comes in contact with the cool end of the pipes, it releases the transferred heat and condenses back into a liquid, flowing back along the pipe to the warm end again to repeat the process. Heat pipes provide a simple and effective way to move heat away from electronics without the need for mechanisms that may require maintenance over time. Inside the modules of the ISS, the respiration of the astronauts includes water vapour. Because the module walls are sometimes cold, condensation can form in the interior. Heaters consisting of nickel chrome wire embedded in silicon rubber are spaced around the interior of the pressure shell of most modules to keep hardware warm and protect the inside of a module from condensation.

Active Thermal Control System

Within the crew cabin, ventilation ducts and fans circulate air. Air is directed over fluid filled heat exchangers. The 'active' system consists of flowing fluids to carry waste heat energy to external radiators. Pumps circulate the fluids continuously. 'Active' fluid lines collect and transport excess heat to the external radiators. Inside of modules in the US segment the internal cooling fluid is water; water is used because it has a high heat capacity, low viscosity, and it is non-toxic. inside the Russian segment it is a water and ethylene-glycol mixture. Outside of the US segment modules, 99.9% pure anhydrous ammonia is used. Ammonia is used because of its high heat capacity and it remains liquid to -78°C (-108°F). Purified ammonia is toxic to humans and therefore use is restricted to areas outside of the ISS pressurized modules. Outside the Russian segment modules, a silicone organic fluid is used. Most of the larger modules on the US segment, the US Laboratory, Node 2, Node 3, the Japanese Experiment Module and the ESA Columbus Module each have a separate internal TCS. The Joint Airlock (Quest) is cooled by an extension from the US Lab internal TCS system. The internal TCS can be operated at different temperatures to support different cooling needs. The colder cooling loop operates at 9.4˚C (48.9˚F). The higher temperature operates at 17.2˚C (63.0˚F). Heat is removed from hardware using coldplates. A coldplate is a broad, flat plate with fins extending from it. The coldplate fins interleave with fins extending from the electrical component which is the heat source. The heat from the electrical component fins is conducted into the coldplate fins. This coldplate finned plate is bonded to a stainless steel flow plate, which is sealed to allow water or ammonia flow through the plate. The fluid conducts heat from the coldplate fins and is transferred with the flowing liquid. Water in the TCS inside the modules is pumped through the plumbing to a heat exchanger outside of the modules. Ammonia in the TCS outside is pumped through a heat exchanger that is in contact with the water filled heat exchanger. Heat is transferred from the water to the ammonia through the heat exchanger.

Nadir side of the ISS, showing the S0, S1 and P1 truss segments with the Radiator Beam Truss Structure and main TCS radiators. The white panels of the Russian segment TCS are also visible.

This is a high angle view shows the Cupola mounted on Node 3, both backdropped against the main TCS radiator panels and solar arrays beyond.

The excess heat is carried by the ammonia to external radiators which reject the heat overboard. On each side of ISS, there are three Thermal Control System (TCS) radiators. Each of the six radiators consists of eight radiator panels, each 3.4 x 2.7 m (131.25 x 107 in). The thermal radiators have a combined surface area of 2046 sq. m (22,025 sq. ft.). Each set of three radiators has a heat rejection capability of 35 kW. The radiators are mounted side-by-side on both sides of the ISS truss, on a rotating beam called the Radiator Beam Truss Structure. They are mounted at slight angles to one another. The beam is mounted on the Thermal Radiator Rotary Joint (TRRJ). The TRRJ can rotate the 3 radiators through 105° in either direction and tries to keep the radiators' edge-to-the-Sun, maximizing radiative cooling. Each radiator panel is honeycomb aluminium carrying 22 stainless steel tubes through which ammonia flows. Each honeycomb panel is sandwiched between white or silver colored aluminium sheets. Heat from the ammonia radiates off the panels into space.

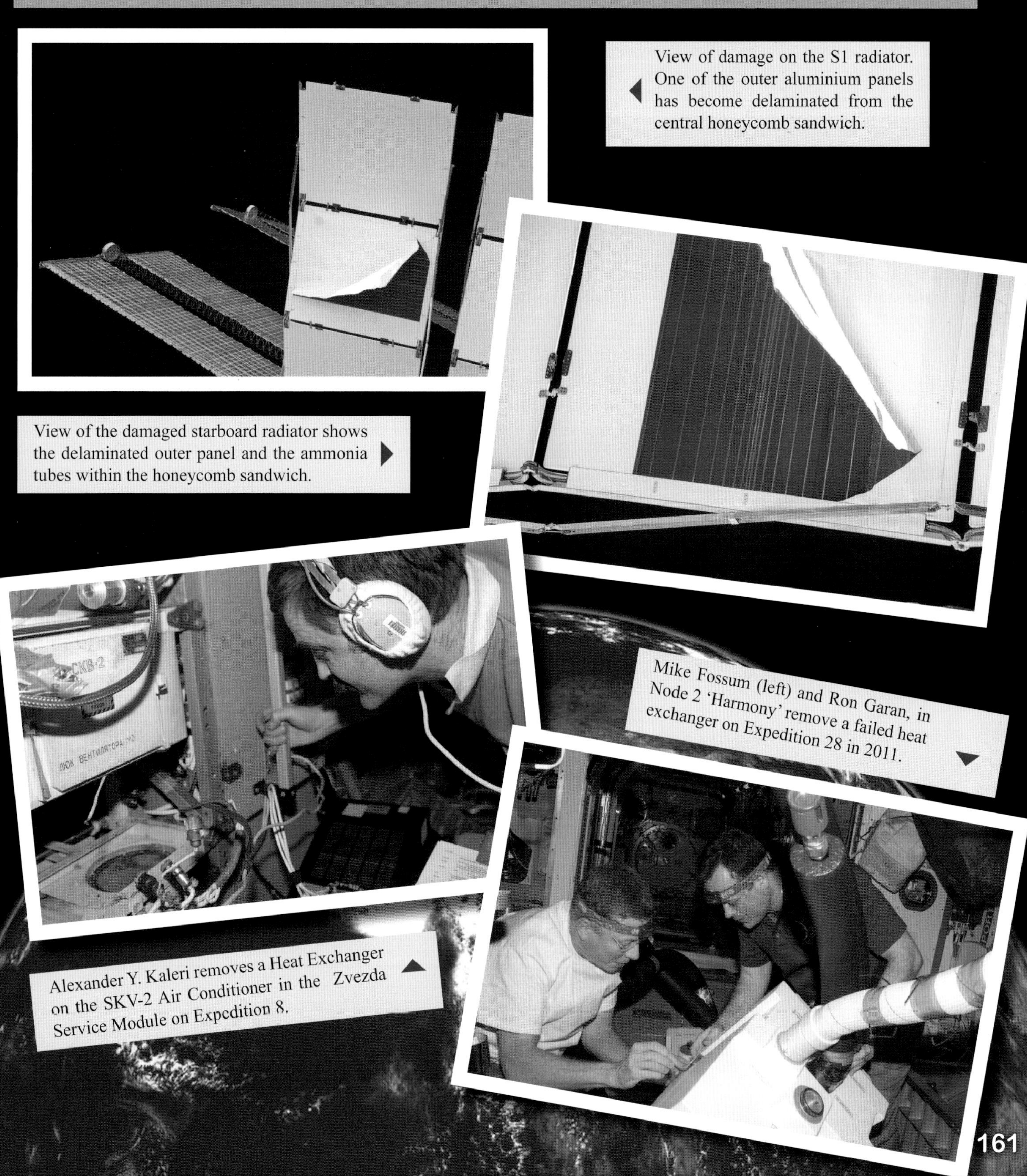

View of damage on the S1 radiator. One of the outer aluminium panels has become delaminated from the central honeycomb sandwich.

View of the damaged starboard radiator shows the delaminated outer panel and the ammonia tubes within the honeycomb sandwich.

Mike Fossum (left) and Ron Garan, in Node 2 'Harmony' remove a failed heat exchanger on Expedition 28 in 2011.

Alexander Y. Kaleri removes a Heat Exchanger on the SKV-2 Air Conditioner in the Zvezda Service Module on Expedition 8.

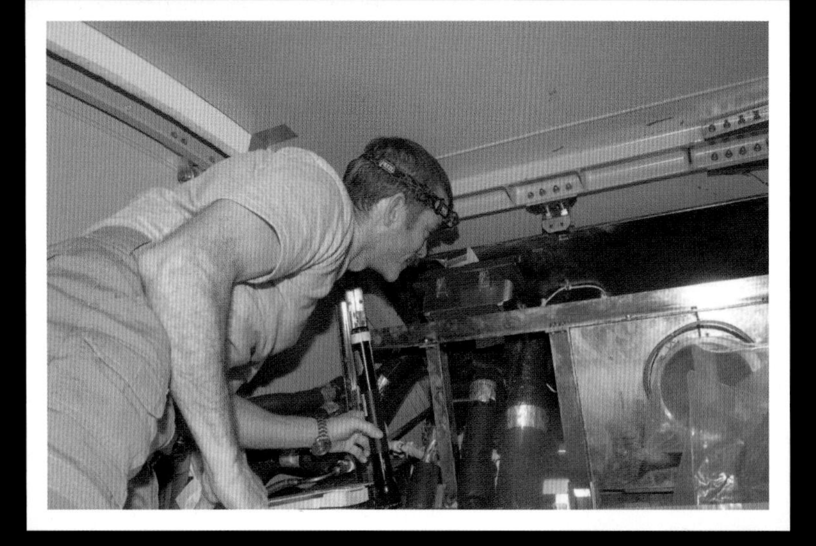

Chris Hadfield replaces a Heat Exchanger inside the Joint Airlock 'Quest' on Expedition 34 in 2013.

Data Management System (DMS)

Virtually every aspect of every system on the ISS is controlled by computers. The Data Management System (DMS) computer network detects and commands systems functions through Multiplexer - Demultiplexers (MDMs). 46 MDM computers inside and outside of the ISS modules control functions such as electrical power switching, the alignment of solar panels, control of the mobile transporter, disposal of excess heat, turning lights on or off, re-orienting the station, and transmission of signals to or from the ground. The system started with only two MDMs but has steadily expanded as the ISS has grown. The different MDMs have common software for controlling systems with common functions in different parts of the ISS, but unique for functions specific to only a particular MDM. Intel 386 processors form the core of most MDMs. When the ISS DMS was being designed in the 1980s, the Intel 386 was state of the art. The entire system was designed, tested and certified around these processor chips. Only in a few instances have processors been updated to newer, faster or higher memory chips in order to keep pace with updated systems requirements. The crew can review and command ISS systems through the MDMs using the Portable Computer System (PCS). Up to eight PCS computers can be connected to the system simultaneously. A separate group of portable or laptop computers, called the Station Support Computer (SSC) system, is used for non-critical functions that are independent of ISS systems. The SSC is used for payload and experiment operations, astronaut schedules and timelines, and other independent functions. The PCS and SSC computers use similar hardware and can be swapped if required, though their software, connectivity and functions operate differently. The ISS uses 1.8 million lines of code, including test and simulation programs and over 40,000 lines of software for 16 of the computers that 'talk' with more than the 2000 sensors, effectors and embedded 'smart' controllers. Two computers are dedicated just to keep the station in the proper orientation as it orbits the Earth.

Rick Mastracchio on Expedition 39 in 2014, replaces the Enhanced Input/Output Control Unit Circuit Card in the External Multiplexer/Demultiplexer (MDM). The MDM will be taken outside of the module for installation during a spacewalk. This photo was taken in Node 2 Harmony.

During Expedition 31 in 2012 the crew watch the launch of a Dragon commercial spacecraft on a Station Support Computer (SSC) in the U.S. Laboratory Destiny. In this image are, clockwise (from bottom right), Oleg Kononenko, Don Pettit, Andre Kuipers, Joe Acaba, Sergei Revin (obscured) and Gennady Padalka.

Caution and Warning

Systems software tracks every ISS module and reports when conditions are abnormal. Detectors are located in every element around the station. Abnormal conditions are reported by illuminated lights on instrument panels and displays, various audible tones, and text and graphic messages on personal computers. The PCS is specifically designed to provide intuitive, easy-to-use software. It was essential to develop a network of information systems that could provide alerts to warn of impending danger or system failures and which enable the crew to understand and deal with each situation. There are four categories of Caution and Warning in place aboard ISS. Class 1, Emergencies are life-threatening and require immediate action, such as fire, depressurization, or toxic contaminants. Class 2, Warnings, requiring immediate corrective action to prevent loss of station or crew, such as guidance, attitude control and high atmospheric pressure. Class 3, Cautions, are not necessarily time-critical but could develop into a Warning like impending failure of part of the guidance or communications systems. Class 4, Advisories, non-critical, provide status of systems operating abnormally.

Alexey Ovchinin uses a PCS computer and is photographed during Emergency Soyuz Descent On-Board Training while in the Svezda Service Module, while Jeff Williams looks on during Expedition 48 in 2016.

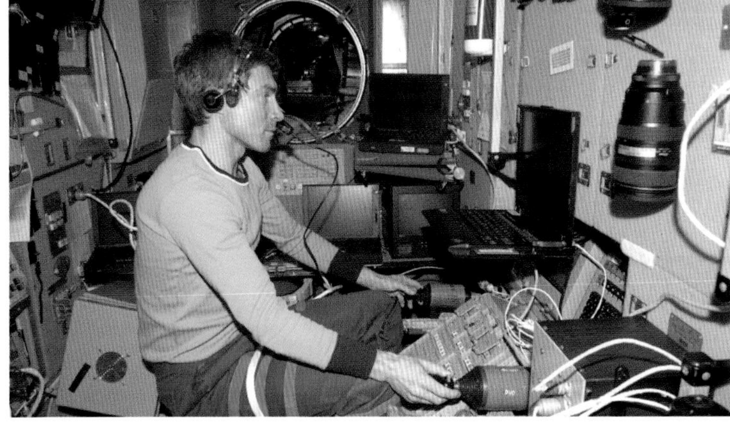

Sergei Krikalev on Expedition 11 works a hand controller on the TORU teleoperated control system in the Zvezda Service Module (SM) in preparation for the arrival of the Progress 18P spacecraft. The Simvol-TS screen and hand controllers, could manually fly and dock the Progress to the ISS in the event of a failure of the Kurs automated docking system.

Guidance, Navigation and Control

The Russian Motion Control System (MCS) and the US Guidance, Navigation and Control (GN&C) both guide and orient the ISS. The two systems exchange data and are complementary, except for propulsion which is provided solely by the Russians. The navigation portion of the US GN&C primarily relies on Global Positioning System (GPS) data to determine the ISS position, velocity and orientation. The Russian MCS can determine station orientation by tracking the stars, the Sun and the horizon. The system can also use a Russian version of GPS. Russian gyrodynes, similar to the US Control Moment Gyros (CMGs), can also be used to control orientation but are used principally when the Russian module flies independently.Orientation data is also drawn from two US-provided Rate Gyro Assemblies (RGAs) mounted in the S0 truss. Each RGA consists of three ring laser gyroscopes. These are mounted at 90° to one another in order to sense movement in three axes of rotation. They use variations in laser light beam lengths to sense attitude change and the rate at which the change is occurring. A US navigation system senses the ISS orientation and sends attitude data to most systems on the US segment. It keeps the US solar arrays aimed at the sun, and the S-band and Ku-band antennas pointed at their communication targets. The US segment capability to reorient the ISS is done using Control Moment Gyroscopes (CMG). The CMGs are located on the Z1 truss, berthed to Node 1 zenith port. This is done using electricity, which is readily available and replenishable from the ISS Electrical Power System. The four CMGs, are each a 98 kg (216 lb) steel flywheel, spun by electric motors at a constant rate of 6600 revolutions per minute. The system has a combined mass of 1197.5 kg (2,640 lb.). The CMGs typically are fully in control of ISS attitude positioning during day-to-day operations. They give the ISS stability and minimize required attitude adjustments. The fly wheels spin on gimbals that permit the CMGs to reorient to any position. As a spinning fly wheel is reoriented, it exerts a small torque, between 10 to 30 N-m (7 to 22 ft-lbs), about as much force as a person could exert by pushing against one end of the ISS truss. In orbit the ISS gently rocks back and forth. The slight rocking is due to aerodynamic drag which is countered by the CMG gyros gently nudging the ISS back towards its normal orientation. Infrequently, large perturbations, caused by venting, or by movements of massive objects on ISS, like the Canadarm2 or spacewalking astronauts, require a rocket motor firing to maintain orientation.

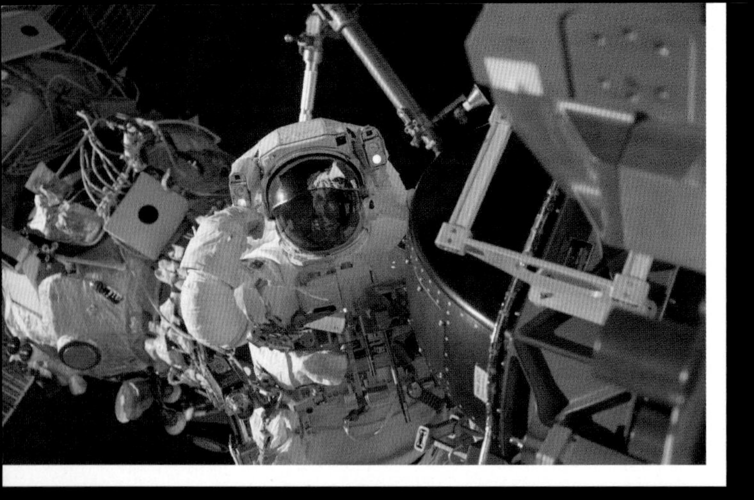

On STS-118 Richard Mastracchio removes a failed Control Moment Gyroscope (CMG), the black wheel, on the Z1 Truss.

Dave Williams anchored to the foot restraint on the Canadarm2, prepares to replace a faulty control moment gyroscope (CMG) in the station's Z1 truss in 2007. Here Williams is taking the spare CMG off of a stowage platform outside of the Joint Airlock Quest. Four CMGs are used to control the station's attitude in orbit.

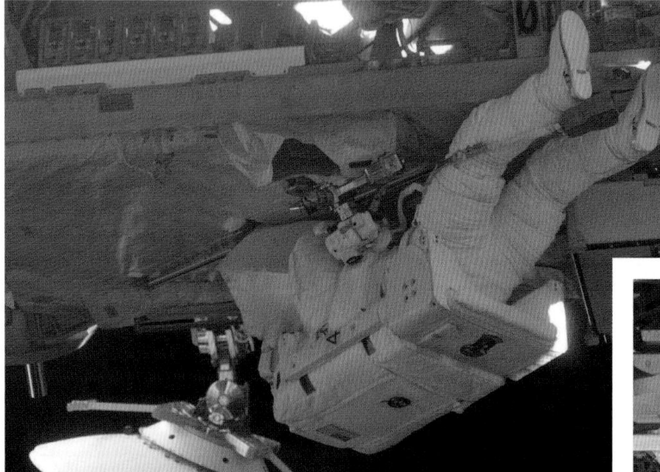

Christer Fuglesang, installs a GPS antennae on the S0 truss. During Expedition 20 in 2009.

Global Positioning System (GPS) Antenna on the S0 Truss.

Propulsion

Even though the ISS is in space, its low Earth orbit (LEO) does encounter molecules of Earth's atmosphere. Aerodynamic drag is small but constant, and will cause the ISS orbit to drop 25 to 50 m (82 to 164 ft) per day. The station's solar arrays are responsible for most of the drag, but can be 'feathered' to reduce the amount of propulsion required for reboost, however this has the effect of limiting the amount of electrical power that can be generated. Propellant consuming rockets must be used to reboost and maintain the ISS orbit. Fuel must be resupplied from Earth. Maintenance of the ISS orbital altitude depends on Russian propulsion systems. This was first provided by the FGB (Zarya), until the arrival of the Service Module (Zvezda) in 2000. The presence of the Service Module effectively blocks the use of the FGB engines. The Service Module subsequently took over responsibility for reboost. Propulsion has also relied upon visiting Progress and ESA ATV logistics vehicles and docked US Space Shuttles. Propulsion can also be called upon when the ISS needs its orbit changed to avoid orbital debris or to relieve the CMGs if these need to be de-spun for maintenance or if they have become 'momentum-saturated'.

During a spacewalk on Expedition 3 in 2003, Vladimir Dezhurov is photographed on the Service module Zvezda just beyond the module's attitude control rocket thrusters.

The aft end of the International Space Station (ISS) Service Module Zvezda and a Progress unmanned logistics spacecraft in 2000. Visible on the aft ends of both the Progress and the Service Module are the rocket engine nozzles used for re-boost to maintain the ISS orbit.

A view during a re-boost rocket firing. Ground controllers at Mission Control Moscow ignited the thrusters of a Progress spacecraft docked to the station's Service Module Zvezda for 14-minutes, raising the altitude of the station by about 3 km (1.9 mi) during Expedition 6 in 2003. This view is from a window in the Service Module.

Communications and Tracking System (C&T)

Two-way audio, video and command communications is provided between Mission Control, crewmembers inside the ISS, those outside performing spacewalks, Earthbound scientists and control centers around the world. Unmanned and manned spacecraft that visit the ISS communicate with the ISS and directly to ground stations and through communications satellites. Astronauts on-board the station closely monitor incoming vehicles and can "wave-off" any vehicle that might prove a hazard. Virtually all functions on board the ISS can be commanded by ground controllers through the C&T system. Ku-band, S-band VHF and UHF frequencies are all used by different elements of the system. As the station was assembled and expanded, multiple communications links were required across numerous frequencies and with different ground and space-based networks. A range of wavelengths were important for communications on board the space station, to and from ground stations and other spacecraft, and between ground facilities on Earth. Different communications frequencies are used for different systems and operations. There are relationships between the frequency and the amount of information a signal can contain. Higher-frequency radio waves can carry more information but are more susceptible to signal degradation transmitting through Earth's atmosphere. The US segment uses NASA Tracking and Data Relay Satellites (TDRS) for transmission of many signals between ISS and Earth. Using two or three TDRS satellites, nearly continuous communications can be maintained in the S- and Ku-wavelength bands. Commands can also reach the ISS through Russian ground stations. But they are only in range of ISS for a portion of most orbits. Russia also has "Luch" tracking and data relay satellites, and the Japanese can also provide geosynchronous communications satellite relays at times. Information is continuously transmitted providing status of ISS systems configuration and operation, science experiment data, information from visiting vehicles, alarms and warnings, and two-way audio and video. The development of technologies such as Global Positioning Systems, mobile/satellite phone and smart phone technology, the internet, the expansion of email and social network systems and the miniaturization of computers have all aided communications functions in the ISS program over its three and a half decades.

The S1 Truss with Canadarm2 in the foreground. An S-Band Antenna Subassembly (SASA) (highlighted) is mounted above the truss. The S-Band system is the primary means of sending telemetry data and voice communications from the ISS to the ground and for communicating with other approaching spacecraft.

Steve Bowen behind the Ku-band antenna on the Z1 truss during an EVA. Ku-band is used to transmit payload data, 2-way voice communications, 2-way video, Email, internet and telephone calls and for ground operators to log onto the ISS computer system. It requires accurate pointing of the large dish antenna towards TDRSS satellites.

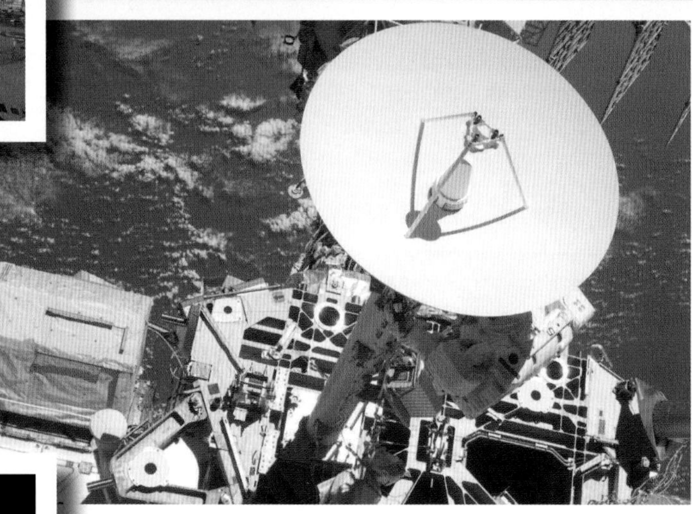

Garrett Reisman, anchored to Canadarm2, holds a replacement antenna for high-speed Ku-band transmissions on Expedition 23 in 2010.

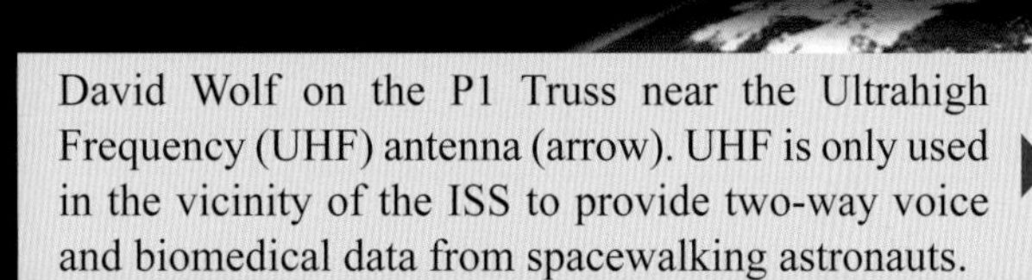

David Wolf on the P1 Truss near the Ultrahigh Frequency (UHF) antenna (arrow). UHF is only used in the vicinity of the ISS to provide two-way voice and biomedical data from spacewalking astronauts.

Michael E. López-Alegría (left - red stripes) and John B. Herrington on the P1 truss. The Wireless Video System (WVS) External Transceiver Assembly (WETA) is mounted on the truss to the top of the picture (arrow).

Nikolay Budarin in the Equipment Lock of the Quest Airlock. wearing a Liquid Cooling and Ventilation Garment (LCVG) and the lower half of an Extravehicular Mobility Unit (EMU). Over his shoulder are three Audio Terminal Units (ATUs). ATUs are intercoms located throughout the US segment of ISS and are used by the crew to talk to control centers on the ground. They have speakers which are used to annunciate emergency, warning and alarm tones.

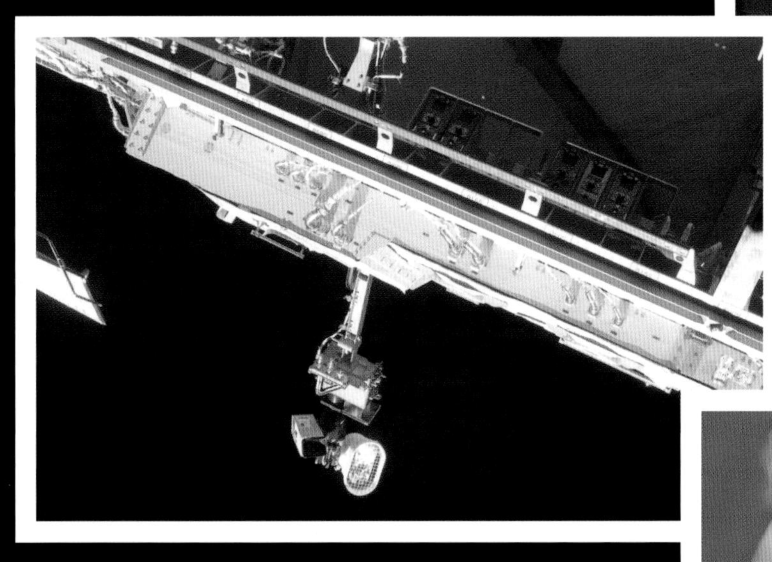

View of the S1 Truss External Television Camera Group (ETVCG). Standard definition ETVCGs are mounted on the exterior of the ISS and can be panned, tilted, and zoomed. They also have lights for use during night operations.

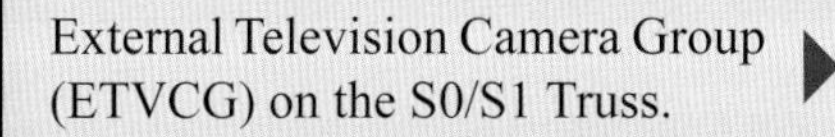

External Television Camera Group (ETVCG) on the S0/S1 Truss.

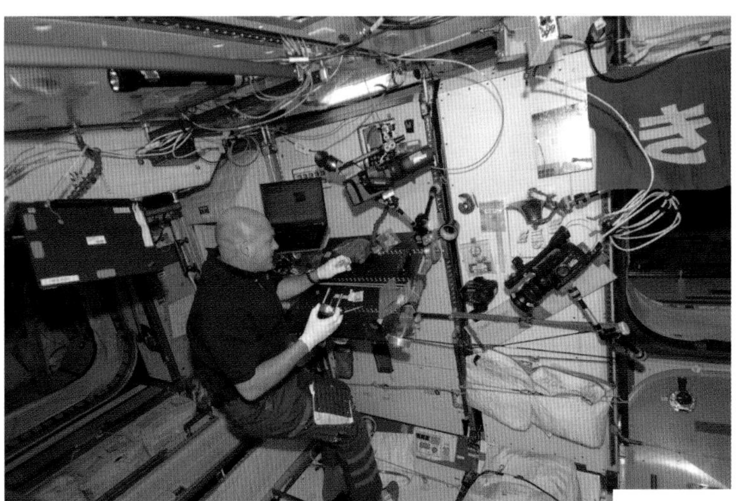

Andre Kuipers with several video cameras inside the Node 2 'Harmony' on Expedition 30 in 2012.

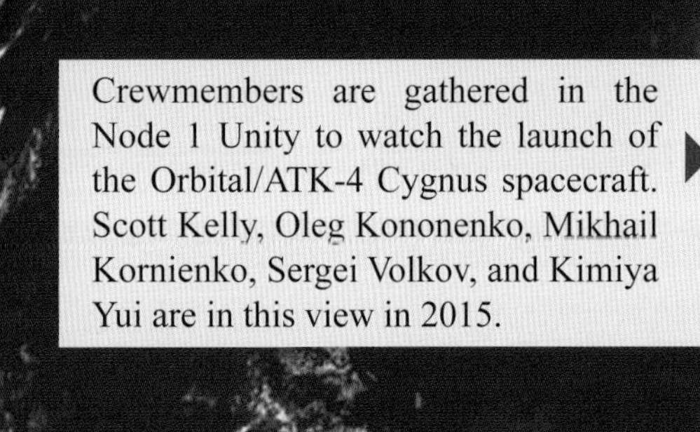

Crewmembers are gathered in the Node 1 Unity to watch the launch of the Orbital/ATK-4 Cygnus spacecraft. Scott Kelly, Oleg Kononenko, Mikhail Kornienko, Sergei Volkov, and Kimiya Yui are in this view in 2015.

Environmental Control And Life Support System (ECLSS)

The Environmental Control and Life Support System (ECLSS) provides and controls oxygen, nitrogen and other atmospheric gases for breathing; it maintains temperature and pressure and removes water vapor, carbon dioxide and other contaminants. ECLSS monitors the atmosphere with fire detection and suppression systems. Early in the ISS orbital phase, the Russian elements, FGB (Zarya) and Service Module (Zvezda) provided the ECLSS functions. The Russian system could maintain a crew size of three. As the US segment was established with its own ECLSS components, the US began providing additional oxygen, nitrogen, pressurization and water reclamation capabilities. The US system can maintain a crew size of about seven. In the US segment, the ECLSS consists of the Atmosphere Control and Supply System (ACSS); the Temperature and Humidity Control System (THCS); the Atmosphere Revitalization System (ARS); and the Fire Detection and Suppression System (FDSS). These systems provide and maintain atmospheric pressure, internal temperature, humidity and breathable air monitoring, maintaining a shirtsleeve environment inside the ISS. Although cabin air flows between all modules of ISS, and there is capability for transfer of water between modules, the Russian segment and US segment of the ISS ECLSS operate largely independently, providing redundancy and together maintaining the ISS atmosphere.

Rex Walheim, during a spacewalk from Shuttle STS-122 in 2008, prepares to replace a Nitrogen Tank Assembly (NTA) stored in the P1 truss.

Russian and US ECLSS Design Philosophies

The U.S. and Russian ECLSS design philosophies are different. The Russian approach to ensuring that a capability is maintained, even in the case of malfunctions, tends toward having backup capabilities provided by a different type of system. For example, to maintain the supply of breathing oxygen on ISS, if the 'Elektron' electrolytic O_2 generator fails, the backup is either oxygen stored in pressurized tanks or perchlorate chemical candles which are burned to create oxygen. In the case of a malfunction, the primary system is restored to functionality by the system's replacement on a regular resupply mission. The U.S. approach to ensuring that a capability is maintained tends towards using redundant equipment; i.e., having two identical units with one reserved for emergency use, or alternatively, operating multiple units at less than full capacity.

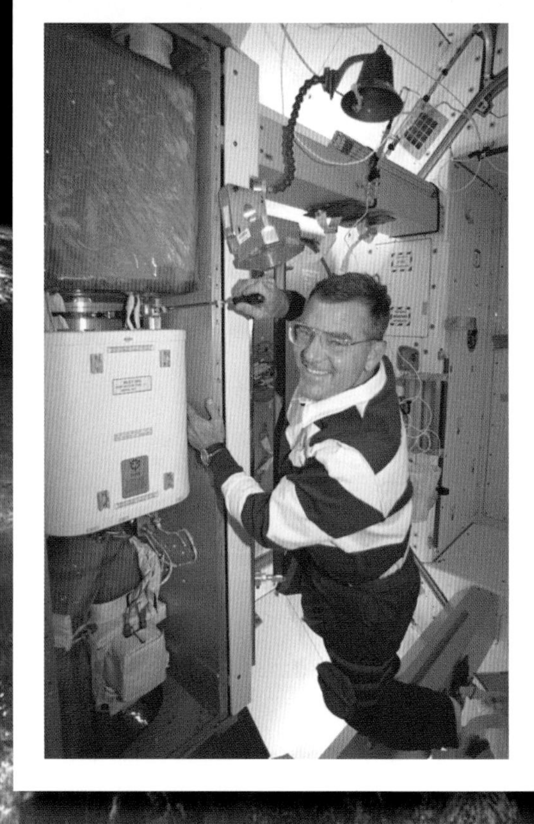

James S. Voss works on a fan at the inlet of hardware installed in the floor in Node 1, in 2000.

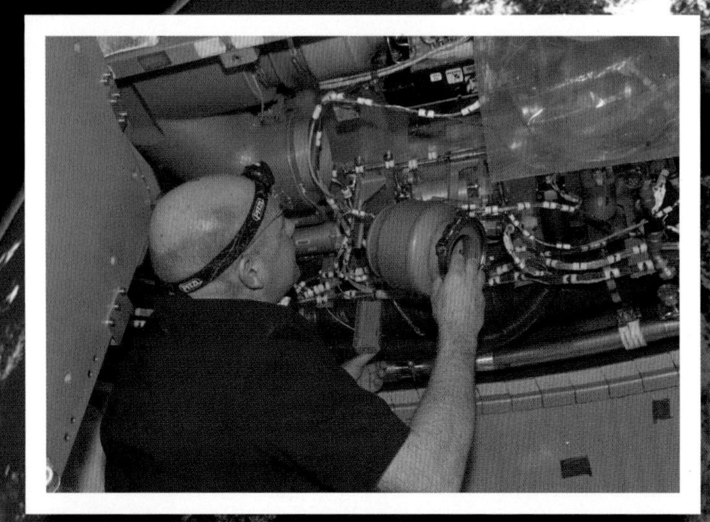

Andre Kuipers on Expedition 30 inspects and cleans ventilation ducts, screens and blowers in the ESA Laboratory Columbus in 2012.

Susan J. Helms works with a water container in the Service Module Zvezda in 2001. Inter-Module Ventilation (IMV) valve and supply and return hoses in the hatch leading from the Node 1 Unity are visible.

Atmosphere Control and Supply (ACSS) and Atmosphere Revitalization System (ARS)

Oxygen and Nitrogen Replenishment

ECLSS provides a breathable atmosphere at normal Earth pressure and comfortable temperature and humidity. The system provides oxygen (O_2) and nitrogen (N_2) at the same ratio as on Earth: about 21% oxygen, 78% nitrogen. Pressure on the ISS is maintained between 724 to 770 mm Hg (~9.66 bar, 14.0 to 14.9 psi), temperatures between 22ºC to 26ºC (72ºF to 79ºF) and humidity at around 45% for crew comfort and to minimize condensation on interior surfaces. Water is removed from the cabin air as humidity condensate at the approximate rate of 1.5 kg/person/day (3.3 lb/person/day). Contaminants such as carbon dioxide (CO_2) and other impurities are removed. The ECLSS is relied upon to maintain a healthy atmosphere in the cabin. The types and concentrations of materials in the atmosphere are critical to human life and yet the crew cannot readily sense the concentration of oxygen or most other materials in the air they breathe. Therefore, the crew must rely upon instruments to verify that the materials in the atmosphere are being maintained in a safe balance. If levels of O_2 or N_2 need to be adjusted, it can be done by mission control or by the crew on-board. As the crew breathes, O_2 is consumed and needs to be replenished. In a perfect system, N_2 would never need to be resupplied, but a small amount of leakage occurs on the ISS, and so supplies are carried to maintain the correct proportion. Tanks mounted outside the Joint Airlock 'Quest' store O_2 and N_2 for use when needed. An additional O_2 tank is stored on the truss. Logistics resupply vehicles carry O_2 and N_2 in their atmospheres. They can also carry high pressure storage tanks from which the ISS storage tanks can be refilled or the tanks can be vented directly into the ISS atmosphere. Both the U.S. and Russian systems use electrolysis of water for generating oxygen though there are significant design differences. The U.S. approach was to design hardware to be serviceable in orbit. Components of the U.S. Oxygen Generation System (OGS) are designed as orbital replaceable units (ORU) and are readily accessible for replacement. The Russian approach does not require that components be individually replaceable, however the entire system, called the Elektron, is designed for removal and replacement and to fit inside the Russian logistics vehicles. The 'Elektron' water electrolysis system provides the primary source of oxygen in the Russian segment, with supply based on an average human consumption of 20-25 liters (35.2- 44 pints) per hour per person. An elektron unit electrolyzes water to produce up to 5.13 kg (11.31 lbs.) / day of oxygen. If the system detects a partial pressure variance of plus or minus five millimeters of mercury relative to this nominal value, the elektron commands changes to the electrolysis current accordingly, resulting in an adjustment to the oxygen flow to return to nominal levels. The elektron unit in the Service Module served as the primary source of oxygen supplied to the ISS crew until systems were orbited by the US. The Russian system can support a crew size of three.

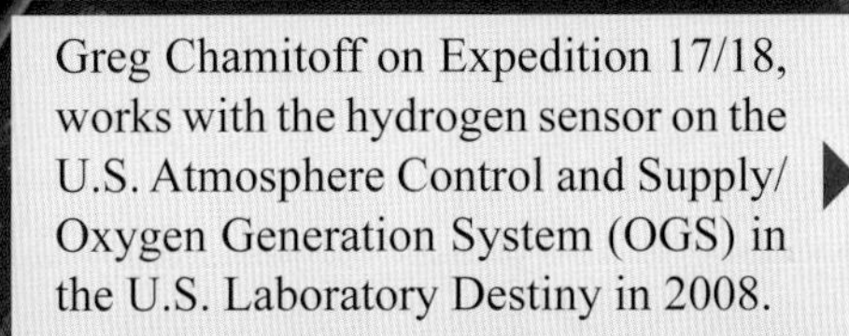

Greg Chamitoff on Expedition 17/18, works with the hydrogen sensor on the U.S. Atmosphere Control and Supply/ Oxygen Generation System (OGS) in the U.S. Laboratory Destiny in 2008.

One back-up to the Russian Elektron is the Solid Fuel Oxygen Generator (SFOG) and generates O_2 by burning perchlorate candles (lithium perchlorate was used on Mir and potassium perchlorate is used on ISS). Typically the candles are used on ISS only after other supplies have been consumed.

Carl E. Walz works on the Elektron Oxygen Generator in the Service Module Zvezda in 2002. In a failure the entire unit is typically replaced.

Andrey Borisenko (left), Alexander Samokutyaev (center) and Sergei Volkov are pictured with Russian Elektron oxygen generation system VM BZh Liquid Units. An old BZh Liquid Unit was being replaced in the Service Module Zvezda.

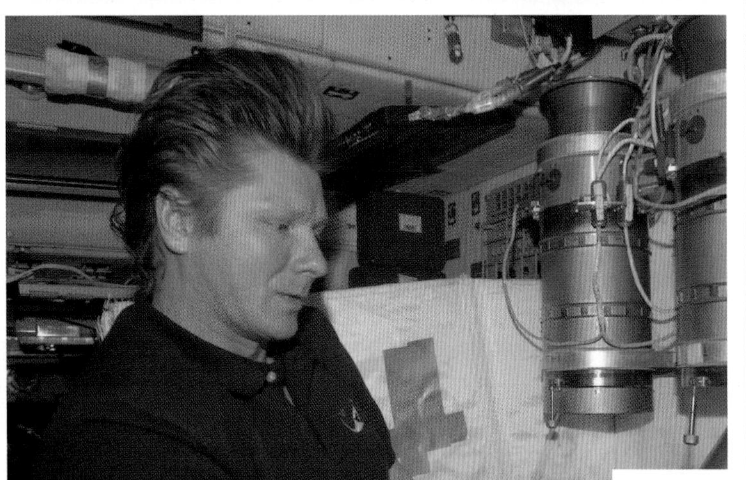

Gennady Padalka performs a check on the Russian POTOK-150 micron air filter unit of the Service Module Zvezda's SFOG (Solid Fuel Oxygen Generator) candles while the US Oxygen Generator System (OGS) was not functioning.

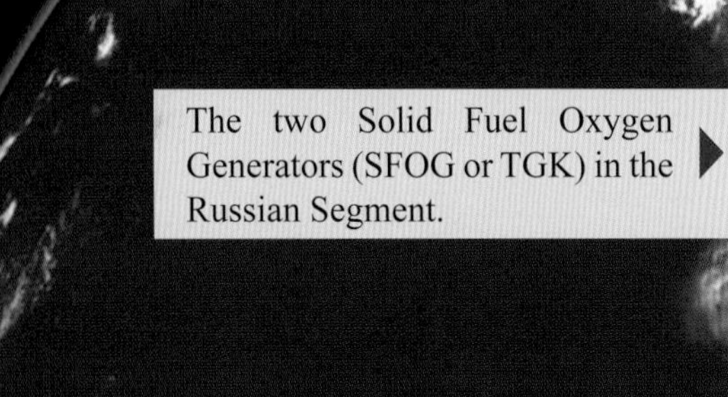

The two Solid Fuel Oxygen Generators (SFOG or TGK) in the Russian Segment.

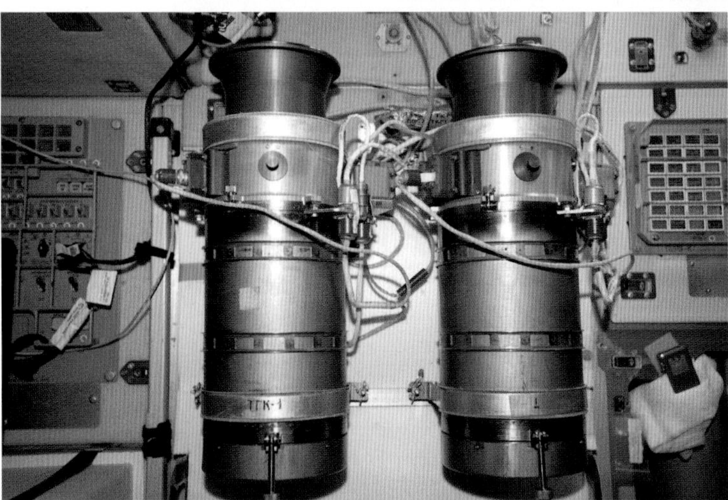

Carbon Dioxide

After O_2 and N_2, CO_2 is the next-biggest atmospheric concern. Besides replenishing oxygen and maintaining O_2 and N_2 in the correct ratio, the ECLSS must remove carbon dioxide and water vapour. Exposure to higher concentrations of CO_2 can lead to headaches, increased respiration, reduced performance or depression of the central nervous system. Sensitivity to CO_2 appears to change in weightlessness. On Earth, gravity and convection help to circulate air. Hot air rises and cool air falls. But in space there is no convection and therefore air must be actively circulated throughout the interior volume. Air is diffused through the two ceiling stand-off vents into the module cabin. Air is captured for recirculation through return air ducts in the floor stand-offs. In closed areas where air might otherwise be more static, such as the astronauts' sleep compartments, fans are used to circulate air to ensure that carbon dioxide (CO_2) does not become concentrated in proximity to the astronaut's mouth or nose. In the early years of ISS operation, the primary way of removing CO_2 on the USOS was via the Carbon Dioxide Removal Assembly (CDRA). The CDRA uses a zeolite and silica gel molecular sieve to desiccate the incoming air and adsorb the CO_2. Air and water are returned to the cabin. Two zeolite beds are used in alternating sequence. As CO_2 from the first zeolite bed is desorbed and vented overboard into space, a second zeolite bed is used to absorb CO_2. The venting of CO_2 to space results in a net loss of O_2. The CDRA in the U.S. segment of ISS and Vozdukh from the Russian segment of ISS, both remove CO_2 from the cabin. The Russian Micro-purification Unit (BMP) removes atmospheric contaminants. All of these vent some air overboard during their normal operation.

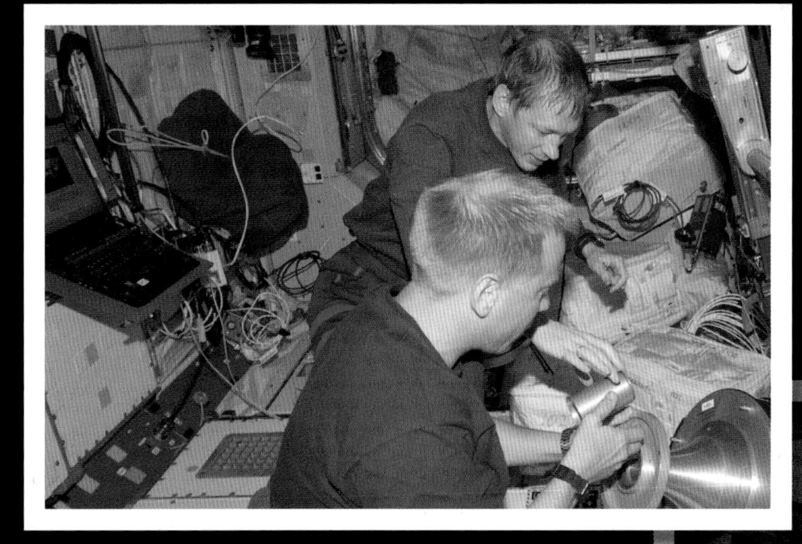

Tim Kopra (foreground) and Frank De Winne, on Expedition 20 in 2009, work with a carbon dioxide removal kit adapter in the Node 1 Unity, troubleshooting a problem with the heater of the Carbon Dioxide Removal Assembly (CDRA).

Doug Wheelock on Expedition 24 in 2010, works on the Carbon Dioxide Removal Assembly (CDRA) in the Air Revitalization 1 (AR1) rack in the US Laboratory Destiny.

Michael Fincke, during Expedition 28 in 2011, performs maintenance on a desiccant/sorbent bed on the Carbon Dioxide Removal Assembly (CDRA) in the Japanese Laboratory Kibo.

Later in the program, in 2010, a Sabatier CO_2 methanation reactor was added. With the Sabatier reactor, CO_2 from the astronauts' metabolic processes reacts with H_2 from the Oxygen Generation Assembly (OGA) to produce methane (CH_4) and water.

Koichi Wakata on Expedition 39 in 2014 works in the Node 2 Harmony to reassemble the desiccant adsorbent on a Carbon Dioxide Removal Assembly (CDRA).

Scott Kelly (left) and Terry Virts during Expedition 43 in 2015 perform maintenance on the Carbon Dioxide Removal Assembly (CDRA) in the Japanese Laboratory Kibo.

Temperature and Humidity Control System (THCS)

The Common Cabin Air Assembly (CCAA) circulates air within most modules. The CCAA blows the returned warm, humid cabin air into a heat exchanger. In the heat exchanger the air is cooled and water vapor is condensed to a liquid. A centrifugal liquid separator removes condensed liquid water from the air stream. As water is removed it is sent to the Water Recovery and Management Subsystem.

Water Recovery and Management

Humans rely more on water, by weight, than on any other consumable. Water is required by humans for consumption, for drinking, rehydrating food, personal hygiene, and for equipment cooling. Human metabolic processes convert water into perspiration and urine. On long duration missions like the ISS, considerable weight can be saved if consumables are regenerated using a sophisticated closed-cycle ECLSS. When the ISS ECLSS works, the use of consumables like water, air, oxygen, or nitrogen, is minimized and a small supply can last a long time. This minimizes the quantities of consumables required to be launched from Earth for resupply. Wastewater and urine is collected from the commode. Plumbing transfers pre-treated urine from the Waste Management Compartment to the Water Recovery System racks. Humidity condensate is received as air circulates through the cabin and collected in a condensate storage tank. The Water Recovery System (WRS) electrolyzes the waste fluids and processes it through a series of filter beds and a high-temperature catalytic reactor to filter out gasses and solids and produce O_2 for breathing and potable water. A Process Control and Water Quality Monitor verifies that the water is suitable for human consumption. The unit is designed to process and supply approximately 93 kg (205 lb.) of water per day to support a crew of 7 people.

Soichi Noguchi with Sabatier methanation reactor during operations in the Japanese Laboratory Kibo on Exedition 23 in 2010.

Doug Wheelock on Expedition 25 in 2010 installs the Sabatier system in Node 3 Tranquility. The Sabatier reactor creates water from by-products of the station's Oxygen Generation System (OGS) and Carbon Dioxide Removal Assembly (CDRA). The hardware is actually owned and operated by the Hamilton Sundstrand company, which sells water to NASA and the ISS Program.

Samantha Cristoforetti on Expedition 42 in 2011, manually purges the Sabatier reactor accumulator in preparation for the Sabatier compressor change out.

Michael Fincke on Expedition 18, works on the Water Recovery System (WRS) in the Destiny US Laboratory in 2008.

On Expedition 19 in 2009, after NASA's Mission Control gave the crew a "go" to drink water that the station's new recycling system had purified, the three celebrated with a 'toast'. Pictured are crewmembers Gennady Padalka (center), Mike Barratt (right) and Koichi Wakata, holding drink bags with special commemorative labels in the Destiny laboratory.

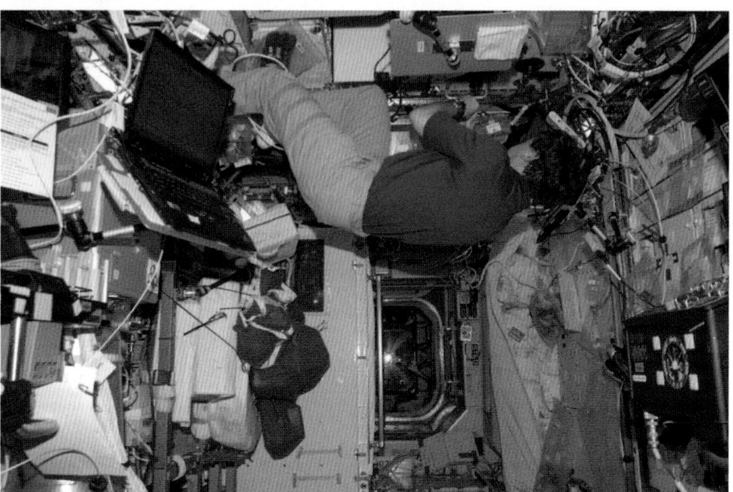

Michael Barratt on Expedition 19/20 collects a water sample from the Water Recovery System (WRS) Potable Water Dispenser (PWA) in the US Laboratory Destiny in 2009.

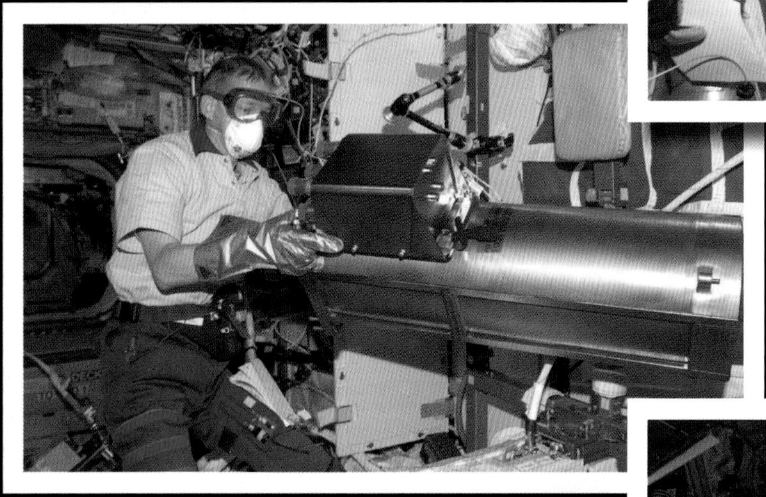

Frank De Winne on Expedition 20 in 2009, works with the Water Recovery System, supporting the test fill of a failed Recycle Filter Tank Assembly (RFTA) in the US Laboratory Destiny.

Jeffrey Williams on Expedition 22 in 2010, installs a Urine Processor Assembly / Distillation Assembly (UPA DA) in the Water Recovery System (WRS) rack in the US Laboratory Destiny.

Mike Fossum on Expedition 29 in 2011, works with the Water Recovery System (WRS) Fluids Control and Pump Assembly (FCPA) in the US Laboratory Destiny.

Mike Fossum on Expedition 29 in 2011, installs the Advanced Recycle Filter Tank Assembly (ARFTA) tank top assembly at the Urine Processor Assembly / Water Recovery System (UPA WRS) in the US Laboratory Destiny.

Dan Burbank, Expedition 30 in 2012, in the Node 2 Harmony, replaces a catalytic reactor in the Water Recovery System (WRS).

Control of Contaminants

The ECLS must be relied upon to identify and remove toxic materials or trace contaminants. Several kinds of unanticipated contaminants are possible in the closed environment of the ISS. Contaminants can include undesirable solids, liquids, or gases. These can originate in materials used during manufacture or during off-gassing after manufacturing or from decomposition from aging. Contaminants can result from maintenance, leaked propellants, leaked cooling fluids, by-products of experiments or payloads, or from over-heated motors or overheating of non-metallics, from crew effluents or detritus, or from hygiene consumables, food, or other unanticipated sources. The ISS environment is closed and sealed and has been for about two decades. Gases and fluids are reused over and over again. Contaminants can remain in the cabin atmosphere and become concentrated over time. On Earth, contaminants are often ignored because of ventilation; at worst they usually only cause temporary discomfort. In a long duration spacecraft, contaminants can become concentrated, and can create unanticipated difficulty with the functioning of environmental control hardware. As the contaminant concentrations increase, changes to the system design may be required to preclude and eliminate danger to the crew. Composition of the atmosphere in the ISS is measured by the Major Constituent Analyzer (MCA), a mass spectrometer that measures O_2, N_2, CO_2, H_2, water (H_2O), and methane (CH_4) in the atmosphere. In the U.S. Lab, the Volatile Organic Analyzer (VOA) uses a gas chromatograph/ion mobility spectrometer (GC/IMS} to identify violations of contaminant concentration limits. In the case that limits are exceeded, the crew is alerted and ventilation is commanded to off or standby.

The Trace Contaminant Control System (TCCS) removes and disposes of gaseous contaminants. It consists of an activated charcoal adsorption bed and thermal catalytic oxidizer with post-sorbent bed.

Ron Garan on Expedition 27 in 2011 works with the Major Constituent Analyzer / Data & Control Assembly (MCA DCA) in the US Laboratory Destiny.

Terry Virts on Expedition 42 in 2014 installs components of the Major Constituent Analyzer (MCA) in the US Laboratory Destiny.

The activated carbon is impregnated with phosphoric acid for ammonia removal. Ammonia is a concern because of its toxicity and its use in the ISS thermal control system. The high-temperature catalytic oxidizer converts carbon monoxide, methane, hydrogen, and other low molecular weight compounds to carbon dioxide, water or other compounds which can be more easily utilized. A sorbent lithium hydroxide (LiOH) filter removes acidic products of catalytic oxidation.

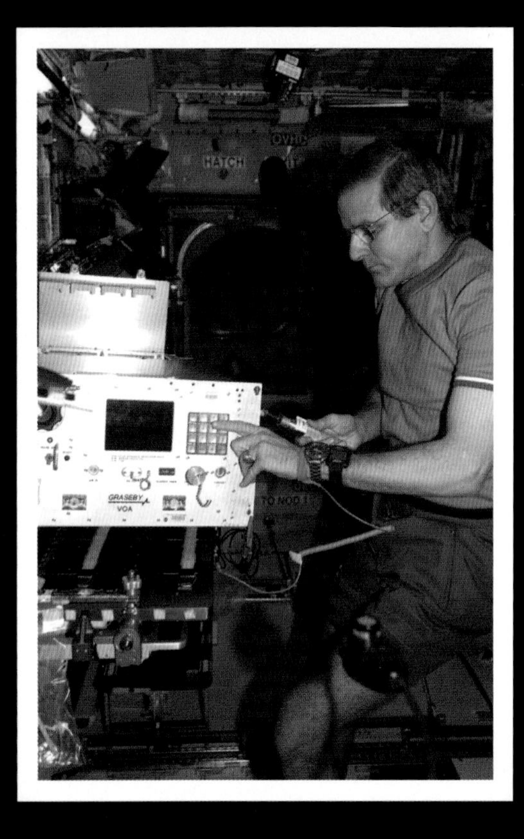

▲ William S. McArthur Jr. troubleshoots the malfunctioning Trace Contaminant Control Subassembly (TCCS) in the US Laboratory Destiny on Expedition 12 in 2005.

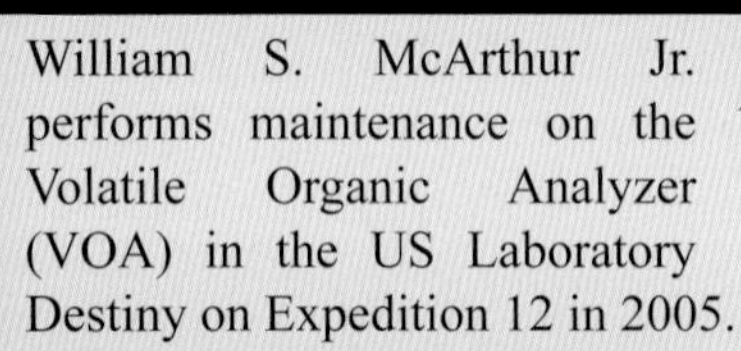

William S. McArthur Jr. performs maintenance on the Volatile Organic Analyzer (VOA) in the US Laboratory Destiny on Expedition 12 in 2005. ▲

Reid Wiseman during Expedition 40 in 2014, works with the Total Organic Carbon Analyzer (TOCA) while Alexander Gerst (background) gets a workout on the advanced Resistive Exercise Device (aRED) In Node 3 Tranquility. ▶

Cabin air captured for recirculation through return air ducts in each module's floor stand-offs, pass through high efficiency particulate air (HEPA) filters that capture particulate and micro-organism contaminants. Crew members manually vacuum the filters periodically to dispose of the contaminants. An element in the Russian Elektron water reclamation electrolysis system subsystem is the Harmful Contaminants Filter, which removes contaminants after a fire. It has long been known that crewmember's bones demineralize and weaken in zero gravity. What was unanticipated by ECLSS designers was that the minerals exited the crewmember as elevated calcium sulfate precipitate. The calcium sulfate interfered with proper operation of the water distillation and purification system. Other unforeseen contaminants, the origins of some which are still uncertain, have also required environmental control system changes. U.S. spacesuits were originally designed for use on short-duration Space Shuttle missions. They were intended to be returned to the manufacturer after every Shuttle flight for cleaning and refurbishment. On the ISS, the same space suit remains in orbit for years. Facilities are not available on the ISS for complete cleaning or refurbishment and over time, contaminants can increase. In one instance, increased particulates slowed the circulation of environmentally controlled fluids. Fluids backed up in the suit and became a breathing hazard for an astronaut during a spacewalk.

Fire Detection and Suppression System (FDSS).

There are cabin fire and smoke detectors throughout the ISS modules, as well as fire indicator panels. Additional smoke sensors are located in specific racks. Fire suppression is provided by several portable fire extinguishers; gas masks are provided throughout the interior of ISS.

Habitation - The Human System

The space environmental can be inhospitable and threatening. People on the ISS can be isolated. Crew performance and safety depend upon the design of the ISS and the facility has to provide for all of the human's needs. Physical and environmental conditions can induce physiological, biomedical, and psycho-social stressors on the astronaut that can affect crew performance and safety. For these reasons, more so on the ISS than any previous space vehicle, the vehicle's design emphasized habitability. For long-duration habitation, consideration must be given to the full range of human activities and functions. All functions must be accommodated; all should be designed into the system from the outset. For early, short-duration spacecraft, the emphasis was on engineering, mission operability, mass constraints, and cost mitigation. Early spacecraft were designed to be operated, not lived in. For ISS, however, habitability functions were a priority and design requirements were established early in the program.

Kjell Lindgren during Expedition 44 in 2015, wears protective breathing apparatus that would be used in the unlikely event of a fire or hazardous chemical leak inside the pressurized modules of the ISS. Familiarization with safety and emergency equipment is standard practice onboard.

"A large part of what we do on the space station is invent, and then test and then prove space-qualified hardware and design. The carbon dioxide removal system, the trace gas contaminant removal system; the waste recycling system, the low and high temperature loops for you to remove condensation and what fluid you put in those loops; how you generate electricity; the type of battery design; thermo-regulation design on those batteries. All of those are legacy designs that have never been tested in the past and are now immensely more tested and proven as a result of what we've done for ISS. So those are staging posts for going further on into space." - Chris Hadfield

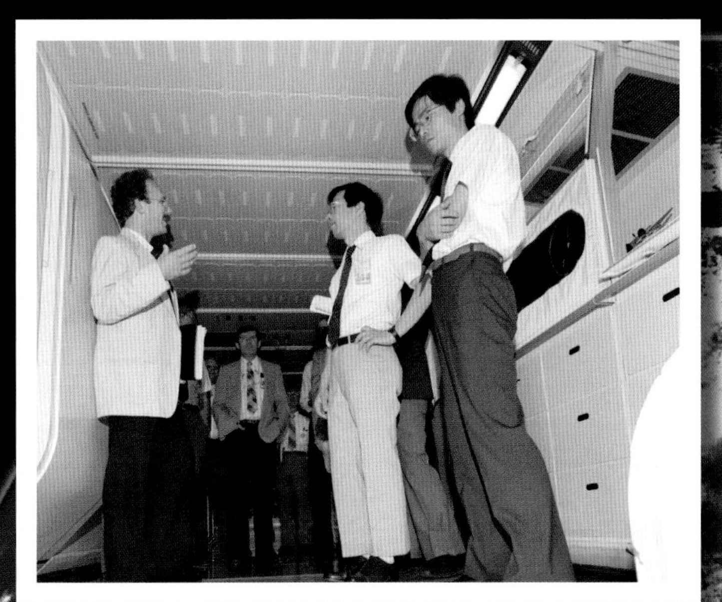

During a Crew Station Review in the Man-Systems Laboratory at the Johnson Space Center, architectural control manager Gary Kitmacher discusses the configuration of the US Laboratory and Habitation modules and standard racks with Japanese engineers. Kitmacher originated the Cupola concept and was one of its leading proponents.

During a Crew Station Review in the Man-Systems Laboratory at the Johnson Space Center, designer Jay Cory discusses with Japanese engineers access to utility runs in the corner stand-offs of the US Laboratory and Habitation modules. Cory was a leading proponent of modular architecture on ISS.

In 1988, during a Crew Station Review in the Man-Systems Laboratory at the Johnson Space Center, Cupola architect Laurie Weaver reviews the early design of the Cupola with US and Japanese engineers and architects.

In 1987, an early configuration of the Cupola is assessed for accessibility and visibility in the Man-Systems Weightless Environment Training Facility at the Johnson Space Center.

Early Crew Provisions

Prior to the first long duration expedition crew boarding the station in November 2000, visiting crews used the Space Shuttle for personal hygiene, eating, sleeping and off duty periods. However, with the addition of the Russian Service Module (Zvezda), many of the required habitation facilities accommodated the ISS crew members.

Upgrading the habitation

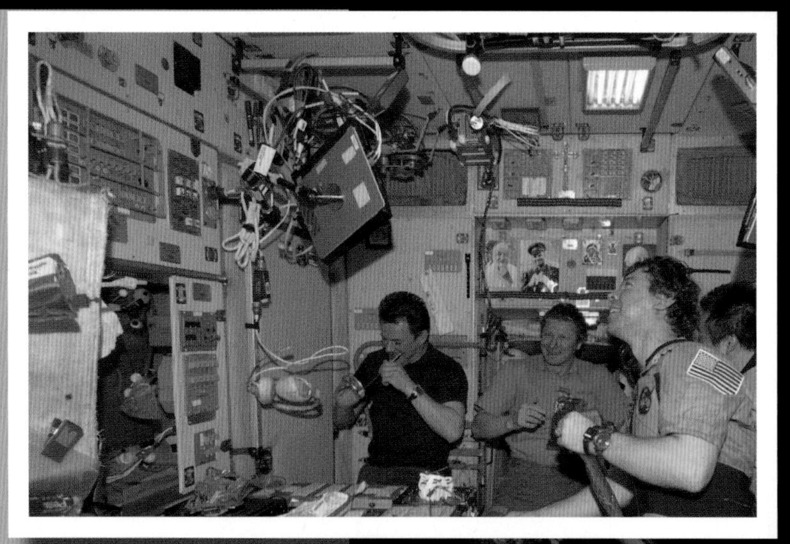

Expedition 20 crew shares a meal at the galley in the Service Module Zvezda in 2009. Pictured (from the left) are Roman Romanenko, Gennady Padalka, Michael Barratt and Koichi Wakata.

As the number of long duration ISS crew members increased, so did the need to house them. By 2008, the station was nearing completion. In 2008, Shuttle Endeavour carried fixtures to enlarge the habitation facilities to support six long duration crew members.
Additional facilities for the larger crew size created a more relaxed environment. This was important for the well-being and productivity of the crew. It helped to increase crew efficiency, which in turn raises productivity and leads to more and better results from each increment. It was also important for the program to maintain a fully functional and serviceable orbiting facility over a period of multiple decades and to balance those needs with the ability to support scientific investigations, maintenance and housekeeping.

Galley

The galley provides food warmers, a refrigerator and a water dispenser. The original galley is located in the Russian Service Module 'Zvezda'. A second galley is installed in the US segment and has been placed in Node 1 'Unity' and the US Laboratory Module 'Destiny' at different times. The food warmers heat food packages. The water dispenser can supply either heated or unheated potable water. The station's refrigerator measures 42 cm deep, 26 cm wide and just 17 cm high (16.5 x 10.2 x 6.7 in).

During Expedition Two in 2001, Susan J. Helms (left), Yury V. Usachev and James S. Voss sharing a meal at the table in the Service Module Zvezda.

Nicole Stott and Roman Romanenko, and Robert Thirsk (mostly out of frame at right) on Expedition 20/21 in 2009, at the galley in Node 1 Unity.

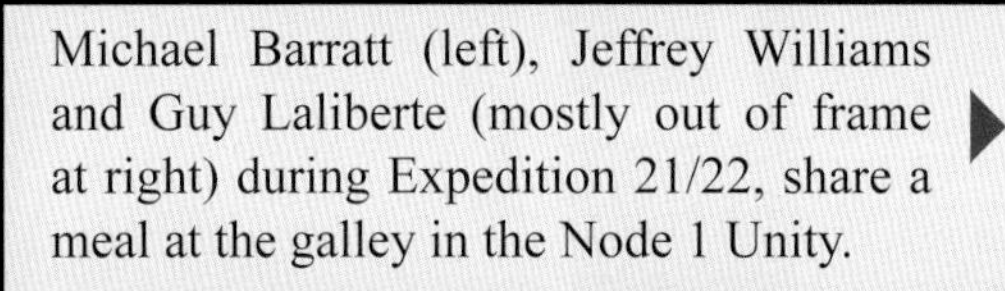

Michael Barratt (left), Jeffrey Williams and Guy Laliberte (mostly out of frame at right) during Expedition 21/22, share a meal at the galley in the Node 1 Unity.

Crews from Expedition 21 and Atlantic/STS-129 gather for a meal at the galley in the Node 1 Unity in 2009. From left to right: Frank De Winne; Mike Foreman, Charles O. Hobaugh, Randy Bresnik, Robert Thirsk, Leland Melvin (top of frame), and Robert L. Satcher Jr.

View of the Waste and Hygiene Compartment (WHC) in the Service Module Zvezda during Expedition 7 in 2003.

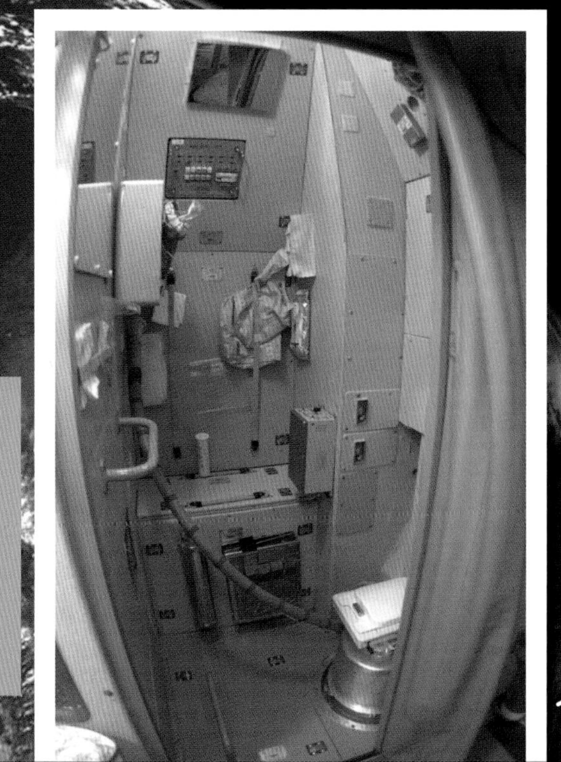

Waste and Hygiene Compartment (WHC)

The original Waste and Hygiene Compartment (WHC) is located in the Russian Service Module 'Zvezda'. A second WHC was installed in the US Laboratory 'Destiny' and later moved to Node 3 Tranquility. The US segment WHC incorporates a Russian-built toilet essentially identical to the commode in the Russian segment. The design incorporates separate channels for liquid and solid waste. Any solid waste goes into a holding tank, while urine is carried to the three-rack Water Recovery System (WRS) for conversion to potable water.

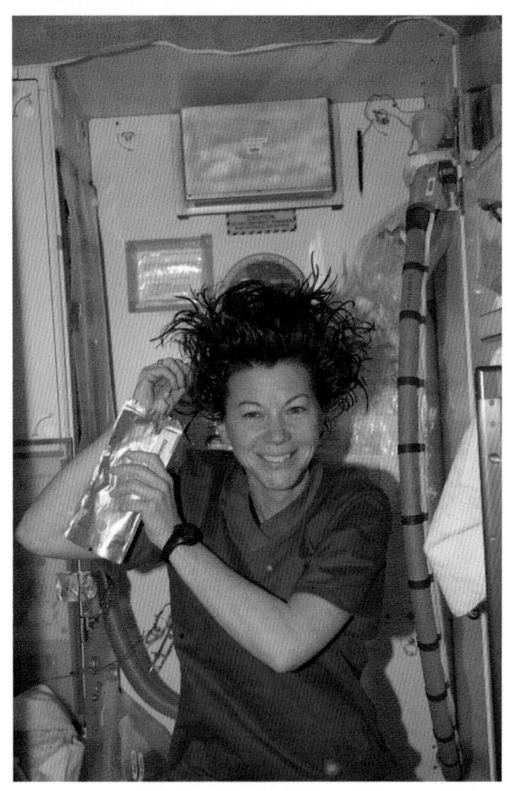

Catherine (Cady) Coleman washes her hair in the Waste and Hygiene Compartment (WHC) in Node 3 Tranquility during Expedition 26 in 2010.

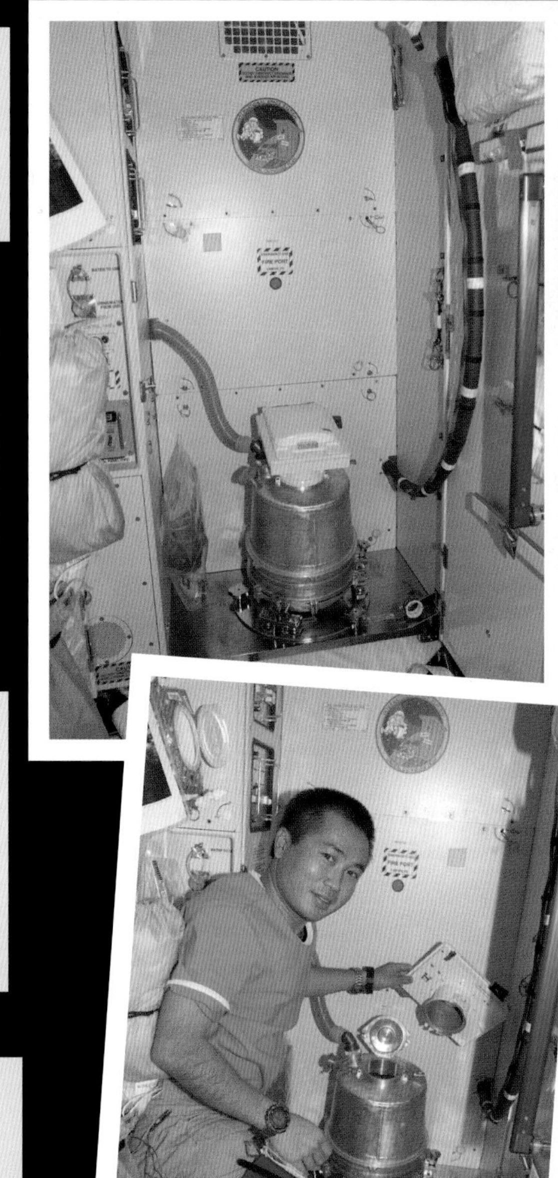

View of the Waste and Hygiene Compartment (WHC) in the US Laboratory Destiny during Expedition 19 in 2009.

Koichi Wakata performs the daily ambient flush of the potable water dispenser in the Waste and Hygiene Compartment (WHC) in the US Laboratory Destiny during Expedition 19/20 in 2009.

Jeffrey Williams removes a mostly full EDV-U urine container from the U.S. Waste and Hygiene Compartment (WHC) in the US Laboratory Destiny during Expedition 21 in 2009.

Aki Hoshide, during Expedition 33 in 2012, removes and replaces the water pump and pre-treat tank in the Node Tranquility. The tank contains five liters of pre-treat solution, a mix of H_2SO_4 (sulfuric acid), CrO_3 (chromium oxide, for oxidation and purple color) and H_2O (water). The pre-treat liquid is mixed with water and used for toilet flushing.

Crew Quarters

A 'private' crew quarters sleep and personal compartment is provided for each crew member. Personal belongings and clothing can be stored out of the way in stowage pockets and compartments. A sleep restraint is attached to a wall. Connections for portable laptops facilitate email communication and entertainment, supported by audio players and DVD/disc players. The compartment also has room to display family photos and mementos on the walls. Ventilation ensures that air is moved to prevent a hazard due to carbon dioxide build-up. Compartments in the Russian segment provide portholes for viewing the Earth.

Yury V. Usachev, during Expedition Two in 2001, works at a laptop computer in his crew compartment in the Service Module Zvezda.

Vladimir N. Dezhurov, during Expedition Three in 2001, takes a break in his crew quarters in the US Laboratory Destiny.

Jack Fischer in his Crew Quarters (CQ) in the Node 2 Harmony during Expedition 51 in 2017.

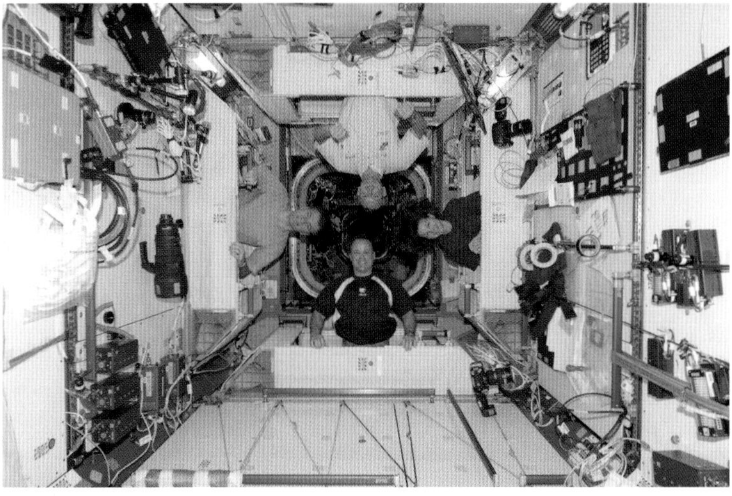

Ron Garan (bottom), Cady Coleman, Paolo Nespoli (left) and Alexander Samokutyaev in their crew quarters in the walls, floor and ceiling of Node 2 Harmony during Expedition 27 in 2011.

Karen Nyberg sewing in her Crew Quarters during off-duty time during Expedition 36 in 2013.

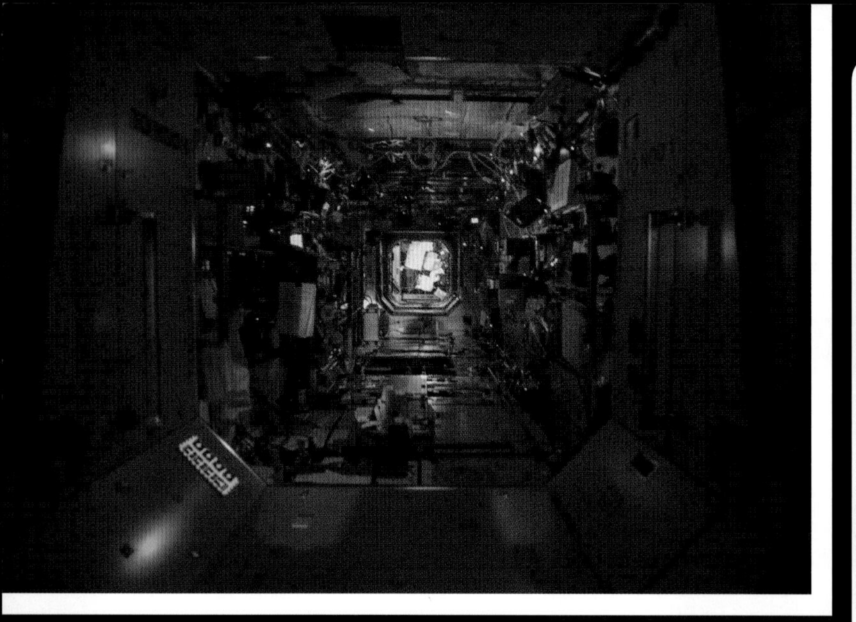

This view from Node 1 Unity into the US Laboratory Destiny, during Expedition 40 in 2014, shows the inside of the space station while the crew is asleep.

"At night, in the dim safety lighting, (on Salyut 7) it was difficult to make out where one of the crew was hanging in the air wrapped in a sleeping bag, and where a bag of clothing or equipment was tied. One morning, when everyone reluctantly got out of their sleeping bags, Solovyov, smiling, quietly complained, "I was beaten up today." "How were you beaten up?" I didn't understand, and thought he must be joking. "It's very simple, I was beaten up with fists!" It turned out that Igor Volk, woke up early and decided to do his exercise regime. Before he had time to wake up properly, he swam into the main compartment, noticed the bags that were hanging there with the equipment, picked the one which seemed to be softest, and began to beat it like a punching bag. That would be OK, but the bag was Solovyov's sleeping bag, in which he was curled up. "You understand," Volodya said with a smile, "I'm asleep and I didn't understand whether it was in my dream or whether someone was really beating me up. And I didn't want to wake up! Then I felt this business was starting to get serious. I stuck out my head from the bag and there's Volk attacking me." We all laughed at this but Solovyov began to choose his sleeping place more carefully." - Svetlana Savitskaya

Crew Health Care System (CHeCS)

Three sub-systems comprise the Crew Health Care System (CHeCS). The Health Maintenance Subsystem (HMS) provides medical equipment and consumables to monitor and mitigate illness or injuries. HMS also provides preventative health care and stabilization, and supports emergency patient transport between vehicles, including restraints and life support capabilities in the ISS and crew return vehicles. Environmental health subsystem (EHS) monitors the quality of the interior environment. The subsystem provides equipment for monitoring radiation levels, surface microbial contaminants, water microbiology, and toxicology. Exercise Countermeasures Subsystem (ECS) provides physiological countermeasures and enables several exercise capabilities. CHeCS enables an extended human presence in space by assuring the health, safety, well-being and optimum performance of the ISS crew. The crew's fitness is evaluated and relevant countermeasures are provided, enabling the crew member to counteract any degradation of fitness by using on board exercise procedures and protocols. A resistive exercise device (RED) simulates weight-lifting on Earth; appropriate restraints on the unit make it possible to conduct a workout that includes bench presses, dead lifts, sit-ups and squats. A bicycle ergometer permits the crewmember to pedal using either feet or hands.

"We've proven you can get to Mars just fine, healthy and good to go, in order to land on Mars and start exploring Mars. We've proven it over and over again. We've been putting people in space for six months and longer since the 1970s. We know there are risks and trade-offs and things you have to do in order to keep people healthy that long, but nobody is coming back from six months on the space station incapacitated. They're reduced for a while, but it's not perfect. The complexity of spinning a spaceship is far, far, far harder than keeping people healthy for six months. But that's all based on engine design and lack of ability to control gravity. Three hundred years ago we had no idea how to control electricity, we didn't know anything about it, and we sure really didn't know how to control steam and we didn't have internal combustion engines, the ability to develop and control those has altered life forever and the comfort and human quality of life. Right now on space stations we are limited by the fact that we have chemical rockets so the cost of getting every kilo of stuff is high enough that we have to husband them really carefully and so the right trade-off is not to put huge amounts of metal up there so we can safely spin the whole thing just to make life a little easier for the astronauts inside. We can always revisit the idea but from an engineering point of view it's completely impractical." - Chris Hadfield

"The three main risks, obstacles to space for humans, is radiation, variable gravity, I would argue variable not microgravity. I once did parabolic flight and you can handle microgravity. Switching from microgravity to regular gravity is really unpleasant and when you land you go up to like seven or eight G's, so after you've been in microgravity and your bones have got weak, is it a good idea to have an 8 G entry? So there's that. And then the third one that we haven't really touched on is the isolated confined extremes of the environment of space. The psychological factors are really important. I once spoke to this astronaut after they came back from space and I asked if they hung out with the other people and they said, no, not really. Then there were these other Russian and American astronauts who came back and they were like a married couple. They were finishing each other's sentences. They were hysterical and you could just tell they genuinely liked each other. They were buds. Space is a unique environment. There is a culture of people who do understand what it's like in space. They have a shared experience in space; but they don't necessarily have a shared experience here on Earth. So how do they create a culture that helps them to work better and learn the aspects of that culture that are healthy? All the aspects of that culture that can be unhealthy? It can be unhealthy but if they work together... so rather than characteristics that make it work better, what can we do to make it work better?" - Nicole Buckley

"The International Space Station is an unprecedented undertaking in scientific, technological, and international co-operation... Collaboration among our international, industrial and academic partners will ensure that the benefits from ISS work are felt across the global spectrum of public and private interests." - Daniel S. Goldin

SPACE STATION APPROVAL AND INTERNATIONAL INVITATION

On 25 January 1984 in his State of the Union address, US President Ronald Reagan gave the go-ahead for the space station. At this time, Reagan offered the international space community the opportunity to participate in the program.

For several reasons including political and financial support, NASA leadership decided that international cooperation would assist in developing the station. Even before the program was formally approved, NASA approached Canada, the European Space Agency and Japan. These partners were all already working with the US on the Shuttle program. Canada had supplied the Shuttle Remote Manipulator System (RMS or Canadarm) the robotic 'arm' used for deploying or grappling hardware in the payload bay and supporting astronauts on EVA, ESA had supplied the Spacelab pressurized sortie module and support pallets, and a free-flyer called Eureka, and Japan would sponsor a dedicated Spacelab mission and other payload work on ISS.

▲ The London Economic Summit of June 7-9 1984 during which the space station was a major topic of discussion. President Ronald Reagan, Prime Minister Margaret Thatcher, Foreign minister Graf von Lambsdorf and Prime Minister Yasuhiro Nakasone.

In the wake of the Reagan approval of ISS, at the London Summit which took place in June 1984, President Reagan formally invited other countries to participate in the Space Station Program. The following years saw the creation of the initial partnership. In January 1985, ESA held a Directors meeting and accepted the Columbus science laboratory as its primary contribution to the station. Canada formally joined the program in April 1985, followed by Japan in May. The same month, a space station international user workshop was held in Copenhagen, Denmark.

In 1985 and 1986 the configuration of the station was being refined. It started as a 122 meter 'Power Tower'. In 1986, a new configuration emerged; a 96 by 46 meter rectangular truss structure called the 'Dual Keel'. 1986 also saw the partners agree on their primary roles. The US would provide modules and the truss sections, and launch the elements on the Shuttle. Additional pressurized laboratory modules would come from Europe and Japan. Canada would develop a more advanced version of the robot arm already in use on the Space Shuttle.

Following the loss of Challenger and her crew of seven, NASA refined its plans. The truss for the station would start with a single long structure to which all the major elements would be attached and deferred the upper and lower keels until a later assembly stage. Challenger had another impact on the station and that was to rethink the approach to supporting astronauts in the case of catastrophic accidents or emergencies. Prior to Challenger, the plan had been called 'Safe Haven'. In between Shuttle flights, astronauts would have to remain on-board the station until the next Shuttle arrived to recover them. This meant that in the case of the loss of any one module, adequate resources would need to be supplied to maintain the crew in orbit for at least six months. After Challenger, a decision was made to begin the development of a Crew Emergency Return Vehicle (CERV) which would always be available to allow astronauts to return to Earth within a matter of hours.

"Even before Reagan's 1984 State of the Union address, we had international involvement. The Japanese and Europeans were just like anybody else. There was really big interest by the French, and the Germans, and the Canadians. It was in late fall, and we were trying to get more detail into it." - Luther Powell

In 1988, the Shuttle returned to orbit. The space station program was officially named by President Reagan on 18 July, '*Space Station Freedom*'. On 29 September, 12 nations, including the US, Canada, Japan and 9 European countries, formally signed an agreement to participate in the Space Station Freedom program. In 1989, on the occasion of the 20th anniversary of the Apollo 11 moon landing, President George H.W. Bush announced a new Space Exploration Initiative to return astronauts to the Moon on a permanent basis and develop the systems to support a mission by humans to Mars. In conjunction with the new requirements for a station emergency return vehicle, NASA asked for increased funding for human space flight. In the fall of 1990, the US Congress reduced the budget for *Space Station Freedom*. NASA was told that, over the period 1991 to 1996, it would have to cut $6 billion from the program. NASA began to scale back plans for the space station to meet the newly imposed budget limitations. The main truss was shortened from 150 to 108 meters. The pressurized US Laboratory and Habitation modules were reduced in length, from 13 to 8 meters. The reduction in size of the modules was in part due to the mass capacity of the Shuttle taking the modules to the station orbit. In order to reduce the number of Shuttle launches, the modules would need to be more heavily loaded and the philosophy turned towards pre-integration of both modules and truss elements to require fewer on-orbit shuttle assembly missions and fewer EVAs.

"The National Commission on Space met in 1984 and 1985. It was still a time of tension between the United States and the Soviet Union. We talked about everything from going back to the Moon to going to Mars. There were people like Carl Sagan advocating a joint US-Soviet manned mission to Mars. I suspect Ambassador Kirkpatrick was put aboard to bring a grownup voice of realpolitik wisdom to any such deliberations, and some more sober, wise experience from her point of view. One guy was a really keen and passionate advocate for much more international cooperation. 'Use it actively as a tool, be more expansive, it should be an explicit part of US foreign policy.'" - Kathy Sullivan

"After Reagan's State of the Union message in 1984, after we pushed Station through the Congress, there was a meeting of the G5 big democracies. He took the model of the Space Station with him. There is a picture with Margaret Thatcher, Prime Minister of the United Kingdom, and Reagan standing at the table with the Space Station. There were discussions about an international partnership. When Reagan made the decision to proceed, he said that we want to work with our friends and allies. About two days after the meeting when we expected Reagan's decision, he didn't say, "We're going to go do it." He called James Beggs, the NASA Administrator, and Beggs called me into his office and said, "Reagan is on the phone, I want you to listen." And, so I listened. What Reagan said was, "Can we do this thing internationally?" Jim said, "Of course." That was the only discussion. STS-9 and working with the Germans on the first Spacelab set the stage for an international partnership That was very, very important. If we ever get to Mars, it's going to be international for sure. It's a big deal." - Hans Mark

"Spacelab was developed for short missions, but its design requirements were not related to the number of missions. We developed the systems for the crew safety. The technologies we needed to develop for ISS we learned from the Spacelab; the thermal control, the structure and mechanisms. Different modules had different areas we had to stress. For instance the logistics modules had to be designed for carrying mass to and from orbit; cargo mass was the key driver for its design. But when we modified one of the logistics modules to be a permanent module, the protection from meteorite and orbital debris, the external protection, is much more demanding in terms of capability for preserving the pressurized environment. The noise level is much more important. The noise in the permanent logistics module is the lowest noise in the whole ISS and sometimes the crew was sleeping in the logistics module because it is really quiet. One of the experiences we got from developing the MPLM, was the cooperation in development of flight procedures with NASA. We developed the software which drives the management of the operation. It made EUS more competitive for commercial elements. The experience of multiple flights allowed us to trim our designs to make them more efficient and save power." - Walter Cugno

"Spacelab, in the 1970s, was a tremendous example of international cooperation. It worked cooperatively very well. In Spacelab it was clear that NASA was the boss, and that ESA was the underling, and we were the only two agencies involved. When the invitation came out to do the space station, ESA jumped on it, because we were going to be working with our old mates, NASA. But when we got involved in station, we could hardly recognize NASA. The relationship that we'd built up during Spacelab was like it had never occurred. NASA's way of doing business on the station was completely different from the way they did it on the shuttle and Spacelab. The personnel were completely different. Station was being led by JSC, whereas on Spacelab, our interface had been with Marshall Space Flight Center. NASA was also changing program managers annually, which made it very difficult. " - Alan Thirkettle

"Spacelab was a campaign mission. If you had a failure you don't repair, you switch to redundant systems. All activity was purely utilization-oriented, as opposed to ISS, which is a mix of system and utilization. The other part of the story is design. The design was used in orbit, but integrated on the ground. The mission being so short and also constrained by logistics. So the applicability of the Spacelab mission to a long term mission was very different. We did learn how to build human-rated equipment for safety and design guidelines; so our entrance fee into the world of manned space was designing Spacelab, but it relied on U.S. technology and help." - Bernardo Patti

Inside the Spacelab Science Module, the crews of STS-71, Mir-18 and Mir-19 pose for the traditional inflight picture. Anatoly Y. Solovyev at bottom center. Clockwise from Solovyev are Gregory J. Harbaugh, Robert L. Gibson, Charles J. Precourt, Nikolai M. Budarin, Ellen S. Baker, Bonnie J. Dunbar, Norman E. Thagard, Gennadiy M. Strekalov (angle) and Vladimir N. Dezhurov.

Preparing Spacelab 1 at the Kennedy Space Center for its flight aboard Space Shuttle Columbia (STS-9) in November 1983. ESA developed Spacelab as a manned scientific laboratory to fly on the Space Shuttle.

"After we had sold the Space Station, President Reagan said
about the Russians

At Congressional urging, NASA management visited Russia to assess the purchase of Russian elements for the station program. During a summit in Moscow on 30-31 July, 1991 Presidents George H.W. Bush and Mikhail Gorbachev signed agreements to initiate a new program to fly US astronauts on the Russian space station *Mir*, and Russian cosmonauts on the US Space Shuttle.

In 1992, the Russian Soyuz was evaluated as a crew rescue vehicle for the space station. It had a 25-year record of orbital operations and using it would save time and funds for Freedom, avoiding the expense of developing a totally new CERV. NASA also assessed the use of a Russian docking module and airlock for the *Shuttle-Mir* and *Freedom* programs. On 18 June, following three months of negotiations, NASA and the Russian Space Agency finally agreed on the possibility of utilizing the Soyuz and Russian docking systems in the *Freedom* program. By October, the two agencies had agreed that an American astronaut would launch on a Russian rocket and complete a three-month mission to Mir, to be picked up by a Shuttle docking to Mir in 1995.

"In President Ronald Reagan's 1984 State of the Union speech he invited friends and allies to join in building the station. The framework for the partnership was set up where NASA was the senior partner and the primary stakeholder." - Melanie Saunders

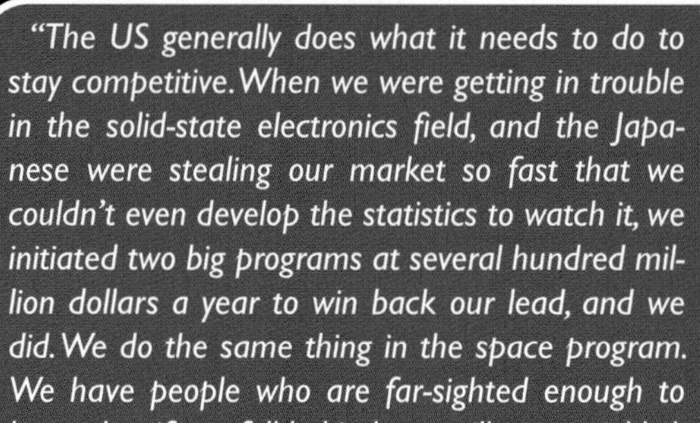

"The US generally does what it needs to do to stay competitive. When we were getting in trouble in the solid-state electronics field, and the Japanese were stealing our market so fast that we couldn't even develop the statistics to watch it, we initiated two big programs at several hundred million dollars a year to win back our lead, and we did. We do the same thing in the space program. We have people who are far-sighted enough to know that if you fall behind, you will very quickly be overtaken. It was a motivating factor for support for the Space Station. President Reagan saw it as a national leadership issue. He believed it was something that would enhance the way in which the world viewed the United States. When we talk to people around the world about technology, they believe, and it's like an article of faith that if the United States decides to do something technological, we will do it and we will succeed. When I talked to the French or the Italians or the English or the Germans or the Japanese and said, 'We're going to do this,' they believed it. And in the next breath, they say, 'We want to be a part of it.'" - James Beggs

"The stabilizing influence of international involvement has been borne out many times. There have been times that we were very helpful to NASA and helped them get their budget because of the international commitments, and, correspondingly, times when Canada and the other partners have had problems with their budget, and NASA was able to help us out and wave the international commitment flag and keep the program on track." - Bryan Erb

"I joined ESA in 1986 on the Columbus program study phase. That lasted about four years and was getting nowhere. We were kind of going in a circle. At the time of Freedom the funding was not clear, and especially the baseline was not clear, and the logistics was only relying on the shuttle. So there were a number of open points that were never truly addressed. Crew rescue and all that. All that was not existing." - Bernardo Patti

"There was an existing government-level agreement signed in 1988, called the Intergovernmental Agreement [IGA] that was signed by the partner governments. NASA also had an agreement with each partner. This bilateral Memoranda of Understanding [MOU] between NASA and each space agency identified what each contributed and what rights each got." - Melanie Saunders

"The international effort of the Space Station was not switched on. It did not go from zero to 100% all at once; it grew with time. I have witnessed it from various roles. The work helped to develop a mutual understanding and mutual trust. We still have barriers between the various partners; there are language barriers and cultural barriers; but we are learning. Everybody who has participated has grown. We grew to become one community all aiming together at the same goals. There were different ways to reach those goals. It might not look understandable for the individual to see how any one person or organization was doing it. But trust has developed because we know that each has the best intentions; each just does it a little bit differently. We learn from each other. Each participates in and helps to develop a narrower, better coordinated operational scenario. That has been one of the greatest rewards for me working as an astronaut." - Hans Schlegel

In 1993, a new US President, Bill Clinton, occupied the White House. NASA revealed that a growth of $1 billion would be required to support the *Freedom* program. The claim was not well received by the new administration. The President directed the agency to redesign the station again, within 90 days, and restrict annual spending to $2.1 billion. On 10 March, 1993, a re-design team identified three options: A (Alpha), B (Beta) or C (Charlie). The *Alpha* design was a slightly reduced *Freedom*; it looked the most promising and the revised *Freedom* began to be referred to as *Space Station Alpha*. None of the new designs that emerged on 7 September, 1993, met the strict budget restrictions imposed by the new President. Even before the NASA redesign effort had settled on the *Alpha* approach, another group working with the Russians had recommended merging the *Freedom* design with the planned Russian *Mir-2*. *Mir-2* would have been composed of modules already in work to support a follow-on to the *Mir* already in orbit. The merger was a sound idea. The approach would provide the needed financial support for the Russian space program; the Russian modules were all designed for autonomous flight and using the Russian modules would reduce costs by eliminating the need for critical US guidance, navigation and propulsion systems, and also by deferring the need for the full US power and habitation systems. The approach should have enabled a crew to man the station earlier than the Shuttle assembly sequence would permit. And use of the Russian *Soyuz* would eliminate the need for another new spacecraft, the US CERV. The core module of the *Mir-2* was essentially identical to the core module of *Mir* which had been in orbit for six years by this time. It was designed as the Russians main habitation module and could be resupplied by unmanned *Progress* spacecraft. So the new station could be continually crewed once the core module was in place without needing the presence of a docked Shuttle. The addition of the Russian modules made the station larger. Because each Russian module had its own power supplying solar arrays, they provided 42.5 kilowatts more electrical power. The station would now be able to host a crew of six instead of four.

One mitigating factor was the need to change the station's orbital inclination from 28.8º to 51.6º. The lower inclination was ideal for the US Shuttle launching from Florida and maximized its capacity of payload taken to orbit. Because the Russian launch site was at a latitude of 51.6º and because their modules were designed for their *Proton* launch vehicles, the new orbital inclination would have to be 51.6º. It would increase the amount of territory the station would fly over but also increase the number of required launches. On 2 September, 1993, five days before the emergence of the *Alpha* design, a US/Russian agreement was signed to merge the *Freedom* and *Mir-2* programs, with the launch of the first element planned for May 1997. The Clinton administration agreed to funding for ISS of $10.5 billion, at a rate of $2.1 billion a year for the period FY 1994 through 1998. The new cost was set at $17.4 billion, which was derived from the $19.4 billion that *Alpha* would have cost, less an estimated saving of $2.1 billion by bringing in the Russians as a partner. These new figures were not set in stone. The new program was renamed the *International Space Station* (ISS). In 1993, NASA changed its management structure for the program. *Freedom* had been managed by a NASA team in Reston, Virginia, and four different NASA centers had major 'work packages', each with a separate contractor. With the new design, the NASA Johnson Space Center in Houston was delegated all program implementation responsibilities and Boeing was named as the US prime contractor. The other contractors were subbed to Boeing and later, in 1996 and 1997, Boeing bought out the other contractors.

"In May of 1989, Congress decided they wanted to cut the Space Station budget. We went through three or four budget cuts. We formed a team of people at Langley Research Centre who tried to determine 'how we would save this money? What are we going to do?' We messed up; we decided that we didn't need our international partners to participate; we would do it all on the NASA side. The International partners were not happy." - Richard Kohrs

"We paid a lot of attention to the Soviet space program. We had a considerable number of collaborative programs with them. We would fly on some of their science satellites and they would fly on ours. In 1982 or '83, the space treaty we had signed with them way back during Apollo was due for renewal. Roald [Z.] Sagdeev, the academician who was the civil head of their space program said, 'The Soviet Union badly wants to renew this treaty.' But the White House didn't want to renew it. We were right in the midst of the serious arms-control negotiations. This went on until the Soviets' collapse. It waited until the collaboration on the Space Station in the Clinton-Gore days. Gore was very anxious to do that." - James Beggs

"In September or October, 1991, Congress asked why we weren't considering the Soyuz as a rescue vehicle for the Space Station. George Bush was still President. Dick Truly was the NASA Administrator. Senator Barbara Mikulski held a hearing. Truly and Yuri Semenov, the chief designer and head of the NPO Energia in Kaliningrad were witnesses. Mikulski asked why, with Freedom wanting a rescue vehicle, why we weren't considering the Soyuz? Semenov thought it was a great idea, and Truly said, 'You know, we probably can do that.' This was just after the time when the Berlin Wall came down and the Soviet Union came down. There got to be quite a lot of interest in the United States in terms of engaging or exploiting or taking advantage of what we could do with the Russians that might be of benefit to us. In January, 1992, we took about seven or eight people to Russia to talk to Energia about using the Soyuz. The Russian, systems and equipment is rugged; it will work in any environment, it is very capable mechanically, and strong and durable. By comparison, United States equipment has tended to be designed to eke out every bit of performance. It's like a fine watch. It's highly refined, it's very elaborate. It may also have the potential for more difficulties. During Apollo-Soyuz, in the early 1970s, when we would meet with the Soviets in Moscow, we would always meet in some rented facility downtown. We didn't know where they actually made things. The day we arrived in 1992, we were driven straight out to Kaliningrad, to a huge factory, a huge brick wall around it, through a great gate, and right into the office that used to be Sergei Korolev's. The people that we met there were the same people we worked with in the early seventies. It was immediately obvious you could use the Soyuz for a rescue vehicle. We reviewed every system. It was a several-day review. The only problem we found was that the limit on the length of time a Soyuz could be left in space was six months. There weren't any other significant changes to be made to the Soyuz." - Arnold Aldrich

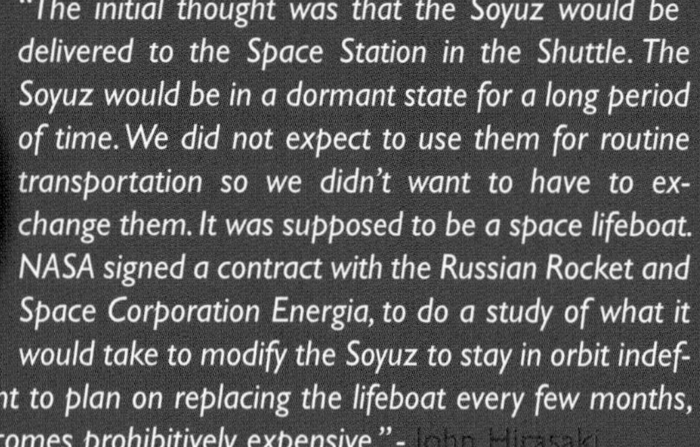

"The initial thought was that the Soyuz would be delivered to the Space Station in the Shuttle. The Soyuz would be in a dormant state for a long period of time. We did not expect to use them for routine transportation so we didn't want to have to exchange them. It was supposed to be a space lifeboat. NASA signed a contract with the Russian Rocket and Space Corporation Energia, to do a study of what it would take to modify the Soyuz to stay in orbit indefinitely. You don't want to plan on replacing the lifeboat every few months, because then it becomes prohibitively expensive." - John Hirasaki

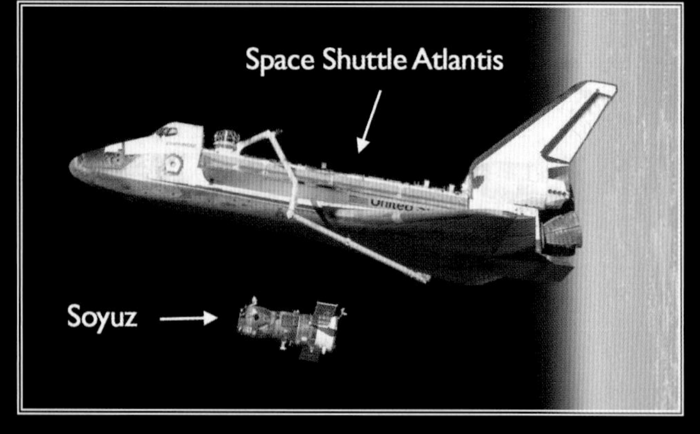

▲ A composite image showing Soyuz and the Space Shuttle to scale.

"President Clinton came on in 1992. We were having major technical and management difficulties in the Freedom Program. The Space Station was on the chopping block; Mr. Goldin, the NASA administrator pleaded with Clinton to allow him to redesign the station to get it under better control as opposed to cancelling it. We had three different options that we looked at, A (Alpha), B and C. Alpha started as a U. S. option. The Russians were in the Washington, D. C., area during the redesign and we had some discussions with them. But we had not been cleared to involve the Russians. It was not until after the Option A decision was made in August or September, 1993, that we looked at what roles the Russians might play. I went to Russia in September-October 1993. Later in 1993, the Russians were brought into the program." - Chester Vaughn

"We started talking with the Russians in 1992 when Boris Yeltsin met with President Bush. They talked about doing a rendezvous of the Shuttle with Mir. Administrator Goldin went to Russia and I went with him. We suggested a Russian cosmonaut fly on the Shuttle. Ryumin didn't want that. I told Goldin, 'Remember, it's the golden rule. We've got the gold so we make the rules.' Once the top Russian political leadership said 'Let's do it,' everybody fell in line. Some Russians were not too happy to work with Americans, but the main thing they liked was the money." - Thomas Stafford

"Before Russia was incorporated into the station program, station was to fly at a 28.5-degree inclination. We did a landing site survey looking at where a Soyuz could land. We found that there were a very limited number of landing sites. We looked at Australia, south Texas and Florida? They were all within range." - John Hirasaki

"In 1993, Freedom was to fly due east out of the Cape at 28.5-degree inclination. It would fly over as high a latitude as Florida but no higher. A number of the launch sites in the other parts of the world are at much higher inclinations. It was very hard for them to come down to a Freedom orbit. So if you wanted to launch from Baikonur, or launch sites in Japan or China, it would take a very large vehicle and be very inefficient. It seemed to me the Space Station was international, was going to be permanent, and it would make more sense to be accessible from launch sites anywhere in the world and to overfly much more of the populated part of the world on a regular basis. I recommended we change the inclination of the Space Station to a 51.6-degree orbit, the latitude of Baikonur. This created a problem. The Shuttle is most efficient, when you fly due east. If you fly at a higher inclination, it takes more propellant to deliver the same cargo. So we had created a situation in which the Space Shuttle couldn't fly the missions to Space Station with the heaviest cargoes." - Arnold Aldrich

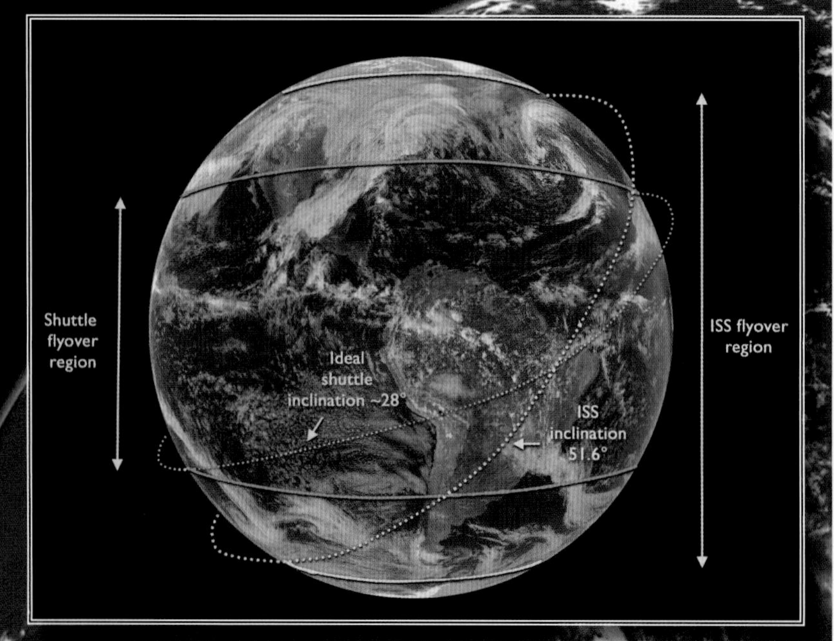

▲ The difference between the normal shuttle orbit and the proposed ISS orbit.

"Daniel Goldin came in as the Administrator in 1992 and the deal was made to add the Russians. The Russians wanted to keep their space program, and they didn't have enough money to fund it. A guy in Reston is going through a presentation, and we see the Space Station with Mir attached. Someone asked 'How are we going to do that?' And the response, 'you need to go figure that out.' It was apparent someone had worked with the Russians, and that was the first time anyone had seen that." - Denny Holt

"The three big players, ESA [European Space Agency], NASDA [National Space Development Agency of Japan], and the Canadians, were very much in the mix in the structure that I was leading. They all had offices here and they came to staff meetings. They had seats at the table. But we tried really hard to keep them out of the critical path. A lot of the leadership in those aerospace agencies, were kind of the have-nots from a human spaceflight standpoint. They'd all put up some, but none of them had ever been involved in a human adventure like this." - Neil Hutchinson

"In '93, Bill Clinton came into the White House and said that he wanted a new station. A lot of people got cold feet in Europe. We were having problems with our own member states funding the Freedom program. We were challenged to downsize the Columbus program. We decided we needed to be able to launch the Columbus module with all its payload racks inside the shuttle. The original Columbus module was twice as long, and a lot heavier than the one that we eventually produced; we couldn't launch the whole thing together. We had to buy additional shuttle flights to launch the payload racks and then integrate them in orbit. So we looked at reducing the size of the Columbus module, and got all the racks on during launch. In the redesign, we had a meeting with the Japanese and the Canadians to agree on a common position with respect to NASA, when there were things that were going on that we didn't like from an international partner point of view. There was quite a bit of solidarity. Through that I think we developed lots of very close relations with our Japanese and Canadian colleagues. Crystal City is an actual place just down the road from the Pentagon. We rented a complete floor; I think the 13th floor of a tower block in Crystal City. That was where we worked from 8:00 o'clock in the morning to 10:00 o'clock at night with pep talks from Dan Goldin and visits from George Abbey. The Freedom team were not allowed into the building unless they were invited. They were actually conducting a CDR at the time we were doing the redesign. In the end, we went back to the Freedom design. It was called Space Station Alpha before it became ISS. Alpha ended up costing about the same amount, and was rather less performing because they did things like scrap the entire data management system, which was considered to be overly complex. They ended up using Honeywell MDMs which were basically 1950s technology. There was a ministerial council towards the end of 1993, which pegged the funding of the program back because no one thought it was going to end in a good conclusion. There was a lot of doubt amongst member states that this was going anywhere." - Robert Chesson

"Station went into a big redesign effort to get the program under control. It took place at Crystal City, Virginia, near Reston, which was where the Freedom Program was located and close to NASA Headquarters. We kept hearing rumors of an office tower meeting with the Russians, but we didn't know what it was about. We came up with a re-design when we got a late-breaking input that the design had to include the Russian Functional Cargo Block (FGB) tug. We said, 'What's a Russian FGB tug?' We didn't know what it was. Then they provided some information on the FGB tug. In one weekend, we redesigned the assembly sequence and the Station interfaces based on the input. That was what we wound up doing." - William Reeves

The Russian FGB being assembled at the Krunichev plant in 1993.

"The Crystal City, Virginia redesign activity was broken into three different teams, concepts A, B, and C. 'A', 'Alpha', was the one selected. The Alpha configuration was technically marginal, but it met the cost and schedule box. Space Station Alpha had no Russian involvement at all. But, then the Zarya Functional Cargo Block (FGB) and the Zvezda Service Module were added. Russian involvement helped the technical aspects of Alpha. The Russian elements gave us time to restructure the program." - Laurie Hansen

"The US announced that we were inviting the Russians into the program. Legally, we couldn't just unilaterally do that. We had to negotiate with Canada, Europe, and Japan the formal multilateral invitation to the Russians to join the International Space Station program. It was a fait accompli at higher political levels, and our partners all knew that. They weren't happy. The original partners, were very worried about Russia being a partner that had such significant contributions that it would make ESA, the Canadians and the Japanese more like junior partners. They were counting on the U. S. to be strong on their behalf. The original partners wanted to ensure that there would not be substantial changes to the international agreements, because it had taken years of negotiations to establish these. The agreements were signed. The agreements were well established. The partners didn't want a whole bunch of changes. They didn't want to just reopen all the negotiations when another partner came in. It was a challenging process because the way the agreements were set up, quite deliberately at the beginning, was one multilateral agreement at the top, at the State Department, foreign ministry, inter-governmental level. That was mostly the legal terms and conditions and just the general framework. But all of the NASA level memoranda of understanding were bilateral. Bilateral with Canada, bilateral with Europe, and bilateral with Japan. To make an agreement with the Russians, I first met with Canada, Europe, and Japan, their space agencies, and we hashed out what I was allowed to change, what could they support. Then I would take that version of the agreement to Russia. They'd accept some, they'd reject others. I had to make sure that every bilateral agreement stayed consistent on all the common terms and conditions. Then things happened, like the Europeans coming in and wanting to change a bunch of stuff; I'd have to see if that was okay. The round-robin of iterative updates to the agreement was a very complex process. I was new to Space Station agreements. I did not start working on them until the Russian invitation. I would read the words and just take that what they said and what I assumed they meant were the same. There were times when I would think something sounded perfectly logical and made sense, then one of the partners, usually Europe, would say, 'You can't possibly change that, we spent years arguing over that clause.' Then I'd get a history lesson on why that was important, what it meant. When we went from the original round of international agreements to the second round of negotiations, which included the Russians, Canada and Europe both wanted to negotiate down their contributions, because they could no longer support the original level of funding. Europe reduced the size of their laboratory. Originally the Japanese lab and the European lab would have been the same size. The European one is now smaller. Japan stayed at 12.8 percent I think and Europe went down to 8.3 percent. Canada also needed to negotiate a reduction in its contribution for the robotic system." - Lynn Cline

"In the Freedom Program, we tried to keep our international partners out of the critical path. In other words, if they didn't show up, the U. S. part of it could continue and we'd have a lab module to do science. Getting in with the Russians put them into the critical path. It was for a good reason but it adds complexity. The decision was made, that was what we had to do to make the ISS work. When historians write about the Space Station, they'll say this is what the government had to do to make it successful, to keep it going. I don't know if that's right or wrong, but that's where we are, so everybody ought to support it." - Richard Kohrs

"NASA realized that they could do something sustainable and reasonably fast only if they had the Russians on board and avoided expenditures they were facing. That's the reason the Service Module, the FGB, and all the crew transportation were taken on by the Russians. That allowed NASA to concentrate on their segment. The design for Columbus in the Freedom era was more of a paper study without ever getting to cutting metal. It was also a learning curve in Europe. It was the first time after Spacelab, which was a program where we had a lot of American help; Columbus, was European industry doing the core job. Our industry today is totally mature because of Columbus, because of ATVs, because what we have done; the late 1980s were a learning curve we had to go through." - Bernardo Patti

"In 1993 the Russians were brought into it. The NASA Administrator said if he hadn't brought the Russians in, we wouldn't have a program. I was disappointed in the politicization of the program. We had been a US national program. We were running with partners, but there was no politics. Bringing the Russians in was political. The whole thing was centered around keeping Russia from transferring technology that would destabilize world peace. In return, we would give them money and American companies would be permitted to explore for oil in parts of Russia. It was a political deal to prop up the Russian economy. The Russians had done a lot of things we haven't. They do automatic dockings. They've welded in space. The little mechanical arm they designed to move the modules around on Mir was just one of the most clever things I've ever seen. They decide they want to do things and they go do them. Technically, they're sharp people. But the Russian technology is certainly not leading-edge. The Russians believe in the adage 'If it isn't broke, don't fix it.' It's a different approach to technology, but it works. In some areas the Russians are backwards-looking. They were despondent that their program doesn't have any money. In the Soviet system, they got a lot from the government. The Russian people looked up to Yuri Gagarin who was damn near worshipped. But now that is gone. When the Mir went down, it was a blow to their national pride. We've had successes. Those successes were based on forming personal relationships. When you get invited to their home and you invite them to your home, then the relationship becomes personal. When you have that kind of relationship, you can count on them." - Henry Hartsfield

"When we started working with partners who were formerly our enemies, not only do you realize that they really are not enemies, they're just like you and I. They have the same hopes and dreams for personal achievement and family life, everybody wants to take care of their family. Technically they are as good and their judgments about what is right and wrong are not very different. Russian design principles are quite different than the principles the US uses. They rely more on physical testing and making the design robust in terms of the safety factor, versus the US approach of making the design very sophisticated, trying to extract a maximum amount of performance, and more extensive analysis. The Russians prefer to have a robust system, tolerant of types of failures that may not have been anticipated. They are very thorough in their testing. They are very conservative when it comes to changing things. Because if it worked last time, why change it? They may not get the same performance, but in terms of safety and robustness, it's a much better approach. We did a comparison of their processes and standards to establish equivalencies to our processes and standards. Then we accepted their process as meeting the requirements. As long as the end result is equivalent in terms of safety, then they've got their standards, we've got our standards. There are similar principles in pressure integrity, fault tolerance, and redundancy. The numeric values and methodology is different but both of them meet the objectives." - John Hirasaki

"To work through the agreements with our international partners, we had what was called a Multilateral Consultative Working Group, MCWG. It was basically the MOU negotiation teams from NASA, the European Space Agency, the Japanese Space Agency [National Space Development Agency of Japan], and the Canadian Space Agency. Those four agencies would get together, and then we would tell them what we were generally planning to show to the Russians. Then we would go meet with the Russians, and then come back and meet with them again." - Melanie Saunders

"When the Russians joined the ISS program, ESA had much more experience working with them than the Americans did. We had worked with them on Euromir. We had a number of astronauts that flew on the Mir station. And they had learned an awful lot. We got quite a bit of experience developing the control center to operate the onboard systems for Euromir." - Robert Chesson

"For crew rotation, we kept our people on the Shuttle. The Shuttle would come and go five to six times a year. The Russians would do their crew rotation on six-month increments with the Soyuz. The other international partners, who have smaller contributions, it took them awhile to accrue a whole increment for a crewmember.... they wanted to have as much flexibility as possible.... So they were interested in using the Shuttle for rotation. Their astronauts knew the Shuttle. They were comfortable coming to the U. S. for training. Other than ESA, they hadn't built up the infrastructure for training in Russia; there were issues of living accommodations, transportation logistics; they had not set these up." - Melanie Saunders

"When the Russians joined the Space Station, I thought this was great from many perspectives: technically, operationally, and politically. It was a wonderful decision to have the Russians join because the space program became a role model of international cooperation across the political, cultural and personal boundaries." - Hans Schlegel

"The US and the Russians had totally different ideas about the purpose of a space station. The Russians were there to man an outpost, to fly in space and to be in space. It was about being there and being the leader. It had little to do with research. The U. S. was focused on research; it was the reason for the station. We were there to do research in microgravity." - Melanie Saunders

"The Russians came into ISS having had experience working with Europe. They had their own bilateral relations with Europe. They did not have the same level of experience in working with Japan or Canada. Many times in bilateral meetings with the US, the Russians made comments, 'they are your problem. They're your partners. Canada? They're at 2 percent. We don't care what they demand.' They were not respectful comments. But when we were in a multilateral session none of that showed. Then the Russians were very polite. They were supportive, seconding some of the partner's positions. With all of the changes, we now had ATVs and HTVs and Arianes and other things in the agreement. If you look at the Freedom agreements and the later agreement, there were some substantial changes in the content of the program. One of the challenging things was keeping the Russian list accurate, because the Russians kept changing what modules they were going to provide." - Lynn Cline

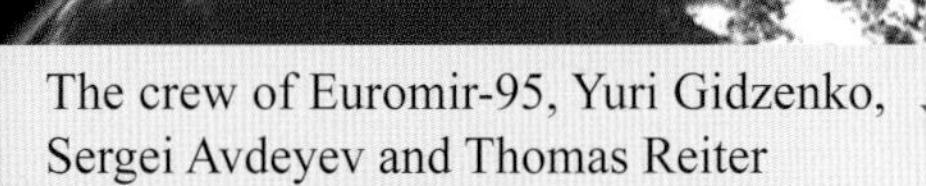

The crew of Euromir-95, Yuri Gidzenko, Sergei Avdeyev and Thomas Reiter

The Final Hurdles 1994-1997

A new assembly sequence emerged on 28 November 1994, with the inclusion of the Russian elements from Mir-2. According to the new plan, the first element in the program would be the Russian Functional Cargo Block, or Control Module, called Zarya. Zarya would provide guidance, control, power and propulsion to the nascent station. In a unique arrangement, Boeing contracted directly with Khrunechev for Zarya, bypassing both US and Russian governments. The module would be paid for by the US but become a part of the Russian segment after launch. The US dollars supplied to prepare Russian ISS elements and keep their own programs going during times of great economic, social and political upheaval in Eastern Europe was welcomed by most of the Russians.

The first American element, Node 1, would be attached to Zarya. Together, Zarya and Node 1 would form a core from which expansion could take place from both sides. In the new assembly sequence, the launch of the first elements slipped by six months, to November 1997, with assembly expected to be completed by June 2002. Work was to begin on the Zarya module almost immediately.

One Shuttle-Mir docking mission was agreed upon in 1992. In 1993 the program expanded to include a series of Shuttle dockings to Mir, and modification of two Russian modules, Spektr and Priroda, to become US laboratories on Mir. These missions would fill the gap until ISS station assembly would begin and the new Shuttle and Mir activities were incorporated into the ISS Program as 'ISS Phase I'. Zarya, Spektr and Priroda were all similar FGB-style modules. Since the same teams of engineers and technicians would prepare all three, preparation of Zarya and the subsequent core module, later named Zvezda, would be deferred until after Spektr and Priroda had been completed and launched in 1995 and 1996, respectively.

By 1996, over 70 tons of US hardware had already been built. The primary structures of Spektr, Priroda, Zarya and Zvezda had all been completed years earlier and had been prepared for outfitting in the case that any would be needed as back-ups for the original Mir, the first element of which was launched in 1986. But as should have been anticipated, by 1995 neither Zarya nor Zvezda were yet being worked on as the Khrunichev and RSC-Energia teams were focused first on Spektr, and then on Priroda. Each would gain attention as personnel were available. Outfitting of the Service Module Zvezda, critical in the assembly to allow a resident crew to live on board the station without a Shuttle being attached, was even further behind the FGB-style modules. Even more problematic, the Russian government withheld funding for the ISS Program and the Russian companies were continuously short of funds.

The delay of the Zarya/FGB would delay the start of ISS assembly. The delay of the Zvezda/Service Module would delay the start of manned long duration operations. The delays had the effect of increasing costs and increasing the duration of time the early modules would need to fly tended only by visiting Shuttles. NASA grew concerned and began to prepare a US module that could serve in lieu of the Russian modules to provide guidance and propulsion to maintain the ISS in orbit, if it were required. The US module was called the Interim Control Module (ICM). The ICM was designed around the propulsion module of a classified military satellite. The addition to the program further exacerbated US problems, increasing cost.

The launch of the first element slipped from May, 1997 to November, 1997 and subsequently to June 1998. In April, 1997, NASA requested that the US Congress add a 'new line item' to the agency's budget, as an insurance against further Russian delays. Called the Russian Program Assurance (RPA), Congress approved for FY1997 a further $200 million to finance the construction of the ICM and other contingency options, in case the Russian delays affected the assembly sequence.

The Russian delays did allow further ground testing of the US elements, several of which had experienced serious issues during testing. During the first pressurization of Node 1, the structure was damaged in the area of the radial hatches and required a redesign and the addition of new structural elements. During the initial testing of the Thermal Control System (TCS) cold plates, it was discovered that welds were porous and the cooling fluids leaked, necessitating re-construction of hundreds of components. So in some ways the Russian delays alleviated other costs and delays that likely would have surfaced as integration problems were discovered later, perhaps during assembly in orbit, which could have been far more difficult to overcome.

Russian government funding continued to be a significant concern and ultimately delayed the launch of the first element. In January 1998, the Russians informed the partners that that they would launch the Zarya FGB on 20 November 1998 but that the launch of the Zvezda Service Module, crucial to establishing an expedition crew on the station, would slip further. Zvezda was rescheduled to April 1999 at the earliest.

"The partners have had two levels of obligations. One level is to provide their elements. Those were one-shot with some continuing sustaining engineering. Each partner built their module, the FGB or Kibo or Columbus. The second level of obligation, each partner is responsible for their share of ongoing operations costs. These turned out to be extremely onerous. The operations costs had to be estimated over the number of years. Each partner has to contribute. It is a difficult thing to share responsibilities. It might be easier if a partner were to agree it was their responsibility to run mission control for example, and not to expect repayment. In the future that might be easier. Or perhaps they have to anticipate there is going to be ongoing overhead and that an accounting mechanism will track the costs they will have to reimburse through the life of the program. The magnitude of that burden was not appreciated at the outset but it needs to be for the next program." - Dan Jacobs

International Program Administration and Implementation

The International Space Station requires a truly international organization of unprecedented scale and complexity. The size and complexity of the support team is often overlooked. It stretches around the globe. Different time zones, cultural differences, physical distances and technical and political processes all have to be factored into the program. The management infrastructure, myriad construction and testing facilities, launch and flight support networks, communications and tracking systems, mission control centers and recovery facilities, as well as the selection, training and post-flight care of crew members are all layers of complication. In his book, '*Beyond the Planet Earth*' (1920), Konstantin Tsiolkovsky spoke about the need for international cooperation in the exploration of space.

The ISS program evolved essentially from the US national space station program, which was born from the superpower rivalry of the cold war and the arms race of the 1950s, competition in the space race and Moon race of the 1960s, and the political, social and international changes across the globe in the 1970s-1990s. The station effort was officially sanctioned and evolved in the 1980s into the *Freedom* Space Station. Almost as soon as it was approved, international cooperative agreements were pursued. The study for a NASA Earth orbital laboratory was established by NASA Associate Administrator for Spaceflight, Robert Seamans, in 1963. The study addressed international influences on future large space stations; and the national security aspects of international involvement.

In the 1960s, American space station development turned towards what evolved into Skylab, but the space station remained of secondary priority until Apollo's Moon landing goal was achieved in 1969. Forecasting the direction of the multi-national cooperative effort of the current ISS, the State Department noted in 1969 that a large space station would be of 'great international value' and that the development of reusable logistics spacecraft providing frequent and routine access to space would be of greater value than prolonged exploration of the moon.

The International Space Station brought together 16 nations and five space agencies, cooperating in space for a common goal.

"When the Service Module was delayed by the Russians we realized that delays by one partner meant all the partners experienced the delay. We had to put in fixes so that even if one partner's element was late or didn't show up at all, the rest of the program could continue without them. We didn't appreciate that during the early design periods. You cannot do away with having partners on the critical path if they are to have a meaningful role; but it needed to be recognized upfront and efforts made to minimize the impacts." - Dan Jacobs

"International partners can save you money and they can cost you money. There may be a savings if you don't have to build all the components; but there can be a cost in terms of coordination and integration. Sometimes they have problems when you're not and that slows you down, and sometimes it takes longer to get to a decision or a consensus. But they typically have problems at different times, so you have some redundancy where you have some give-and-take. Somebody can help you when you're having a problem, and you help them when they're having a problem. The political safety net that international partners provides is huge, and it's a huge inhibitor to the whims of an administration and Congress. It can help you through transition and those different changes of focus and desire and people's visions and their legacy and all of that. NASA can end up changing directions constantly and never get anything done, which has been a problem. Having the international partners really mitigates that." - Melanie Saunders

"At KSC we put together a team that goes out to the manufacturing sites to understand the hardware and the processes before we prepared it for launch. We tried to add value from the processes being used at all the factories, including the internationals. When we got the ISS flight hardware to Florida, we started getting it ready to fly. There were few active interfaces between the ISS elements and the Shuttle. Shuttle was just delivering an object to space that would be activated in orbit." - John J. "Tip" Talone

"November, 2018, marks 20 years since we launched the International Space Station's first module, Zarya, into orbit. Once Zarya was in place, it took more than a dozen years and 40 separate launches to assemble the station and 200 successful Russian, European, Japanese, and American NASA and commercial launches to keep it crewed and supplied. Today, the space station has been continuously occupied for more than 18 years by 232 people from 18 countries. More than a 100,000 people from space agencies and contractors in 37 US states and 16 countries have worked on the program. Since it has been in orbit, we've accomplished an unprecedented amount of work in space with more than 2600 scientific investigations conducted by researchers from more than a hundred countries. In the future, we expect the cadence of flights, the number of crew on-board, and the amount of scientific research conducted on the orbiting laboratory to increase. The space station is evolving, and the pace of its evolution is speeding up." - Kirk A. Shireman

"Most of the international policies and the agreements, at least at the Memorandum of Understanding [MOU] and the program level, had already been established at the beginning of the program, at the very onset. So when I started working in Space Station, around 1998, it was more of a case of implementing the agreements than it was establishing them. There's an intricate amount of detail that was negotiated in the original agreements. They talked about common operations costs and what was going to be provided and what was negotiated hardware. As an example, the U. S. got 50 percent of the resources in Columbus laboratory module being built by the European Space Agency (ESA). So where they had eight payload racks in Columbus, the U. S. owned and would use four of those racks. We had to identify which four racks we wanted in the Columbus module. Then we had to negotiate with them how those particular racks would get installed, checked out, certified and verified, and documented to meet both US and European needs. We had similar arrangements with the Japanese. We had bartered with Japan for them to provide more hardware than ESA would provide. The Japanese had a larger share of the common operations costs." - Rick Nygren

"When the Columbia shuttle accident happened in February 2003, a small part of the station was in orbit – it was manned with Americans and Russians, and the Americans couldn't get to it. The US was utterly dependent on the Russians, who had the only operational transportation system. So all of a sudden, NASA, the leader of the venture, was the beggar in chief, and had to very, very quickly kowtow to the Russians, to develop contractual arrangements with them to buy launches, to buy transportation. The US solicited the help of the other partners – the Japanese, the Canadians, and the Europeans – to help them. The US forgot their introspection and their internal machinations, because there were Americans in orbit and they had to save them. That was a catalyst for change in the relationships and for a change in the way NASA did business. They changed personnel, and the new personnel that came in, both on the shuttle side and on the station side, were just fantastic. The best of the best. We went from a situation, almost overnight, from having a rotten relationship, really poor, to a good and constructive relationship. It took a disaster for that to happen." - Alan Thirkettle

"We nearly lost the program with the Columbia accident. At the 2005 ministerial council, I had to go and beg for money to continue the program. They told me I was dreaming; that ISS would never get finished. We hadn't launched Columbus. All they had was the U.S. lab and enough solar arrays to be able to run it. Our delegations told me that NASA would decide to stay with that and not bother to develop the rest of ISS. I got a budget which was effectively, I think most delegations thought it was to cover the run-down of the program. But the Japanese and the Canadians really pushed back. It was a battle for quite a time to keep the funding and the enthusiasm going. But once ESA got the run-down budget, then we had enough to keep us going until the ministerial council in 2008. Then they continued the funding, and then we had enough to be able to get to the launch of Columbus and ATV." - Robert Chesson

"It happens from time to time because it is such a long term program. Governments change, priorities change, your ability to sustain the funding changes over time, or we run into technical difficulties that increase the amount of funding needed. All these things can affect the relationship. But, if you look at the successful international collaboration, it far outweighs these exceptions where one partner or another had a big difficulty, but the problems are the ones that get the headlines that people remember." - Lynn Cline

"At the outset there was not a strong appreciation of the challenges we would face because of the length of time the program would take to develop and operate and be in existence. All five major partners have gone through several administrations, several changes in priority, several changes in budget support and levels and several changes in national policies." - Dan Jacobs

"You have to be resilient in a program like ISS because it requires a lot of spending over a long time period. So other programs say that you are spending too much and that you will never get to launch. People argued that we spent a lot of money, but we were still on the ground and no one has seen any benefit. You get bombarded by scepticism the entire duration of the program. An engineer by nature must be positive; he wants to make things happen. If he hears that he is wasting the taxpayer's money and his own time, you need a thick skin to get through that. We had a number of crises. We had one shuttle flight accident. But in the end, the whole station was delivered and it looks very close to the artistic impressions that we had in the 1990s. Many people said it was too complex and that it would never work." - Bernardo Patti

United States

In the US, the National Aeronautics and Space Administration (NASA) manages the ISS. NASA was created in October 1958, in the aftermath of Sputnik. Since then, NASA has maintained technical superiority in human space flight operations, from the one-man Mercury and two-man Gemini missions in the 1960s, through the historic lunar landing missions of Apollo and the long duration missions of Skylab, to the fleet of reusable Space Shuttles that NASA operated between 1981 and 2011.

NASA and the US provided the truss structure, the solar arrays, the first connecting Node, *Unity* and the US/Joint Airlock, *Quest*. The US Laboratory, *Destiny*, the cupola and many of the station's sub-systems are managed by the American space agency. The Control Moment Gyros (CMG) which maintain station orientation are also NASA supplied. Node 2, *Harmony* and Node 3, *Tranquility* and components and assembly of the Cupola were performed by ESA in barter/exchange for US/NASA services.

Many of NASA's field centers have been involved in the space station effort, from the earliest space station plans of the 1960s, up to operational support for the facility that is now in orbit. NASA Headquarters in Washington DC, provides managerial oversight of the field centers and is responsible for the political and financial links to the US government.

Lyndon B. Johnson Space Center (JSC)

JSC in Houston, Texas, is the center of the human space flight effort. It is the home of the ISS Program Office which manages all aspects of the US ISS Program. Several JSC institutional directorates support the ISS and other Programs. The principal supporting directorates are: Engineering, Health and Human Performance, Safety and Mission Assurance, and Flight Operations. The Astronaut Office and offices associated with Mission Control and Flight Integration are housed within Flight Operations.

Mission Control

Ground control of the ISS is provided from facilities in most participating countries. With the ISS systems and laboratories operating simultaneously aboard the station, control centers handle the operations and scientific activities for all on-board facilities. Additional control center support is required for launches and for flights of the logistics/resupply vehicles.

The Houston Mission Control Center (Building 30), named the Christopher C. Kraft Mission Control Center, is the primary control center for American human space flight. This center has been in use beginning with Project Gemini in 1965, and continuing through Apollo, Skylab and Space Shuttle. Two historic flight control rooms used since 1965 were augmented by new facilities that were brought on line in 1995. The new facilities are located in a new five story building (known as Building 30-South or 30S) and feature three control rooms designated Red, White and Blue. The multiple control rooms can support ongoing flight operations and training or simulations.

Houston Mission Control operates 24 hours a day 7 days a week by teams (or shifts) of certified Flight Controllers. With the advent of continuous mission operations aboard ISS in 1998, NASA was faced with operating Mission Control continuously for decades. If the primary MCC facility became incapacitated a back-up would be required. Prior to 2007, the role of the back-up ISS control center was filled by Mission Control in Moscow, Russia. For example, Russian mission control was used to back up MCC-H during Hurricane Rita in September 2005. However Moscow restricted ISS contact to occur only while ISS was passing over Russian ground stations. Based on capabilities and costs, it was determined that a more effective approach would be to establish a back-up control center at Marshall Space Flight Center (MSFC) in Huntsville, Alabama.

Marshall Space Flight Center (MSFC)

The Huntsville Operations Support Center (HOSC) at MSFC provides the primary ISS Payload Operations Center (POC). The POC also supports the function of back-up US ISS control center. Normally, the MSFC POC operates 8 hours per day, five days a week but can support continuous mission operations. The facility features ground systems and flight operations hardware that mirror the control room in Houston. In order to provide the back-up capability, the Huntsville facility upgraded voice services and S-band command and orbital communications.

The Back-up Control Center originally was intended to ensure safety of the crew, vehicles and on-board systems. Support of more intense activities such as docking and EVA were not originally intended. However in 2008, the Houston Mission Control Center was shut down due to Hurricane Ike, at a time when a Russian Progress was docking to the ISS. JSC Flight Controllers supported the operation from the Huntsville back-up control center.

In its role as the ISS Payload Operations Center (POC) the HOSC controls US experiments aboard the ISS. The POC links controllers on the ground with the ISS crew and international researchers. This is the American 'Science' Mission Control. The facility was originally developed to support *Spacelab* missions during the Shuttle program and was expanded to handle the science activity on ISS.

The POC is the primary US center for ISS payload operations and coordination with the international partners' centers in Korolyov, Moscow Russia; Tsukuba, Japan; and Oberpfaffenhofen, Germany. Though each partner nation staffs their own facilities, POC is responsible for coordinating and synchronizing payload activities among the partners, to ensure that payload operations are conducted safely and to optimize on-orbit resources and crew time. POC also represents the United States (and NASA) interests in science research aboard the station. Planning for science activities can occur days, weeks or even months before they are implemented.

US Telescience Support Centers

ISS Telescience Support Centers manage the US ISS research portfolio and perform remotely operated science experiments in the American segment of the ISS.

Each Telescience Support Center focuses on different fields of space research:

- Marshall Space Flight Center, Alabama: material sciences; biotechnology; microgravity and space product development
- Ames Research Center, California: gravitational biology and ecology
- John Glenn Research Center, Ohio: fluids and combustion research
- Johnson Space Center, Texas: human life sciences, physiology, behavioral studies and crew heath and performance.

John F. Kennedy Space Center (KSC)

KSC in Florida is the main US launch center. Space Shuttles, Dragon and most planned crew and cargo logistics spacecraft launch to the ISS from KSC. The center served as the focal point for processing, testing and preparing all of the US-launched ISS elements prior to flight. ISS pre-launch processing and payload integration coordination with the Shuttle and commercial launch providers have been supported. Processing of the Space Station elements and hardware is handled in the Space Station Processing Facility (SPPF).

Space Station Processing Facility (SPPF)

The SSPF has an office/laboratory area and a High Bay area. The office and laboratory area has three main floor levels. The offices and support rooms provide biological laboratories, checkout rooms for payload racks, areas for management of flight crew equipment, and off-line processing labs. The High Bay provided accommodations for many of the ISS assembly elements as they were prepared for launch on the Space Shuttle and is still used to support preparations for logistics flights to ISS.

The High Bay facilities are:

- High Bay: a full height area containing eight designated work areas used for staging, experiment integration, payload integration and verification. It is also employed as a post-landing disassembly facility.
- Airlock: provides vacuum hook-ups for cleaning and decontaminating each ISS element.
- Hardware Inspection Area for inspecting each element of the station
- I-Bay: includes power distribution boxes, work tables and data testing areas. At the end of this area is the Hazardous Fluid Servicing Area, for working on elements of the IUSS Thermal Control System containing ammonia.
- Low Bay: used for processing experiment racks.
- Rack Room: used for experiment rack processing.

The off-line laboratories are used for processing smaller experiments. One room is devoted to fabricating and repairing multi-layer insulation blankets for the station.

The Mission Cargo area is used to prepare cargo for each logistics flight to ISS. Food, clothing, tools and scientific experiments are packed for flight in Cargo Transfer Bags (CTB). Protective padding is designed and cut for packing around sensitive hardware. Flight crew equipment including crew personal items, EVA equipment, small hardware items, emergency escape gear and cameras is processed and stowed for flight. It is expected that the processing facility will be used in support of future programs using the KSC launch site. All major ISS US segment elements and payloads have been processed through the SSPF, including those of ESA, Canada, Japan and Russia. Testing of systems operations is a routine part of pre-launch processing. ISS posed a unique problem in that some systems would never be completely integrated on the ground. Some elements would already be in orbit before new elements would complete manufacture and arrive for launch processing.

Following the retirement of the Shuttle the SSPF had a smaller role to play in supporting the ISS by preparing cargo for commercial resupply spacecraft, while leasing out other smaller areas of the facility.

The Multi Element Integrated Testing (MEIT) program tested flight hardware as it was prepared for flight. Electrical, data and fluid lines were connected. To the extent possible, testing was conducted as integrated units. Emulators simulated any elements missing from the ground test. Many problems were identified and corrected prior to hardware being launched into space.

US ISS Prime Contractor

The prime US contractor for ISS is the Boeing Company. Several thousand Boeing employees are directly involved in ISS development, operation and sustaining engineering. Additionally, Boeing sub-contracts to numerous other companies throughout the United States. In support of ISS, Boeing has had important support facilities in Texas, Alabama, Florida, California and Oklahoma. Boeing produced many of the US elements, including the US Laboratory, interconnecting nodes and structures, power system, data management system, environmental control and life support, and associated critical hardware and software.

The Boeing facility in Huntsville, Alabama, fabricated many of the major US modules. This included the payload racks, the *Quest* airlock, cargo systems and components of the Cupola. The facility was responsible for the construction of the Environmental Control and Life Support (ECLS) System, Cupola windows, and control systems and subsystems associated with these elements. Boeing's Huntington Beach site in California was responsible for the design, development, construction and delivery of the integrated truss structure, the Pressurized Mating Adapters (PMA) and the Mobile Transporter (MT). This facility also developed the Communications and Tracking (C&T) System; the Guidance, Navigation and Control (GN&C) System; the Command and Data Handing (CCDH) System; and the Thermal Control System (TCS) and assisted in outfitting the Cupola.

Other leading US contractors include:

- Rocketdyne Propulsion & Power in Canoga Park, California, supplied the ISS Electrical Power System (EPS).
- Lockheed Martin, Bethesda, Maryland (solar arrays, communications systems);
- Space Systems/Loral, Palo Alto California (batteries, electronics and other items);
- Allied Signals, Torrance, California (gyros and other navigational gear);
- Honeywell, Phoenix, Arizona (command and data systems, gimbal motors);
- United Technologies, Windsor Locks, Connecticut (pumps and control valve assembly).

Aerial Photo of Johnson Space Center near Houston Texas.

"I'm a certified instructor for all the new crew members that go to the ISS to operate Canadarm2. We have a mockup of the robotics workstation that we train them on and everything that's running on the screens is totally simulated but they get to interact with the hardware that they would have up there and they've all come back and said it's great in terms of training and it's quite realistic. Nothing quite beats the cupola views you have on the ISS but we try and get as close as possible." - Kristen Facciol

A full-size simulator of the space station Cupola ▲ shows the station's Canadarm2 working alongside a space shuttle. The simulated image is projected onto a hemispherical dome.

Neutral Buoyancy Laboratory (NBL) operations and facilities in the Sonny Carter Training Facility at the Johnson Space Center (JSC). Overall view looking down into the pool with various International Space Station (ISS) mockups visible at the bottom .

Cygnus cargo module being prepared for launch.

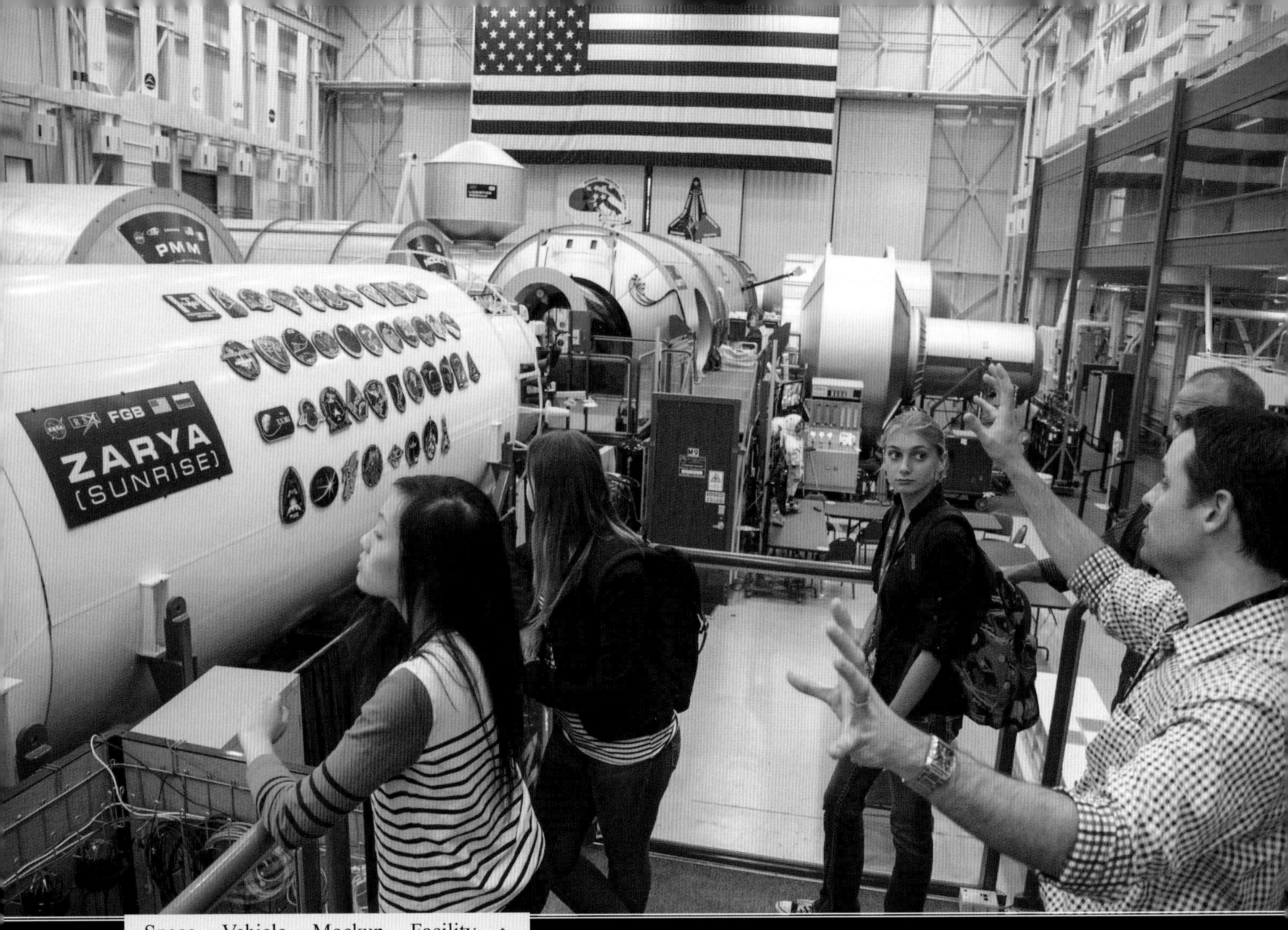

Space Vehicle Mockup Facility (SVMF) in JSC Building 9, interns during tour of ISS mockups and racks.

▲ Expedition 52 crew members Jack Fischer, Fyodor Yurchikhin, Randy Bresnik, Sergey Ryazansky and Paolo Nespoli during Emergency Scenarios training at SVMF in JSC Building 9.

Expedition 55 crew members Scott Tingle, Anton Shkaplerov, Norishige Kanai, Drew Feustel, Ricky Arnold and Oleg Artemiev during Emergency Scenarios training in SVMF. ▼

Expedition 55 crew members Scott Tingle, Anton Shkaplerov, Norishige Kanai, Drew Feustel, Ricky Arnold and Oleg Artemiev during Emergency Scenarios training in SVMF.

Expedition 56 crew members Drew Feustel, Ricky Arnold, Oleg Artemiev, Alexander Gerst, Jeanette Epps and Sergei Prokopev during Emergency Scenarios training in SVMF.

Johnson Space Center ISS Mission Control Center Building 30

Expedition 58/59 (57S) crew members Anne McClain, David Saint-Jacques and Oleg Kononenko during Routine Operations training in SVMF.

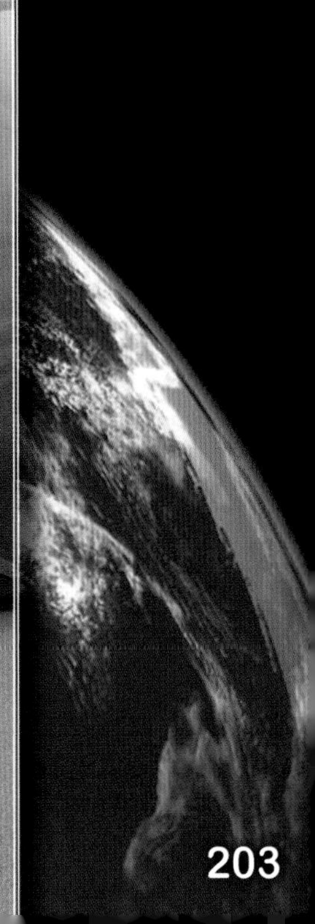

Routine Operations training in the mock-up of the Harmony module at Johnson Space Center.

Node 2 and Kibo In preparation for flight at the Kennedy Space Center (KSC) Space Station Processing Facility (SSPF).

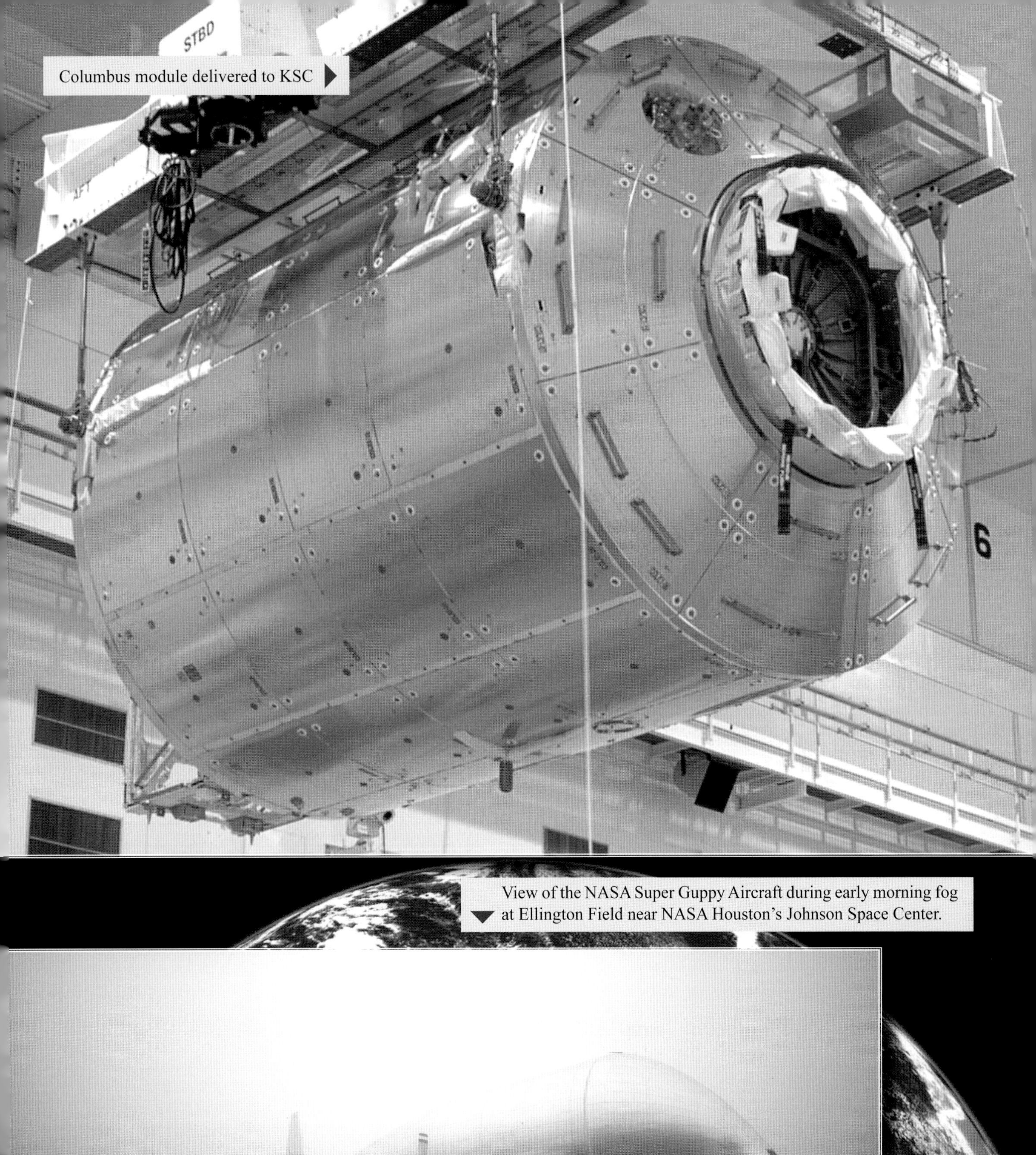

Columbus module delivered to KSC

View of the NASA Super Guppy Aircraft during early morning fog at Ellington Field near NASA Houston's Johnson Space Center.

The trainer for the Space Shuttle, positioned on a special cradle for travel, is loaded onto a Super Guppy aircraft at Ellington Field near the NASA Johnson Space Center.

Cupola being mounted on the Node 3 Tranquility for launch at the KSC SSPF.

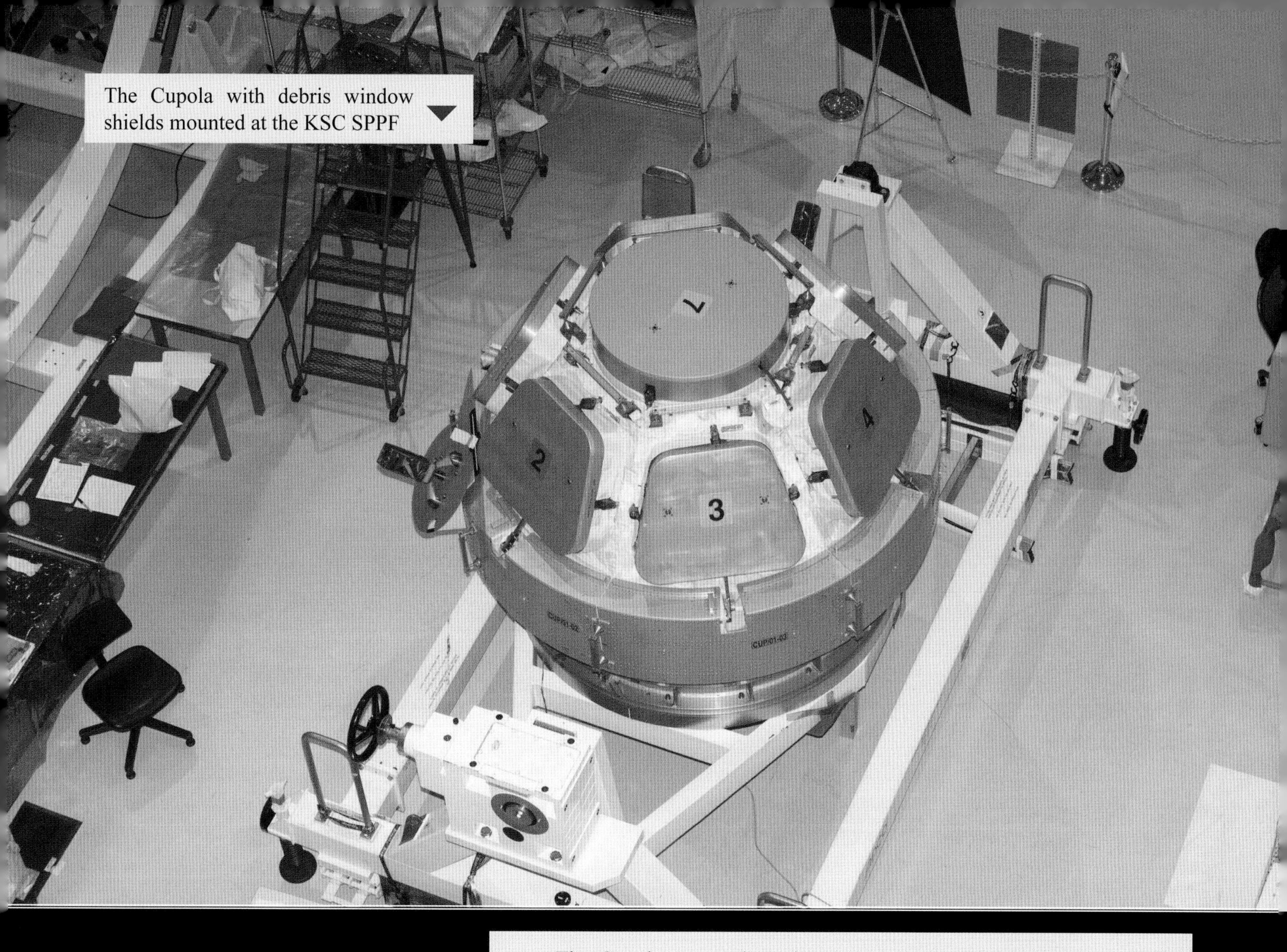

The Cupola with debris window shields mounted at the KSC SPPF

The Cupola mounted on the Node 3 inside the Shuttle payload bay in preparation for launch. The Shuttle Remote Manipulator System Canadarm1 is in the foreground towards the top of the view.

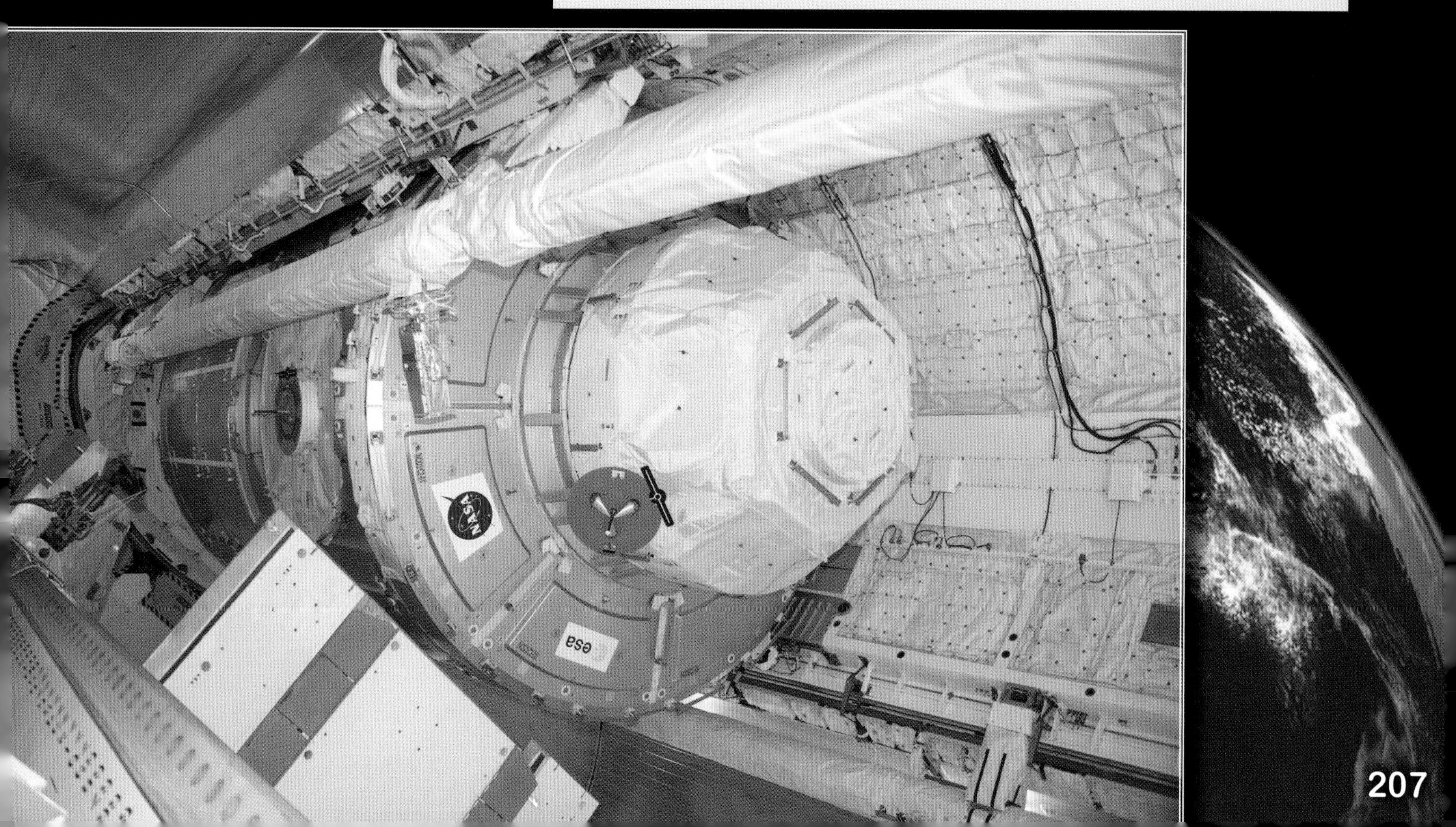

CANADA

The Canadian Space Agency (CSA) was established in 1989, though Canada had an active space program beginning in the 1960s with its first satellite launch. In the 1970s, Canada was an important contributor to the Space Shuttle Program, developing the Shuttle Remote Manipulator System (RMS) or Canadarm which was critical in the assembly of the ISS. Canada has conducted a broad program of operations and research, including flights of Canadian astronauts on the Shuttle and ISS and payloads on Shuttle, Mir and ISS. CSA provides the ISS Mobile Service System (MSS) drawing upon its earlier Shuttle RMS experience to develop the more sophisticated ISS Space Station RMS (SSRMS), Canadarm2, and the Special Purpose Dexterous Manipulator (SPDM), Dextre.

Canadian centers involved with ISS are:

- John H. Chapman Space Centre, Saint-Hubert, Quebec, which serves as the CSA headquarters, overseeing the management of the national program and its involvement in the ISS program, as well as Canada's cooperative ventures with NASA, ESA, Japan and Russia.
- The David Florida Laboratory, Ottawa, Ontario, serves as a major engineering facility.
- The Mobile Servicing System Operations Complex (MOC), located at the Chapman Space Centre in Saint-Hubert, Quebec, was the primary facility for developing the Canadian ISS systems for flight. It also acts as the primary astronaut training facility for these systems. Support for Canadarm2 operations is also located at this facility.

"The Canadians were the closest thing to having one of the international partners in the critical path, because they had the arm and we knew that we weren't going to be able to get the thing assembled without the arm." - Neil Hutchinson

"Goddard Space Flight Center had Martin Marietta out of Denver building a robotic device. When it came down to cutting budgets, the Canadians were building their robotic device and they supplied their own money to build it. We had to make a choice and we wound up taking Goddard out of the Space Station Program, and cancelled the contract." - Richard Kohrs

"The Space Station started with a NASA robot arm that was being developed by Goddard Space Flight Center, the Flight Telerobotic Servicer. When the Canadians joined the program with a robot, the US telerobotics was cancelled. That saved NASA a billion dollars that became the reserve funding for the program." - Denny Holt

"The fact that we have a Canadian robotic arm, has made the ISS possible. Without our arm we wouldn't be where we are. It has done amazing things and it continues to do amazing things even with the visiting vehicles, without the arm we couldn't have captured the HTV, Dragon, or Cygnus cargo vehicles. They provide a lot of the cargo and the payloads and science that are going on the ISS now. The arm has played a huge role and Canadians should be proud of that." - Charladean Smith

The headquarters of the CSA is located at the John H. Chapman Space Centre in Saint-Hubert,Quebec.

"The CSA-NASA partnership is evident at every level. The Canadian ISS element is part of the US Operating Segment. Our robotics workstation talks directly to the US Command and Control computers. Our robotics flight control team is made up of Canadians working in partnership with Americans. We work together in a seamless, integrated fashion as one team." - Tim Braithwaite

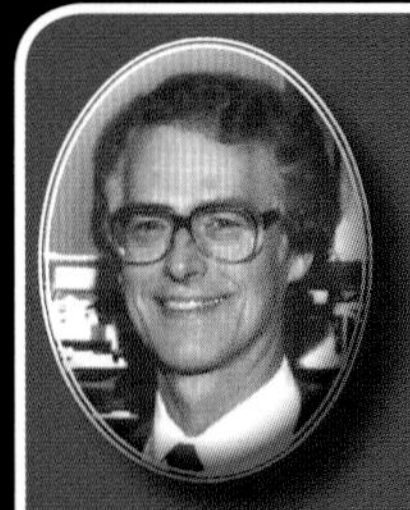

"When space station was starting up, Senator Jake Garn was looking for a vehicle to promote U. S. competitiveness, and he had latched in on automation and robotics and artificial intelligence as key areas that, if the United States was going to be increasingly competitive, he felt this needed to be pushed and he selected the space station as the vehicle to push it. Garn said, 'If we promote the use of automation and robotics on space station, that will drive the technology.' It will make space station better, and at the same time drive U. S. industry." - Bryan Erb

Canadarm2 in preparation for flight

"In my visits to NASA and its contractors I found a little known optimal appendage to the orbiter. It was called the Shuttle Attached Manipulator System (SAMS). It was an anthropomorphic (human like) arm, attached to the shuttle and could be used to deploy and retrieve satellites from the shuttle's cargo bay. It was largely unheard of as no firm requirement had been established for it. The SAMS was 50 feet long and 15 inches in diameter and behaved much the same as a human arm. We started to work with NASA's Johnson Space Flight Centre in Houston on its definition and also with Rockwell who had become the space shuttle orbiter prime contractor." - John MacNaughton

"Canada's role would be to design and build the robotic systems needed to help the astronauts to construct the space station, and then to maintain and repair it. Canada's role was and still is critical, as it is part of the core of the station. Canada's financial contribution ultimately would be some $1.5 billion, more than 10 times the Canadarm investment. The station's planned lifetime was 30 years, so if the Canadarm experience is any guide, the return on investment on the station should be at least five times the cost." - Phil Lapp

"The difference between cultures at NASA, ESA and CSA is huge; because you think about all the different member states that are part of ESA and they all have different percentages of how they work and who gets hired and that part of the politics is interesting. Even just working with the different cultures. In the hallway you'll hear multiple languages, but then in the meeting room it's all English." - Charladean Smith

"I tend to start with, 'Have you heard of the International Space Station?' because there are a surprising number of people who haven't. Then I get into how it's the largest laboratory orbiting the earth and because it's such a harsh environment we have to have robotic systems. Then I describe the Shuttle and how we had the original Canadarm on the five shuttles. And I describe the Station arm as a giant robotic inchworm that allows us to move to different work sites across the International Space Station with which we can access all the various components that we need to get to, and so that we can minimize the number of times we have to send people out into a very harsh environment. That's the very basic explanation." - Kristen Facciol

Canadarm2 during development, here being placed into a chamber for environmental testing.

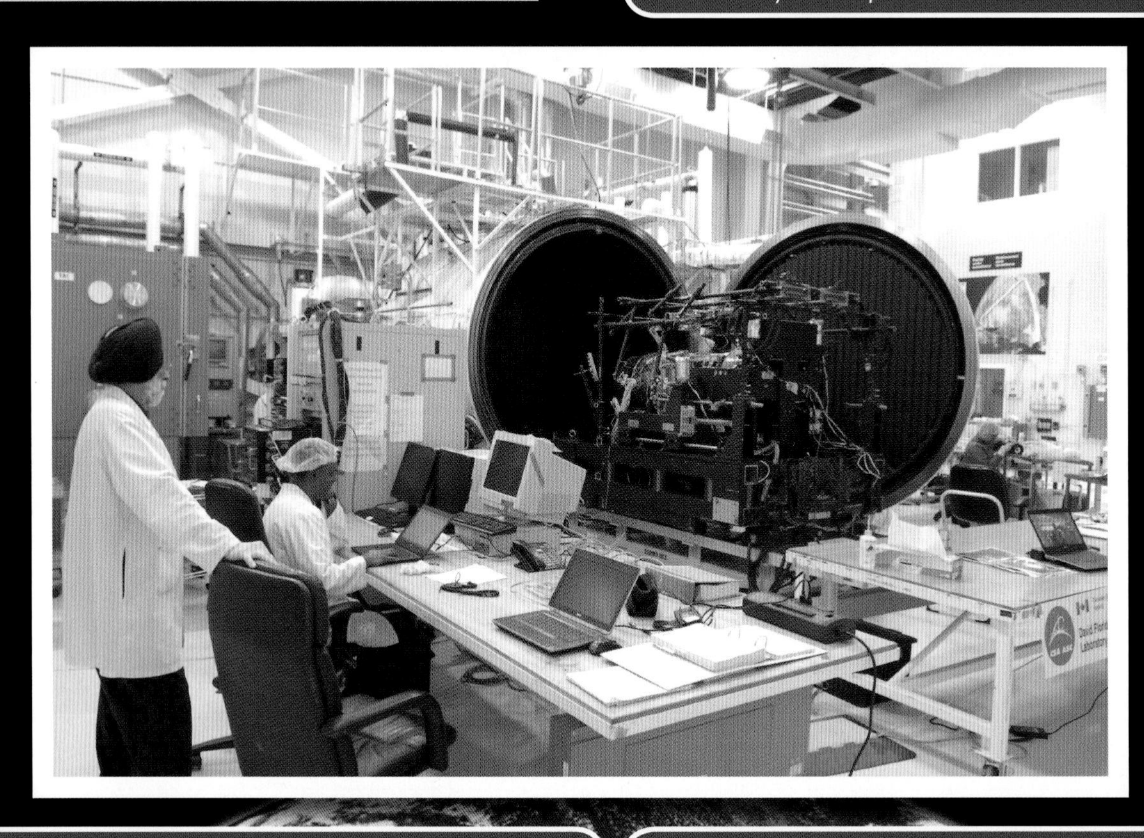

"Dextre was originally for fine repair maintenance. But the different ways that it has been used could not have been thought of when it was started. If it's there it will get used in some way. Things often do not go the way we expect and plan. Curveballs get thrown at us and it takes people thinking outside the box to come up with work-arounds and to put together a plan to achieve our objectives. Again and again people and teams do come up with workarounds to make the mission successful. We are on the path to automated assembly. I think there's still some gaps that we need to work on but we've come far even from when the ISS was being built. When ISS was being assembled the crew did most of the arm operations that we had pre-planned for them. They followed the procedures. Nowadays we do so much from the ground for the ISS. The crew now only does free flyer captures. The majority of the other operations are all from the ground. We are working towards automation. On the Deep Space Gateway there is a big push to do a lot of the tasks automated. It would be pre-planned on the ground then we send up a script and the arm executes it with intelligence and assesses its own performance. We still have steps to get there but we trust that automation will get the job done." - Charladean Smith

"The SSRMS systems are becoming more modularized, like plug and play. It's not fully automated. We still do ground commanding of all the robotic systems. We support a lot of the EVAs with robotics. But in terms of repairs we've actually done quite a few recently both robotically and using astronauts. The way the system is designed it is a bunch of orbital replacement units that can be taken out and fully replaced. For example we replaced an end effector on Canadarm2 and that had to be done by an astronaut. But, there was a failed camera on the arm and we were able to use Dextre or SPDM, the smaller robot with the two arms and the ability to rotate its body. We picked that up with Canadarm2 and then robotically changed out these things. So we were able to do that completely from the ground. It was not automated; it was planned on the ground then the sequence was commanded by flight controllers. We are working towards more modularization. It is much better to replace one component than to replace an entire system. We have spares in orbit for the majority of the components and whenever required we're able to either robotically or using the EVA, swap those out." - Kristen Facciol

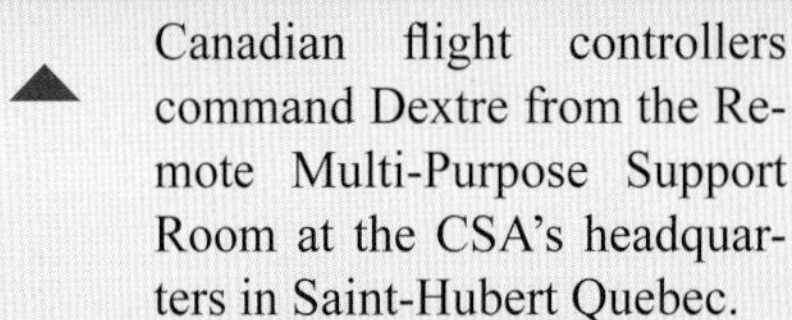

Canadian flight controllers command Dextre from the Remote Multi-Purpose Support Room at the CSA's headquarters in Saint-Hubert Quebec.

"I believe our greatest achievement from a Canadian standpoint was to show that the space station shows that we can play with the world technologically, scientifically, and professionally, astronauts, engineers, everyone, to make human space flight possible. That we can play with the rest and we can hold our own." - Nicole Buckley

"From 1999 to 2003, I was the Shuttle RMS lead at Titan. Titan was a contractor that hired me to work in engineering robotics on the simulations for the SRMS and SSRMS. We did an analysis for that so we made sure that the simulation had all the parameters they needed. Based on the different missions, we would plug in the payloads that the arm would be moving around and do an analysis to make sure that the loads and trajectories fit in the parameters of the capabilities of the arm. We supported the virtual trainers and desktop trainers that the crew would use to train different aspects of the mission. We would make sure that the simulations ran as they needed to run. From feedback from the crew they always said that they were surprised at how similar the simulations were to actually flying the arm in orbit. Later I was hired on to the Canadian Space Agency in 2003 and I got hired as a robotics flight controller. At any given time there have been 7 to 10 CSA robotics flight controllers. There is a group of 30 to 40 NASA robotics flight controllers as well. We're considered part of their group. We were considered part of the NASA MOD robotics flight controllers team even when we're back here at CSA." - Charladean Smith

"When I was younger I wanted to be an astronaut. Every child dreams of being an astronaut. I knew my route to that was going to be through medicine or engineering. I always had in the back of my mind that that was going to be where I was going to end up. And I made sure I performed well in science and math. I went to Space Camp in Laval, Quebec when I was 11 years old, that was when I was first introduced to the Canadarm and we got to do simulated repairs on the Hubble space telescope. I thought, 'This is so cool! There's this giant robotic arm up in space doing things.'" - Kristen Facciol

"For Canadarm2 everything that we do is simulated on computers. We have full robotic planning simulators that allow us to look at the trajectories we have to plan. Once that is simulated it goes through dynamics analysis to make sure that the proper parameters and loads are used appropriately in that assessment. It's one thing to say 'we can move the arm this way', but can we physically do it is a different question. We don't have any actual ground-based hardware that does it now. What we train on now is as accurate as it can be. The tools that have been developed over time based on what we've learned in orbit are incredibly robust and we trust them because we've been able to prove their capabilities for so long. What we've learned in space has created a platform for the direction we're going in from the medical perspective as well. A lot of what we've learned is going into nuclear robotics as well. We absolutely did start with a blank slate and paved the way to make a name for Canada in robotics not just in space but in other remote environments and difficult to access situations. When we send people to Mars the kinds of maintenance and repair and replace functions would probably be managed and programmed on-board the spacecraft rather than back at mission control. It is much more difficult to deal with the programming when you have to deal with time delays in communications. Just sending the sequences of commands the way we do now, we wouldn't be able to deal with those delays." - Kristen Facciol

Dextre being prepared for flight.

"I think the Canadian built robotics are beautifully designed and continually evolving paradigms of what space robotics will be. And when we have robots on the surface of the moon, when we put the base there, or the robots that we tend to strap to the outsides of long duration zero-g ships and interplanetary ships and such, they're all going to be based on Canadarm1. It is great, precedent setting Canadian-designed and built." - Chris Hadfield

"Dextre was originally going to have a limited set of things it was going to do. At the outset it was mainly a demonstration program that was going to assist with some maintenance. But once it was demonstrated, all of a sudden everyone needed it to be used for science that the Canadians had never thought of when they built it. But scientists think of it when they see what it can do. So now we need a new agreement with the Canadians to get more spares from them because it's going to be used much more than we thought. It is just another example of the benefit of having our partners provide these different things. You end up with different capabilities that no one thought of, but now everyone recognizes how beneficial they are." - Dan Jacobs

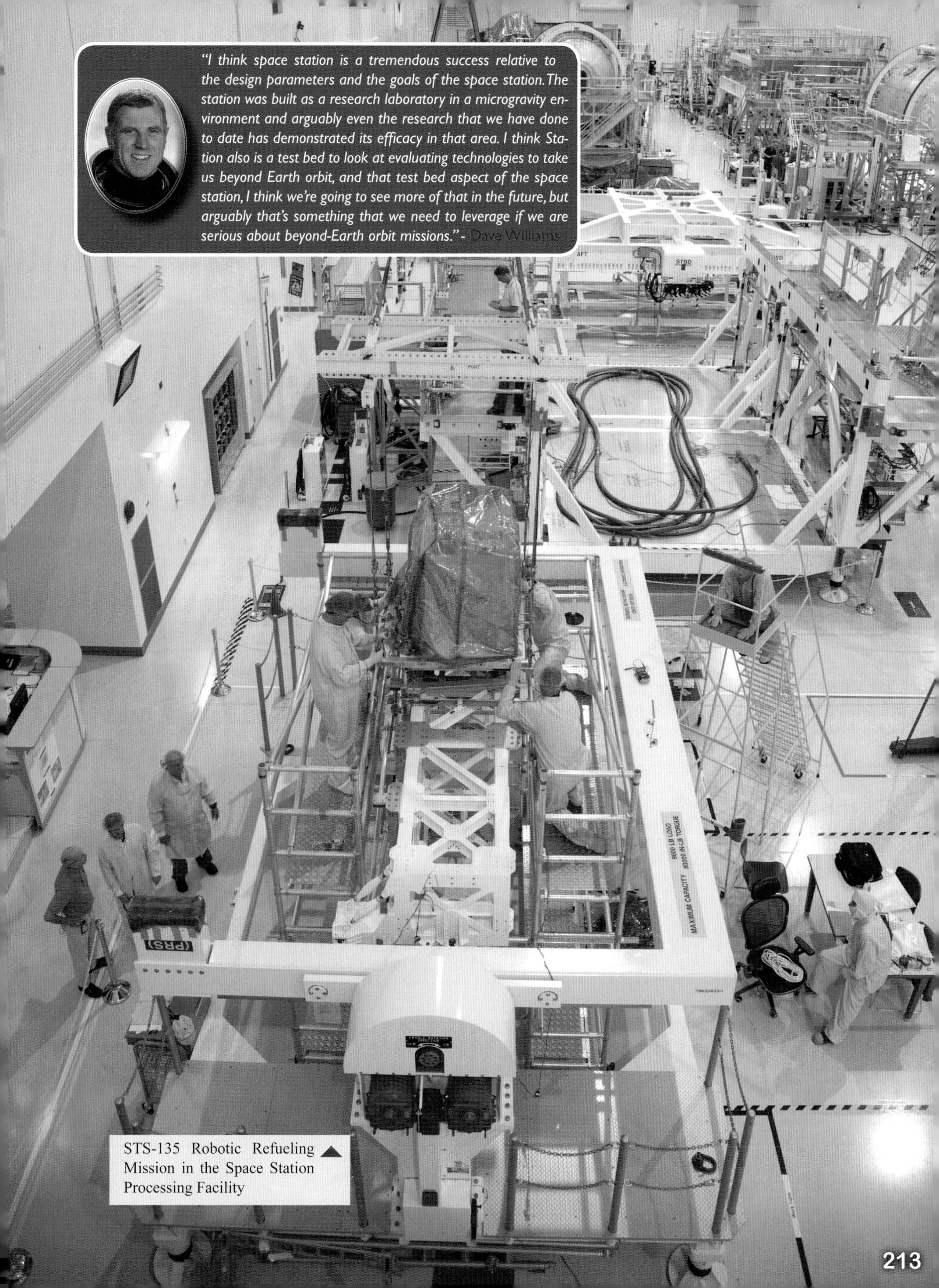

"I think space station is a tremendous success relative to the design parameters and the goals of the space station. The station was built as a research laboratory in a microgravity environment and arguably even the research that we have done to date has demonstrated its efficacy in that area. I think Station also is a test bed to look at evaluating technologies to take us beyond Earth orbit, and that test bed aspect of the space station, I think we're going to see more of that in the future, but arguably that's something that we need to leverage if we are serious about beyond-Earth orbit missions." - Dave Williams

STS-135 Robotic Refueling ▲ Mission in the Space Station Processing Facility

Located at the Canadian Space Agency's (CSA) headquarters in Saint-Hubert, Quebec, the Mobile Servicing System (MSS) Operations Complex is the main facility for planning and monitoring Canadarm2 and Dextre's activities on board the International Space Station. All astronauts and flight controllers who operate the Canadian robotic systems on board the station are trained at the MSS Operations Complex.

"In Mission Control Center (MCC) in Houston there is the front room, the room most often seen on TV, with the Flight Directors and an operator from each system. Each of those people has a back room. It was always the Canadian Space Agency's plan to have a back room, a remote Multi-Purpose Support Room (MPSR), in Saint-Hubert. Previously they had been located in Houston. The lead sometimes is in Houston, but whether in Houston or Saint-Hubert, they all see the same data, see the same video, communicate on the same headsets, listen on the same voice loop." - Tim Braithwaite

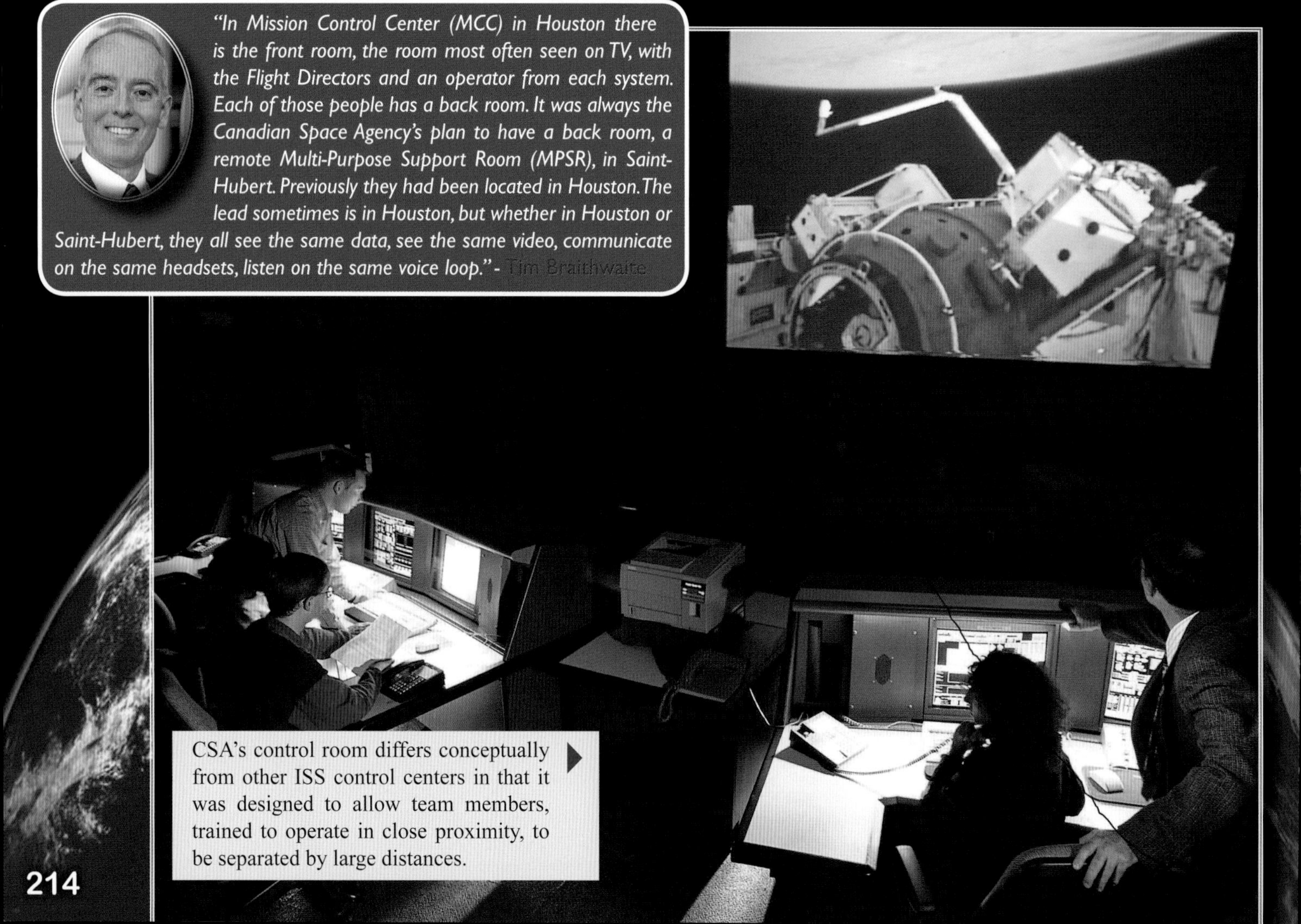

CSA's control room differs conceptually from other ISS control centers in that it was designed to allow team members, trained to operate in close proximity, to be separated by large distances.

Europe

The European Space Agency (ESA) was established in 1974, consolidating the European Launch Development Organization (ELDO) and European Space Research Organization (ESRO). With its headquarters in Paris, ESA manages contributions from 16 member states. For the ISS Program, 12 ESA member states have been participants. Several of the member states continue their own active national space programs in addition to cooperating through ESA.

ESA has worked closely with NASA in human space flight for many years. In the 1970s, ESA developed the Spacelab Module and Pallets which were flown on 22 Shuttle flights between 1983 and 1998. Twenty ESA Astronauts flew on the Shuttle between 1983 and 2011. ESA has also worked with Russia, leading to a series of crew members flying on Salyut, Mir and Soyuz.

For ISS, the European Space Agency (ESA) developed and provided the Columbus laboratory and five Automated Transfer Vehicles launched between 2008 and 2014, two of the Nodes (2 and 3, through the Italian Space Agency) and the Cupola. A European Robotic Arm is also under development for attachment to the Russian segment. The Italian Space Agency also provides the Multi-Purpose Logistics Modules (MPLM) that were used for Shuttle logistics flights to ISS and one of which was converted to a permanent ISS storage module. They also build the pressurized compartment of the Cygnus commercial logistics vehicle.

Prominent ESA facilities are:

- ESA Headquarters in Paris, France, is the central facility for oversight and management of programs and budgets.
- The European Space Research Center (ESTEC) in Noordwijk, The Netherlands is where most of the programs are developed and implemented.
- The Automated Transfer Vehicle Control Center (ATV-CC), Toulouse, France, is the controlling facility for each flight of the ATV logistics vehicle.
- The European Astronaut Center, Cologne, Germany, is the central facility for ESA astronaut training.
- The European Automated Transfer Vehicle ATVs were controlled from European sites in coordination with MCC-Houston and other international partner control centers.

Launch Site

The Guiana Space Center (GSC), Kourou, French Guiana, was originally the national launch facility for France. It has become the main launch center for ESA programs, including Ariane 5 ATV missions to ISS. The 'European Spaceport' in French Guiana is located in South America, close to the equator, which gives the maximum possible assistance to vehicle launches owing to the west to east rotation of the Earth. Following a decision by the French to create an independent national launch facility in 1964, Kourou was mainly used for sounding rocket programs. In December, 1979, the first Ariane orbital launch took place. A new launch complex and payload support facility is used for Ariane 5 operations, including the launch of ATV vehicles to the ISS. Russian Soyuz-ST booster rockets have also been launched from Kourou since 2011.

"President Reagan invited international partners to join in a space station program. ESA decided that they wanted a piece of this. It seemed to be a natural succession after the Spacelab program. I worked on phase A. We didn't call it phase A, but that's basically what it was, for two or three years. But it didn't seem to be getting off the ground very quickly. I also did some study work on transfer vehicles that were the forerunner of the ATV. I was working for the ESA director of launchers. it was decided within ESA to swap the director of launchers with the director of manned space flight. That person was Jorg Feustel-Buechl. He was somebody that I absolutely wanted to continue working for. He was just a brilliant director. He really was the father of this work in Europe. In ESA, whenever we want to get a major program approved, we had to go to the member states at the agency and make a proposal to them. We were going in 1995 to a ministerial conference. We were going to propose that Europe participate in the ISS program, to do the Columbus module, to do an ATV, to develop all the operational capabilities, and to have a utilization program. It was a package that added up to about 2.4 billion Euros. About 18 months before this ministerial conference, the Columbus module was about 10 meters long, a length of eight racks internally. Industry said it was going to cost 1.4 billion to develop, and it would take two shuttle launches, each of which would cost 200 million to get it up there. We would have launched the module empty apart from its systems and then the racks would have to go into the logistics module that the Italians had been developing for NASA. The 1.4 billion version of Columbus had an external payload platform and an antenna on the back of it that was going to interface with the European data relay satellite. We couldn't afford it. There was no way in the world that we were going to get that kind of money. Feustel said, 'if we can get the Columbus module for 650 million, we can get the program approved.' Industry was saying it was 1.4 billion, so I had to cut it in half. After about a year of very, very intensive work we got them to agree to a price of 657 million. To do that, we halved the length of it, literally. The module had to be cut in half at 5 meters long. We packaged it in such a way that it could all be launched in one shuttle launch. So the launch costs were cut in half as well. So they sent us a bid for 670– 657 million that was conditional upon us not interfering too much, and was non-negotiable. They said, 'if you try and negotiate a penny out of this, then we withdraw the offer,' which was a very unique situation. I was quite sure I could get at least 10% out of their price. Feustel told me, 'if they say it's non-negotiable, that means that they've got fat in it and you'll be very happy as a project manager that your prime contractor has got fat.' He was absolutely right. We ended up hardly having any overrun on Columbus, even though we had a four year extension by the time we finally flew." - Alan Thirkettle

European Mission Control

The Columbus Control Center, (COL-CC) in Oberpfaffenhofen, Germany, is the ESA Mission Control for ISS Columbus operations, and the primary operations center for European experiments on the station. The control center for the European Columbus scientific module is based at the German Space Operations Center (GSOC) of the German Aerospace Center (DLR) in Oberpfaffenhofen, near Munich, Germany. A team of 75 scientists and engineers supervise and coordinate all ESA activities associated with the ISS and specifically the activities of astronauts in the Columbus laboratory. The COL-CC uses the call-sign 'Munich' during radio communications.

GSOC has been used since the 1970s and since 1983 for manned space flights. In 1983, during the first Spacelab mission on Shuttle STS-9, GSOC was linked to NASA's MCC-H and in 1985 the center became the Payload Operations Control Center (POCC) for the (German) Spacelab D1 mission, successfully demonstrating the capabilities of the center with responsibility for the scientific aspects of the mission while NASA handled flight operations. This was the first time that an American mission had been controlled, in part, outside of the United States. A new building was commissioned and constructed in the 1990s for Spacelab D2 and the ESA missions to Mir. GSOC also participated in the STS-99 mission, with German astronaut Gerhard Thiele on board during the Shuttle Radar Topography Mission (SRTM) in 2000.

A new purpose-built center was certified operational in April 2005 for the ISS Eneide mission of ESA Italian astronaut Roberto Vittori. The Col-CC team worked with launch team at Baikonur, the astronauts in space, the control teams in America and Russia and partner teams in Italy and the Netherlands. From July 2006, Col-CC supervised all activities of the ISS Astrolab mission of German ESA astronaut Thomas Reiter, the first European long duration mission on ISS (7 months) during Expedition 13/14. In this operation, Col-CC coordinated activities with the European Astronaut Center (EAC) and the DLR Institute of Aerospace Medicine in Cologne, including monitoring the state of health of the astronauts. The ground crew at Col-CC also coordinated the on board experiments operated during the Astrolab mission.

In 2008, Col-CC began to supervise ISS-based tasks of the ATV, including the transfer of supplies from the resupply vehicle to the space station. They became part of the communications network while ATV was docked to station. With the attachment of the Columbus laboratory in 2008, the control center had the responsibility for looking after the European scientific laboratory in addition to coordinating the scientific program.

ATV Operations Control Center (ATV-CC)

A separate control center for ATV was created in 2002 at the Toulouse Space Centre, France, and is managed by the French space agency CNES on behalf of ESA. Prior to flight, the ATV-CC created the mission plan and then prepared and validated control system inputs. The control center oversees all phases of the ATV mission, including the stabilization of its orbit, transfer maneuvers, rendezvous and docking, and end of mission activities: undocking, de-orbit and atmospheric entry. During the flight the ATV-CC coordinates with control centers in Moscow and Houston.

"People have made the comment to me a number of times, 'Well, the Russians must have been the toughest negotiators.' No question, the Russians are very skilled negotiators. But I found the most frustrating and most difficult negotiators to be the Europeans. The reason for that was Germany needed X, France needed Y, Italy needed Z. None of those could be breached. The lead negotiator for the Europeans basically had virtually no flexibility to compromise." - Lynn Cline

"In 1996, ESA member states said, 'we are now serious about paying for a human exploration program.' And there was a package deal at the ministerial meeting in Toulouse in 1996. The Germans supported the French with Ariane 5. France supported the Columbus and utilization. Hermes program was discontinued. There was a constraint that we were to design to cost of the Columbus. We were told by industry, 'this is what we have. Tell us what we can do with it' and that was when Columbus became much shorter because it was reduced to cost. There was a reduction of the number of launches. We repackaged but provided the same utilization volume in a much more compact configuration. JEM, stayed the same, but was then twice as long. JEM is very much luxury in term of packaging, and accessing equipment. Our equipment packaging is not very user friendly for the astronauts exchanging equipment. We launched Columbus in one launch where JEM/Kibo was launched in three missions." - Bernardo Patti

ESTEC in Noordveig, ESA research and technical center at which ATVs were developed.

"ESA told NASA, 'rather than paying you money to launch Columbus, why don't we give you the nodes instead? They'll be done by the European consortium that's doing Columbus, and the structure will all be based on exactly identical work to that of Columbus, which is based on Spacelab, so you can trust it. The MPLM, the logistics module that the Italians were in a bilateral doing with NASA, also had the same pedigree as far as the structure was concerned. So we said, 'we will do the mechanical part of the nodes. NASA will provide the avionics and the life support equipment, because that is tied into the rest of your system.' So we built the nodes. That way we didn't have to pay for the launch. That was a Feustel out of the box way of doing business. NASA jumped on it. For NASA it was perfect because nodes 2 and 3 were going to cost them a lot more than $200 million from Boeing. NASA was happy, and they trusted the Italian company that was doing all of the mechanical stuff with this. We were happy because it meant that our taxpayers' money would go back to European industry, but also, we could do it really quite cheaply in spacecraft development terms. These nodes didn't have to be requalified for our structure because they were very similar in size to Columbus, to the MPLM, and to Spacelab. So we had essentially a prequalified piece of hardware with a track record and a lot of the analysis. So this barter was a considerable help in our overall contribution to the station and our ability to control the cost of the station. The nodes did not change the percentage of research time ESA would get, because the nodes were a NASA responsibility. NASA was still paying for the nodes. When NASA agreed to this, first of all, that we would do the nodes, they said, 'the nodes are going to be done by the company in Italy that's doing the logistics module, and the logistics module's being done in the relationship between ASI, the Italian space agency, and NASA. They had nothing to do with ESA.' NASA said 'we like this relationship we have with ASI. We want that same relationship for the nodes. All ESA, has to do is to provide the money.' Feustel said, 'if that's what they want, it's okay. ESA will have no responsibility. Industry is doing the work, and ESA doesn't need to get involved.' But every time NASA wanted to make a design change, and NASA makes a lot of design changes, ASI accepted the changes, passing them on to industry. Industry rather naively assumed they were going to get paid for these changes. ASI didn't put sufficient discipline into it. They weren't at all good at cost control, or cost management. They got themselves into a large cost overrun, because they kept accepting change after change. Industry asked ASI for money. ASI said, 'we haven't got any money. It's ESA that pays for this.' Industry went to ESA and said, 'you have to pay for these changes,' and ESA responded, 'what changes? We don't know anything about changes. We're hands off.' ESA had to help negotiate an agreement with NASA, ASI, and the Italian company, and our member states for NASA to pay for the large number of changes they had made. ASI contributed some of the money. Industry had to contribute some. And ESA had to contribute some as well, in exchange for which ESA took over the management. ASI became observers. ESA took over the management, and now managed it in exactly the same way that we were managing the rest of the program." - Alan Thirkettle

The International Space Station Node 2 module is shown in a processing facility. Under contract of the Italian Space Agency (ASI), Alenia Spazio in Turin, Italy led a consortium of European sub-contractors to build Node 2. The module was built for NASA under a barter agreement with the European Space Agency (ESA) in exchange for the launch of the European Columbus Laboratory by the space shuttle to the International Space Station. Node 2 provided a passageway between four International Space Station science experiment modules: the U.S. Destiny Laboratory, the Kibo Japanese Experiment Module, the European Columbus Laboratory and the Centrifuge Accommodation Module.

"NASA was developing the cupolas. ESA needed more shuttle transportation, and so we told NASA that ESA was prepared to take over development of the cupolas. Alenia in Torino, Italy, was the prime contractor. It was a competitive bid. They gave us a proposal the likes of which we had never seen before. Alenia had done the logistics modules, they were doing nodes 2 and 3, they were doing Columbus, and they were doing the module for the ATV. They were the pressurized habitable environment element developers for the ISS, and they would not be outbid by anybody else. They put together a really tremendous proposal for that and developed it. Initially, there were supposed to be two cupolas but then NASA decided in their wisdom that they only needed one. NASA said to us, 'as you're only going to deliver one, we won't give you as much transportation.' ESA responded: 'no—no, we signed an agreement, transportation for cupolas. We are going to give you two. If you don't launch the second one that's up to you, but we want the transportation we had agreed to. We will not debate further about it.' NASA finally came to their senses and agreed." - Alan Thirkettle

"The cupola barter granted ESA the launch of some equipment and the on orbit resources for such equipment so that ESA could begin utilization on ISS early, before Columbus was there. And the cupola is also a kind of flagship for Europe, because that's a beautiful piece of equipment, which is used and seen by everyone, where the astronauts are taking pictures, and all of it was built by European industry." - Bernardo Patti

"In another barter, ESA needed about another $10 million or $12 million of transportation. NASA was having difficulty transporting their station elements to KSC. They either had to go by barge or they would have to go by truck. Those are not great transport environments. They really wanted to fly the things. But the elements of the station are so big that you need a special airplane like the guppy. NASA has a guppy, but it was falling apart. But I think Airbus had a spare guppy they had used for transporting elements of the A380 within Europe. Airbus now had a jet powered guppy. The NASA guppy was an old Boeing Stratocruiser turboprop. Feustel said to NASA, 'I will buy you a guppy in exchange for which, you will launch science to the space station for me.' The guppy still had a certificate of airworthiness and was still being maintained as a back-up though they knew they would probably never use it again. They also had a warehouse full of spares. ESA offered that they could have all of that. A couple of NASA guys came over, they inspected this thing to be sure they were happy with it, and indeed we sold a guppy to NASA." - Alan Thirkettle

The European Columbus laboratory settles into a new home in the Space Station Processing Facility at NASA's Kennedy Space Center

"There were other barters. We had to launch and retrieve science; samples from experiments; they have to go up and down. At the time in the 1990s, only the shuttle could carry things up and down. Shuttle was unique at that time. ESA would have to pay a large amount of money for this and ESA didn't want to pay money. So we said, 'we can take over things that you, NASA, are responsible for.' So ESA designed, developed, and delivered a minus 80 degree freezer to keep things like biological samples very cold in orbit, waiting for the logistics module that's going to bring them down. We did a couple of freezers for NASA and we did a science glove box in exchange for logistics transportation." - Alan Thirkettle

"About year 2000, the NASA program costs had gone out of control and the office of management and budget, the treasury of the U.S., which was run by Sean O'Keefe, decided ISS ought to be scrubbed, or canceled. The U.S. government looked at how they could control costs and get rid of some of the operational costs. They took about 2½ years to do this. They cancelled the hab module and the centrifuge module. We were not too worried about the centrifuge module, although we had thought of the possibility of trying to do it for ourselves. The hab module, we weren't really too worried about either, because it was seen as a bit of a luxury. We were concerned about them trying to turn Columbus and the Japanese module into part-time gymnasiums for the crew. That disturbs the local microgravity environment, and at that time, we expected a lot of material science work to be going on, and that would not survive with people jumping around in zero g. During this time, the international partners were out on a limb and were being ignored. We observed meetings between NASA and the government, to see the questions that were being asked. The U.S. government decided it had a problem and it was going to solve it, and if there was collateral damage in the form of international partners, that was tough. ESA's member states said, 'Well, if the Americans are so interested in themselves, why don't we pull out? We can use the money for a different space program.' We came very, very close to canceling our participation in ISS. What stopped us was the intergovernmental agreement that had the status of a formal international treaty. So we would have to break a treaty. Politically there was an imperative that we had to be really, really certain that cancellation was what we wanted to do. We came quite close." - Alan Thirkettle

▲ Automated Transfer Vehicle ATV in the acoustic test facility

▼ ESA's Node 3 in preparation for shipment to KSC

ESA's Columbus Control Center in Oberpfaffenhofen

"ATV was like any other vehicle to us, but with the big difference being that we could launch a lot from an ESA point of view with that vehicle. The ATV operations were done from Toulouse. They had a separate control center for ATV there. But ATV was always a big impact for us because we got a new bunch of ESA equipment delivered with" - Alexander Nitsch

▼ ESA Training Academy

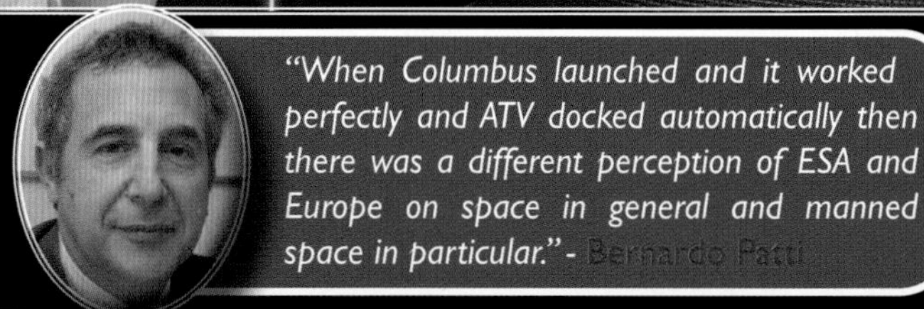

"When Columbus launched and it worked perfectly and ATV docked automatically then there was a different perception of ESA and Europe on space in general and manned space in particular." - Bernardo Patti

▼ ESA's main control ESOC room

Ariane 503 transfer from assembly building to launch pad at Kourou

"The Columbus launch was very exciting; we had worked for so many years on the Columbus module; to actually see it going up from Cape Canaveral was quite something." - Robert Chesson

Ariane 5 launch from Kourou

JAPAN

Japan established a National Aeronautical Laboratory (NAL) in 1955. Later the Institute for Space and Aeronautical Science (ISAS) formed. In 1969, Japan created the National Space Development Agency (NASDA), which became the center of its national space program. NAL, ISAS and NASDA merged to form the Japan Aerospace and Exploration Agency (JAXA) in 2003. The Japanese space agency JAXA developed the largest scientific laboratory on the ISS. It is used in conjunction with an experiment logistics module and a platform that allows the placement of experiments outside the Japanese module. An experiment airlock and an integrated remote manipulator are used for placing external experiments. The JAXA H-II booster rocket H-II Transfer Vehicle (HTV) supplement the logistics resupply of the station.

Several Japanese astronauts have flown in space. The first Japanese to fly in space was a journalist who flew on a Soyuz to the Mir Orbital Station. Several Japanese flew on the US Shuttle. Japanese participation in the space program expanded with the arrival of the Kibo laboratory and the use of the Japanese HTV logistics vehicle to support ISS.
JAXA Headquarters in Tokyo is responsible for overall management of Japanese involvement in ISS.

Tsukuba Space Center (TKSC), developed the Kibo laboratory module and houses the Kibo control center.

The TKSC facilities:

- Space Station Integration & Promotion Center (SSIPC), The Space Station Test Building (SSTB) was used for Kibo system testing and now houses on-orbit operations support for experiments and research performed aboard Kibo.
- TKSC Astronaut and flight controller training facilities train the space station crews in Japanese objectives, research, experiments and operation of Japanese hardware.
- The Kibo Systems Integration Facility, houses the Kibo Engineering Model (EM) and the Proto-Flight Model (PFM). Kibo systems checkout equipment, a station interface simulator, and a Manipulator Functional Test Facility provide a main arm simulator. Functional and performance tests of the Japanese RMS and demonstration tests of the PM/EF berthing mechanisms were conducted in this facility.
- The Space Experiment Laboratory (SEL) provides support for developing and testing experiments to be performed aboard Kibo. Smaller laboratories support different experiment types. The Material Science Experiment Support Technology Development Laboratory evaluates compatibility between different materials, their containers and potential contaminants. The Telescience Technology Development Laboratory simulates systems to verify the operations and procedures. The Life Science Experiment Support Technology Development Laboratory evaluates and prepares life sciences experiments for the station.
- The Material and Life Science Laboratories assess prospective experiment programs in fluid physics and biotechnology. The Ground Reference Experiment Laboratory verifies procedures, prepares reference samples and gathers control data for comparison with subsequent space-based operations. Where necessary, ground experiments are conducted in conjunction with on board experiments to compare results.
- The JAXA Astronaut Training Facility (ATF) and Weightless Environment Test Building (WET) are used for crew training. WET is a weightless simulation facility using a large swimming pool to simulate EVAs.

Launch Site

The H-II rocket launches the HTV from the Osaki Launch Complex for mid-size rockets range at the Tanegshima Space Center.
The primary Japanese launch site was constructed in the south-eastern corner of a long and very narrow Tanegshima island in 1966. The location facilitated launches over the ocean on trajectories to the east or the south, as close to the equator as possible (31°N Latitude) from within Japan. The site was constructed mindful of the small but critical local off-shore fishing industry. The site has several ranges, which are used for different launch vehicles.

Japanese Mission Control

Once the Kibo Laboratory became part of the ISS, the Japanese established a separate control room to support Japanese scientific operations on the station. A team of 50 flight controllers handles Kibo operations. They are divided into teams covering ten disciplines under the direction of a Flight Director. The JAXA Mission Control Room in the Space Station Operations Facility at Tsukuba Space Center is manned by three shifts over each 24 hour period. The Japanese team manages Kibo laboratory operations. The flight control team monitors the status of commands, the downlink of data, and the performance of systems, payloads and experiments. They can make real time decisions or change plans. The team can communicate directly with the crew inside the Kibo laboratory or operating facilities outside the pressurized sections. The Japanese team also has direct links with the other international partner control centers. The flight control team prepares and evaluates all plans and procedures relating to Kibo tasks. They conduct regular training simulations of all aspects of the Kibo program, including off-nominal and contingency procedures, and use this training to troubleshoot any problems or anomalies aboard the station. A JEM Engineering Team (JET) provides technical support to the flight control team. The JET evaluates real time data and performs pre- and post-flight analyses.

HTV Flight Control Room (HTV MCR)

HTV flights are monitored and controlled by the HTV Mission Control Room (HTV MCR), located at the Space Station Operations Facility (SSOF) in JAXA's Tsukuba Space Center (TKSC). This facility controls and manages all HTV operations in collaboration with the NASA Mission Control Center at the Johnson Space Center in Houston, Texas. Communications with each HTV is established via the TDRSS network after the HTV separates from the launch vehicle. HTV systems are nominally activated automatically, vehicle attitude is stabilized, and internal self-checks are conducted. After communications is established, the HTV MCR issues commands to control subsystems and vehicle maneuvering. The NASA MCC begins monitoring HTV operations when the spacecraft is one orbit or 90 minutes from a point 5 km (3.1 mile) behind ISS. From that point, the HTV MCR and MCC-H jointly cooperate to control the HTV mission at ISS.

"In the 1980's, the focus of the Japanese space programme was in domestic technologies, in launch vehicles and satellites. I, myself was interested in international co-operation, so I proposed to my manager to work on the International Space Station. At first I worked not on space station but the space shuttle utilization programme on Spacelab-J. This and other shuttle missions were a very good opportunity for Japan to learn about manned space activity. On Spacelab J, we had 35 experiments on three racks in the module for life sciences, material sciences and fluid science many experiments in a small space, a very Japanese way. That was the first time we fabricated equipment for space with humans with a lot of safety requirements, a very hard experience for our engineers, which helped us a lot to develop Kibo on ISS. At first, many engineers did not understand English and we required translators for the technical co-ordination meetings. That was very hard. At that time we did not have very good computers, so we exchanged drawings by facsimile for 100 pages. That was very hard. So when the Marshall Spaceflight Center staff got to the office in the morning they always complained because the fax machine tray was full." - Shigeki Kamigaichi

"I worked with the Japanese bringing in the requirements during the early days of Space Station. The Japanese contribution to ISS is still 100 percent of what it was in the beginning. They built a laboratory module. They built the Exposed Facility. They built a manipulator arm. They added an HTV logistics vehicle. Language is a terrible problem for the Japanese. The Japanese are too proud to ask for interpreters. They've been formally trained in English, but most words, colloquially, just go by them. They miss a lot. There are also problems beyond the language, in a cultural sense. The Japanese do not make decisions the way NASA makes decisions. NASA gets a dozen people around a table, and discuss, argue, and then decide 'this is what we'll do.' Meetings in Japan are held for photo opportunities. If the Japanese have a Program Manager who wants to meet with the US Program Manager, the Japanese do not expect to do any work at the meeting. The work has already been done. They only expect to consummate the agreement, or managerially validate what was already decided. They work by consensus at a working level. They are prone to go along with whatever NASA wants to do, so they often just let NASA make the decisions." - Carl Shelley

"In negotiations, the Japanese will push you really hard. The most important thing in dealing with Japan is if your position is one you cannot bend on, you just need to state it over and over. Be patient; be firm. Don't start trying to say it in three different ways, because maybe they didn't understand the first time. No, they understood the first time. If you start changing it, then they assume, 'Ah, she can modify that, we'll try and exploit that opening.' I learned to stick to it until you can't stand saying it anymore. You just had to keep at it." - Lynn Cline

"Japan has a space center at a town called Tsukuba. It looks like JSC. All the buildings are there; the facilities are there. They've got water tanks. They've got simulators. They have a control center that looks like the one at JSC. They've got all the things you'd find at JSC. What the Japanese do not have is the people to use all of their facilities. We advise them on establishing the teams of people to interface with the JSC Mission Operations and ISS Program Office." - Carl Shelley

"The most exciting time during my life was to work with NASA from 1995 to 2009. And I had two greatest achievements. One is, of course, is to conduct the first EVA by first Japanese person during STS-87. It was quite a challenge but I enjoyed it very much. Also, the second achievement was to install the first Japanese manned facility to the International Space Station. And I was the first Japanese person to enter this Japanese-made space facility and I was very honored to do that." - Takao Doi

"Japan had no experience in astronaut training prior to joining the ISS programme so our training in Japan required coordination with our international partners. So our programme of training developed from that cooperation. In Japan, English is taught in school as an introduction not in-depth, so it was very tough for Japanese astronauts and engineers to master even basic language skills. For the Russian language, it was even harder and we had to provide language teachers and translators for our people when they travel. For Japanese astronauts, it was very tough because of the language to fly on Soyuz than the Shuttle but in recent years, their experience has improved so they are able to manage understanding Soyuz systems in Russian easier." - Chikara Harada

"For Japan, the only way to do space science was to fly on the shuttle or join the space station programme, we learned a lot about International co-operation by working with the NASA people. Before we participated in the ISS programme, most of the other countries believed Japanese space technology very poor. At that time the head of the Japanese space agency could not meet with the head of NASA only through the US foreign affairs department director so we were not an equal partner, but now we are seen as an equal partner with NASA, so today, Japanese space technology, through the ISS programme and Kibo, is seen very differently around the world." - Shigeki Kamigaichi

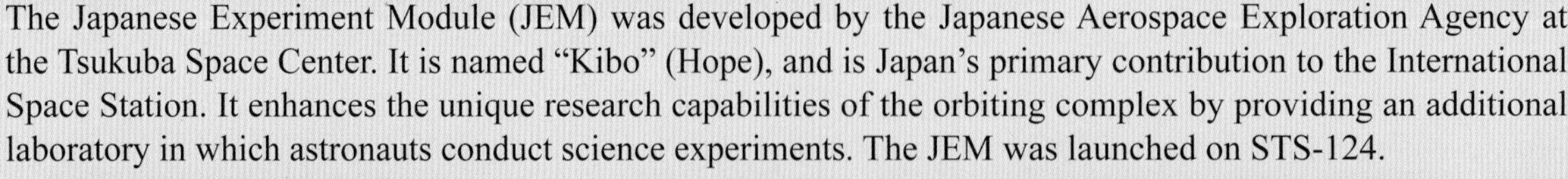

The Japanese Experiment Module (JEM) was developed by the Japanese Aerospace Exploration Agency at the Tsukuba Space Center. It is named "Kibo" (Hope), and is Japan's primary contribution to the International Space Station. It enhances the unique research capabilities of the orbiting complex by providing an additional laboratory in which astronauts conduct science experiments. The JEM was launched on STS-124. ▲

"Constructing the ISS and operating ISS is the biggest international collaboration in human history and I'm very happy that Japan could participate in it as a country from Asia – only one country from Asia. And Japan also learned how to conduct a long kind of mission, ISS mission. And, before that, Japan had only shorter duration missions and acquiring the new technology one-by-one by conducting shorter duration missions. But participating in the ISS program, Japan learned how to conduct a very long space mission. And Japan made a big commitment to continue this new ISS program for more than 10 years, now 20, whatever." - Takao Doi

The Space Station Operation Facility (SSOF) is responsible for controlling Kibo operations. At the SSOF, operation of Kibo systems and payloads are supervised and Kibo operation plans are prepared in cooperation with NASA's Space Station Control Center (SSCC) and Payload Operation Integration Center (POIC). ▼

"Due to the diversity of the space station programme, we need, even on the ground, a variety of people from various backgrounds, cultures, age groups and experiences and through this diversity we make the team stronger. We have to work together, even if it is a challenge sometimes, to achieve a shared goal allowing for the maturity of the programme and ensuring its aims remain peaceful. I believe that the Kibo laboratory was the turning point for the Japanese space programme from a developing country in terms of human spaceflight to a developed country as one of the five equal partners of the international space station." - Chiaki Mukai

The JAXA Flight Control Team (JFCT) consists of flight directors and more than 50 flight controllers. The team monitors and controls the Japanese Kibo laboratory of the ISS around the clock in a three-shift per day schedule. The flight director oversees and directs the team. Flight controllers possess specialized expertise on all Kibo systems. ▲

The JAXA Flight Control Team (JFCT) monitors the status of command uplinks, data downlinks, system payloads and experiments aboard Kibo. The team has the capability of making real-time operations planning changes, and can communicate directly with the crew aboard Kibo and the various international partner mission control centers located around the world. The team troubleshoots problems or anomalies that may occur aboard the Kibo during flight operations. ▲

"We spread our experiments across the crew when there is not a Japanese astronaut on-board. The very first complex life science experiment just happened to be done by a Japanese astronaut which was very lucky. The schedule can be changed a lot on ISS even scheduling Japanese experiments during the time when a Japanese astronaut is on-board can change. But, in many cases, a NASA astronaut can do the experiment. Today, Kibo is just a module for experiments it doesn't have any life support capabilities to sustain a human crew or independent sub systems but in the future, possibly with Kibo II, we would like to develop a module with those facilities. Of course this depends upon JAXA, the Japanese Government and cooperation with international agencies discussing the issues, but one of the targets would be cis-Lunar, we would like to put a station around the moon." - Shigeki Kamigaichi

Kibo operations are jointly monitored and controlled from the Space Station Operations Facility (SSOF) at the Tsukuba Space Center (TKSC), in Japan, and the Space Station Mission Control Center (MCC) at the NASA Johnson Space Center (JSC), in Houston, TX, where the overall operations of the ISS are controlled. ▲

"It becomes kind of routine to conduct these kind of space activities onboard the International Space Station, specifically for industries. They could benefit by developing scientific modules. But after they completed constructing those modules, those engineers have to work on different projects. So in a sense, it's better to continue that kind of development once every 10 years. But after we finished developing the Japanese experiment module in 2004 – so more than 10 years – we didn't have any new development craft for human space activities. Now that Japan announced that they would work with the United States to go back to the Moon, to explore the Moon in 2020s. And then that is quite exciting to me, not only for myself but also all the students, industries, or the scientific communities." - Takao Doi

View of a tool interface located on Micrometeoroid Orbital Debris (MMOD) Shield of the Japanese Pressurized Module

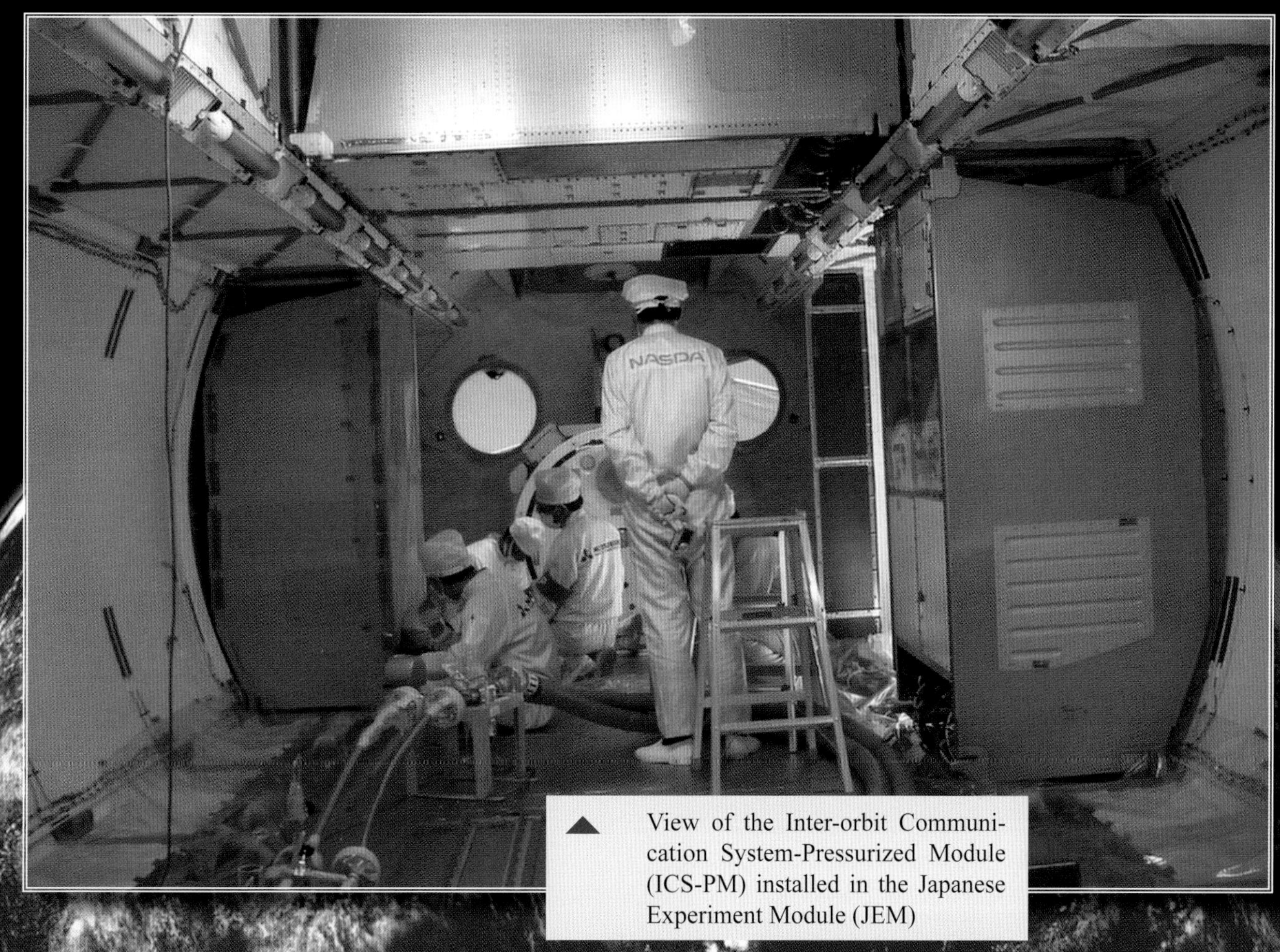

View of the Inter-orbit Communication System-Pressurized Module (ICS-PM) installed in the Japanese Experiment Module (JEM)

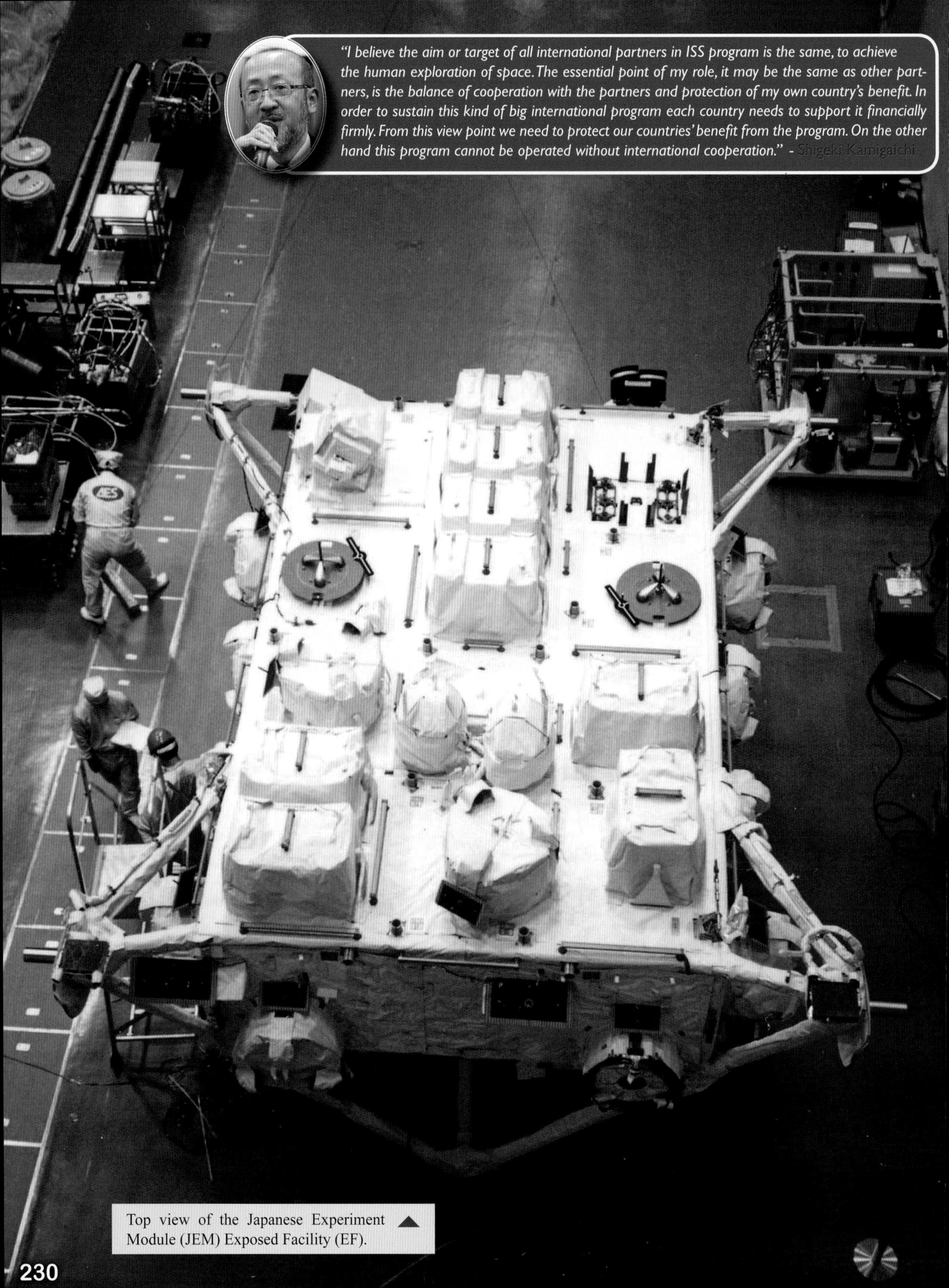

"I believe the aim or target of all international partners in ISS program is the same, to achieve the human exploration of space. The essential point of my role, it may be the same as other partners, is the balance of cooperation with the partners and protection of my own country's benefit. In order to sustain this kind of big international program each country needs to support it financially firmly. From this view point we need to protect our countries' benefit from the program. On the other hand this program cannot be operated without international cooperation." - Shigeki Kamigaichi

Top view of the Japanese Experiment Module (JEM) Exposed Facility (EF).

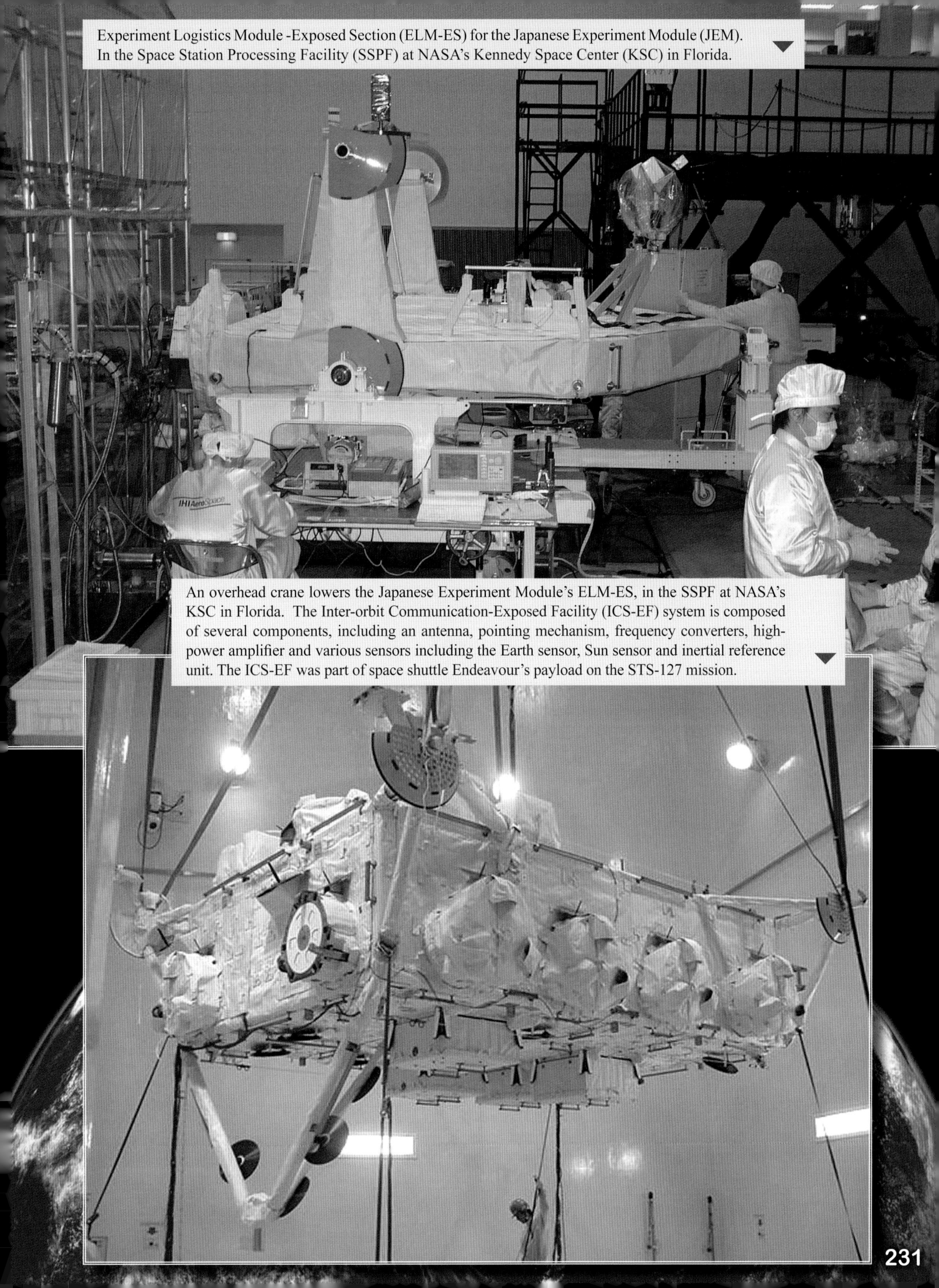

Experiment Logistics Module -Exposed Section (ELM-ES) for the Japanese Experiment Module (JEM). In the Space Station Processing Facility (SSPF) at NASA's Kennedy Space Center (KSC) in Florida.

An overhead crane lowers the Japanese Experiment Module's ELM-ES, in the SSPF at NASA's KSC in Florida. The Inter-orbit Communication-Exposed Facility (ICS-EF) system is composed of several components, including an antenna, pointing mechanism, frequency converters, high-power amplifier and various sensors including the Earth sensor, Sun sensor and inertial reference unit. The ICS-EF was part of space shuttle Endeavour's payload on the STS-127 mission.

"At the start of the Kibo operations most of the ground flight controllers were engineers who were involved in the development of the Kibo module. After ten years of Kibo operations they are now senior controllers. Most important is the relationship with the astronauts, so we decided to put astronauts with flight controllers in one unit. The ISS program itself has been changed a lot since its beginning, mainly shrunk. But Japan did not change the basic design of the Japanese Experiment Module. But we also needed to reduce the budget size in Japan. I believe we could do this by efficiency. And now we still need to reduce the budget size more so that we can invest in future manned program. For this purpose these days we encourage industry involvement in the ISS program." - Shigeki Kamigaichi

▲ Tsukuba Weightless Environment Training Facility (WET)

A view of the Space Station Operations Facility (SSOF) which is responsible for monitoring and controlling Kibo operating systems, monitoring and controlling Japanese experiments onboard Kibo, implementing operation plans and supporting launch preparation. ▼

"Japan showed the world that they could make a big commitment to the international collaboration of humanity. And I think that -- Japan had never done that before. So I think it made Japan ready to go on, not only on the human space program but on to other things."- Takao Doi

RUSSIA

Russia and the Soviet Union achieved many significant space firsts:

- first space satellite, Sputnik 1, 1957
- first living creature in space, the dog Laika, Sputnik 2, 1957
- first probe to reach the moon, Luna 2, 1959
- first man in space, Yuri Gagarin, 1961
- first woman in space, Valentina Tereshkova, 1963
- first crew of more than one person, Voskhod 1, 1964
- first EVA, Alexei Leonov, 1965
- first soft landing on the moon, Luna 9, 1966
- first automated lunar sample return, Luna 16, 1970
- first soft landing on Venus, Venera 7, 1970
- first space station, Salyut 1, 1971
- first soft landing on Mars, Mars-3, 1971
- first continuously used, multiple expedition space station, Salyut 6, 1977-1982
- first modular, 'permanent' space station, Mir, 1986-2001

Russia brought its experience of three decades of manned space station operations to the ISS program. From the outset, the Russians sought to extend long duration space operations as a stepping stone to traveling to the planets. The Russian space program was first established as competing design bureaus for developing rockets, spacecraft, and space stations. Several principal Russian organizations have been involved in the ISS. The Russian Space Agency in Moscow provides managerial oversight of the ISS Program and is responsible for the political and financial links to the Russian Federal government. The establishment of the Russian Space Agency was relatively recent and did not occur until after the decision to join the already established station program that would become the ISS.

S. P. Korolyov Rocket and Space Corporation Energiya

RSC-Energiya (RSC-E) was established as Experimental Design Bureau-1 in 1946 by its Chief Designer, Sergey Korolev. For its first 10 years, it focused on development of ballistic missiles for military use and then on an Intercontinental Ballistic Missile (ICBM). As the ICBM looked as though it would prove successful, Korolev sought and gained permission to launch a military reconnaissance satellite that could also be used for carrying a human. Subsequently he was directed to launch the first satellites. RSC-E continues to build Soyuz rocket boosters based on the first ICBM, and Soyuz spacecraft, first flown in 1967. The manned Soyuz continues to carry ISS crews to and from orbit and the unmanned version serves as the Progress logistics carriers. Derivatives have been used to carry the small Russian ISS modules. RSC-E has also served as the lead designer and integrator of the heavy modules DOS and TKS, used for the earlier Salyut and Mir stations and used currently on ISS as the FGB, Nauka MLM and Service Module.

Khrunichev State Research and Production Space Center.

Together with its OKB-23 Salyut Design Bureau, located in Moscow, Khrunichev has been the group responsible for the construction of all Soviet and Russian large space station modules, such as FGB, Service Module and Nauka. Khrunichev also builds the Proton rockets used to launch these large modules. Khrunichev has a close working relationship with OKB-1 RSC-Energia. RSC-E acts as the program management, integration and engineering arm while Khrunichev is responsible for construction of the large modules.

The Gagarin Cosmonaut Training Center (GCTC)

The GCTC has been the home and training center for Russian cosmonauts since 1960. This is also where the international crew members complete their training on Russian systems, including Soyuz, Progress and Orlan EVA operations. GCTC began as a Soviet Air Force Special Training Center. In 1965 it became interdepartmental and a year later expanded its work to include training for research test pilots as well as cosmonauts. In 2009 the training facility was moved from military supervision to be within the purview of the new Russian Federal Space Agency, Roscosmos.

"In 1991, the Soviet Union collapsed. The Russian Space Agency was established in 1992. Station was almost canceled in 1993 by the US Congress because of funding issues. Here was a case of the U. S. not being the ideal partner from the perspective of our existing international partners. We had a legally binding, existing agreement with Canada, Europe, and Japan for Space Station Freedom. Then the U. S. decided that we should invite Russia into the program. Part of this was because of the change in administration and the budget cuts, and we almost lost Station. But if we could bring Russia in with the substantial infrastructure contributions of their very mature, established human spaceflight program, then that would compensate for things the U.S. could no longer afford to do. This was the time period of the Gore-Chernomyrdin Joint U. S. -Russian Commission on Economic and Technological Cooperation. Gore was given the lead in the U. S. government for a whole range of things with Russia. Space was one of them." - Lynn Cline

Institute of Medical and Biological Problems (IBMP)

The IBMP is the central facility for Russian human life science research on ISS. In addition to specialists across the medical and physiological disciplines, IBMP maintains a mission control room that features a TV-video conference link to TsUP and a Team Leader, with up to 5 specialists on consoles. The IBMP control room is used as required. Flight controllers responsible for life science and medical activities work at TSUP on a continuing basis with additional specialists called up during EVAs.

Russian Mission Control

The Russian Federal Space Agency Mission Control Center (TsUP) is located in Korolyov, Moscow Region. The Russian Mission Control is called Tsentr Upravleniye Polyotom (the Center for the Control of Flight). This is normally abbreviated to TsUP (and pronounced 'tsoup'). Russian cosmonauts in orbit typically address the TsUP as 'Moskva' ('Moscow'). The TsUP facility is one of the major research departments of the RSA Central Research Institute for Machine Building (TsNIIMash) and was not originally a control center. Its conversion began in 1970 and it became operational with Soyuz 12 in September 1973. Since that time, it has been used for all manned space flights, although a duplicate room was intended for the aborted Buran space shuttle program. The layout of TsUP is in the form of five rows of 24 consoles. The main control room accommodates approximately 35 flight controllers responsible for systems performance and mission operations. Ancillary control rooms provide work stations for controllers managing vehicles operating independently of the ISS, and for science payload and experiment operations management. Since 1973, the role of Lead Flight Director has been filled by former cosmonauts. The flight control position responsible for talking with the cosmonaut crew in space is also normally filled by fellow cosmonauts; called 'Glavniy Operator' this is the equivalent of the American Capcom. The Russians are less restrictive of who can speak with the crew in orbit and routinely call in specialists as the need arises. Flight controllers typically work a single 24 hour long shift, followed by 3 days off. Then they come in for 2 planning days and on the third day conduct another 24 hour long shift. Small Mission Support Rooms are adjacent to the main control room. Representatives of NASA and ESA are frequently hosted. Throughout NASA-Mir and later during the early stages of ISS assembly, NASA had a Houston Support Group that resided at TsUP to provide support and control assist Russian cosmonauts and facilities. These groups served as back-up to their respective control centers to ensure continuous and safe orbital operations. With the completion of ISS assembly there is less need for the international partner groups in the control centers and most dialogue is continued remotely through communications channels. MCC-M is located in close proximity to the RSC-Energiya main center which facilitates calling in developers and suppliers of space hardware, payloads and supporting equipment.

Launch Site

Baikonur Cosmodrome in Kazakhstan has been the main launch facility for space flights during the years of the Soviet Union and now for Russia. It launches most of the country's manned and unmanned space missions and is the processing and launch center for all Russian-launched ISS elements, as well as the Soyuz crew transport and unmanned Progress resupply craft. Baikonur Cosmodrome was originally constructed in the mid-1950s as the first ICBM launch site. It is located close to the southern limits of the former Soviet Union, in what is now the Republic of Kazakhstan. The network of launch pads is linked by railway lines to the assembly and administrative areas and to the factories where the rockets and spacecraft are produced.

"From ASTP, we got the androgynous docking mechanism from the Russians that is used on Station now. That was one of the things that benefited both countries and is benefiting the Space Station now. If we gave the Russians anything, it was probably process and techniques about how to plan a mission; not that they hadn't planned a hundred more than we have through the years; and most of them successfully. I think we taught them about the way to set up working groups and do a logical approach to a mission plan and technical agreements, more structured than they were used to working. We learned some things from the Russians; some technical, and some philosophical engineering things; like, the U.S. probably goes overboard on testing, especially hardware testing. The Russians were forced by budget constraints to not do all these elaborate tests. Basically, they build their hardware and do some physical testing, but they don't do a lot of analytical testing. When they get it in orbit, and if it turns out they weren't right, they fix it. If it's not broke, you don't fix it. I think following more of the Russian approach on Space Station, because it's just so expensive to test everything or analyze it to death with all possible failure modes, it's just there's a point of diminishing returns. I think some of our people have learned that the Russians do some things pretty cleverly. We learned a lot about the Russian character, too, at least the guys we worked with. They were upright, patriotic people. Most of them have a great sense of humor too." - Kenneth A. Young

"The Russians came into the ISS program; they were already operating Mir. They had their own operating system and it was totally different from what we had been planning. Getting each side to appreciate that there were benefits to be gained by speaking with the other side and either incorporate or at least tolerate what was absolutely necessary to make the program possible. The benefits of decentralizing functions needs be taken into account in the future." - Dan Jacobs

View of the outside of Russia's Federal Space Agency Mission Control Center in Korolev, Russia, located on the outskirts of Moscow.

"An interesting dynamic was that at the outset, the Russian Space Agency was a brand-new agency. It was a very small agency that suddenly was put in charge of all of the state-run companies like Rocket and Space Corporation (RSC) Energia and Khrunichev [State Research and Production Space Center. I think the companies resented having this new government entity above them because the companies were the ones with the experience. It was difficult for some at NASA because the US and Russian systems didn't match. NASA had expertise in the field centers and in the contractor community. The Russians had a tiny government agency and huge companies, and the companies were more like our Johnson Space Center (JSC) in Houston. The NASA people wanted to treat the Russian companies like contractors. It was often confusing." - Lynn Cline

"The Russians and the US had two different cultures; each was very proud; each had a history in space. The Russians really didn't respect anybody in space except NASA. NASA were the only ones who had been to space other than them. But there were also feelings of jealousy and competition. The Russians had a great sensitivity to being in a true partnership. They did not want to be dictated to by NASA. That was a real challenge to overcome in terms of US attitude towards the Russians; initially the US attitude was that: 'It is our Space Station, this is the way we're going to do things.' We were able to overcome that attitude by taking a step back." - Randy Brinkley

"Openness and information sharing was one of the key cultural clashes that we had with the Russians. Americans are information rich; we have documents, at least in NASA, we freely pass information back and forth. In Russia, under the old Soviet system, knowledge was power. If you wrote out what you knew, they could replace you. In the U.S. we focus a lot more on documenting processes. The Russians had a lot of it in their heads. We were constantly asking the Russians, time and time again, 'Let me see your drawings. Let's see your documents.' We kept getting the run-around. During Phase I, we had a contract and had money set aside for the partnership; we were paying them to deliver documentation. Money talked to the Russians, and that helped facilitate a lot of the data transfer that we did get." - Charles Lundquist

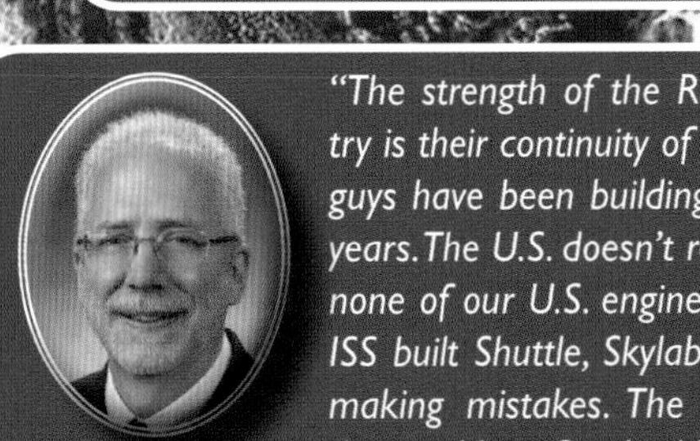

"The strength of the Russian space industry is their continuity of experience. The same guys have been building spacecraft for thirty years. The U.S. doesn't really have this. Almost none of our U.S. engineers or technicians for ISS built Shuttle, Skylab or Apollo. So we are making mistakes. The Russians don't have that problem. On the NASA side, a lot of our senior people have retired or passed away, including the astronauts. The Russians were of the opinion that they knew how to run a space station; they didn't need us intruding on their work. It took a lot of effort to justify NASA's need to know and understand details of the Russian segment to make sure that it was an integrated station and that from end to end, it works together." - Mark Geyer

"I was recommended to the commission by the army general, then commander in chief of the Air Force, Pyotr Stepanovich Deinekin as an independent expert. The work of the commission initially was not easy: political forces dominated each side. All of us were very fortunate to have such outstanding personalities with great authority as Vladimir Utkin and Thomas P. Stafford. The first is the creator of the famous intercontinental ballistic missile SS-18 Satan, and the general designer of one of the most powerful space vehicles - the Zenit rocket, later the director of the main research institute of the space branch, the Central Research Institute of Machine Building. The second is a most famous astronaut, one of the humans who visited the moon. The key issues were financing the ISS deployment and ensuring its security. Of course, the station was not threatened by collision with other space vehicles (like a car on a highway), but from so-called "space debris" - particles of technogenic origin, sometimes rather large, so it needed to be protected. The Americans, in accordance with their methodology, imposed on us the creation of an expensive system of protection against "space debris" for the service module of the Russian segment of the ISS. Our own methodology had been clearly demonstrated in the form of a full-scale experiment - which had been successfully functioning for about 13 years by that time, the Mir orbital station. And when we began to consider the predicted period of its "non-destructive" existence (i.e. before depressurization), using the American method we came to an amazing estimate: a little more than a year! This paradoxical result was due to the fact that the Americans assumed all of the particles of a certain size would be piercing, not distinguishing them from "passing", the energy of which, of course is much less. But one of our representatives, who participated in the negotiations, signed the relevant binding document, and we had to follow it. So the question of the need to develop international legal norms regulating the participation of cosmonauts and astronauts of various countries in work on orbital complexes is long overdue." - Vladimir Kovalyonok

"US astronauts carried their suits to the ISS on the Shuttle. Then they carried them back to Earth where they are repaired, replaced by elements, separate nodes, if necessary. And they are also fitted on the figure on Earth. We decided to leave our space suits on the station ever since the time of "Salyut". The design allows it to fit it on any astronaut in orbit. Repairs take place on the station, but a Russian space suit lasts up to three years. The Americans brought their spacesuits to the ISS, they worked once, but that's it. Then there was nothing to return them to the Earth. The shuttle did not fly for 2.5 years. So they were forced to work in our "clothes" even on the American segment of the ISS." - Isaak Pavlovich Abramov

"We were the first group to ever go to Baikonur [Kazakhstan], what we called at that time Tyuratam, and see all of their launch operations and their training operations at Star City, and then ended up being a capcom for the ASTP mission. It was interesting. It was great working with the Russians. I found out that the cosmonauts—a pilot's a pilot the world over. We got along very well. I still have some friends that we interact with there." - Robert Crippen

"There are a lot of similarities with the cosmonauts. Most of them were pilots. Most of them had done the same things that we've done. Their system is different. Star City is typically just cosmonaut training, and they don't participate in the development of their programs to the same extent that we do. That was a big difference. The society they lived in Soviet times was quite different from what we have here. The methodology of training, comparing the two different cultures, is about the same. I think the people have different cultures, but we all come from the same roots. It is not surprising to see that instruction and training really look alike, and mostly, now that we work together, it's even more evident." - Jean-Loup J. M. Chrétien

Prior to Expedition 47 Jeff Williams dons his Russian Sokol suit with some help from NASA Crew Surgeon and Star City Fight Doctor James Pattarini, on the final day of Soyuz qualification exams at the Gagarin Cosmonaut Training Center (GCTC) in Star City, Russia.

"When I started working with the Russians in 1993, we had to come to grips with some new concepts. We would provide the Shuttle with launch and return capabilities. They had never had return capability before. The US was going to focus heavily on scientific payloads and through the Shuttle and Spacehab Projects, I had worked to streamline the documentation and review processes since the early 1980s. The Russians had not focused on science and payloads, and they had several discrete organizations that often did not communicate with one another and did not share information. In the initial contract discussions we agreed that we would not put a limit on the mass of US hardware in the Russian modules because two-way traffic would allow hardware to come and go. Instead we agreed to work cooperatively to agree on what would fit. Through the first three years working together we came to agreement with Russians from Energiya, GCTC and IBMP on the documentation that would be required. We formatted this into 'blank books' that the hardware providers would fill-in. The completed documents would be shared and reviewed for safety, training and technical integration. This was similar to what I had done previously on Shuttle and Spacehab, but was a new concept for the Russians. I think the Russians appreciated this documentation integrating their organizations and helping to establish formal processes. This worked well in Phase I. We formed a team that worked together and most of us developed great respect for one another. The Russians continued to use the same processes and documents later in ISS, but the US documentation after Phase I was very different and was still being refined 20 years later." - Gary Kitmacher

"A significant difference between the Russians and NASA, we had huge turnover in our personnel. The Russians had many veterans. Many Russians had worked Apollo-Soyuz, 20 years earlier. NASA personnel in the field centers learned that personal relationship with their Russian counterparts was extremely important. They needed to build trust, respect, and understanding." - Lynn Cline

Expedition 53-/54 crewmembers Alexander Misurkin (left) Mark Vande Hei (center) and Joe Acaba (right) walk through Red Square in Moscow in front of St. Basil's Cathedral as they prepared to lay flowers at the Kremlin Wall where Russian space icons are interred in traditional pre-launch ceremonies. ▲

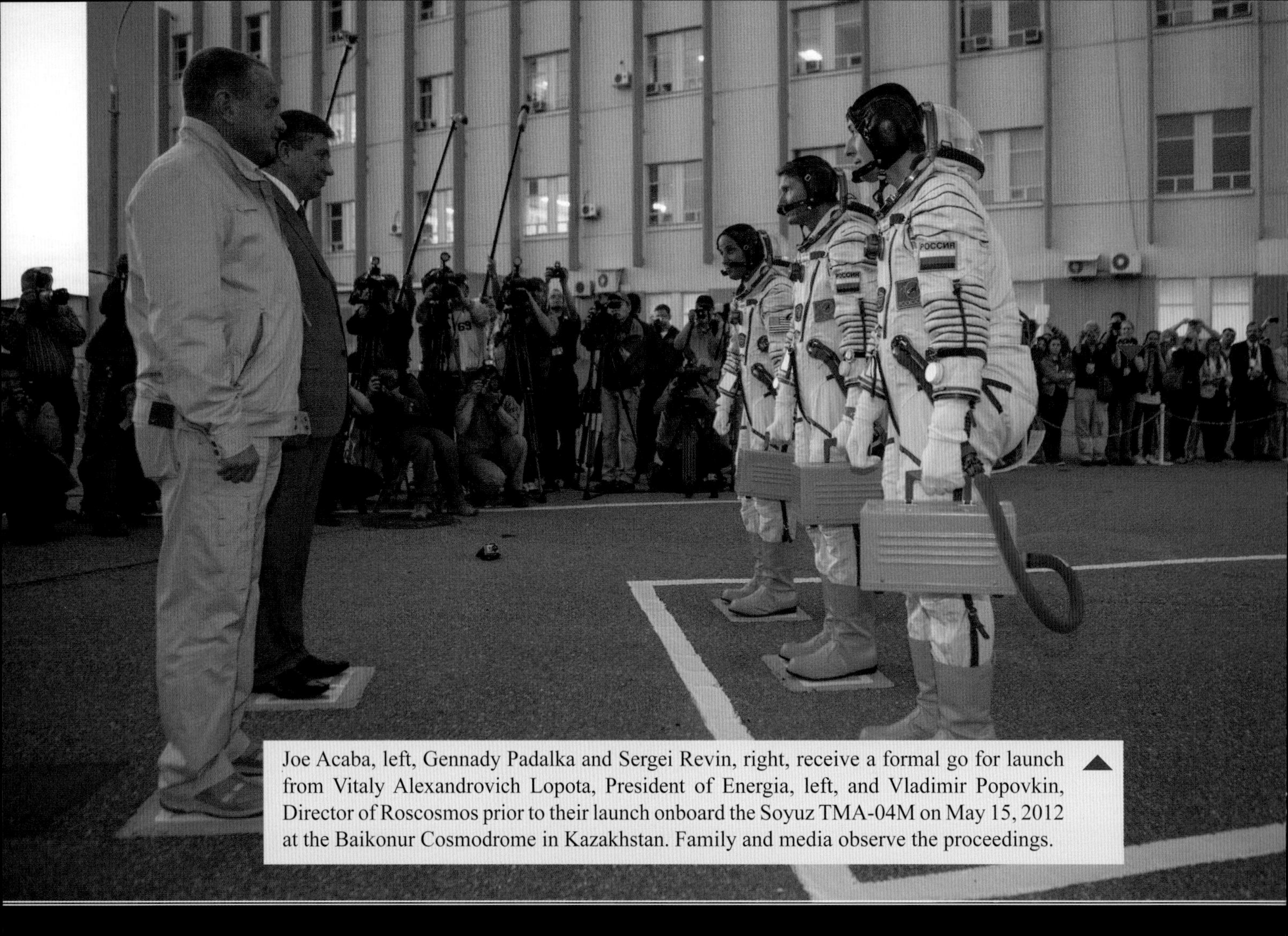

Joe Acaba, left, Gennady Padalka and Sergei Revin, right, receive a formal go for launch from Vitaly Alexandrovich Lopota, President of Energia, left, and Vladimir Popovkin, Director of Roscosmos prior to their launch onboard the Soyuz TMA-04M on May 15, 2012 at the Baikonur Cosmodrome in Kazakhstan. Family and media observe the proceedings.

The Expedition 34/35 crew members are seen from behind as they speak to their families after having their Sokol suits pressure checked prior to their launch to the International Space Station, on Wednesday, Dec. 19, 2012, in Baikonur, Kazakhstan.

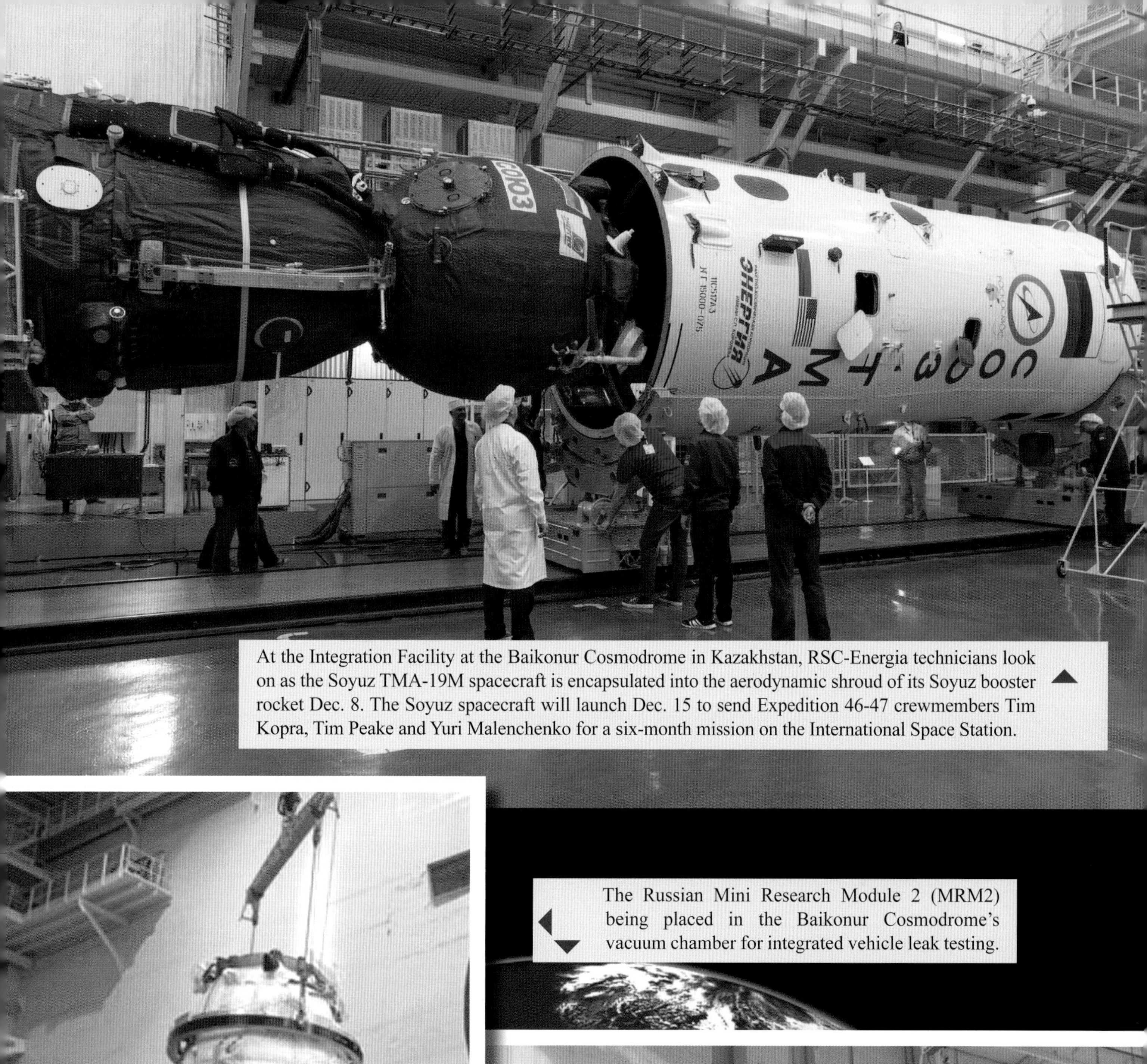

At the Integration Facility at the Baikonur Cosmodrome in Kazakhstan, RSC-Energia technicians look on as the Soyuz TMA-19M spacecraft is encapsulated into the aerodynamic shroud of its Soyuz booster rocket Dec. 8. The Soyuz spacecraft will launch Dec. 15 to send Expedition 46-47 crewmembers Tim Kopra, Tim Peake and Yuri Malenchenko for a six-month mission on the International Space Station.

The Russian Mini Research Module 2 (MRM2) being placed in the Baikonur Cosmodrome's vacuum chamber for integrated vehicle leak testing.

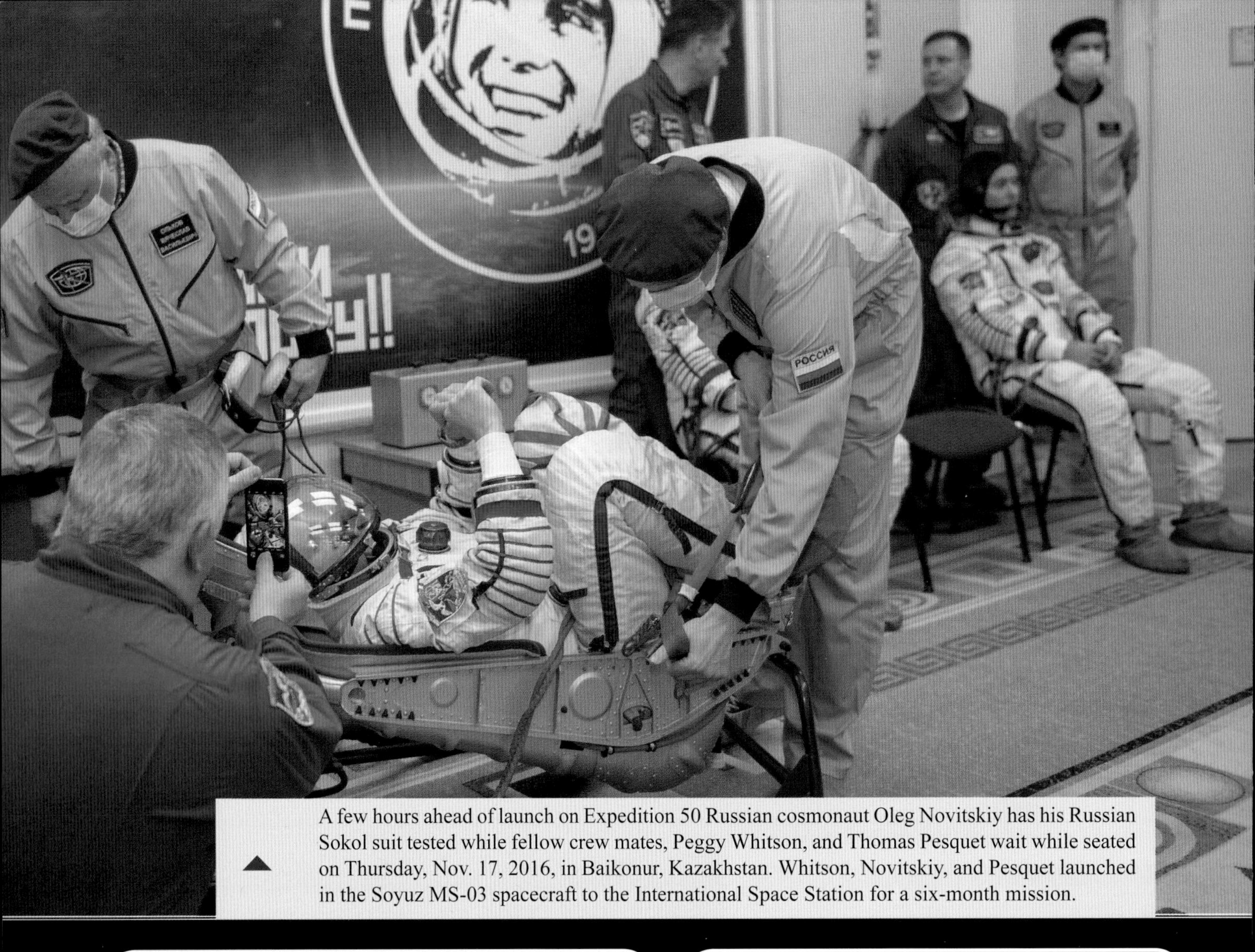

A few hours ahead of launch on Expedition 50 Russian cosmonaut Oleg Novitskiy has his Russian Sokol suit tested while fellow crew mates, Peggy Whitson, and Thomas Pesquet wait while seated on Thursday, Nov. 17, 2016, in Baikonur, Kazakhstan. Whitson, Novitskiy, and Pesquet launched in the Soyuz MS-03 spacecraft to the International Space Station for a six-month mission.

"It took us a while to get used to the Russian training regimen and teaching methodology. They used chalkboard and posters in the classrooms. The astronaut transferred information from the blackboard to a notebook. In the beginning there were no textbooks, workbooks, videotapes, or computer-aided instruction. The Russians train differently. They don't really fly the Soyuz. The young cosmonauts didn't necessarily have a lot of flight time. The cosmonauts might have between 500 and 1,000 hours of jet time. Most of our pilot astronauts had thousands of hours of jet time. US astronauts would fly routinely to Cape Kennedy and to Edwards Air Force Base in the Shuttle Training Aircraft. The Russian cosmonauts would fly a training flight maybe every three months because it wasn't viewed as being relevant to their training. All of us did lots of simulations. Possibly because the Shuttle is a very complicated vehicle, the US simulations always often integrated simulations with Mission Control. The Russians were pretty much the cosmonauts, the simulator, and a simulator instructor. Only rarely would they plug in the control center. Their control center was in another city; Star City is one place, and Mission Control Center is another part of the Moscow region. The Russian organizations that were involved were very different. At the beginning, the military did Star City, and civilians did the Mission Control Center. They were different organizations, and we found it unusual that they didn't train together often." - William Readdy

"We had to learn to work with the other international partners. The Canadians, are an associate member of ESA, so that worked fine. The Japanese were a joy. You have to accept their behavior and how they go about making decisions, which is a slow process, but once they've made a decision, and they've made an agreement with you, you can absolutely take it to the bank. They are so honorable it's ridiculous. They are a joy to work with and they have been all the way through the program. The Russians; they have capabilities that were far greater than anything that had been done in the west. It is not just a matter of Sputnik and Gagarin and their being first. Their system engineering was and is fantastic, and far better than the Americans, the Europeans, or anyone else's. The Russians design things simply because they can't afford to design things complicated. Their technology is way behind everyone else's. Their computing technology, avionics technology, materials technology, are all crude, but their system engineering and their pragmatism are an order of magnitude better than anyone else's. The Russians came across as arrogant; like there was nothing they could hope to learn from anyone else. For a long time we did things their way because they refused to accept that in a program of this magnitude cooperation means give and take." - Alan Thirkettle

"The entire Soviet people were proud and convinced that space is not just a waste of money, but an investment. During Yeltsin, and in the first years of Putin's rule, we did not pay attention to this, that's why the opinion was formed that we do not need space. In fact, this is not so. In the remote village where I was born, there was neither a river nor a lake. And I developed a system for observing the world's oceans. It's very simple: where there is plankton - there are fish. According to the most conservative estimates, these observations provided savings of about 200 million Soviet rubles. The government then came to the conclusion that it was possible to launch commercial satellites. The minister, who was supposed to prepare the documents, said: "Why do I need this?" In other words, the situation has arisen when we ourselves were cutting the branch on which we were sitting. So, we must develop space technologies." - Vladimir Kovalyonok

"Putting on an Orlan suit, for everything it takes 2-3 minutes. without help. The Americans cannot do this. Their suit it takes about 15 minutes. The Russian Orlan-M weighs no more than 95 kg, while the American's weighs up to 115 kg. The science fiction writers used to use the phrase: "I've put on a spacesuit." Today you would say: I did not put it on, but I went into it." - Isaak Pavlovich Abramov

Jeff Williams, prepares for his final day of Soyuz qualification exams at the Gagarin Cosmonaut Training Center (GCTC) in Star City, Russia

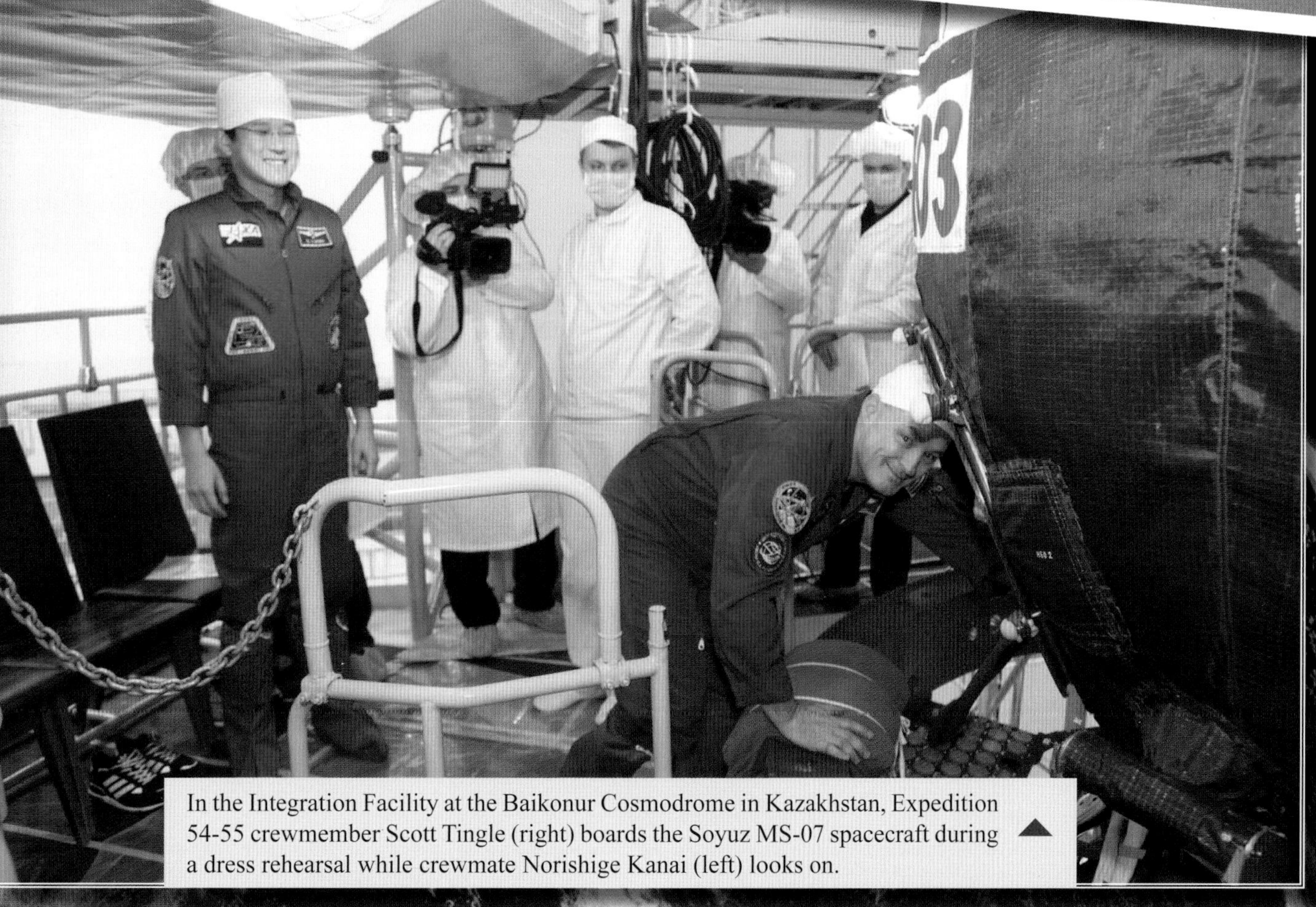

In the Integration Facility at the Baikonur Cosmodrome in Kazakhstan, Expedition 54-55 crewmember Scott Tingle (right) boards the Soyuz MS-07 spacecraft during a dress rehearsal while crewmate Norishige Kanai (left) looks on.

"In preparing for flights to the ISS, all are quite serious about learning the language, since one part of the station's systems is of Russian production and the other is American. Usually we speak the language that is most convenient for each particular situation. At one time the question was discussed for a long time in which language to print all the documentation. From my experience I understood: it is best to read in the original language. No matter how the translation is prepared, there are still a lot of inaccuracies. We speak Russian with the Russian Mission Control Center, and English with the American one. And we communicate with each other in the language that is best suited at the moment. If a colleague does not understand me, I switch to English, and if I do not understand him, he chooses the words in Russian. During joint flights, we also often develop our own slang language. An untrained person is unlikely to understand anything by reading the onboard documentation. Sometimes we invent words for convenience. For example, when the ISS was being assembled, we had a tool called "Makita electric wrench." In order not to say the long phrase "give me a Makita electric wrench", we simply said: "primakitit" or "otmakitit." And everything was very clear." - Sergei Krikalev

Expedition 14 crew Michael E. López-Alegría (standing), and Mikhail Tyurin, participate in a training session in an International Space Station Zvezda Service Module mockup/trainer at the Gagarin Cosmonaut Training Center in Star City, Russia.

The Soyuz rocket and TMA-10M spacecraft are assembled at Building 112 at the Baikonur Cosmodrome on Sept. 22, 2013 in Kazakhstan.

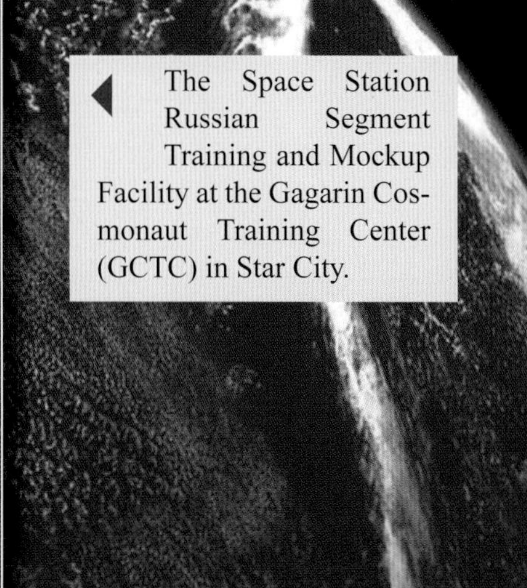

The Space Station Russian Segment Training and Mockup Facility at the Gagarin Cosmonaut Training Center (GCTC) in Star City.

"These modules were brought together for the first time in space, and that's a huge testimonial for the reliability and the attention to detail of the aerospace engineers in all of the various countries who were part of the space station program; because it was fit checked for the first time in space and it all came together and worked." - Dave Williams

Spektr and Priroda Modules, destined for the Mir Orbital Station, and the Functional Cargo Block (FGB) and Service Module, destined to become part of the ISS, at the Khrunichev factory in Moscow in 1994.

Khrunichev technicians install and test hardware inside of the Service Module at the factory in Moscow.

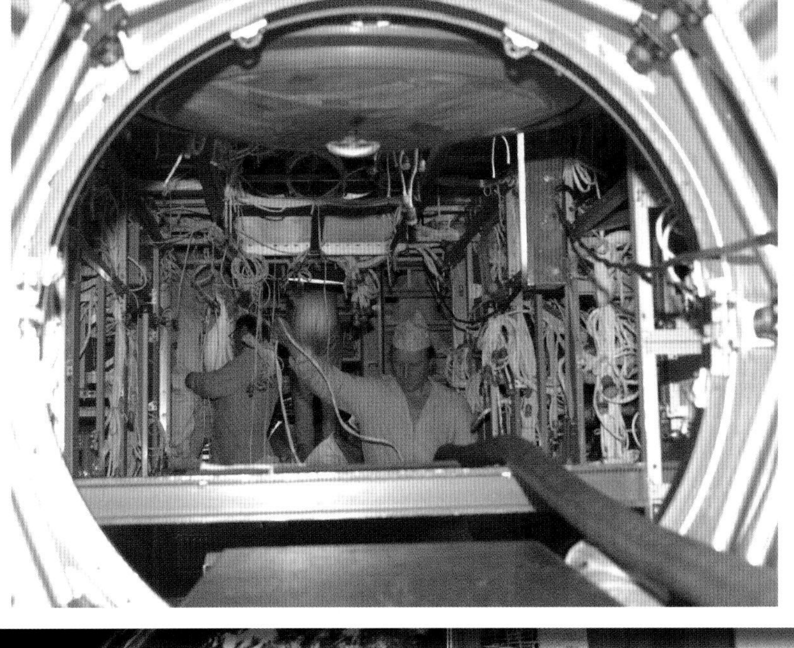

ISS Service Module roll-out. at RSC-Energia near Moscow, Russia during preparation for shipment to Baikonur Cosmodrome in Kazakhstan.

The Russian Docking Compartment and airlock, named Pirs. It serves as a docking port for Soyuz or Progress spacecraft arriving at the station. Here the Pirs is under construction at Energia near Moscow.

Overall view of the Hydrolab pool, used for training on the Russian ISS segment, with mock-ups visible underwater and in the background. The Hydrolab is at the Gagarin Cosmonaut Training Center (GCTC) in Star City.

STS-101 crew wearing US Extravehicular Mobility Unit (EMU) spacesuits, with safety divers, train on the Service Module mock-up in the Hydrolab pool at Gagarin Cosmonaut Training Center (GCTC) in Star City.

Gennady Padalka shakes hands with Deputy Head for Cosmonaut Training, Gagarin Cosmonaut Training Center (GCTC) Valery Korzun, after the successful completion of qualification exams for Expedition 43. Mikhail Kornienko, at right, looks on.

Expedition 47 backup crew Shane Kimbrough, left, Sergei Ryzhikov, center, and Andre Borisenko, enter the Soyuz simulator for their Soyuz qualification exams, at the Gagarin Cosmonaut Training Center (GCTC) in Star City, Russia as media look on.

Expedition 43 backup crew members: Sergei Volkov, left, Alexei Ovchinin, center, and Jeff Williams answer questions from the press outside the Soyuz simulator ahead of their Soyuz qualification exams, at the GCTC in Star City, Russia.

Expedition 27 crew from top, Andrey Borisenko, Ron Garan, and Alexander Samokutyaev wave farewell from the bottom of the Soyuz rocket prior to their launch to the International Space Station from the Baikonur Cosmodrome in Kazakhstan, on April 5, 2011.

The Soyuz TMA-18 spacecraft is rolled out by train to the launch pad at the Baikonur Cosmodrome, Kazakhstan, March 31, 2010. In preparation for the launch with Alexander Skvortsov, Mikhail Kornienko and Tracy Caldwell Dyson.

The Soyuz TMA-02M spacecraft is seen just after being raised into a vertical position at the launch pad at the Baikonur Cosmodrome in Kazakhstan on June 5, 2011. The rocket was being prepared for launch June 8 to carry the crew of Expedition 28 to the International Space Station.

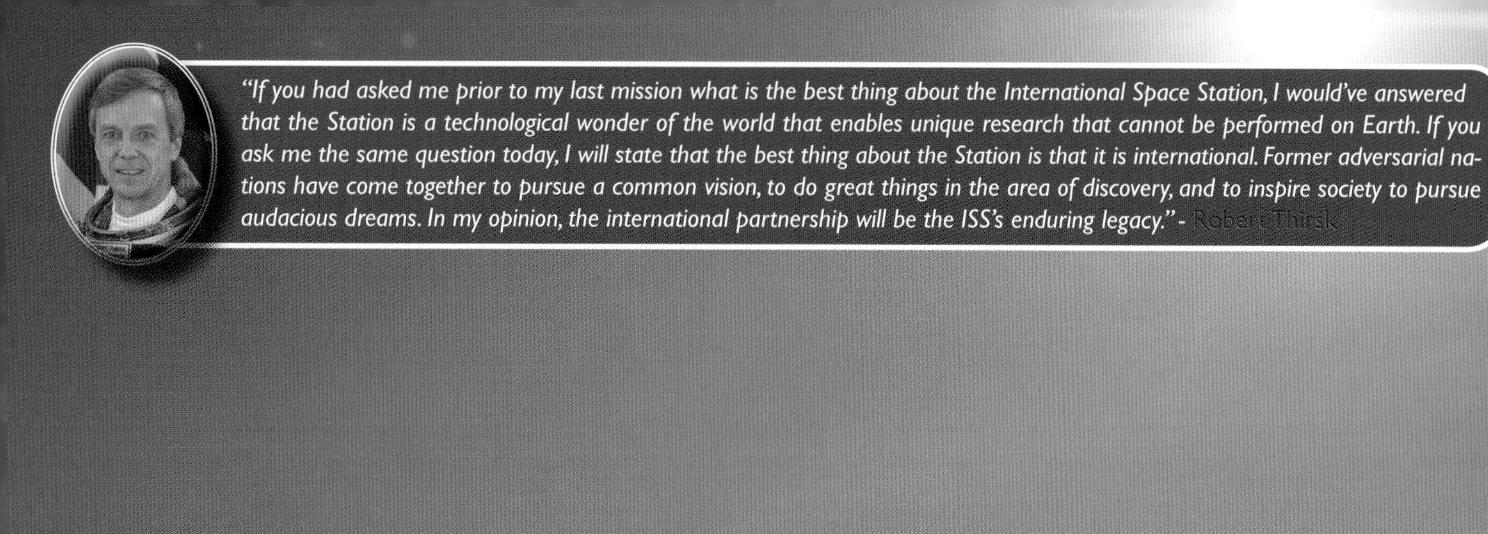

"If you had asked me prior to my last mission what is the best thing about the International Space Station, I would've answered that the Station is a technological wonder of the world that enables unique research that cannot be performed on Earth. If you ask me the same question today, I will state that the best thing about the Station is that it is international. Former adversarial nations have come together to pursue a common vision, to do great things in the area of discovery, and to inspire society to pursue audacious dreams. In my opinion, the international partnership will be the ISS's enduring legacy." - Robert Thirsk

▲ Members of the media photograph the Soyuz TMA-04M launch from the Baikonur Cosmodrome in Kazakhstan May 15, 2012. Heading to the International Space Station onboard the Soyuz are Expedition 31 crew Gennady Padalka, Joseph Acaba and Sergei Revin.

▼ The Soyuz TMA-19 spacecraft is raised into vertical position at the launch pad of the Baikonur Cosmodrome in Kazakhstan, Sunday, June 13, 2010.

The Soyuz rocket with Expedition 33 crew Oleg Novitskiy, Kevin Ford and Evgeny Tarelkin onboard the TMA-06M spacecraft launches to the International Space Station on Oct. 23, 2012, in Baikonur, Kazakhstan.

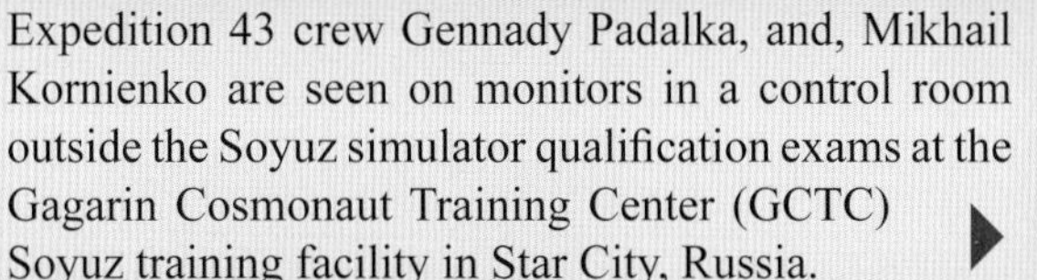

Expedition 43 crew Gennady Padalka, and, Mikhail Kornienko are seen on monitors in a control room outside the Soyuz simulator qualification exams at the Gagarin Cosmonaut Training Center (GCTC) Soyuz training facility in Star City, Russia.

View of the Flight Control Room at Russia's Federal Space Agency Mission Control Center in Korolev, Russia, located on the outskirts of Moscow.

Expedition 49 crew Andrey Borisenko, front left, Sergey Ryzhikov, front center, Shane Kimbrough, front left, with Kate Rubins, back left, Anatoly Ivanishin, back center, and Takuya Onishi, back right, on a screen at the Moscow Mission Control Center in Korolev, Russia a few hours after the Soyuz MS-02 docked to the International Space Station on Friday, Oct. 21, 2016.

The Soyuz TMA-17M spacecraft is seen after it landed with Expedition 44/45 crew members Oleg Kononenko, Kjell Lindgren and Kimiya Yui near the town of Zhezkazgan, Kazakhstan on Friday, Dec. 11, 2015.

"Soyuz really saved the space station." - Vladimir Kovalyonok

▲ A photographer stands atop the Soyuz TMA-05M spacecraft long after it landed with Expedition 33 crew Sunita Williams, Akihiko Hoshide and Yuri Malenchenko in a remote area near the town of Arkalyk, Kazakhstan on Nov. 19, 2012 after four months onboard the International Space Station.

In 1984, President Ronald Reagan authorized NASA to assemble a large space station. At that moment in time NASA had completed only a single three-hour demonstration EVA or spacewalk from the Shuttle. The design of station was ongoing, so its exact construction or assembly requirements were not yet defined and the quantity and nature of the EVAs that would be required was still unknown. Background work had been started some years earlier. Space Station related EVA studies began in the 1970s and continued during the Space Shuttle's first decade of operations. Several space structures were fabricated to test processes and designs. In 1978, the Massachusetts Institute of Technology (MIT) had begun a program of human factor's engineering studies of techniques for assembly of structures in space. The studies focused on the productivity of an EVA crew during assembly operations. MIT developed a program of seventeen underwater tests. These tests were run in the Neutral Buoyancy Simulator at Marshall Space Flight Center, in Huntsville, Alabama. The evaluations focused on the assembly of erectable trusses.

Jerry L. Ross (left) and Sherwood C. (Woody) Spring are photographed as they assemble pieces of the Experimental Assembly of Structures in Extravehicular Activities (EASE) device in the open payload bay. At the left of the frame is the IMAX camera gear and to the right is a Getaway Special (GAS) canister. Behind the EASE device are the closed cradles for the two satellites that were deployed during this mission.

Astronauts in Extravehicular Mobility Units (EMU) working on both Experimental Assembly of Structures in Extravehicular Activity (EASE) and on Assembly Concept for Construction of Erectable Space Structures (ACCESS) in the payload bay of the orbiter Atlantis during STS-61B. Sherwood C. Spring works on ACCESS. Jerry L. Ross is partially visible behind the stowed EASE truss.

This work was used during Shuttle mission STS-61B in 1985, as the Experimental Assembly of Structure for EVA (EASE) and Assembly Concept for Construction of Erectable Space Structures (ACCESS). The tests evaluated the ability of the astronauts to fabricate truss structures in orbit and validated the underwater test results. The STS-61B crew showed that the three-sided pyramid-shaped EASE required far too much free floating activity to make it a suitable method to construct a space station. The conclusion from the ACCESS experiment confirmed the feasibility of EVA space assembly of erectable trusses, and played a role in the decision to initially baseline the Space Station truss as a 5 m (16 ft.) erectable structure.

"At the end of 1987 I was appointed to the Space Station Advisory Committee. At our first meeting we were asked to provide recommendations on needed technology, management and policy issues. As the station design matured the amount of time the astronauts would need for EVA was growing rapidly; a troubling development seen by some members of Congress as an argument to cancel the program. We visited various labs that were developing robots and concluded that the robotic technology was not advanced enough to help in the immediate future so we recommended that NASA reexamine the design of the many elements to reduce the amount of EVA time needed." - Donald Beattie

EVA became a central issue. The initial success of the station would require successful assembly during EVA. At first EVAs would be conducted from the Space Shuttle. As EVA systems were incorporated in the ISS, the EVAs could be conducted independently of the Shuttle.

The number of EVAs required was projected to be a substantial obstacle to overcome before the next phase of station operations could proceed. This became known as the space station's 'Wall of EVA'. As requirements were forecast, a related effort was undertaken to assess the potential for robotics' support of assembly operations. The requirements were incorporated into the design of the Mobile Servicing System that Canada would develop as their contribution to the program.

In December, 1989, Charles R. Price, headed the team as Chief of the JSC Robotics Systems Branch. Thirty-three people contributed, including seven astronauts. The assessment group examined ten primary components of the station; looking at the design, assembly, expected system failures, the number of Orbital Replacement Units (ORUs) and expected requirements for post-assembly maintenance.

Projecting their findings across the predicted 35 year lifetime of the space station, the group determined that an average of 507 maintenance actions would be required at the station each year. This translated into about 625 work site hours a year, or 3,276 hours of EVA time. This was equivalent to 273 two-person EVAs annually, or more than 5 two-person spacewalks each week. During the 4.5 assembly years forecast to build the station, an estimated 1,752 maintenance actions equated to 11,517 hours of EVA, or 960 two-person spacewalks. This would require more than 4 two-person spacewalks every week.

After assembly was complete, the predictions identified 190 two-person spacewalks a year for a minimum amount of required maintenance when the station was still at its newest. This was 3.5 two person spacewalks a week. As the station grew older and systems deteriorated, it was anticipated that the maintenance requirement would grow to a peak of more than 10 two-person EVAs each week.

"The Space Station Freedom Program started with a design that didn't work. We had one investigation by [William F.] Bill Fisher and Charlie [Charles R.] Price. Charlie was out at JSC Engineering; he was a robotics guy. Bill Fisher was in the Astronaut Office. They were called the Fisher-Price Team.

They assessed the maintenance demand. The reason the maintenance demand was so high was because they had let the Reliability and Maintenance guys just go run the numbers, based on specification numbers, and nobody paid much attention to what they were. It looked like they were pretty high. Everybody said, "It can't be that high."

And it wasn't, but at the same time, the numbers were in public record, and there were a lot of people starting to pick them up and saying, "Well, wait a minute. If this is the case, you won't ever do anything but EVAs [Extravehicular Activities] to fix or maintain your Space Station.

The complexity factor got lost in the noise. NASA had done a flight experiment in 1985. Jerry [L.] Ross and [Sherwood C.] Woody Spring did a beam building experiment on STS 61-B, two flights before Challenger. They built a beam in the Shuttle bay. It took quite a while, and nobody factored in the amount of time that it took them, because everybody had fallen in love with the design. Because everyone wanted a big Space Station." - Denny Holt

If all the assembly, maintenance and utilization EVA requirements were added together, annual EVA time could reach 2,864 hours in 1992 and rise to an annual rate of 4,013 hours by 2001. This was a staggering amount of EVA time. It became almost impossible to imagine how all of the EVAs could be accommodated or how the program would have adequate hardware and capability for support.

In the five years between finalizing the station design and the beginning of ISS assembly in orbit, planning for this extensive EVA program was conducted. The EVA plan blended in as an element of the assembly sequence.

During the man-tended phase of operations, and in the early station increments, EVAs would be conducted by Space Shuttle crews operating out of the Shuttle airlock. Once EVA facilities were incorporated into the Station, EVAs would begin to use the station facilities. Both Russian and US segment airlocks would be available after the first two years of assembly. The Canadian Space Station Remote Manipulator (SSRMS) and its mobile base would become important for supporting spacewalks further away from the docked Shuttle and the habitable elements. The US and Russian airlocks and the Canadian SSRMS would all be incorporated into ISS within about a year of the first ISS increment crew taking up residence.

In November 1992, NASA announced that EVAs would be assigned to a number of Shuttle missions to train astronauts and flight controllers, and to develop space construction techniques.

The first of the training missions was STS-49 in 1992. The astronauts evaluated the Assembly of Space Station by EVA Methods (ASEM), which was a follow-on from the EASE/ACCESS assembly activities on STS-61B. The astronauts assembled a prototype truss. But by the time STS-49 flew, NASA had already changed the station design to use pre-assembled truss segments for the station.

EVAs on STS-54 in 1993 demonstrated the ability of an astronaut to carry and move a substantial bulky mass and to use large tools. STS-57 in 1993 demonstrated moving large massive payloads while riding on the RMS. Simplified Aid For EVA Rescue (SAFER), a jet powered back-pack for returning an astronaut to the station in case they got detached, was flight tested on STS-64 in 1994.

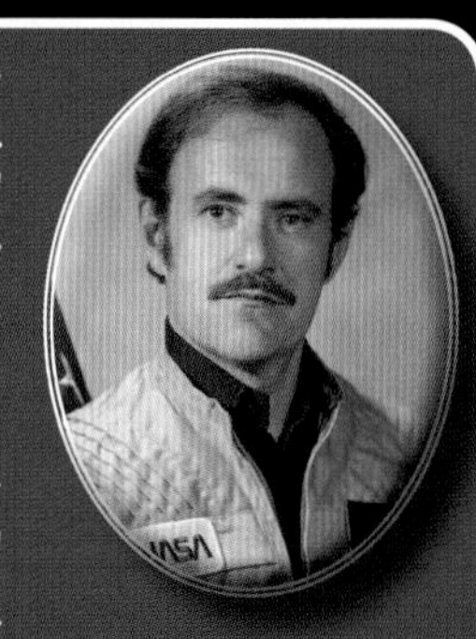

"I got assigned to the early conceptual design of Space Station. It started with dual keel. It had a hangar for fixing satellites. After the return to flight, after Challenger, there was more emphasis on Station; we all became more realistic. Station kept getting smaller and smaller. We discarded the dual keel as being impractical. The big concern was how many spacewalks it took, and a big concern was there was no way to do that many spacewalks. I think in the end we've probably done more spacewalks than we said were undoable, but at that time people were very reluctant about it." - John "Mike" Lounge

Kathryn C. Thornton (foreground left) releases a strut during the first test of space station assembly methods. Assembly of Station by Extravehicular Activity (EVA) Methods (ASEM) in Endeavour's payload bay. Thomas D. Akers, positioned on the far side of the truss, waits for Thornton to hand him the strut. The two astronauts are building the ASEM structure during mission's fourth EVA. In the background are the shuttle's vertical tail, and the orbital maneuvering system (OMS) pods.

On STS-54 Gregory J. Harbaugh, (with red stripes, left), carries Mario Runco, Jr along the payload bay starboard side. The objective of this exercise was to simulate carrying a large object and evaluate the ability of an astronaut to move about with a "bulky" object in hand. This EVA was the first to broaden experience in preparation for Space Station.

"EVA is a coordinated effort between the people inside and the EVA crew members outside. It is required because of the complexity of the mission; there is frequently interaction; things may have to be commanded from inside while the crewmembers are working on things outside. It happens today on every one of these missions that we fly to the Space Station, where we're doing these complex EVAs." - Richard Covey

STS-60 in 1995 was the first intended flight of the EVA Development Flight Test (EDFT) series; but the EVA was cancelled when an airlock hatch jammed. Beginning with STS-63, and then continuing with STS-72 and STS-87 in 1996 and 1997, a series of simulated EVA assembly tasks were conducted. On STS-87 the crew evaluated a telescopic crane boom, simulating the replacement of ISS solar array batteries. The flight also evaluated the free-flying Autonomous EVA Robotic Camera Sprint (AERCam Sprint), a 35.5 cm (14 in) diameter sphere resembling a soccer ball that featured twelve small nitrogen thrusters, a pair of TV cameras and avionics system that could be remotely flown from the Shuttle's aft flight deck. The next year ISS flight operations would begin. In the series of EVAs conducted on the Shuttle with the EMUs, and on the NASA-Mir missions using the Orlan suits, it was revealed that the two suit systems were reliable and ready to use for ISS EVA by either Shuttle or space station crews.

▲ Bernard Harris and Michael Foale (above) conduct an EVA on the Space Shuttle Discovery during STS-63. Foale's feet are seen secured to a work platform attached to the Shuttle's robotic arm. Below Foale and the arm is nose of the Shuttle and The Spacehab module in the Shuttle's payload bay.

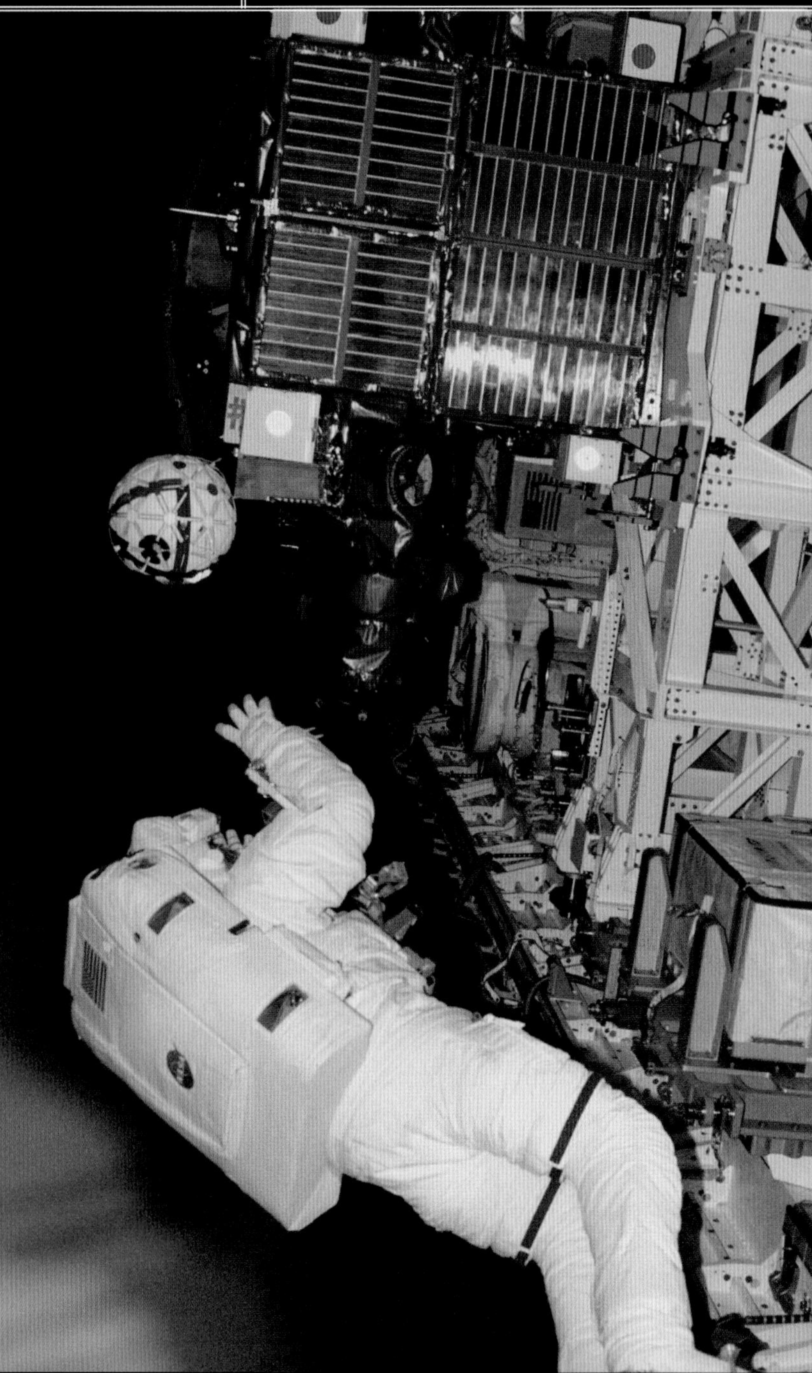

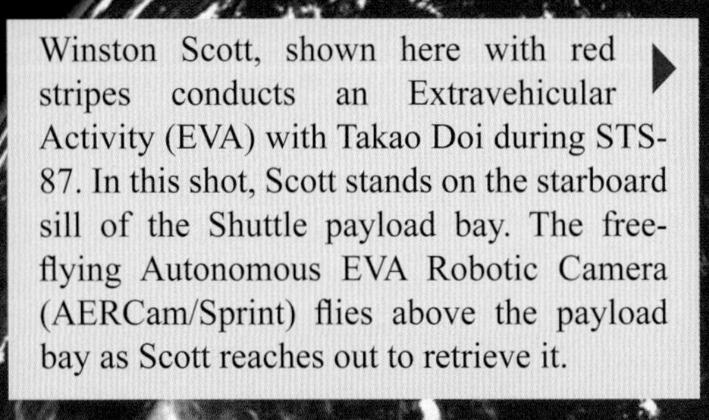

Winston Scott, shown here with red stripes conducts an Extravehicular Activity (EVA) with Takao Doi during STS-87. In this shot, Scott stands on the starboard sill of the Shuttle payload bay. The free-flying Autonomous EVA Robotic Camera (AERCam/Sprint) flies above the payload bay as Scott reaches out to retrieve it. ▶

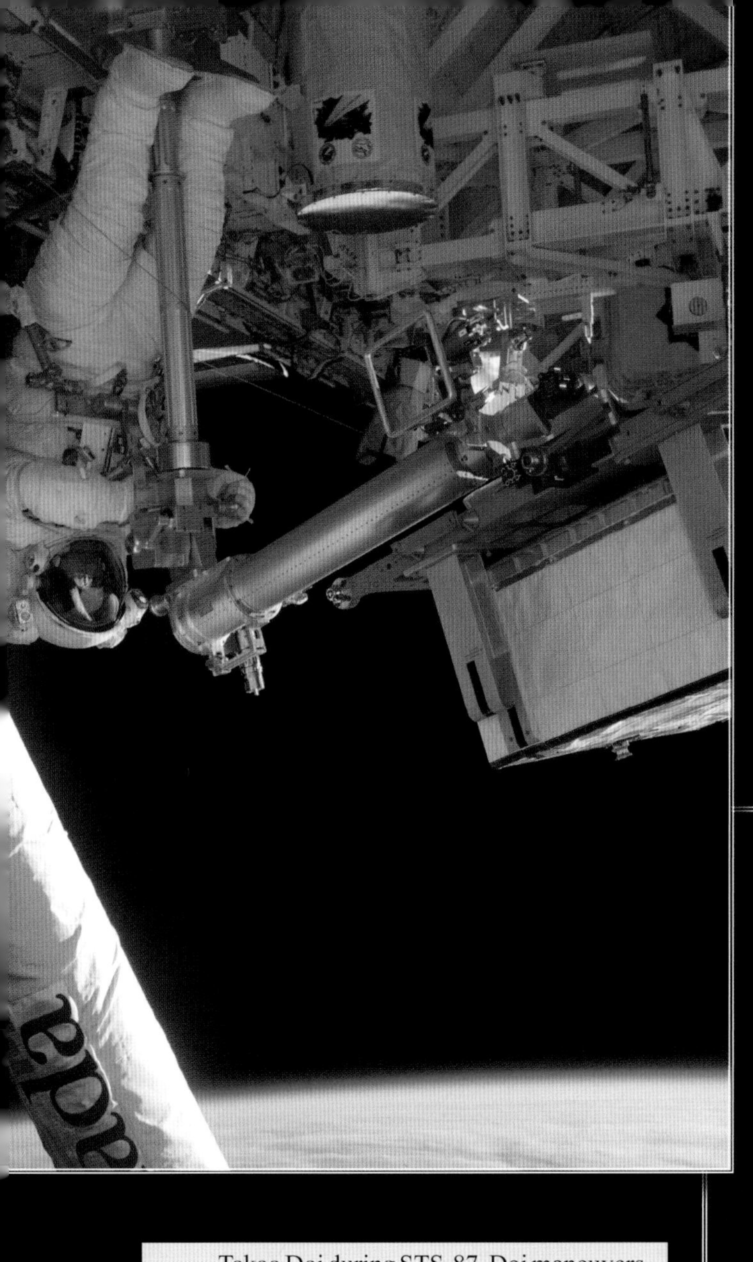

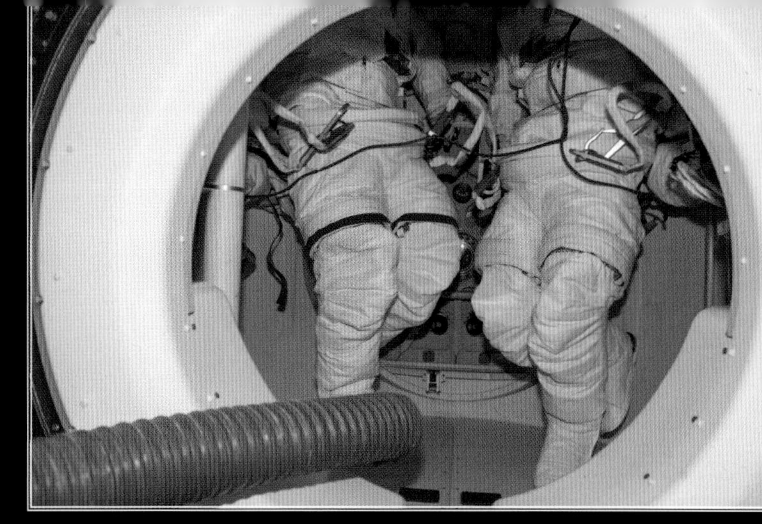

The bottom portions of the extravehicular mobility units (EMU) for STS-98's two space walkers are seen in the airlock aboard the Space Shuttle Atlantis. Thomas D. Jones is on the left (red stripe), and Robert L. Curbeam on the right.

Takao Doi during STS-87. Doi maneuvers the Translation Device crane with a large mockup ISS battery attached.

On the first Shuttle ISS assembly mission, STS-88, Jerry L. Ross (right), is perched on the end of Endeavour's remote manipulator system (RMS) arm. The solar array panel for the Russian-built Zarya module can be seen along right edge.

Scott Parazynski is anchored to the Articulating Portable Foot Restraint on the Orbiter Boom Sensor System (OBSS), being maneuvered to the P6 4B Solar Array Wing (SAW).

"One event that really stands out during Expedition 16, while STS-120 was docked. We were redeploying a solar array to its final location out on P6. On STS-116 they had a problem retracting the solar array. When we did the redeploy, the first half deployed normally, but the second half ended up tearing the solar array before we could stop the deploy. It ended up being probably about a five-foot tear. The folks on the ground came up with a very elegant solution. I call it the Apollo 13 moment, because they had to use only what we already had on-board. The ground figured out a way for us to build what they called "cuff links". Then it had to be attached on both sides of the tear. Attaching the two together would hold the array stable and not allow that tear to propagate any further. We used the Station arm with the Shuttle inspection arm that we normally used for inspecting the bottom side of the Shuttle Orbiter looking for damage. For me, it was the most stressful time, because we knew that if that array didn't get fixed, then we would have to jettison the array. If the array was only partly deployed, then the Shuttle could not undock. If we had to jettison the array, then the next Shuttle was supposed to bring the ESA Columbus module up, but we wouldn't have enough power for Columbus. It could potentially impact that whole Station assembly sequence." - Peggy Whitson

Rick Mastracchio performing construction and maintenance on the International Space Station. During the spacewalk, Mastracchio and Clay Anderson (out of frame), relocated the S-Band Antenna from Port 6 (P6) to Port 1 (P1) truss, installed a new transponder on P1 and retrieved the P6 transponder.

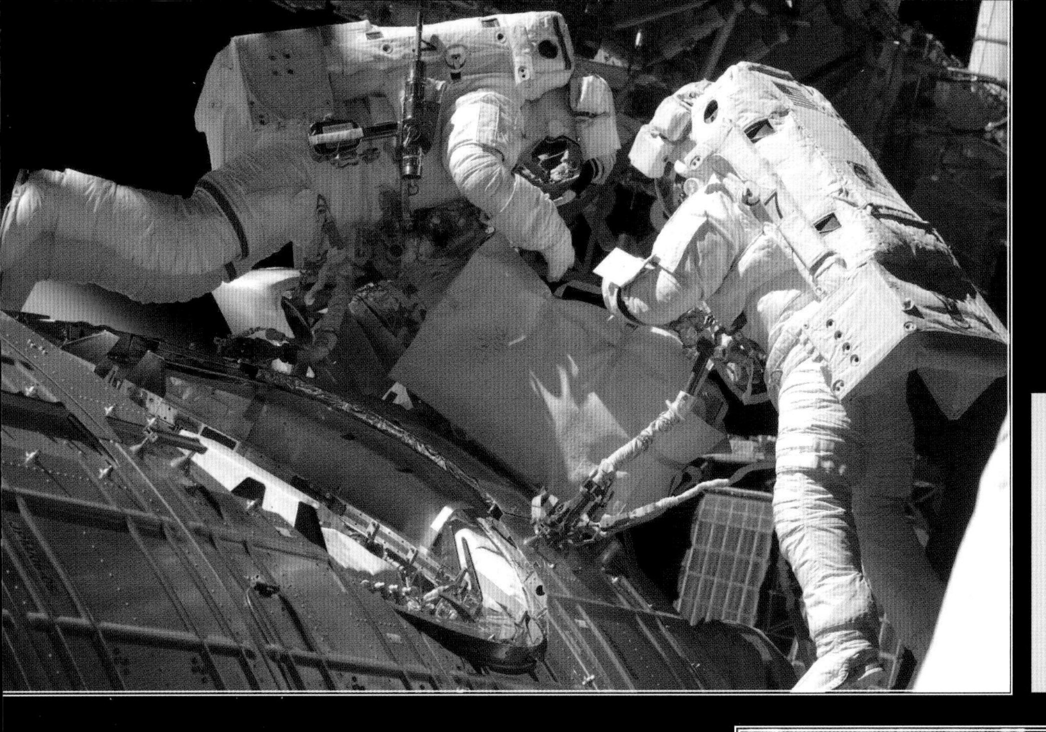

Mike Fossum (left) and Ron Garan, participate in an extravehicular activity on the International Space Station. Fossum and Garan installed television cameras on the front and rear of the Kibo Japanese Pressurized Module (JPM) to assist Kibo robotic arm operations.

Mike Fossum works on the outside of the Japanese Module Kibo

Jack Fischer works outside the Japanese Kibo Laboratory module repairing insulation at the connecting point of the Japanese robotic arm during the 200th spacewalk in support of International Space Station maintenance and assembly.

A wide shot features the extravehicular activity of William McArthur and shows most of the long robot arm or remote manipulator system (RMS) attached to the cargo bay of the Space Shuttle Discovery. McArthur holds onto a handrail on the RMS. A pressurized mating adapter (PMA) is at top of the frame.

"EVA has been a big part of the Space Station success. Most of the EVAs to assemble the Station were done by Shuttle crews. They had the advantage of getting a lot of training runs in the Neutral Buoyancy Lab (NBL) in the months leading up to a flight. With the long duration expedition crews performing the EVAs we don't have that luxury. We're more dependent on the ground teams to develop the EVA plans. For the astronauts, EVA has always been the most challenging activity that we do; it also is the biggest reward. I had the opportunity to do a Russian EVA in an Orlan spacesuit. It is a completely different system." - Jeffrey N. Williams

"EVA is a coordinated effort between the people inside and the EVA crew members outside. It is required because of the complexity of the mission; there is frequently interaction; things may have to be commanded from inside while the crewmembers are working on things outside. It happens today on every one of these missions that we fly to the Space Station, where we're doing these complex EVAs." - Richard Covey

Thomas D. Jones grabs a hand rail while working on the International Space Station (ISS) during a space walk.

An Internal EVA is activity in an unpressurized area inside of the space station that requires use of an EVA suit. The first EVA at the station, without a Shuttle docked to the complex, occurred inside the Russian segment on 8 June 2001 during the second expedition. This was inside the depressurized Zvezda node, preparing it for the arrival of additional modules. The Russian EVA airlock, Pirs (meaning Pier), arrived in September 2001. Since that time all of the Russian EVAs have been supported by crew members wearing Orlan suits.

In January 2010, a second Russian EVA airlock/docking port arrived. This one was called Poisk (search, seek or explore). Conducting a spacewalk can take some preparation time. It is also the responsibility of several crew members to assist and support EVA operations. So most EVAs on station by increment crews take several days of preparation. Ground training for EVA takes place in 1g mock ups, suspended jig devices, parabolic aircraft flights, underwater or using virtual reality 3D glasses. The pioneering EVAs (1998-2001) at station were all conducted from the airlock of the docked Shuttle orbiter as the installation of ISS airlocks would not occur until well into 2001.

The primary objective of the first dozen years of ISS operations was the assembly of the facility. With the completion of the assembly in 2011 and the retirement of Shuttle, attention turned to a more 'routine' program of housekeeping and maintenance, and the support of external experiments and investigations. On STS-128, one spacewalk focused on swapping out an older ammonia tank for a new one. With a mass of 820 kg (1800 lb.), the new tank was the most massive object moved by EVA astronauts. Once the new tank was securely attached to the station, the old one was installed in the Shuttle's payload bay for the return to Earth.

STS-134 was the penultimate Shuttle, and the final mission on which Shuttle astronauts would perform an EVA. On 27 May 2011, as the two EVA astronauts re-entered the Quest airlock and closed the hatch, they symbolically closed the door on Shuttle EVA operations, a point in history that was observed at the end of the EVA by Greg Chamitoff.

"I have the honor to share this last spacewalk with all the folks working on the ground and the thousands of people who helped build this in the Shuttle and the Station programs. We're floating here on the shoulders of giants. The ISS is a pinnacle of human achievement and international cooperation – 12 years of building and 15 countries. And now it is the brightest star in the sky and hopefully the doorstep to our future. So congratulations everybody on Assembly Complete." - Greg Chamitoff

Sunita Williams appears to touch the bright sun during an EVA. Williams and Aki Hoshide (visible in Williams' helmet visor), completed the installation of a Main Bus Switching Unit (MBSU) that was hampered by a possible misalignment and damaged threads where a bolt had to be placed. They also installed a camera on the International Space Station's robotic arm, Canadarm2.

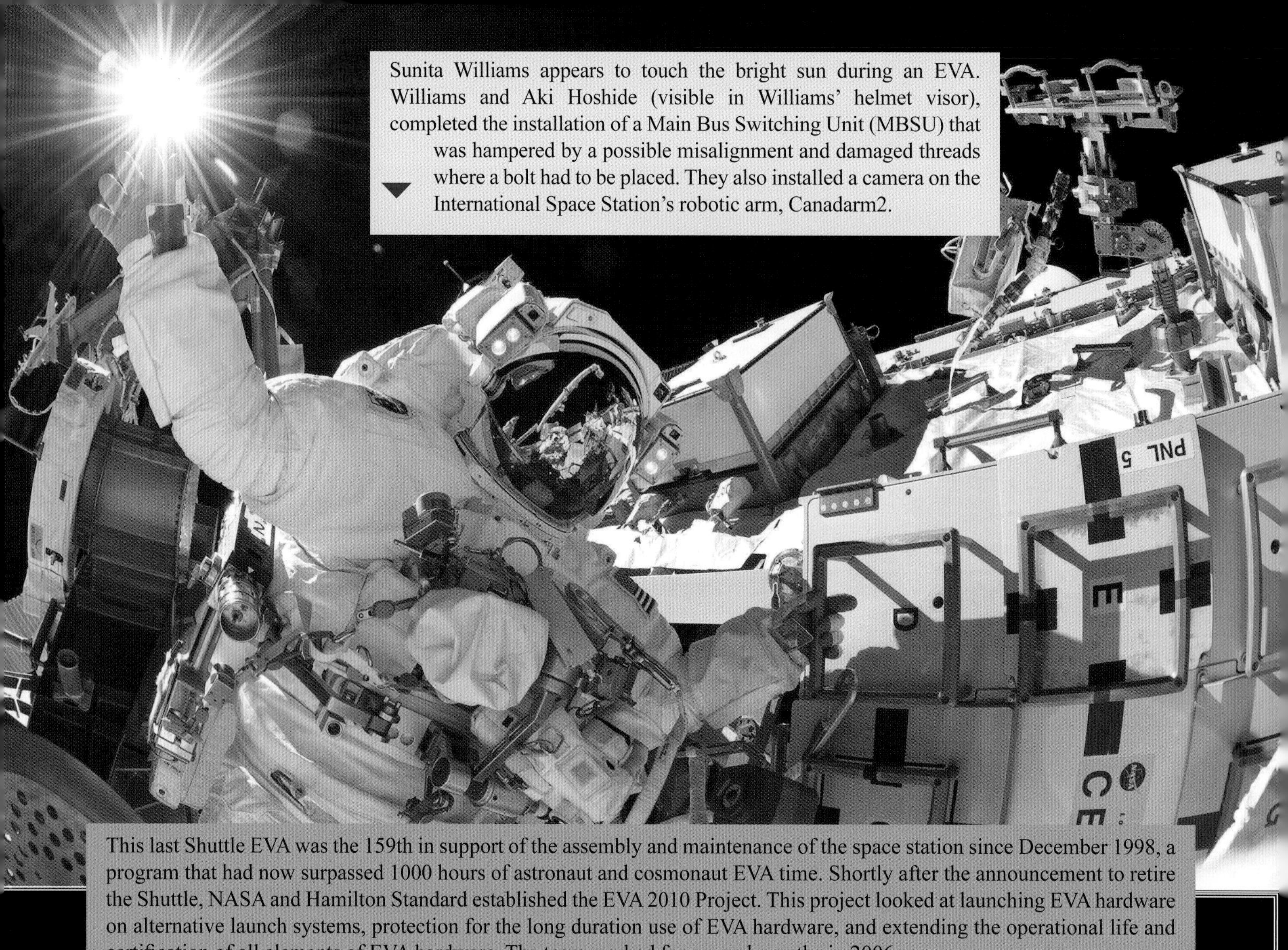

This last Shuttle EVA was the 159th in support of the assembly and maintenance of the space station since December 1998, a program that had now surpassed 1000 hours of astronaut and cosmonaut EVA time. Shortly after the announcement to retire the Shuttle, NASA and Hamilton Standard established the EVA 2010 Project. This project looked at launching EVA hardware on alternative launch systems, protection for the long duration use of EVA hardware, and extending the operational life and certification of all elements of EVA hardware. The team worked for several months in 2006.

Anchored to a Canadarm2 foot restraint, Luca Parmitano during a space walk on Expedition 36.

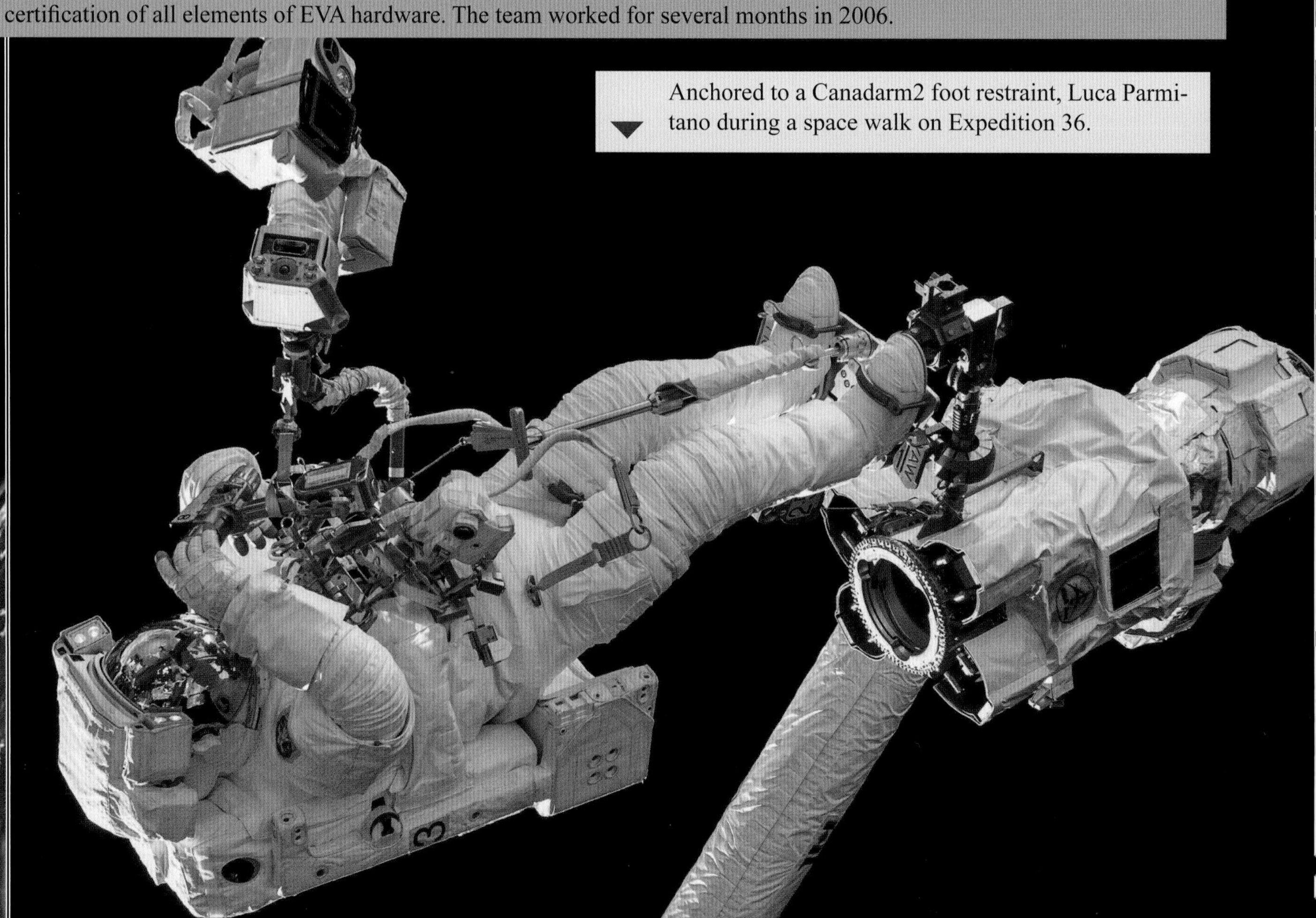

Luca Parmitano in a selfie on Expedition 36.

Reid Wiseman works with Alexander Gerst (out of frame), outside the space station's Quest airlock relocating a failed cooling pump to external stowage and installing gear that provided back up power to external robotics equipment.

Mike Hopkins on Expedition 38 works with Rick Mastracchio to change out a faulty water pump outside of the International Space Station.

The group made several recommendations: a) increase the certification of EMUs from the previously established three years life to 6 years; b) establish new procedures for testing recycled ISS water prior to use in the EMU cooling system until ultra-clean EMU water could be resupplied on resupply vessels; and c) extend the on-orbit certification of EVA tools. The goal was to ensure that EVA from the ISS US segment in EMUs could be supported without the Shuttle. The group identified that limitations of cargo launch mass following the retirement of the Shuttle would restrict the number of EVAs that could be supported each year to eight. They said that on-orbit maintenance could support EMU life beyond 2 years and they felt that 6 years was feasible, but the report concluded that an 8 year interval was 'inconceivable'.

"The program was awfully nervous about spacewalks being very risky. Early on Shuttle, there were several early satellite servicing missions; the Solar Max and the Palapa/Westar retrievals first, and then we had a hiatus before the Intelsat retrieval. I thought we lost some collective memory just between those clusters of servicing missions. There were some glitches in the planning and preparations by the time of Intelsat. To me it represented a failure on our collective part to carry all those lessons forward and to convey them to each other so that we were progressively building. The Hubble experience fed into and supported a lot of aspects of how Station EVA training was done. It created a very robust capability. Once Station assembly is complete the need for lots of EVAs goes away." - Kathy Sullivan

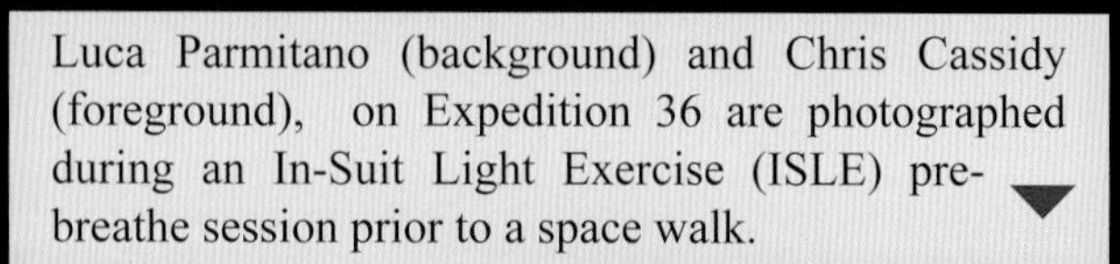

Luca Parmitano (background) and Chris Cassidy (foreground), on Expedition 36 are photographed during an In-Suit Light Exercise (ISLE) prebreathe session prior to a space walk.

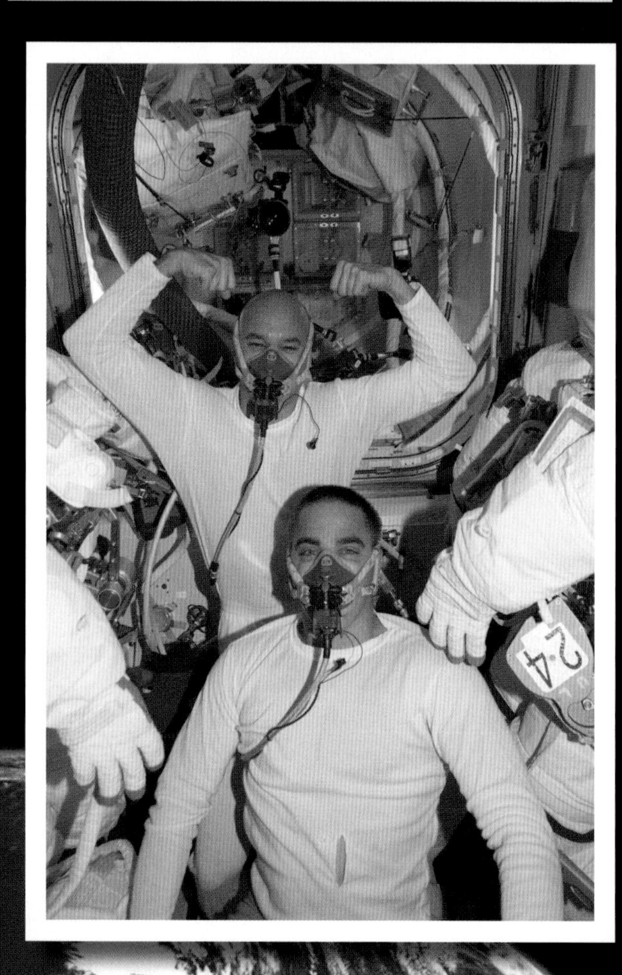

"When we started the ISS assembly, we had what we called the Wall of EVA. Some people thought that we we're never going to be able to surpass this wall of EVA. How are we going to do this? But now that we have finished, look at how well every one of those EVAs went. Even when there were problems, we were able to work around the technical challenges. That's the advantages of having humans in space. Humans are so adaptable. The preparation that we did going into it, to get through all those EVAs, we never had an issue. We had an issue recently. Luca Parmitano had water in his helmet. That was because the suit was not really designed to be maintained in orbit continuously. Up until then, there were technical problems we had to work around, but we completed all those EVAs with great success." - Robert D. Cabana

Scott Kelly routes cables on the International Docking Adapter (IDA).

Scott Kelly works to remove and stow the P3/P4 electrical jumper.

In the July 1958 issue of the British Interplanetary Society magazine Spaceflight, J.C. Guignard wrote about the potential of an extensive spacewalking program supporting the construction of a future space station; he asked: "...but just how easy will such work prove in practice?" During planning EVA at ISS in the 1990s the amount of EVA work to be accomplished was massive and became known as the "Wall of EVA". By the completion of ISS assembly in 2011, 12 years 7 months after it started, the "Wall of EVA" had been scaled. The job was complete. It had been difficult and had required the work of hundreds of skilled personnel, but the job had been completely achievable.

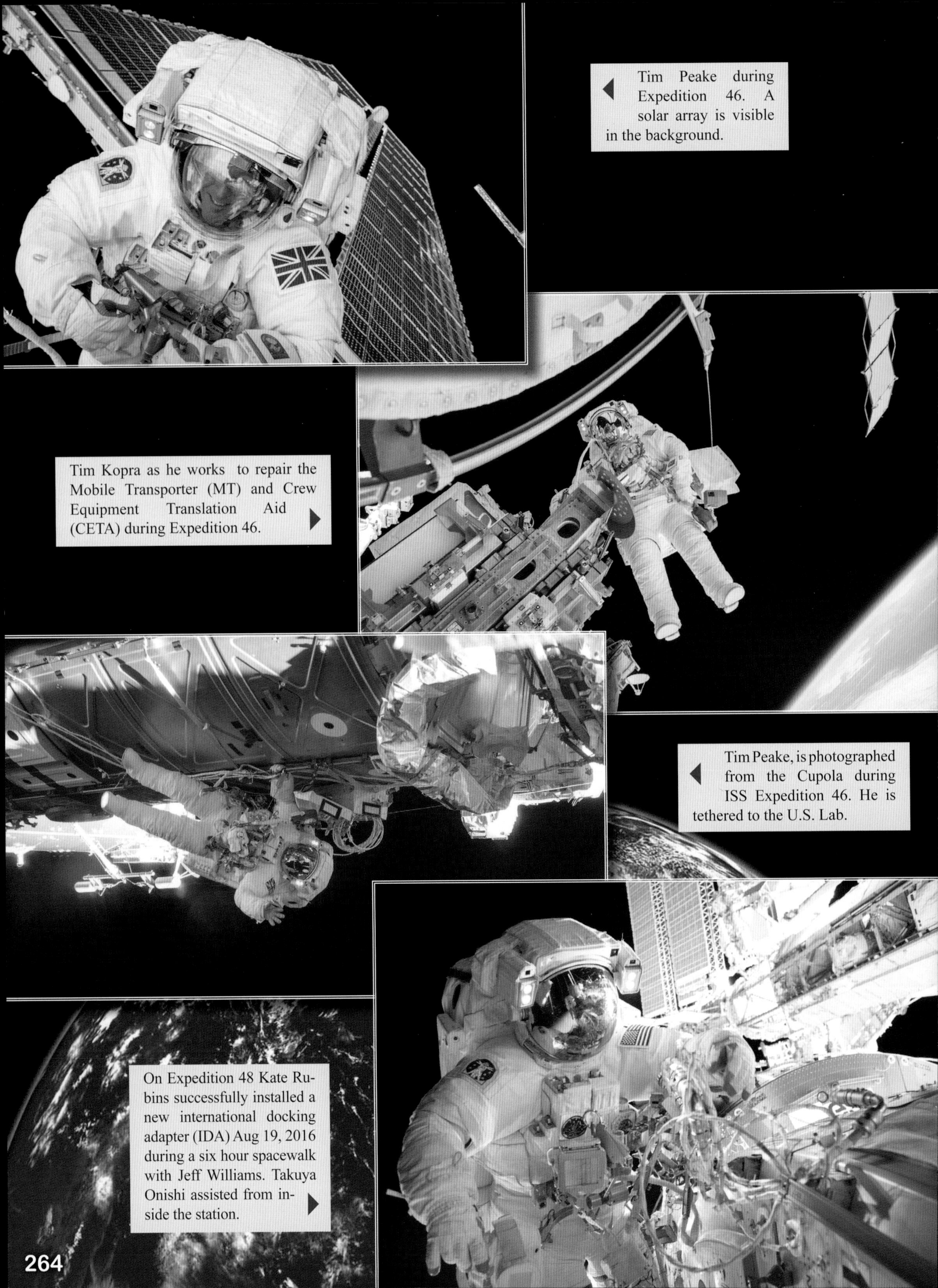

Tim Peake during Expedition 46. A solar array is visible in the background.

Tim Kopra as he works to repair the Mobile Transporter (MT) and Crew Equipment Translation Aid (CETA) during Expedition 46.

Tim Peake, is photographed from the Cupola during ISS Expedition 46. He is tethered to the U.S. Lab.

On Expedition 48 Kate Rubins successfully installed a new international docking adapter (IDA) Aug 19, 2016 during a six hour spacewalk with Jeff Williams. Takuya Onishi assisted from inside the station.

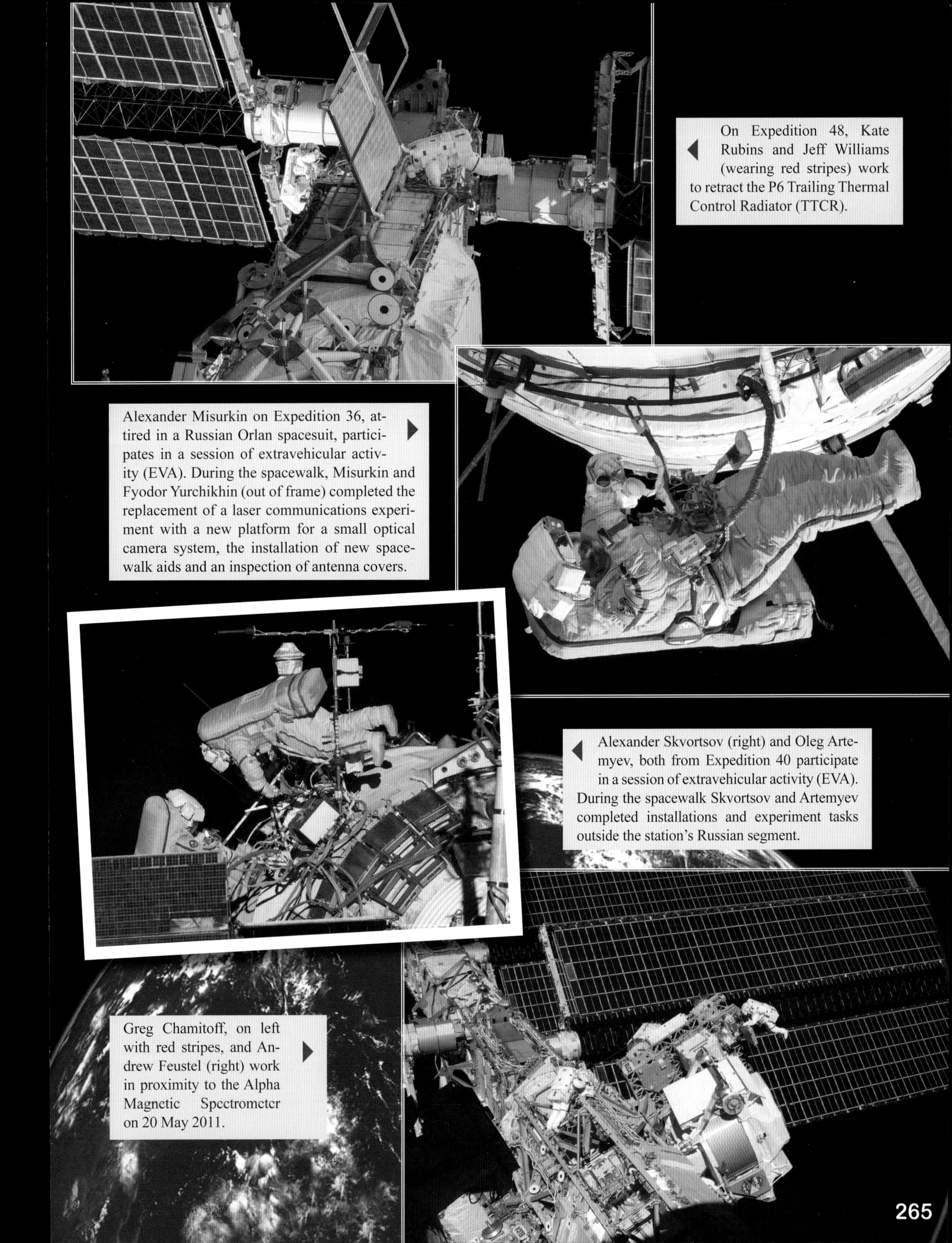

On Expedition 48, Kate Rubins and Jeff Williams (wearing red stripes) work to retract the P6 Trailing Thermal Control Radiator (TTCR).

Alexander Misurkin on Expedition 36, attired in a Russian Orlan spacesuit, participates in a session of extravehicular activity (EVA). During the spacewalk, Misurkin and Fyodor Yurchikhin (out of frame) completed the replacement of a laser communications experiment with a new platform for a small optical camera system, the installation of new spacewalk aids and an inspection of antenna covers.

Alexander Skvortsov (right) and Oleg Artemyev, both from Expedition 40 participate in a session of extravehicular activity (EVA). During the spacewalk Skvortsov and Artemyev completed installations and experiment tasks outside the station's Russian segment.

Greg Chamitoff, on left with red stripes, and Andrew Feustel (right) work in proximity to the Alpha Magnetic Spectrometer on 20 May 2011.

On Expedition 48, Jeff Williams pauses for a photo after installing a hemispherical reflector cover on Pressurized Mating Adapter 2 (PMA-2).

On Expedition 48, Kate Rubins and Jeff Williams (wearing red stripes) work to install a shroud over the retracted P6 Trailing Thermal Control Radiator (TTCR).

Maxim Surayev on Expedition 41 in his Russian Orlan suit, inside the Pirs Docking Compartment 1.

Edward M. (Mike) Fincke during Expedition 9 wearing a Russian Orlan spacesuit, participates in an EVA with Gennady I. Padalka (out of frame). The two spent 4 ½ hours outside the Station swapping out experiments and installing hardware associated with Europe's Automated Transfer Vehicle (ATV) in preparation for its first launch.

"I was on STS-101. Originally it was focused on the Russian Service Module. In the early EVA training in our Neutral Buoyancy Laboratory we didn't have a Service Module mockup to train on for spacewalks. So we had to go to Russia to do the EVA training. It was unique because it was in U.S. spacesuits, and that's the only time that U.S. spacesuits were integrated into the facility in Star City in Russia, to do EVA training. It was a very unique chapter that is sometimes lost." - Jeffrey N. Williams

Sergey Ryazanskiy during Expedition 37 attired in a Russian Orlan spacesuit, is pictured during a spacewalk with Oleg Kotov (out of frame) continuing the setup of a combination EVA workstation and biaxial pointing platform that was installed during an Expedition 36 spacewalk months earlier.

Alexander Skvortsov on Expedition 40 participates in a spacewalk supporting science on the International Space Station.

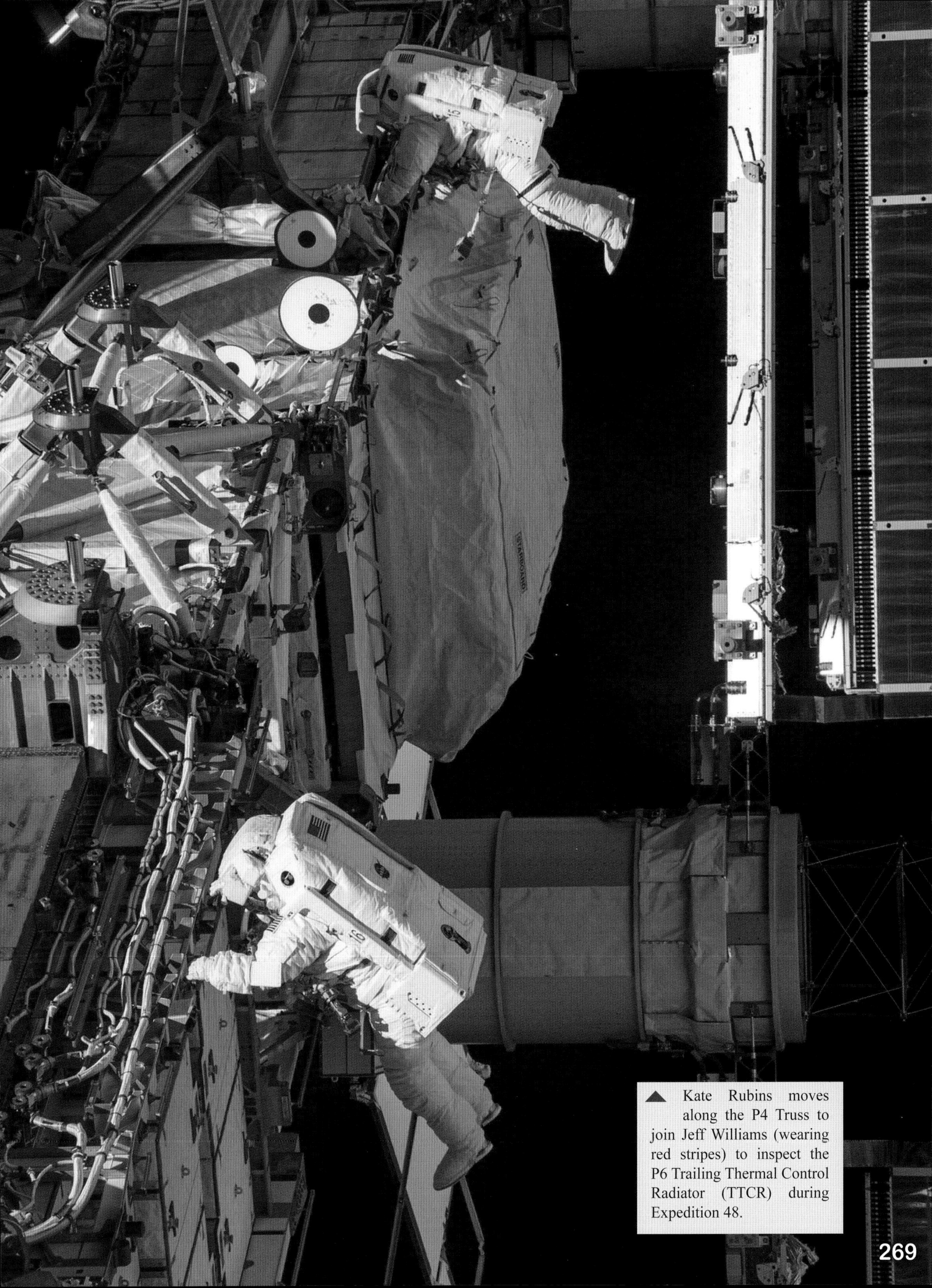

▲ Kate Rubins moves along the P4 Truss to join Jeff Williams (wearing red stripes) to inspect the P6 Trailing Thermal Control Radiator (TTCR) during Expedition 48.

U.S./Joint Airlock (Quest)

NASA/Boeing

Before the arrival of the Joint Airlock (called Quest) on STS-104, all station-related EVAs were conducted via the airlock of the docked Shuttle. The attachment of the Joint Airlock Quest in July 2001, allowed station-based EVAs to be performed without the loss of air. The first EVA through the airlock was by STS-104 astronauts on 20 July 2001 to verify the airlock's performance. The first station-based EVA using the Quest airlock without a docked Shuttle occurred on 20 February 2002, by increment 4 crew members Carl Walz and Dan Bursch. The Quest Airlock was designed for EVAs by astronauts wearing US EMU suits, however it can also support EVAs using the Russian Orlan suits.

The Joint Airlock Quest has two compartments and is berthed to Node 1 Unity on the starboard port. The compartment closest to the Node is called the Equipment Lock and can house two standard racks. The racks house environmental control systems and air and avionics electronics data management equipment. Specially designed storage areas hold batteries, power tools and EVA equipment and supplies. The equipment lock is used for storing and maintaining EMUs. It is also used as a staging area during the preparation for EVA.

The outer compartment of the Quest Joint Airlock is called the Crew Lock. It is narrower in diameter than the Equipment Lock and its design is based on the Space Shuttle Airlock. The Crew Lock features the EVA hatch which provides passage to and from space.

Inside the Crew Lock, an Umbilical Interface Assembly (UIA) supplies water, waste water return, oxygen, communications and spacesuit power for two EMUs while before or after the astronauts leave the airlock. The UIA is located on one wall.

Prior to a spacewalk, the Quest Airlock is used as a 'camp out' facility. The EVA astronauts sleep in the airlock while the atmospheric pressure is reduced from 14.7 psi to 10.2 psi. This purges nitrogen from the bloodstream and helps to prevent decompression sickness (the bends). Once in their EMUs, the pressure in the spacesuits is 4.3 psi.

The US airlock Quest, arrived at ISS via STS-104 in July 2001. In the decade after the arrival of Quest there were 97 EVAs. Many of the EVAs from Quest have included Russian and international crew members.

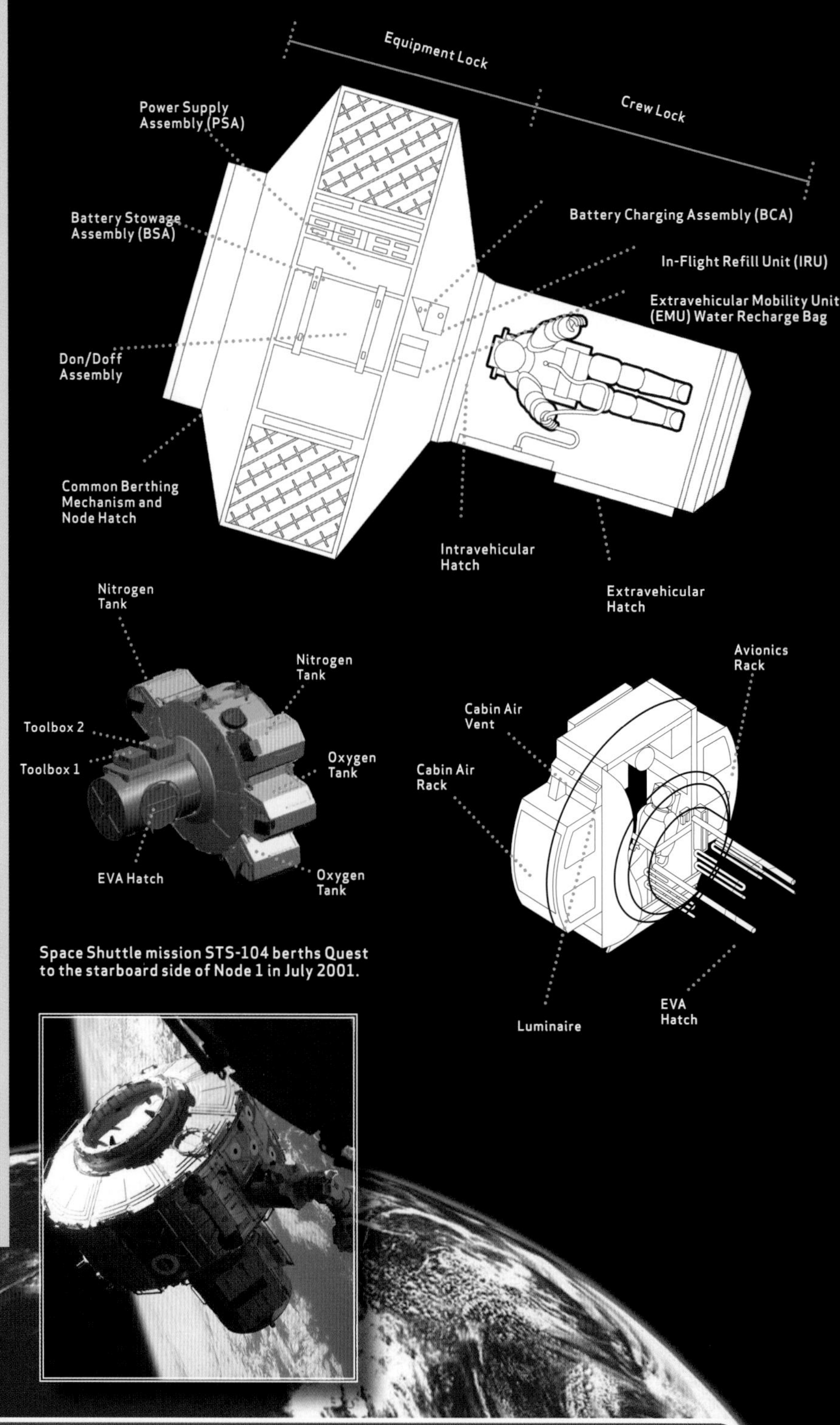

Space Shuttle mission STS-104 berths Quest to the starboard side of Node 1 in July 2001.

"Once I was an experienced EVA crew member, I got assigned to work in the EVA Division. When you go from atmospheric pressure to the spacesuit operating pressure of four pounds per square inch, you can get the bends. We have to spend up to four hours breathing pure oxygen to denitrogenate the blood, to make it safe to depressurize the suit. By depressing the cabin to only 10.2 psi [from the normal 16 psi] when we know we're going to be doing an EVA we can prebreathe for a shorter time. On Space Station there had been an argument between the operations people and the life scientists. The operations people wanted to be able to run Space Station at ten psi, because they knew that we were going to be doing a lot of EVAs, and they wanted to be able to do a 40-minute pre-breathe. Life scientists said no; a purpose of Space Station was to do life science research. They said all of their baseline data was at one atmosphere, and if the station was at other than one atmosphere, then none of their data was going to be valid. So Station was only designed for one atmosphere; you can't reduce the pressure. As a result, if you're going to use a four-psi spacesuit, you're going to have to pre-breathe for four hours. It's a fairly involved and time-consuming activity. Before we started, everybody recognized that assembling the Space Station was going to take a lot of EVA, and sophisticated EVAs at that." - Jeff Hoffman

Docking System Probe
Zenith Docking System (male) and Hatch Entrance to Service Module
Wide-Beam Antenna
High-Gain Antenna
Drain Valve
Attitude Control Antenna
Kurs Antenna
EVA Hatch 2
EVA Hatch 1
Movable Handrail
Pressure and Deposit Monitoring Unit
Interior Orlan Storage
Interior Control Console
Refueling Hydraulic Valves
Nadir Docking System and Hatch Port for Soyuz or Progress

Russian Airlock Pirs and Poisk Airlocks

S.P. Korolev Rocket and Space Corporation Energia (RSC Energia)

The Pirs (pier) and Poisk (search, seek or explore) Russian Airlocks are identical. They are berthed at the zenith and nadir ports of the Zvezda Service Module. Each serve two primary functions; as docking ports for visiting Soyuz and Progress logistics carriers, and as airlocks for EVAs by two space station crew members using Russian Orlan EVA suits. There are two EVA hatches on either side of each pressurized module. They provide redundancy and either hatch can be used interchangeably in case of anomalies.

Pirs and Poisk each have systems that support the flow of fuel, air and water supplies from the tanks of a Progress freighter to storage in the Russian FGB and Service Module.

"The Russian airlocks, Pirs and Poisk, are only compatible with the Russian suits. The joint airlock, Quest, we can take US EMUs or Russian Orlans outside. We could take an Orlan and an EMU out in parallel, but things like communications would be a problem because they are on different radio systems. We might be able to set up a relay. We have three suits of each type on board so we have redundancy. The crew will be trained on both suits, one more as a primary than the other. We have the capability to go anywhere we need to outside. We'll usually take Orlans or EMUs outside, a pair at a time." - Richard Fullerton

On Expedition 45 Scott Kelly and Kjell Lindgren, still wearing their Liquid Cooling and Ventilation Garments (LCVGs) after completing an EVA, pose for a photo in the Destiny U.S. Laboratory.

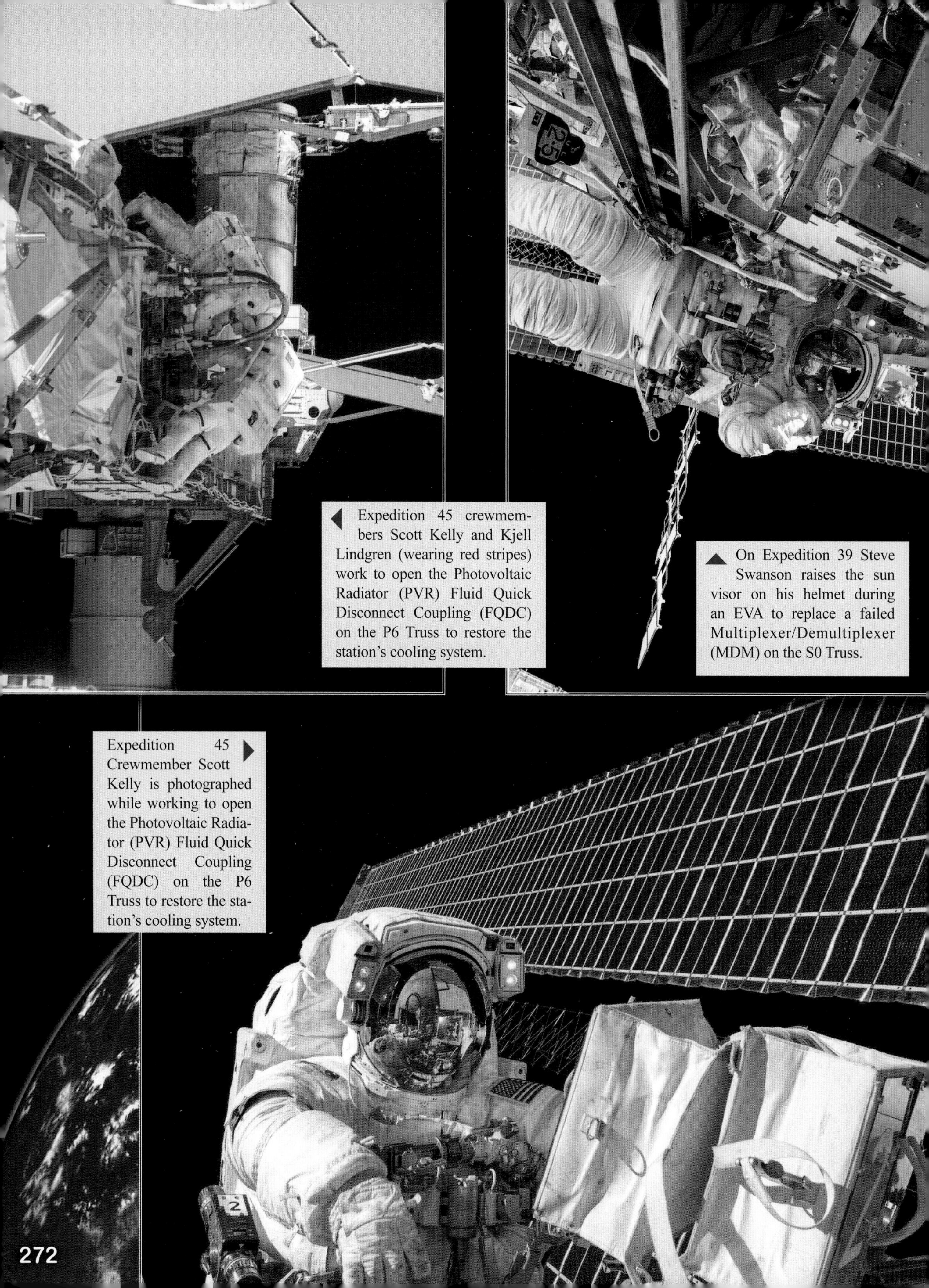

Expedition 45 crewmembers Scott Kelly and Kjell Lindgren (wearing red stripes) work to open the Photovoltaic Radiator (PVR) Fluid Quick Disconnect Coupling (FQDC) on the P6 Truss to restore the station's cooling system.

On Expedition 39 Steve Swanson raises the sun visor on his helmet during an EVA to replace a failed Multiplexer/Demultiplexer (MDM) on the S0 Truss.

Expedition 45 Crewmember Scott Kelly is photographed while working to open the Photovoltaic Radiator (PVR) Fluid Quick Disconnect Coupling (FQDC) on the P6 Truss to restore the station's cooling system.

Spacewalk training is conducted in large water immersion pools. The Neutral Buoyancy Laboratory in Houston is one of the largest operational pools in the world, it holds 23 million liters (6 million gallons) of water. All ISS astronauts preparing for EVA train here. The pool provides underwater full scale mock-ups of a large portion of the ISS modules, truss, solar array rotary mechanism, and robotic arms. Besides being used for the ISS, it is often rented out for use by oil, shipping and other companies to test remotely operated vehicles, other new hardware and assembly techniques.

"The NBL is a hundred feet wide 200 feet long and 40 ft deep, and keep in mind that the space station itself is one longitudinal structure but you can't fit it into the pool like that, so they have the central segment of the space station running width-wise of the NBL and then the two trusses, the port and starboard trusses are running along the 200-foot side of the pool. So that way you can fit the whole Space Station in and, barring the inflection point where you go from the central section to the port and starboard where you have to make a 90-degree turn underwater that you wouldn't be making in space, it's extremely high fidelity." - Dave Williams

A rendering showing the layout of the NBL.

The Weightless Environmental Test (WET) facility pool in the Environment Utilization Center (EUC) at the Japan Aerospace Exploration Agency (JAXA) Space Station Integration and Promotion Center (SSIPC) at the Tsukuba Space Center (TKSC), Ibaraki prefecture, Japan, holds a full size mock-up of the Japanese Kibo laboratory.

The Hydrolab at the Gagarin Cosmonaut Training Facility provides training primarily for crews using Orlan spacesuits near Russian modules of the station.

"When Mission Control plans a mission and during the mission, we become part of the planning process. We provide the input that becomes the EVA time line for the cosmonauts. During the EVA, our EVA expert sits next to the shift flight director in the MCC. We know all of the tasks for that EVA. If there is a situation that requires quick decision making, we are there. We consult between ourselves and then we make a decision and let the flight director know." - Alexander Alexandrov

Extravehicular Mobility Unit (EMU)

NASA/Hamilton Sunstrand/ILC Dover

Hamilton Standard of Windsor Locks, Connecticut began work on the original Shuttle EMU in 1974. The first EMU suits were available in 1981 for contingency use on the first Shuttle missions. In 1998 an 'enhanced' version of the EMU came on line.

The EMU is a multi-layer design; from the outermost layer working in:

1. Thermal Micrometeoroid Garment (TMG) Cover Ortho/KEVLAR® reinforced with GORE-TEX®. This is the outermost covering of the suit.
2. TMG Insulation, comprising five to seven layers of aluminized Mylar®, with more layers added on the arms and legs of the suit.
3. TMG liner of Neoprene-coated nylon rip stop.
4. Pressure garment cover with a restraint of Dacron®.
5. Pressure garment bladder of urethane-coated nylon Oxford fabric.
6. Liquid cooling garment with Neoprene tubing.

The suit normally operates at 0.3 atmosphere (4.3 psi, or 29.6 kPa) and utilizes 100% oxygen. The primary oxygen tank pressure is 900 psi and has an 8 hour maximum EVA capability, although 8.93 hours has been achieved. There is also a secondary oxygen tank available for emergency or contingency use that stores oxygen at 6,000 psi, and which can provide a 30-minute back up supply of oxygen. The mass of the full EMU is 178 kg (393 lb.). The suits are designed to last 30 years.

Communications Carrier
TV Camera
In-Suit Drink Bag
Light
Caution and Warning Computer
Radio
Display and Control Console
Antenna
Sublimator
Water Tank
Oxygen Control Actuator
Contaminant Control Cartridge
Primary O_2 Tanks
Space Suit Assembly (SSA)
Secondary O_2 System
Connection for Service and Cooling Umbilical
Primary Life Support System (PLSS)
Temperature Control Valve
Colored ID Stripe
Liquid Cooling and Ventilation Garment

The main components of the suit are the liquid coolant garment; the hard upper torso assembly; the lower torso assembly; the arm assembly and EVA gloves; the 'Snoopy-Cap' communications assembly and helmet; the EVA visor assembly; the primary and secondary life support systems; and the SAFER unit.

The suits are mostly white in order to reflect solar heat, the EMU suits have identification stripes in red on the arms, legs, helmet and backpack.

The Simplified Aid For EVA Rescue (SAFER) is a nitrogen-powered backpack that allows the wearer to maneuver independently in order to return to the station airlock facilities should he/she become inadvertently detached from the station during EVA.

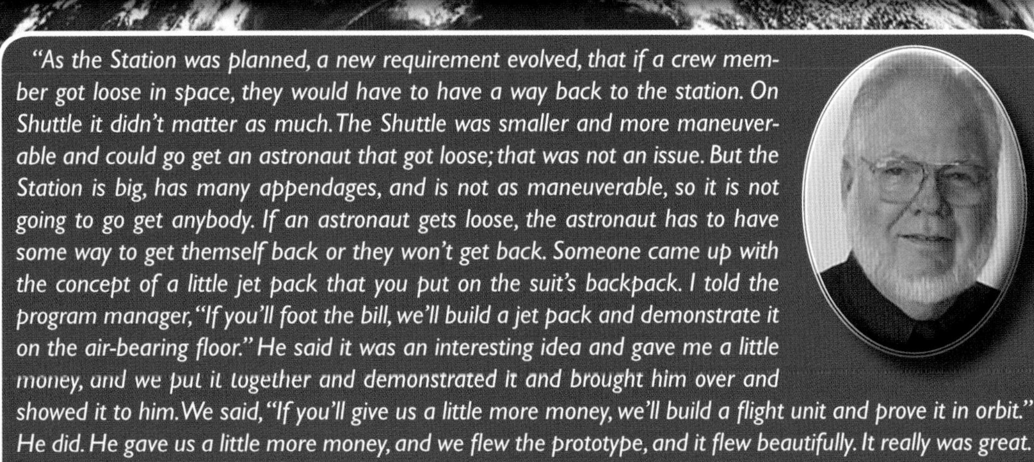

"As the Station was planned, a new requirement evolved, that if a crew member got loose in space, they would have to have a way back to the station. On Shuttle it didn't matter as much. The Shuttle was smaller and more maneuverable and could go get an astronaut that got loose; that was not an issue. But the Station is big, has many appendages, and is not as maneuverable, so it is not going to go get anybody. If an astronaut gets loose, the astronaut has to have some way to get themself back or they won't get back. Someone came up with the concept of a little jet pack that you put on the suit's backpack. I told the program manager, "If you'll foot the bill, we'll build a jet pack and demonstrate it on the air-bearing floor." He said it was an interesting idea and gave me a little money, and we put it together and demonstrated it and brought him over and showed it to him. We said, "If you'll give us a little more money, we'll build a flight unit and prove it in orbit." He did. He gave us a little more money, and we flew the prototype, and it flew beautifully. It really was great. Then I went to the next program manager and told him, "You're probably going to need one of these." He agreed, and we built the SAFER units and they have been flying ever since." – Walt Guy

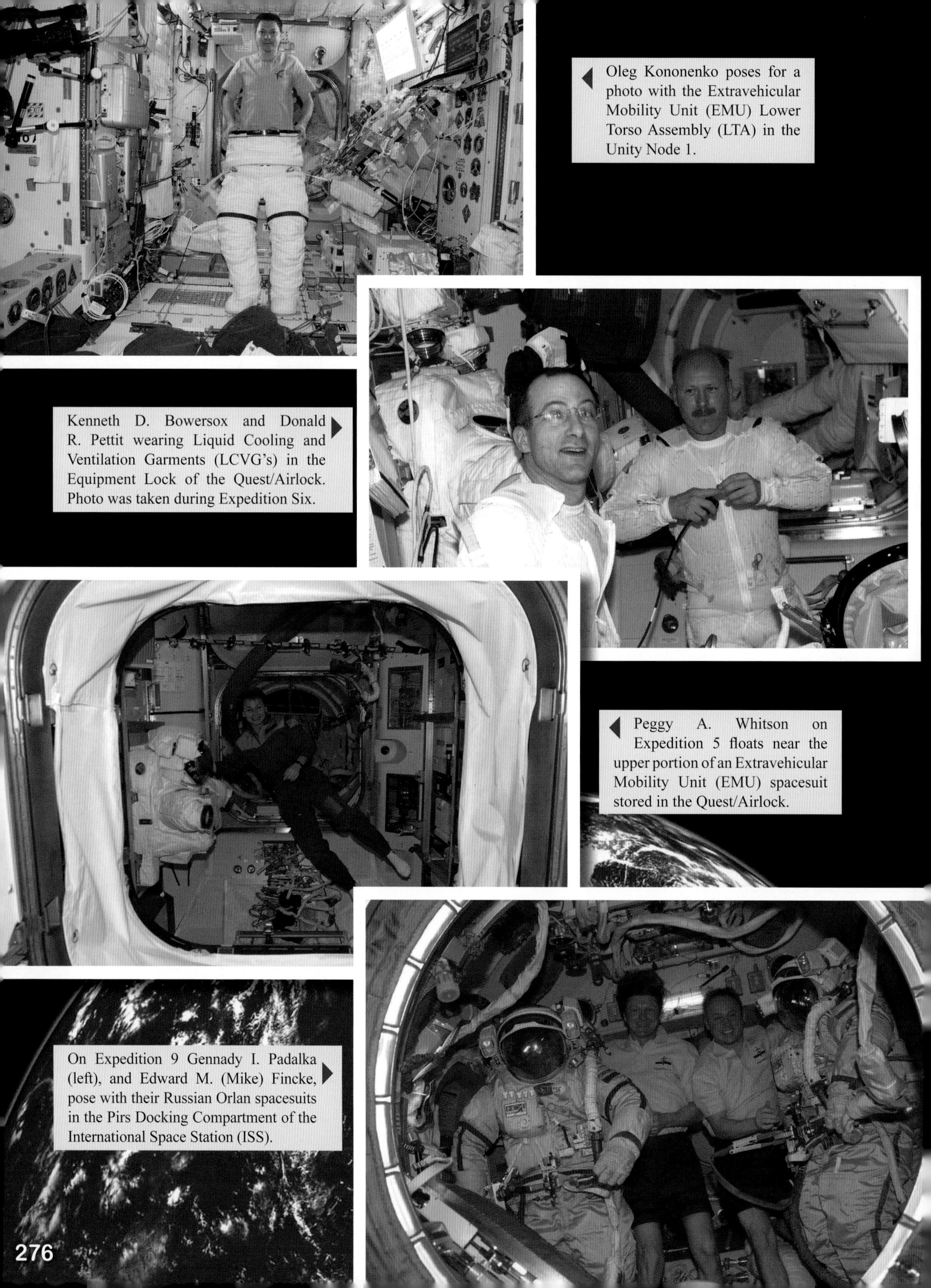

Oleg Kononenko poses for a photo with the Extravehicular Mobility Unit (EMU) Lower Torso Assembly (LTA) in the Unity Node 1.

Kenneth D. Bowersox and Donald R. Pettit wearing Liquid Cooling and Ventilation Garments (LCVG's) in the Equipment Lock of the Quest/Airlock. Photo was taken during Expedition Six.

Peggy A. Whitson on Expedition 5 floats near the upper portion of an Extravehicular Mobility Unit (EMU) spacesuit stored in the Quest/Airlock.

On Expedition 9 Gennady I. Padalka (left), and Edward M. (Mike) Fincke, pose with their Russian Orlan spacesuits in the Pirs Docking Compartment of the International Space Station (ISS).

Chris Cassidy (right) and Luca Parmitano, Expedition 36, are pictured in the Joint Airlock Quest as they prepare for the start of a session of extravehicular activity (EVA). Cassidy and Parmitano are wearing Extravehicular Mobility Unit (EMU) spacesuits.

On Expedition Two Yuri Usachev (right) and Jim Voss (left) in their Orlan suits in the Transfer Compartment of the Service module Zvezda. They are practicing for an Extravehicular Activity (EVA) to be performed later from this compartment.

On Expedition Six Donald Pettit (right) and Kenneth Bowersox (red stripe), both wearing Extravehicular Mobility Units (EMU), conduct Extravehicular Activity (EVA) dry-run activities in the Joint Airlock Quest.

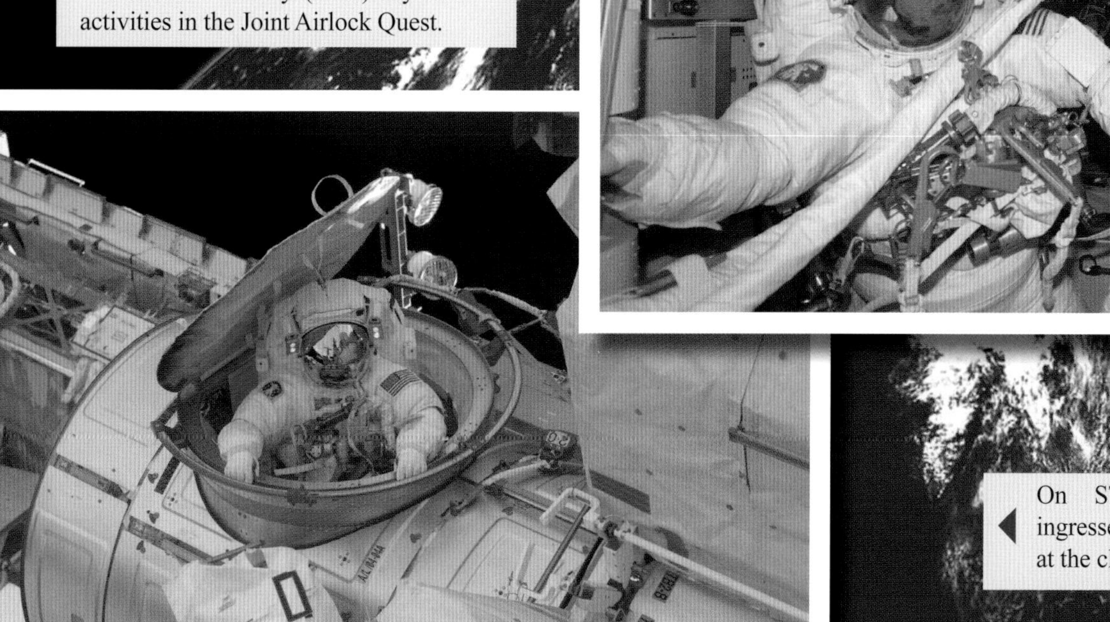

On STS-134 Andrew Feustel ingresses the Quest Airlock (A/L) at the close of a spacewalk.

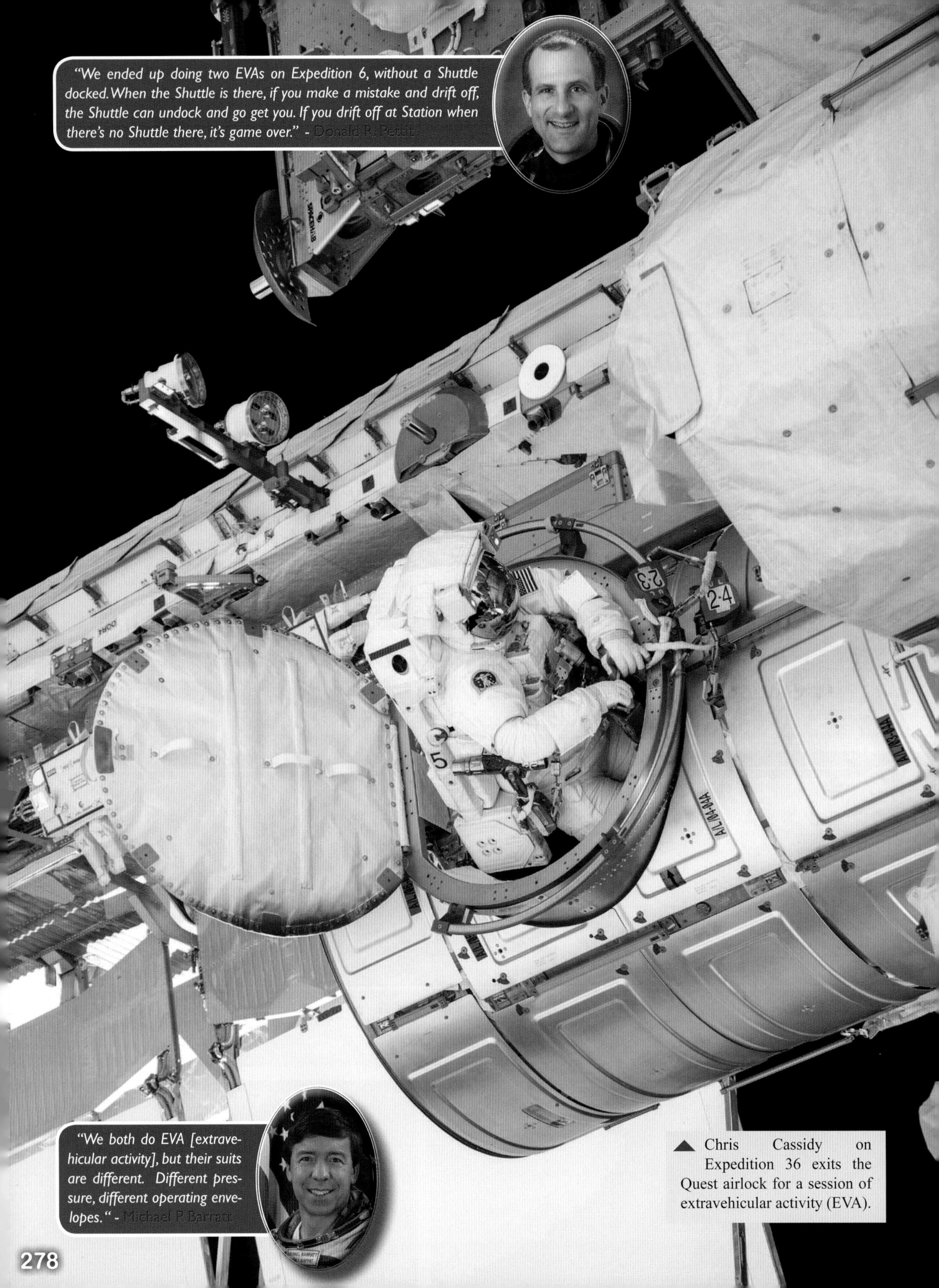

▲ Chris Cassidy on Expedition 36 exits the Quest airlock for a session of extravehicular activity (EVA).

Orlan Spacesuit

Science Production Enterprise Zvezda

The Orlan is the Russian EVA suit system.

The forerunner of the Orlan suit was developed for the Soviet manned lunar EVA program that was cancelled in the mid-1970s. The suit at that time was called Kretchet. The suit was developed and tested between 1967 and 1971 and then modified to support orbital station EVA operations.

Orlan is a 'one-size-fits-all' design, with adjustment straps and fittings that can be sized to accommodate the wearer. The main body and integrated helmet of the suit are constructed from aluminum. The arms and legs are constructed from flexible fabric. A liquid coolant garment is worn inside which features an integral hood and foot units. The backpack is integrated to the suit torso and is hinged to allow entry and exit from the rear of the suit. This allows for the user to be able to close and open the suit himself. The rear suit entry allows for easier maintenance and servicing.

The suit operates at a nominal 0.4 atmosphere (5.8 psi) with a 100% oxygen atmosphere. It can support a maximum EVA duration of 7 hours. The mass of the suit is quoted at 108 kg (238 lb.) and has an on orbital lifetime of 12 years, or 4 years if it is not periodically returned to Earth for major servicing.

Each suit carries either red or blue markings to allow identification of the cosmonaut.

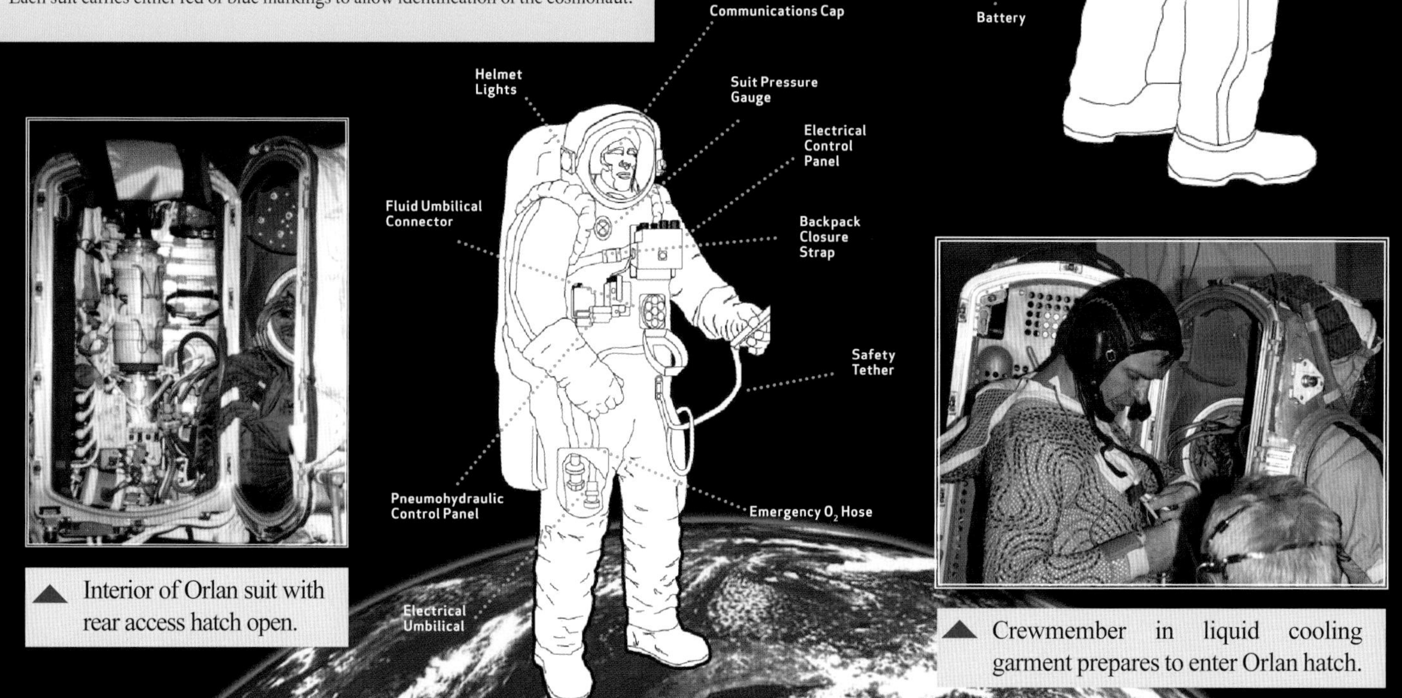

Interior of Orlan suit with rear access hatch open.

Crewmember in liquid cooling garment prepares to enter Orlan hatch.

"Orlan is a very reliable space suit, and the main thing, the comfort, actually, of this particular space suit is that it's almost like a refrigerator. You can open it and walk in there without anybody else's assistance. It's pretty simple to control from the panel, from the cosmonaut's panel. It has a few levels of protection against any off-nominal or contingency situations. It has redundant systems. It has a redundant pump, a water pump. It has a redundant vent. Also, in case of a contingency situation, if both systems fail, it can work in what we call injector mode. It injects the oxygen into the space suit and the cosmonaut can continue working. It requires some modifications as far as the elbow hinges are concerned, to make it more flexible, which we've done, but as far as the capability is concerned, it's a very reliable space suit. And the fact that our cosmonauts can work continuously up to seven hours in those space suits tells a lot about the capabilities of the unit. There are three types of dimensions and sizes for the gloves and for the unit itself, so you can adjust according to the height of the cosmonaut. Each pair of gloves is changed for each EVA. We change the gloves. We bring the gloves back to Earth as a souvenir. We have a standard training plan for each cosmonaut who is training to perform an EVA. First they study and learn the space suit. That's about ten hours. Then they train in the pressure chamber to use the space suits. Then they are trained in the vacuum chambers. They train different procedures for air-lock procedures, and with scenarios of contingency situations. Then on the airplane with zero G [gravity], they also train cosmonauts for the flexibility to ingress or egress for the EVA. We train them in use of the Orlan and the airlock, and the use of restraints. They train to open and close hatches, and to translate along different paths on the station. They train to recover and bring a crewmember inside who may be unconscious or who is incapacitated. We train them to use tools. The training gives them more confidence when it comes to real life. For specific mission tasks, they have only two or three training sessions." - Alexander Alexandrov

Dave Williams standing on the foot restraint of the Canadarm, installs a gyroscope on the ISS during the STS-118 mission.

"The gyros come with electrical connectors and when you're doing a space walk the biggest challenge of all is that you're wearing gloves that are essentially like hockey mitts. So when you're doing up all these connectors it can be a challenge. Sometimes they go right away and other times it takes a little bit more finesse to be able to get them to work but we got the gyroscope installed and then we waited, and you kind of feel like you're waiting for the judges to give you your score, but they powered up the gyroscope and everything worked great so it was another successful space walk." - Dave Williams

Clayton Anderson, poses for a photo in the Cupola of the International Space Station while space shuttle Discovery remains docked with the station.

"WOW! I can't believe it. Here I am, safely in orbit, living on board the International Space Station (ISS). In a day or two, life on the ISS begins in earnest." - Clayton Anderson

Many space explorers have said that preparation for space flight took their whole lifetime. Life experiences and educational qualifications were fundamental to selection as an astronaut. But life experience and formal education was not enough. The astronaut goes back to the class room and begins learning all over again. After years of ground work and patience you hopefully get assigned to a mission. Training then got put to the test. After a short ten-minute rocket ride the dream is fulfilled and you reach space. A whole new adventure begins. Then you realize that all the training was just a guide. The learning process is not over. Everything you know changes. In space, you have to learn to live and work off the home planet. Peggy Whitson flew to the station for the first time in 2002 on expedition 5, and in 2007, returned to the station, this time as Station Commander for ISS expedition 16. In her journals, she wrote about going back to the station.

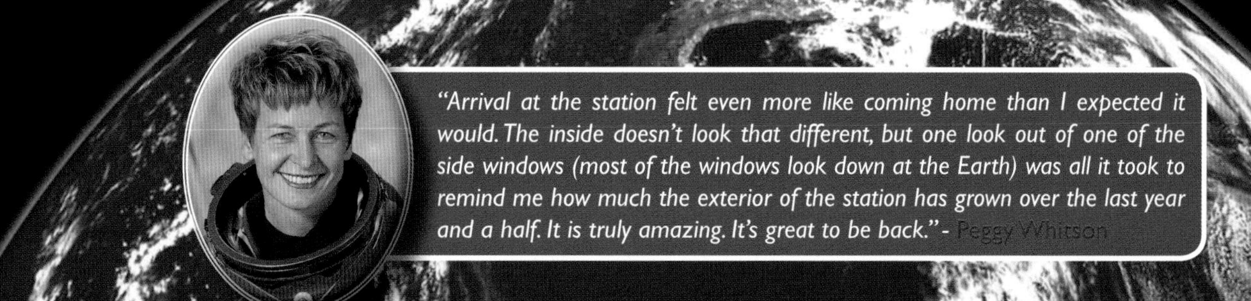

"Arrival at the station felt even more like coming home than I expected it would. The inside doesn't look that different, but one look out of one of the side windows (most of the windows look down at the Earth) was all it took to remind me how much the exterior of the station has grown over the last year and a half. It is truly amazing. It's great to be back." - Peggy Whitson

The crew resident throughout an increment would be called the 'expedition crew'. The dictionary defines an expedition as 'a journey for defined scientific purposes'. Expedition crew members would remain on-board the station for several months, called an increment. The expedition crew was a tradition that began with the Soviet Salyut and Mir space stations. For ISS, the crew consists of a station Commander and one or more Flight Engineers, who regularly maintain and operate the station day after day, week after week for several months. They are then replaced by the next expedition crew, after a short handover period of joint operations. Short duration crews like Soyuz exchange flights, are termed 'visiting' missions, or flights or crews. Shuttle crews were typically referred to as 'Assembly Mission' crews.

Incrs 0-2	Incr 3	Incr 4	Incr 5	Incr 6	Incr 7	Incr 8

"We defined the 'increment' as one of a series flights contributing towards a common goal', the latest in a series of expeditions designed to build upon the experiences and activities of the previous crews. The objective behind this is to create a baseline database of experience, results and confidence over a period of time, learning from and improving upon what has gone before to pursue more ambitious and far-reaching programs.

"One thing that the US had never done before was supplying a space station. Station was going to be up there, continuously manned; we would be sending up new scientific experiments to work on; we had to send up all the food, all the propellant, everything that was needed. Things that were used up had to be brought back. The crew had to go back and forth. Until now we had always planned for just seven- or fourteen-day Shuttle flights; they had a beginning and an end. Now with space station, we had a continuum of time. How do you break that up? How do you do the planning and the crew turnaround?

"That was where we came up with the concept of an increment; it was the period of time from the arrival of one Shuttle to the arrival of the next Shuttle. At that time when the term originated, that was the plan; there would be only Shuttles. That would be the only time that you could change out a crew and bring up stuff that would change the configuration of the Space Station. So we settled on 'increment'. So there would be increment managers who would plan each increment, who would follow the details of each. They would plan what needed to go up and plan the activities to be accomplished. There would be a tactical operations plan; it would encompass two years' worth of increments.

"One of the reasons for taking this approach was that we had international partners. We had Japan and Europe, and each was planning to build their own modules. Canada was going to use the U.S. module. We didn't want the individual partners making plans independently. We didn't want Japan coming up with a plan and saying, okay, this is what we're going to do in our module for this period of time," because we didn't have that many crewmen; it was a limited number of crewmen.

"We were trying to say, the crew has to do everything. They are not representing only their own country's science and only working in their own lab; they have to be able to do whatever needs to be done. For efficiency we are going to have to ignore nationalities and geographic origins on the space station and put together the most sensible integrated time line. We didn't want the partners going off building their own plan; early on that was a big source of tension." - Anne Accola

The Increment and Expedition

In 1999, the first crew was preparing to take up residence on the station. The crew's daily schedule was under final consideration. At that time, it was planned that each crew member would be assigned an individual daily schedule, working about eight hours in each 24 hour period. They would get two days off each week. Which holidays to be celebrated was still being considered. Family conferences were planned to occur once a week. Astronauts could use their free time using some of the personal items each crew member was allowed to take to the station, like books, computer games, videos and music. The working 'day' typically starts with about an hour for breakfast, followed by 30 minutes for personal hygiene and 30 minutes for 'daily coordination' – essentially, reviewing the plans for that day. ISS expedition 8 Commander Mike Foale explained the organization of a working day on station in more detail for his 2003/2004 expedition 8.

"It's according to a pretty much standard plan worked out by, and in concert with the two mission control centers in Houston and Moscow. It begins with a wake-up call 'alarm clock' and two hours of getting yourself together, shaking out the cobwebs, and checking the status of the station. For the status checks, we look at the pressure in the station and check there are no alarms. It is absolutely inconceivable that there would be alarms that the ground would not have reported to us or we would not have heard in the night, but we check the alarms anyway. Then we go through our morning ablutions and breakfast. The first order of business for organizing our activity is what we call the Morning Planning Conference, which is scheduled for about 10 to 15 minutes and ties up with the control centers. Before we start talking, I am looking at a computer laptop screen that summarizes in graphic form the main blocked activities of the day – activity versus time. These run from wake up through sleep for an official workday that goes from 6 am to 10pm. That includes two hours post sleep from 6 to 8 am and two hours of pre-sleep 8 to 10 pm, although we are pretty flexible during those periods. The Conference always began with Houston, discussing the American plan. They very quickly gave me the changes to the flight plan if there were any, or some reminders of what I was going to do that day. Then we were handed over to Moscow and they immediately started talking to Sasha Kaleri about Russian tasks. If we had a European on board then we handed over to the European control center and if I was working on a European experiment in the US lab I would go to a fourth center in Huntsville Alabama. So we were kind of filtered down the centers, but basically the priority goes to Houston, then Moscow then possibly Huntsville and Bremen." - Michael Foale

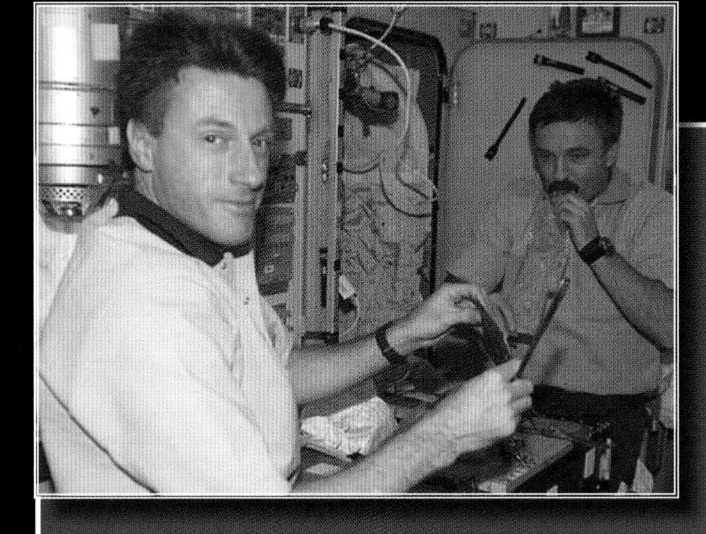

Michael Foale and Sasha Kaleri

"With the addition of the European lab and the Japanese lab, both Bremen and the Japanese control center came on line full time as the station expanded over the 2007-2008 period. The eight-hour period of 'tasks' during the working day could be spaced out with up to 1.5 hours for lunch. Two hours were allocated at the end of the working day for exercise and training, followed by an hour for a review of the day's activities and 1 hour for 'bathing' prior to the sleep period.

In the day's activities, there will almost always be one or two periods of exercise (each an hour long), either adjacent as a two hour block or as separate one hour blocks for each person. There will also be an hour for lunch. In general, as long as we are not restrained by orbital mechanics, lighting, visibility and so on, there are periods where certain activities are more natural for a person during a workday. So exercise is not scheduled over lunch, for example, when the other person is having lunch. Generally (in most crews) there was flexibility in determining the rules for lunch (whether to eat together or separately) before flying.

At the end of the day, we had the Evening Planning Conference, which allowed us to report data observations that were not reported, or were too small in detail to be reported, during the day. We could also tell Houston or Moscow where we had stored specific data files, pictures and so on and it was an opportunity for us to complain if necessary about an activity, so that we could get it changed for the following day. The ground crew also has chance to ask us questions and give us last minute updates of whatever is planned for the next day." - Michael Foale

Astronauts have continuously occupied the space station since 2000. They operated the station throughout the assembly phase, with the station changing configuration with the addition of new elements every few months. When the Shuttle Columbia was lost, in 2003, the Shuttle fleet was grounded for more than two years. During the early expeditions and the post-Columbia period of caretaker 2 person crews (2003-2006), GMT was followed on the station and the emphasis was more towards the Russian side in terms of time zones than the Houston side. Mike Foale explained, that the reason was simply because *"the Russians were much more dependent upon public transport in Moscow, so we had to follow the time when public transport was running for people to get to work. Houston took a bit of a hit on the hours, but they all had cars and could drive to work."*

Coming and Going

For the International Space Station program, while Shuttle was still flying, it was decided that the expedition crew would initially consist of three people. Expedition crews would launch to ISS either aboard a Russian Soyuz OR an American Shuttle. Those expeditions launched aboard a Soyuz would have two Russian crew members, Soyuz Commander and Flight Engineer, and one experienced NASA astronaut, who would be the Station Commander. Expedition crews launched aboard the Shuttle would include two NASA astronauts under the command of a veteran Russian cosmonaut. During the early expeditions, crews would be on-board for between three and four months. This would gradually increase to six-month tours of duty later in the construction process. The best place to test the equipment and enhance the crew training for a space mission is in space itself, so the early expeditions were designed to evaluate the integrity and efficiency of the station and find out about any habitability issues. Regular maintenance and cleaning routines would also be important to help keep the station running at peak performance.

With regular rotations of station crews every four to six months, the system of receiving and incorporating new crews on board the station became a standard operation. With the end of Shuttle flights to the station and the suspension of 'visiting' guest crews or passengers, interruption by short-term visitors was no longer the disruption to the resident crew's routine it once was, although there is still plenty to reorganize and prepare. To accommodate a visiting mission, the resident crew's schedule is changed to incorporate the visiting crew program. This also applied to Shuttle assembly missions, extensive EVAs, and unmanned logistics resupply missions. Though some schedules remain independent of the main crew, all crew members try to assemble for main meal times on the station.

The challenge of dealing with the arrival of new crew members is eased if the resident crew have been able to train with the new arrivals on Earth prior to meeting them in space. A crew member is always more comfortable working and living with new crew members with whom they have trained, or even flown, previously. Still, after several months on the station, it is soon obvious to the resident crews how much they have adapted to life in space when witnessing the antics of newly arrived colleagues.

Some issues had to be resolved through negotiation. An interesting point astronaut Kent Rominger, raised in 2006, was about who would have overall responsibility during the Shuttle-ISS docked phase of a mission when there were both a Shuttle Commander and a Station Commander aboard:

"At one time when the Shuttle showed up, the Shuttle Commander had the responsibility for everything. Now, in 2006, it has been decided that in fact the head of flight crew-operations, which was flying a mission, should set the rules and be in overall charge. Since then, it's never really been addressed. But when I showed up on STS-100 (in 2001) theoretically I think that I was in charge. What you really want is for the Station Commander to be in charge of the Station and the Shuttle Commander to be in charge of the Shuttle. If there is a major emergency, the first ground rule is that you go to your vehicle, if you can. If you cannot, then whatever vehicle you are on, the commander of that vehicle is the overall commander." - Kent Rominger

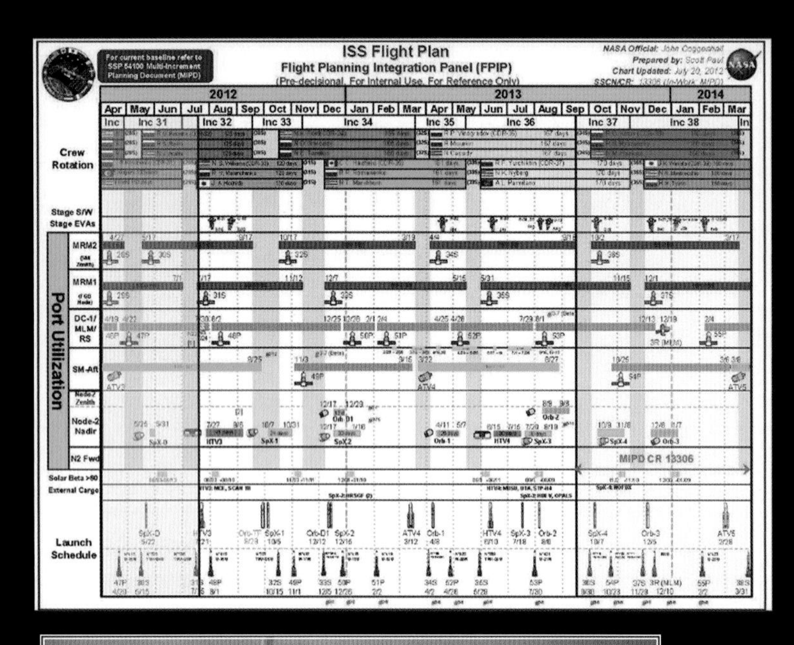

ISS Daily Flight Plan identifies blocks of time for each crewmember to perform specific tasks, as well as identifying day/night cycles, major events such as vehicle approaches and dockings, or EVAs, and communications windows.

The Soyuz rocket with Expedition 33 crew members Oleg Novitskiy, Kevin Ford and Evgeny Tarelkin onboard the TMA-06M spacecraft launches to the International Space Station on Oct. 23, 2012, in Baikonur, Kazakhstan.

Space shuttle Atlantis and its six-member STS-132 crew head toward Earth orbit and rendezvous with the International Space Station. Liftoff was at 2:20 p.m. (EDT) on May 14, 2010, from launch pad 39A at NASA's Kennedy Space Center.

Expedition 43 crew Terry Virts, left, Anton Shkaplerov, center, and Samantha Cristoforetti sit in chairs outside the Soyuz TMA-15M spacecraft just minutes after they landed in a remote area near the town of Zhezkazgan, Kazakhstan on Thursday, June 11, 2015.

According to ISS expedition 15 crew member Clayton Anderson, preparing for the arrival of a Shuttle crew was never a trivial task. It was somewhat reminiscent of having the family over for a holiday, or guests staying over. Preparations on the station included collecting numerous unwanted items and bundling them up for transfer to the Shuttle, as well as cleaning up the station and packing away loose items. Extra room was needed with the expanded crew and the new equipment coming aboard. Sometimes, the station crew prepared equipment and hardware in advance of the arrival of the new crew, saving time when the visiting crew needed to use the items, such as for a spacewalk (EVA). This kind of work was often referred to as 'get-ahead' tasks. Despite the volume of ISS however, an observation frequently made by expedition crews is that there still seems to be insufficient room for orderly storage. Clearing room is often a space age game of chess, a ballet, a house clearance and a spring cleaning all in one. This is one reason why the logistics supply missions are welcome arrivals. Still, Anderson hoped that his station crew would give the Shuttle crew more items to return to Earth than the Shuttle crew were delivering! In 2005, ISS expedition 11 crew member John Phillips wrote about the highs and lows of the pending arrival of a new crew at the station:

▲ John L. Phillips of Expedition 11 participates in a ham radio exchange with students at Albany Hills State School in Brisbane, Australia from the Zvezda Service Module of the International Space Station.

"Sunday July 17. This week has been dominated by preparations for the arrival of the Space Shuttle Discovery and her seven-person STS-114 crew, and then the disappointment of the postponement of the launch, and now a reshuffle of our schedule. We may see Discovery next week."

The circumstances were somewhat unique, as this was the first Shuttle mission scheduled after the loss of Columbia in 2003. NASA was being extra careful and the mission was eagerly awaited by everyone. The delay was both frustrating and challenging. During that week, the ISS-11 crew had celebrated their 100th day aboard Station. They were roughly half way through their planned expedition and had yet to receive any visitors. In his journal, ISS expedition 17/18 crew member Greg Chamitoff described the preparations and expectations for the arrival of a Shuttle crew and the resulting increase in the number of people on board from three to ten as "traumatic." Chamitoff had experienced crew changes with the arrival of the ISS expedition 18 crew members and the departure of the two Russian cosmonauts of ISS expedition 17, who had been his long time colleagues aboard the station. Chamitoff had been concerned about this change, but was amazed how well it worked out. He wrote about it as an energizing event, with so many people around after spending so long with only two other crew members to work and live with. Chamitoff even had a chance to practice his juggling with Space Flight Participant Richard Garriott who had arrived for a short visit. Chamitoff wrote:

"This transition to now being part of Expedition 18 has given me new energy and enthusiasm that might have started to fade just a little before the Soyuz arrival. In some ways, I feel like I'm starting all over again."

The Shuttle that was due to arrive was the one that would be taking Chamitoff home, but he did not feel that his mission was over until he started moving all of his personal items out of the sleep station:

"Preparing for the Shuttle arrival has been a huge undertaking, and it has completely dominated everything for the past several weeks. Most of our work has been in preparing things that need to go down to Earth."

Expedition 44/45 crew members Oleg Kononenko, Kimiya Yui and Kjell Lindgren during emergency scenarios training in the International Space Station training mockups Mar. 11, 2015 at the Johnson Space Center in Houston Texas. ▼

When the Shuttle mission was docked to the station with an expedition crew on board, it could be busy and challenging. Docked days included the welcoming and media celebrations, followed by routine safety briefings and any changeover of crew members. The next few days were dependent upon the payload manifest and EVA program, but were usually the busiest, as most tasks were scheduled for early in the mission in order to get them completed in case there is an unexpected emergency or alteration to the flight plan. The latter days are quieter before the visiting crew leaves. It is usually enjoyable to have a visiting crew there for meals adding to the enjoyment of the occasion. In his 4 April 2011 blog entry Dmitri Kondratyev wrote:

> *"In recent days we have been preparing to meet the new crew. After docking, on April 6, the ISS crew will again consist of six persons. A lot of work, so nobody will be bored, even the newcomers. To meet the new crew we have prepared the cabin on the Russian and American segments, placed a new table in the wardroom so that all six people can be accommodated. The earlier arrival of cargo ships delivered their personal items, toiletries, food stocks. Their washing supplies are already in their correct location. Everything is ready to receive new guests."*

Following a space-to-Earth press conference, members of the International Space Station and Space Shuttle Endeavour crews posed for a group portrait on the orbital outpost. Donald Pettit appears at photo center. Just below Pettit is Heidemarie Stefanyshyn-Piper. Clockwise from her position are Shane Kimbrough, Steve Bowen, Eric Boe, Chris Ferguson and Michael Fincke, along with Yury Lonchakov, and Sandra Magnus and Gregory Chamitoff. ▼

Pictured from left to right, Oleg Kononenko, Yury Lonchakov, Sergei Volkov, Michael Fincke, Richard Garriott; and Greg Chamitoff take a moment for a group photo in the Harmony node of the International Space Station as final preparations are being made for the departure of Volkov, Kononenko and Garriott in the Soyuz TMA-12 spacecraft. ▶

The STS-117 crewmembers give the Expedition 15 crewmembers packages from Earth in the Destiny laboratory of the International Space Station shortly after Space Shuttle Atlantis docked with the station. Pictured are (left to right) Oleg V. Kotov, Fyodor N. Yurchikhin, Sunita L. Williams, Steven Swanson, Lee Archambault and Rick Sturckow.

On ISS Expedition 8 in 2003, Mike Foale and Alexander Kaleri had few interruptions from visiting crews, chiefly because the Shuttle fleet had been grounded. But both took the arrival of their visitors very seriously, as Foale explained:

> *"We prepared for the visiting crews and did as much as we could to accommodate them. In a planning sense, it did significantly disrupt our days, but we went to great lengths to accommodate them and disrupt our days more to make sure that the visiting crew was satisfied."*

For ISS Expedition 3 crew member, Frank Culbertson, the biggest concern when a visiting Shuttle mission arrived was where everyone would sleep and how the logistics could be stretched to support the visiting crew. So, once the training is over and all that stands between you and your mission on the station is the eight minute ride to space and the two day chase to docking. But, there are things that can go wrong and most astronauts will tell you that the launch is probably the most dangerous part of the mission, but after all the slog of training and waiting for your chance, this is the glamorous part of the job!

Space shuttle Atlantis, docked to the International Space Station, is featured in this image photographed by an STS-132 spacewalker during the mission's first session of extravehicular activity (EVA).

A Soyuz spacecraft, which carried the Soyuz 5 taxi crew, is docked to the Pirs docking compartment on the International Space Station (ISS). The new Soyuz TMA-1 vehicle was designed to accommodate larger or smaller crewmembers, and is equipped with upgraded computers, a new cockpit control panel and improved avionics. The blackness of space and Earth's horizon provide the backdrop for the scene.

Mark Kelly (left) and Andrew Feustel are pictured on the aft flight deck of space shuttle Endeavour during rendezvous and docking operations with the International Space Station on flight day three.

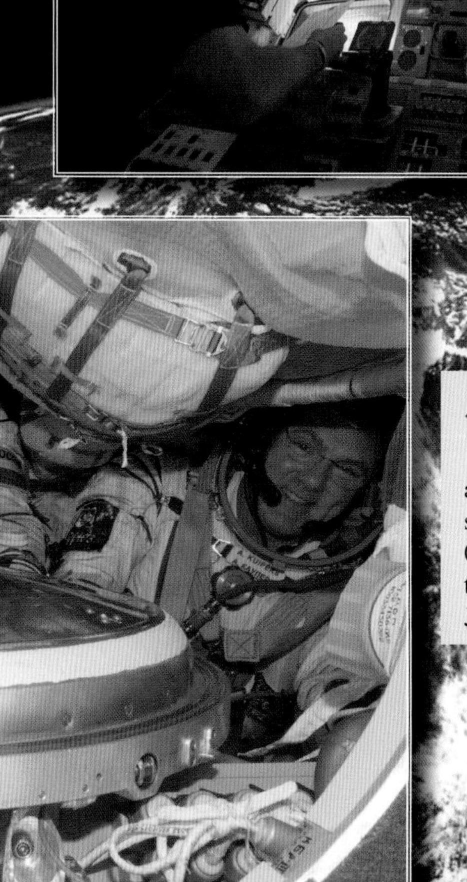

Oleg Kononenko (center), Don Pettit (left) and Andre Kuipers, attired in Russian Sokol launch and entry suits, conduct a standard suit leak check in the Soyuz TMA-03M spacecraft in preparation for their return to Earth scheduled for July 1, 2012.

Attired in their Russian Sokol launch and entry suits, Oleg Kononenko (center), Andre Kuipers (left) and Don Pettit pose for a photo in the Destiny laboratory of the International Space Station.

In November 2011, ISS Expedition 29/30 Flight Engineer Don Pettit was launched to the station aboard Soyuz TMA-22. He posted in his blog of 17 January 2012 about the journey, revealing some of the 'highlights' not normally included in formal press releases.

"For my Soyuz launch I had worn a standard Shuttle diaper with two inserts for extra absorption."

He had found from previous experience that it was advantageous to add a little extra protection, as in weightlessness, urine has a habit of getting into places where it shouldn't:

"We were in our spacesuits for over 12 hours and that's a long time. Even with the extra inserts, my diaper became completely overwhelmed. It leaked really bad; I could feel it happen, and was powerless to control the flood. When the time came to de-suit, I was more than ready to get out of the Sokol suit, but dreaded the impending mess. Fortunately, I was able to cover my stained underwear with a pair of woolen bib overalls."

Pettit's crew in the Soyuz were now starting the two day chase to docking with the space station. But in the confines of the Soyuz, while he may have been able to hide the stains, the aroma inside the spacecraft was far from sweet. Tactfully, Pettit did not detail this part of his journey. But it recalls the story of a Shuttle astronaut whose waste management system leaked during EVA, with the urine being soaked on his liquid coolant garment. Back inside the Shuttle, the smell was so bad that the rest of the crew threatened to put him and his suit back outside for the remainder of the mission!

"On docking day, we put on our Sokol suits again and strapped in about six hours before arriving at Space Station," Pettit continued. "By the time we docked I was tired, dehydrated, hungry, had to use the bathroom, and was still wearing my yellow stained long underwear. My sinuses were a bit congested, with the standard red puff chipmunk face. Our Soyuz cabin pressure

A view photographed from the International Space Station shows the Space Shuttle Atlantis backdropped over terrain as the two spacecraft were nearing their much-anticipated link-up in Earth orbit.

*was at 830 mm, but station is maintained at 740 mm. When we equalized the two, I got a splitting sinus headache. When we opened the hatch we were immediately on camera, downlinked live to the world as we were greeted by the smiling faces of our space station crew mates. All I wanted to do was have a good 'rest stop', get something to drink, and hide in my sleep station (in that order). We were pulled into the Service Module, where we were once again on camera with Russian Mission Control and my family, all anxious to chat. They wanted to know what it was like. I felt like a red-faced, dehydrated, puffy sack of **** (perhaps best left to the imagination). That is what it was really like; I was able to force a smile."*

Adapting from a Shuttle launch to life aboard the station was a rollercoaster ride, according to Expedition 15 crew member, Clayton Anderson:

"The last eleven days have seemed like a category 5 hurricane to me. Once Atlantis docked to ISS, we were greeted by a fire alarm, loss of attitude control and the failure of all 6 of the Russian command and control computers! And that was in the first two days! Now, with the station in a more normal configuration with all computers functioning and the STS-117 crew and Atlantis safely undocked and on their way back to Florida, I feel a sense of calm, as if I am in the 'eye' of the storm, provided by a couple of much needed days off. Yet, I am certain that the eye will pass over us and we will encounter the 'back side' of the storm in a day or two, as life on the ISS begins in earnest."

For first-time space flyer Greg Chamitoff on Expedition17, the ascent to orbit and his adjustment to space was great fun, as he recalled in his expedition journals:

"I can't believe I have been up here for 3 weeks already. It's been overwhelming in so many ways. The past three weeks have been a blur of new sensations, out of this world experiences and hard work. The Shuttle launch was amazing – smooth – it could have been a ride at an amusement park. For those of us on the mid-deck, without windows, it was just a little shaking and rumbling – surprisingly benign. Of course, the Gs picked up to 3 G. Having spent some time in the Russian centrifuge to practice emergency Soyuz descents, which can hit 8 G for up to 40 seconds, 3 G is easy, although the fact that it is sustained

▲ The STS-124 and Expedition 17 crewmembers pose for a group portrait in the Destiny laboratory of the International Space Station while Space Shuttle Discovery is docked with the station. From the left (front row) are Karen Nyberg, Akihiko Hoshide, Ron Garan, Mike Fossum, and Ken Ham. From the left (back row) are Oleg Kononenko, Greg Chamitoff, Garrett Reisman, Mark Kelly and Sergei Volkov.

for a few minutes makes it a little uncomfortable. It was a great ride up. I couldn't believe that a lifetime of wishing for this was suddenly happening all in that moment. It was an unforgettable 8½ minutes. Adapting to zero-g has been a lot of fun. The first sight of the station was so cool."

Each of the crew in space with Chamitoff, including seven from the Shuttle STS-124, and three on Expedition 17, had spent between 8 and 12 years in training. Only two of them were veterans of earlier missions. There was so much international training in the ISS program that they had met or worked together many times on the ground, so when they met in space, it seemed like a gathering of old friends, making the transition to the expedition crew much easier for Chamitoff. While the Shuttle was docked with station for 10 days, everything was just a blur of hard work. But eventually, Chamitoff settled into his planned six-month expedition, once the Shuttle had departed.

After a successful docking, it takes a couple of hours to check the rigidity and security of the seals and to equalize the pressure on both sides of the hatch to allow the hatches to be opened. The new crew members float aboard for a series of 'standard festivities' and television downlinks. Then all of the crew gathers to receive safety briefings from the current expedition crew commander.

A 'ships-bell' arrived on the station with Expedition 1 and US Naval Officer William Shepherd continued a naval tradition of formally signaling the arrival and departure of vehicles and arriving crew and commanders to the ISS. On ISS Expedition 5, prior to the arrival of the Space Shuttle STS-112, Peggy Whitson scrambled to try to get an answer to an important question… What is the appropriate bell-ringing etiquette for an arriving Shuttle? Whitson had rung the bell following instructions from naval officer Dan Bursch as the ISS Expedition 4 and Endeavour crew departed the station. But, not being a naval or military officer, Whitson could not remember the protocol for an arriving vehicle. She soon found out and, armed with the appropriate etiquette updated from

Brent W. Jett (left) and William "Bill" Shepherd ring the traditional Navy bell in Node 1 / Unity during the STS-97 crew ingress into the International Space Station (ISS).

the ground, Whitson rang the bell and called, "Atlantis arriving." She was very proud to get it right and it was only later, when the Shuttle crew came into station, that she found out they had heard neither the call nor the bell: "So much for trying to establish a tradition!" she recorded in her letters home. She made sure they heard the departing call and rings when the Shuttle left!

In 2011, Expedition 27 commander Dmitry Kondratyev explained the updated protocol for taking over the command of the station a decade after the first crews took up long term residence. Kondratyev wrote that, one day prior to the undocking of the Soyuz TMA craft, the official transfer of station command passes from the outgoing expedition commander to the next commander. NASA indicated in 2009 that the expeditions formally ended with the undocking of the returning Soyuz, but evidence in the daily station reports reveals that in practice each commander normally passes over command of the station the day prior to leaving the station, although this can vary slightly depending upon real-time situations. Similar to military-based change of command (and both the American and Russian military etiquette have much in common), the mandatory words are spoken in both Russian and English: "The command of the ISS is passed," and "command of the ISS is taken." This space-age version of the old tradition is itself becoming a tradition on the station, whether the participants are military or not. Hand over activities can take several days, depending on the crews and programs. The hand over period can be filled with briefings and press conferences, and special telecasts and radio briefings to Prime Ministers, Presidents and Royalty, as well as leaders of the various space agencies and, occasionally, celebrities. Soyuz seat liners are exchanged; this marks a critical point at which crewmembers are assigned to their new return vehicles. The return vehicle landing coordinates are updated. Handovers can be hectic, as Leroy Chiao, on ISS Expedition 10, explained: "The handover period is a very busy time, especially since we seem to go into shorter and shorter handovers. Handover between Expeditions 9 and 10 was about ten days. Chiao explained that, following the changeover with the outgoing Expedition 9 crew, they were given a few days of light duties to get used to living and working on the station:

"What the ground does is to try to give you a little bit more extra time for activities. Of course anything critical or time critical is still scheduled, but they do try to introduce other activities a little later on as they know you're going to get more experienced and get used to being on the station. It takes a little while, even after a week of hand overs, to remember where everything is stowed, how everything works and to use different sources, even though you have been trained on it. Actual life on board is a little bit different to the simulator. The crew on board figures out ways to do things that you can't really quite do in a simulator, because they are just so different. And so you get a few light days to get up to speed. The experience of the flight control teams shows here because they pretty much perfectly wrap the delays into our work. They had only about a week for the handover from Expedition 10 to 11, so we were already a little bit short on time. It is a busy time and we didn't really get any down time. We were busy telling the new ISS Expedition 11 crew, Sergei Krikalev and John Phillips, everything they needed to know, showing them things a new crew needed to know just as we were coming off our long duration flight. For the science, it depended on the experiment whether it was shut down during the hand over period of if the outgoing crew member could give tips first hand on the points and peculiarities of each experiment. The new crew needed to know where we had put everything, and how things worked on the station now, as opposed to when Krikalev had been on board previously on ISS Expedition 1, five years earlier. Even for experienced guys like Sergei, the hand over is extremely important. Most of it is just showing the crew where things are located. But we just did the daily things and they were free to change things and do things their way, just as we did."

Hand over days are long and tiring for all of the crew members. Although visiting crew members normally carry on with the individual science programs, independent of hand over activities, for incoming expedition crews, getting up to speed is the priority for their first few days prior to the departure of the out-going crew.

Mike Fincke, on ISS Expedition 9, was thankful for the handover period of 8 to 10 days in order to get his 'space-legs':

> ***"I felt I was starting to get my space legs after about 4 to 5 days in space, which was about 2-3 days on the station, to the point where I started to be really useful or helpful. The first couple of days I wasn't feeling completely myself. It takes about 5 days. The whole period of adaptation to being an astronaut, where you feel completely comfortable in the environment, takes close to a month."***

Crew members making a second or subsequent visit find it noticeable how things have changed since their last visit; often they note how different crews or crew members perform similar operations to those on their earlier visit. When visitors arrive, for the long-duration expedition crew it often seems to be a case of as soon as they are here, they are gone. The arrival of a new crew is an opportunity to update the station crew on news and stories from home, and there are jokes told, personal items delivered and moments of fun. This mixture of fun, work and conversation with new individuals can be a welcome break to the sometimes monotonous routine of life on the station. Of course, there are often consequences to entertaining 'house guests'. ISS Expedition 14 crew member Suni Williams wrote about the departure of the Shuttle crew who had delivered her to station and about adjusting to the long flight ahead:

> ***"It's been about 10 days since the Space Shuttle Discovery left us. They managed to leave quite a mess. After all the house cleaning, we are ready to start the New Year by celebrating."***

Schedule

Once a crew is aboard station and has adapted, a schedule has to be followed to achieve objectives. On early space flights, particularly US flights, the astronaut's schedules were timed to the second and front-loaded with primary objectives for maximum return in the event of any kind of abort. Space Shuttle missions were tightly scripted with a large supporting team on the ground a team, usually of seven astronauts on-board. Space station missions are less tightly scripted, usually with a smaller supporting team. Plus, the astronauts spend a lot of time maintaining the station. On the Shuttle, it came home and was fixed in the hangar.

A real difference between being a crew member on the Shuttle and being a crew member on the station is that you really have to learn everything for station. It's not just a scientific program; it's the robotics; it's the EVA tasks, and much more. All the things that an entire Shuttle crew has to learn, each individual has to know for the station, particularly if you're the only American aboard and have the primary responsibility for the US segment.

The Soviet/Russian long duration station experience did not follow this protocol. Their approach to the work day was more relaxed, in part because the missions were so much longer and had more time to complete objectives. This approach was not adopted by the Americans for the Skylab program, but it was something NASA learned during Shuttle-Mir and have adapted successfully on ISS. The difference between a Shuttle flight and a station flight became evident when the Shuttle docked to the station. For veteran Shuttle astronauts who had gone through the training several times, there was a clear difference, as Steve Smith explained:

Michael J. Bloomfield (foreground), Stephen N. Frick (bottom right), Steven L. Smith (top left), Ellen Ochoa, and Lee M. E. Morin (bottom) gather for an informal photo in the Destiny laboratory on the International Space Station (ISS).

Informal pose of three crew members of the SMEAT, seen inside test chamber prior to start of isolation test. Robert Crippen (left), William Thornton, and Karol Bobko

"On a Shuttle flight, in training, you are all going to be together because you are all going to be in space for 10-12 days together. On a station flight, that's not true. You've got a bunch of people that are only going to be in space about 10-12 days and they are all going (it seems) 1000 miles per hour doing their tasks. Then you get the 2 or 3 in space on the station. They're at a much more leisurely pace, because they know that they are going to be there for a long time. They know they cannot work as hard as you do on a Shuttle flight for six months. You really need to pace yourself." Smith also explained that, in training for a non-docking Shuttle flight, most, if not all of the equipment you will work with is on the ground and all the people you are flying with are attending the same meetings or training sessions, as a team. This is not true on Shuttle flights to the station, because some of the tools you would need to use were already up there, as were some of the people you were to work with. Although the Shuttle crews were instructed on certain tasks involving tools, the actual tools were not available to practice with or become familiar with because they were already in space. So it was all a bit of an awkward moment when we first opened the hatches on 110. The seven of us were ready to go, but Carl Walz, Dan Bursch and Yuri Onufriyenko, on station, had a more slow and steady pace, because they had to work seven days a week for six months."

Astronaut Robert Crippen, in training for many years for the US Air Force Manned Orbiting Laboratory, subsequently a support crew member for Skylab, and later the first pilot of a Space Shuttle, spoke about the compatibility of crew members on short versus long duration missions.

"Missions on the order of a week…probably anybody can put up with anybody for a week. However, you have to put up with everybody for a year, training for that, is more stressful. Now with the International Space Station, where they're up there for much longer, hopefully when we'll have people on the Moon and maybe eventually when they're going to Mars, they're going to be together a lot longer, so compatibility is something that is extremely important. The Navy pays attention to it in things like submarines and other vehicles where you've got a close confinement."

There is not really a 'typical day' on the station. Indeed, each 24 hour period is governed by the dynamics of its orbit. Perhaps the most stressful part of any day is to try to keep to the schedule, particularly for essentials or tasks that need to be completed at a certain time. Having to spend time finding parts, or searching for something which is not where it should be can make it easy to fall behind. It always feels better to be ahead of schedule, and even though the crew can incorporate catch up time into a flexible working list at the start of an increment, there is still pressure to keep to the plan. Achieving the mission's plan is most rewarding, while falling behind can be most frustrating to any crew member. It is usually busy on the station, if for no other reason than to get the maximum return for the cost of each mission. During ISS Expedition 5, Peggy Whitson wrote of the day-to-day operations aboard station towards the end of her first mission in 2002:

"Most days are busy, a combination of the routine things like rebooting laptops, reading daily summaries and messages from the ground, maintaining inventory, and exercise, with a good measure of the "fun" stuff like payloads, robotics ops, EVA prep and occasional R&R (removing and replacing hardware components, not rest and relaxation). My time here has been eventful, but for me not particularly stressful. I enjoy the challenge of learning new things, and being as efficient as possible in my work and I also get a lot of satisfaction out of knowing that I am helping the scientists on the ground with their data collection and research. I think having sat in the investigator seat on the ground makes it easier for me to understand their perspectives, worries and concerns, and try to address them... And of course this 'routine' happens in the novel environment of space."

There are specific tasks or experiments that have to be observed at specific times on most expeditions and the flight plan is color coded to alert the crew to these events on the required schedule. But a day on station is not micromanaged by the ground and there is flexibility in the crew's time on the station. There is always a list of things that 'must' be completed at a particular time, but other items can be moved around in the day, or moved to later in the mission. Mission Control is responsible for organizing the scheduled tasks in order of priority and ensuring that the crew has enough time to complete the program. The crew's task is to complete the schedule and they often make suggestions to optimize things, working with the ground crew in order to improve the schedule as the mission progresses and to provide guidelines for repetition of these tasks by subsequent crews. Lessons about how to live and work effectively in space, and in conjunction with the ground teams, are constantly being learned. Just as valid are the occasions when everyone learns how not to complete an operation, or when a procedure does not go the way it was planned on the ground before the flight.

Peggy Whitson explained that the ground is responsible for prioritizing the crew's objectives, balancing activities to objectives, science with maintenance and housekeeping:

> ***"They have the big picture in mind. Each week, they go and figure out that we are going to do this much maintenance this week, this much science, this much of whatever other activities. They have a big picture plan in mind, which might mean we might do more science during the next three weeks than we will later, because we are going to be really busy prepping for the EVA for instance. So they do that planning on the ground. The crew members can actually implant some things, for instance requesting things to be added to the task list items, and then on their free time they can pursue those things that they feel important as well."***

In her Expedition 18 journal, Sandy Magnus commented on the role the ground team plays in organizing the daily schedule on the station.

> ***"This international team comes together about a year in advance of the increment crew arriving on station. Each partner country or agency has a team of long term planners and together they map out how to fit all the proposed experiments and operations in to the expedition and in which priority they need to be addressed."***

Magnus pointed out that these priorities not only include physically operating the equipment or experiment, but can also include installing new items of hardware or software, maintenance, EVA, robotics and system work.

> ***"The planning needs to ensure that it all synchronizes efficiently, safely and logically, with little interference with other operations. The flight crew and ground support team has to work efficiently and effectively at the same time, coordinating activities on the ground and in space."***

Training and briefing is an integral element of each expedition for the flight, ground and science teams. Before the launch, the flight crew and their ground support has to have a good picture of the overall program of activities and objectives and what is expected each week in order to achieve the mission's objectives by the end of the expedition. John Phillips recorded his working day on the station as part of a 'caretaker' crew of just two during ISS Expedition 11. Operating on GMT, the crew awoke around 06.30 and then spent the next 45 minutes preparing for the day's work schedule. Phillips and his Commander, Sergei Krikalev, had breakfast together, followed by a 'Daily Planning Conference' during which the crew spoke with the Mission Control Centers in Houston, Huntsville and Moscow. These conferences allowed the crew to talk about the planned day and any last minute changes since the previous evening. Control centers in Europe and Japan were later added to the conference when the ESA and Japanese laboratories came on-line.

Lunch is usually taken at about 1:00 PM, depending on the schedule or task in hand. Many crews like to eat lunch together although some crew members prefer to snack alone or try to complete their morning tasks, only taking a short midday break. Then it's back to work in the 'afternoon' until about 5:30 to 6:00 PM, though it is not unusual for a crew to elect to work until 7:00 PM, when a second 15-minute daily planning conference is held. The evening meal is usually taken after the conference. Peggy Whitson said that because she was flying with two Russians on her first mission, her social time with both of them was chiefly at meal times:

> ***"They spent most of their time working over in the Russian segment while I was in the US segment pretty much on my own. It was no problem really; I had the ground to talk to all day as and when I wanted. The meals were our social time; typically in the evening we would spent an hour or two just chatting after we ate."***

Pre-sleep work, such as preparation for the following day's activities, or calling home and answering emails usually starts around 8:30 PM. Bedtime officially begins at 9:30, although no one usually turns in that early. On ISS Expedition 11, John Phillips tried to read for the 30 minutes prior to turning in for sleep at about 10:30 PM each evening. If crew members have nothing else to occupy them, such as emails, photos, or report writing, many are happy to just watch the Earth passing by below them through the windows. Slotted into this working day the crews also find time for exercise periods. As the crew size increases, this schedule essentially remained the same, although the scheduling crew members was more flexible. As well as following the mission schedule and their own individual flight plan, each crew member invariably adopts their own way of doing things on the station. This is a natural human adaptation to what is effectively both a home and workplace. ISS Expedition 17 crew member Greg Chamitoff summed up the adjustment to working on the station in his inflight journals:

"I was flying back and forth through my home up here, putting things away, reorganizing things; feeling good about places I cleaned up and new places I've designated for various things. As I swung from hand-hold to hand-hold, the feeling that suddenly hit me was 'I'm home'. Not, of course, my real home with my family, but it was a strange sense of fully owning this space and being completely comfortable here."

Chamitoff thought that this was probably due in part to the fact that over the previous week, his work on the station had involved preparing both the Progress resupply craft and the European ATV vehicle for departure, filling them with all the trash and unwanted items no longer required on station. He realized that his sense of home was perhaps down to being able to put his hands on almost everything aboard the station, either to work on it, stow it, remove it or relocate it. He wrote:

"Life is good in space. The days are long and a lot of work. Sometimes the works feels productive scientifically or technologically and sometimes it's just manual labor."

Of course, things can and do go wrong during the day, but it is important to realize that such happenings can also work out for the best. It can provide valuable experience and data, not only for future ISS operations, but also for far reaching programs in the decades to come. As each crew is in orbit for a long-time, they are expected to be able to address the many problems they are faced with in real-time, using their own experience and that of previous crews, the on board tools and spares and the redundancy set into the system to achieve their objectives. It might not be exactly with the hardware or within the time line originally envisaged, but at least it has been achieved. ISS Expedition 9 flight engineer Mike Fincke concurred with this:

"That's exactly it. And it's perfect to do it now, when we are only 240 miles above the planet, so we get the really good experience for when we build bases on the moon and we are 240,000 miles away. We will have a lot of these troubles to figure out, but we will be that much more flexible."

When the ISS Expedition 5 crew experienced a two-day break in communications with mission control in Houston because of hurricane threats, work on the US segment had to be adjusted. Whitson recalled:

"We did OK. I actually ended up doing different kinds of activities during those two days that didn't require Ku band capabilities, which most of the US science activities require. They rescheduled my activities to be consistent with the fact that we didn't have that much comm. If I had been doing something more complicated, such as working with the robotic arm it would have been a lot more challenging."

ISS Expedition 6 Commander Ken Bowersox recalled how the time line was planned for his 2002 to 2003 stay on the station and what has changed since:

"The ground is responsible for planning all the activities that need to be done, at least on the scheduled flight plan. Now (2006) there are regular meetings with the crew to discuss what is to be done and the crew can make suggestions about timing or priorities. But then there is unscheduled time when the crew also have a say up on orbit. There are things we call task list activities for the crew to do, other activities that the ground team would like to see done, and that the crew are made aware of."

As station commander, Bowersox balanced the challenge of conducting science, maintenance and housekeeping along with any other operational requirements, including space walks, dockings, and outreach activities:

"The key is that you have to have a good operating station in order to get your science done, so the balance comes naturally. You know immediately, when high priority tasks need to be taken care of. For example, when the bathroom (toilet) requires servicing, everything else stops and you fix the bathroom on board station. So when the tanks are full they have to be fixed and that doesn't show up on the schedule anywhere. You just know that it has to be done. The other priorities are generally taken care of during communication with the ground, building them into the daily ops plan."

Weekends

The schedule for the expedition crew is normally much lighter on weekends. Some crews scheduled a 'movie night' on Friday nights and occasionally on Saturday night as well if they were lucky. Several hours on Saturdays are filled with routine clean-up activities, maintenance and scheduled conferences with the ground. Sunday is normally a free day, although special events can alter the schedule so that the light duty day may be a Friday, the rest day is Saturday and there is a full working day on Sunday. When there's time, Sundays can also be a day to complete a list of catch-up work from the previous week's efforts or to perform get ahead tasks for forthcoming activities. Often, the day is used just to catch up on the emails, read or listen to music. Routine maintenance is usually coupled with report writing or sending messages from the station to post on the NASA web site. Of course, even this 'schedule' doesn't always go according to plan, as Expedition 11 crew member John Phillips wrote on 24 July 2005:

"For some reason I don't understand, I always find myself falling behind my own work plan on Saturday and Sunday, despite the fact that there's not much on it!"

Though Sunday is officially designated a 'day off', the expedition crew is never really 'off duty' on the station. There is always something to do, things that need attending to, unexpected situations, or an observation that needs investigating. There is more time to catch up with things and of course simply being where they are in orbit is often sufficient reason for crew members to conduct their own investigations, experiments or observations outside of the main flight plan, because their inquisitive nature inspires them to do so. As Expedition 17 crew member, Greg Chamitoff wrote:

"Today was a day off, and we seemed to spend much of it chasing icebergs. We had numerous orbits today with perfect lighting conditions as we passed Cape Horn (South America) and got far enough south from there to see the shoreline of Antarctica. The thrill was the discovery of unusual patterns of ice that we've never seen or heard of before. Perhaps the NASA Earth observation folks will be able to explain it to me later."

Holidays and Celebrations

Flying aboard the ISS during the turn of the year offers a distinctive view of the seasonal celebrations we observe here on Earth, During Expedition 12 in 2005, American Bill McArthur and Russian Valery Tokarev had a lot of celebrations. Apart from Thanksgiving, Christmas, New Year and Russian Orthodox Christmas. McArthur also explained that they celebrated the New Year, but not sixteen times to coincide with the day's sixteen orbits:

"Valery and I, not being youngsters and true party animals, celebrated New Year's Moscow Time and shortly after that, we went to bed."

During Expedition 14, Suni Williams wrote in her mission log that the crew dined on turkey, Russian mashed potatoes and onions, dried cranberries, green beans and mushrooms on Christmas Eve. The Christmas day feast included festively decorated Russian sugar cookies. Williams also commented on the celebrations for a Russian New Year celebration:

"New Year is the really big holiday in Russia, then comes Christmas. The Russian Christmas is on January 7th, as per the Russian Orthodox church calendar, so their holiday season has just gotten under way. We are hoping to watch a classic Russian New Year movie this evening, called 'Irony of Fate."

After a three-day weekend for New Year 2007, Williams noted that there was going to be a second one for Russian Christmas just a week later:

"It's finally back to work for everyone after the holidays. Well, not necessary for us all. We actually have another three-day weekend this weekend. We are trying not to get used to them, because they will not be happening here again for quite some time. One of the interesting things is that we get to 'pick' our holidays. We are allotted a handful of days off for holidays throughout the year, so the crew, along with the ground teams, picks which holidays would be the best to take off – dictated by operational constraints etc. Remember, we have our international partners here as well, so not all the holidays are the ones observed in the US. For example, we have January 7th off for Russian Christmas but don't get the day off for Martin Luther King Day. I voted for taking all the holidays off, but that didn't go over too well…"

The New Year's 2011 celebration had settled into the tradition many of us know down on Earth. Expedition 26/27 cosmonaut Dmitri Kondratyev explained in his January 2011 blog from the station:

The Expedition 16 crewmembers pose for a Christmas photo in the Zvezda Service Module of the International Space Station. From the left are Yuri I. Malenchenko, Peggy A. Whitson and Daniel Tani.

"All the main traditions of the New Year at the station had been met, except of course, champagne. In the supplies sent up from Earth there was a container marked 'New Year diet included' "which we opened on December 31. There was a lot of tasty treats for New Year's table. We got some stocks that are set aside for the occasion. Opening gifts from relatives and friends has become one of the most interesting events of the holiday. The holiday on station has become like home. Over many years, the New Year celebrations at the station have accumulated a number of related items and all sorts of Christmas decorations. We found a Christmas tree which, although small, helped to decorate the interior of the station. All the crew celebrated the New Year in GMT, but different countries also celebrated New Year in their own time zone. We noted New Year, Moscow Time. Paolo Nespoli, from Italy, congratulated friends and relatives in Europe time and the American crew members marked New Year with controllers at Mission Control JSC in Central (Houston) Time. The main event was a show of the traditional Russian film 'Irony of Fate'. Our (non-Russian) colleagues found this movie too much, especially with our translations and commentary. Our colleague's joke watching the movie was that it was a good lesson in the Russian language."

Other celebrations include family birthdays and anniversaries, not just of a crew member's own immediate family, but also of the extended family of their colleagues. When a member of the family is not just a distance away, but not even on the planet, celebrations take on a whole new meaning, both for the family on the ground and especially for the crew flying high above them in orbit. Bill McArthur's youngest daughter, son-in-law, and his oldest daughter's husband all celebrated birthdays while he was in orbit. Their celebrations were a lot of fun, because there was so much support from the ground. McArthur observed:

"It was nice, because our workload was significantly reduced right near the middle of the mission, which was a very nice break and a chance for us to get a little rest. I had video conferences with my family associated with all those events. Being in orbit added a little bit of spice to the celebrations, plus I realize how much more I am going to enjoy celebrating them this year because I will appreciate being with my family all the more than I would have before."

Birthdays are observed on the space station. But, a crew member often has to work on their birthday, even if it falls on a weekend. During Expedition 24, two American women were part of the crew: Tracy Caldwell-Dyson and Shannon Walker. On 14 August 2010 Caldwell-Dyson celebrated her birthday, but it fell on a Saturday; it was not a day off, so she spent the day cleaning the station. At least her colleagues shared a great evening meal with her. Some strange traditions are observed in space – such as the Russian tradition of pulling the ears of whoever is celebrating their birthday. On ISS this means being 'pulled' by the ears through the station.

One regular 'celebration' is observed on 12 April each year on the station. This is Cosmonautics Day in Russia. It was the date Yuri Gagarin became the first person to fly into space, in 1961. It is also the date of the first Space Shuttle launch, Columbia on STS-1 in 1981, with John Young and Bob Crippen aboard. In 2011, the celebrations included the 50th anniversary of the Gagarin flight, the 40th anniversary of the first space station, Salyut, the 30th anniversary of the first Shuttle flight, and the tenth anniversary of the exchange of the first and second ISS resident crews. It is often during these relaxed days that the realization of their isolation away from Earth can become more apparent to the astronauts, as Greg Chamitoff observed in his journal:

"While we were enjoying our day off, we've been acutely aware of events unfolding in Texas, especially Houston, as Hurricane Ike blasts through there today. This hurricane has been so huge that it has been difficult to photograph from here. It was interesting, but my first sense of serious isolation up here gradually crept up on me. The first thing to go was the uplink of news broadcasts, newspapers and radio broadcasts. Usually, we get a steady stream of them from a limited number of sources of daily information about what's going on down there. Once Mission Control in Houston decided that they needed to shift operations to a back-up Control Center, the flow of information to us stopped. Though some updates are issued by the back up Control Center, this is limited to only a few items per day, about the storm status. Then all the email syncs were canceled, as the back-up Control Center could no longer support email uplink or downlink. We were essentially cut off, except for the comm link with Mission Control. Of course, we're completely spoiled up here with such modern capabilities and defaulting to comm. only with Mission Control (via TDRSS satellites) is still way better than in the 'old days of space flight'. Nevertheless, it made me realize how dependent we were on the continuous flow of information, audio and video, to enable us to do our work and feel 'connected' to everyone and everything on the ground. I could suddenly imagine what it would be like to lose comm. for an extended period of time due to some system failure."

Time Off

Once the assembly of the ISS was largely completed and the flow of logistics allowed the crew size to increase to six, operations were refocused towards science during the week, with the crews using weekends for recreation, voluntary extra science work or for their own experiments. Expedition 18 crew member Sandy Magnus, for instance, devoted Saturday to cooking. Sundays included a couple of hours of exercise but are mainly private days; often there are private family video conferences. It is also a great day for taking extra photos and completing things that the crew member wants to do before coming home.

Expedition 12 commander Bill McArthur found that running the station as Commander of a two-person caretaker crew in the wake of the Columbia accident, did not seriously hamper his science objectives and activities aboard the station during his 2005 to 2006 mission:

The Expedition Two crewmembers -- Susan J. Helms (left), Yury V. Usachev and James S. Voss -- share a meal at the table in the Zvezda / Service Module of the International Space Station (ISS).

Mikhail Tyurin plays a guitar among stowage bags in the hatch area of the Quest Airlock.

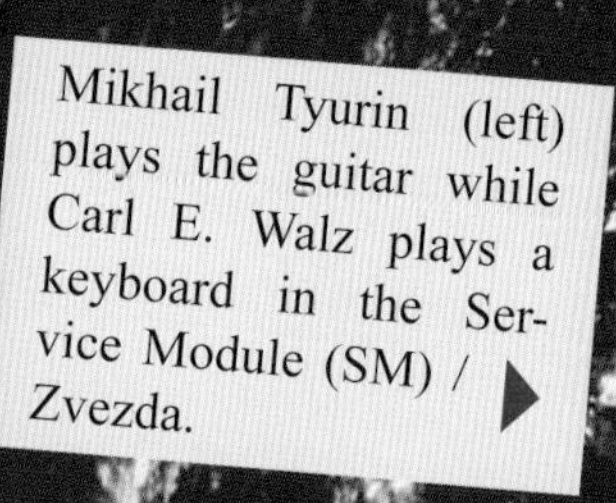

Mikhail Tyurin (left) plays the guitar while Carl E. Walz plays a keyboard in the Service Module (SM) / Zvezda.

Jim Voss is photographed holding a keyboard in the Zvezda Service module of the International Space Station (ISS). He is wearing a brightly colored Hawaiian shirt.

Daniel Bursch smiles for the camera from the International Space Station (ISS) Destiny U.S. Laboratory with a candy cane and a toy floating in front of him.

Peggy A. Whitson (right) and Sergei Y. Treschev share a meal in the Zvezda Service Module on the International Space Station (ISS).

"The challenge is pretty well managed by time lines from the ground. One of the things I did was to try to make myself as available as possible, so that on the weekends there was time available and I could go ahead and do some additional science work. Early on, the folks on the ground asked me how I wanted to spend my free time. I told them that I define free time as those opportunities in the schedule that I got to pick the work that I was doing, instead of having to get the ground to do it for me."

In the early years of space stations, TV was broadcast up to the stations. American football and baseball were followed, as were Russian soccer and ice hockey. Other interests included the Olympics and the soccer World Cup. More recently, on ISS, it has been possible to view such events real-time or recorded via the internet rather than on traditional broadcast television. On Expedition 3, Frank Culbertson's Personal Preference Kit (PPK) included his trumpet. It caused a little problem for him, as crew mate Mikhail Turin joked:

"Frank liked to play his trumpet, but Vladimir Dezhurov and I suggested that he would be better playing it in the airlock where we could close the hatch." Culbertson later explained the incident in more depth: "I'd go into the airlock and shut the door. I did not latch it, but just shut it to keep the trumpet from bothering the rest of the crew. When I'd finished and I was going to open the hatch to go back into the main part of the station, the hatch wouldn't open. I thought my crewmates had locked me in but actually, it was the pressure differential in the ventilation system and it just required a little extra heave to open up."

Tyurin joked that Culbertson was never quite sure whether his colleagues had changed the pressure deliberately just to isolate his trumpet playing for a little longer! Culbertson also spoke about the guitar aboard the station, and that the crew played or listened to music whenever they could. There is also quite a library aboard the station, but it was difficult to find the time to do a lot of reading. Part way though Expedition 3, Frank Culbertson started to receive crossword puzzles from mission control in Houston, scanned from US newspapers. He would often do a crossword each evening before he went to sleep.

Astronaut Background and Training

Culbertson said that his previous experience as the manager of the Shuttle-Mir program helped him to work with the Russians for his ISS mission.

"I felt as though I lived on Mir for 30 months, although it was other people who were actually up there. I was very much involved with them, both during their training and their stay on board, trying to work through the issues that were affecting them. So I was very much aware of what they went through and I knew what to expect to a great extent when I got to ISS. I had a good working relationship with the Russians and understood the language, so I felt prepared to deal with Moscow and all aspects of the Russian program."

Culbertson had flown two short missions on the Space Shuttle earlier in his career. He found the hardest part of preparing for his long duration space station mission was the separation from his family, which he compared to a naval deployment. For Shuttle, although many hours were spent away from the family preparing for flight, most training was done in Houston at the Johnson Space Center near where the astronauts lived. On occasions when astronauts went across country they would often fly their own supersonic jets so were never more than a few hours away. Shuttle training was never far from home. ISS training was different. It included trips to Russia, often for several weeks away from home. Culbertson was an experienced pilot but for spaceflight he tried to understand the science program. From his position as manager of the Shuttle-Mir, he felt he had great familiarity with the science. Bill McArthur said that a broad range of life skills, training and experience helps a crew member during a long flight on the station. He explained how his background in education and his military service (which focused on engineering) helped him to understand his role in commanding and managing a mission to the station:

"Above all, the ISS is a very complex and technical vehicle, and I think that having an engineering and aviation background gave me a familiarity with working on complex systems. I think my whole career was a nice progression, leading up to spending time on ISS. All my Shuttle missions had elements which were, I think, good preparations for being on the station. STS-58 was a Spacelab mission and so that exposed me to conducting research in orbit. STS-74 involved rendezvous and docking (with Mir), robotic arm work and at least preparations for space walks. STS-92 was another rendezvous and docking with ISS, robotics and EVA, plus actual familiarity with the station itself, both interior and exterior. So I think that was all a nice build up to a good ISS expedition. My comfort level of working with complex systems even before the Shuttle missions allowed me to deal with working in an environment in which you really want to avoid mistakes, but at the same time, you do want to bond with your flying machine as well."

Ken Bowersox said that his previous experience in the military and on previous scientific Shuttle missions all contributed to his approach to commanding a long duration station mission.

"I think they're all beneficial. A lab runs different to a battleship, and a space station runs different to a lab, but the key is to be able to integrate all of the things that you have in your background in order to provide an optimal environment in the

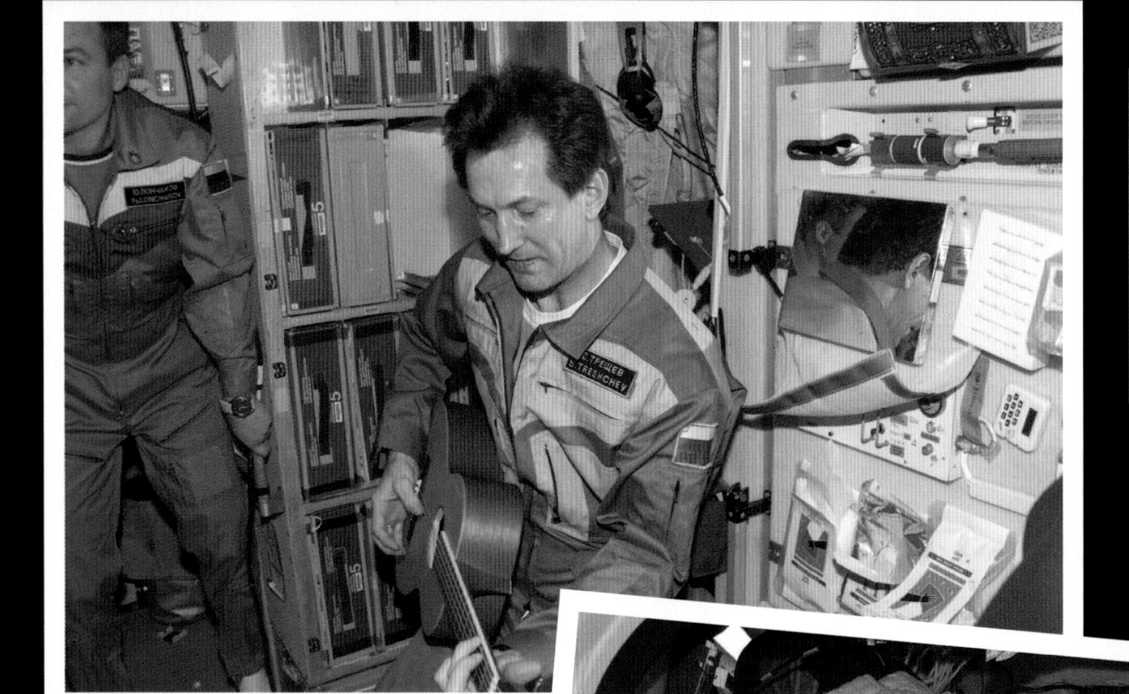

Sergei Y. Treschev, plays host to some "visitors" to the International Space Station (ISS) as he performs on a guitar in the Zvezda Service Module on the International Space Station (ISS). Sergei Zalyotin is partially out of frame to the left of Treschev

Kenneth D. Bowersox (right) and Donald R. Pettit are pictured in the Quest Airlock on the International Space Station (ISS). Bowersox and Pettit are wearing their Extravehicular Mobility Unit (EMU) spacesuits

Umberto Guidoni floats toward the galley onboard the Space Shuttle Endeavour as the spacecraft heads toward a rendezvous with the International Space Station (ISS).

"Of course it is possible to correspond by e-mail, but it doesn't replace ordinary letters. Letters are read in a special way. They don't get opened immediately - you hide it, put it off. And then at the end of the day read it alone, then re-read it. On Mir it was more convenient, the modules were arranged like the petals of a flower, you could fly to the far end, and no one would disturb you, the ISS is a corridor - you can't just retire. Letters that come there are interesting not only in content. After all, your loved ones touched that piece of paper." - Yuri Baturin

circumstances that you are given. So, all three of us on Expedition 5 during 2002 and 2003 worked together to establish the best environment that we could set up on station to help accomplish our goal. I think that is something you learn in a military background. You need to set up the environment that works for you, not necessarily the one that is exactly in the rule book."

When taking over from the Expedition 4 crew in 2002, Peggy Whitson, who flew Expedition 5 and later commanded Expedition 16, thought that the combination of her lack of previous flight experience and a strong desire to ensure that goals were achieved as early as possible contributed to her experiencing "a rapid learning curve" during the changeover before the out-going crew departed:

"Probably I had some insecurities about that. I think that I was very concerned whether or not I had the right training, or enough training of different types, because I didn't know what I would need to know until I got there. Of course I worried way too much, but I was fine."

Whitson thought that a shorter visiting mission may have aided her self-confidence and comfort level about adapting to living and working in space before embarking on the longer mission: "The first time, I think I was just worrying about it. It was just me." She completed a NEEMO underwater mission after her first mission to ISS, which she thought was useful help for unflown members of her dive team:

"I think it's really a good idea to have one flown guy there who can say 'this is just like spaceflight' or 'no, this is absolutely nothing like spaceflight, don't worry about this'."

Whitson also thought that her long association with working with the Russians, starting in 1989, was very useful although one of the biggest hurdles in her training for the first mission was mastering the Russian language. The travel was another such challenge:

"Traveling back and forth is very tiring, fatiguing. You end up spending approximately 4-5 weeks in either country for station training flows. It is pretty challenging."

During the post-Columbia period of "caretaker" two-man crews were assigned lighter duties, tasked with maintaining the critical systems to keep the station functioning, plus housekeeping and managing consumables. The science program was much reduced. The Russians had considerable experience in maintaining a research station with only two cosmonauts, but the Salyut, and even the larger Mir, were considerably different from the complex ISS.

The need to reassign crews in 2003 from three people to the two-person caretaker crew was a challenge. Mike Foale was in training with Valery Tokarev and Bill McArthur when he was reassigned with Alexander Kaleri for ISS-8. The two men's previous experience on Mir helped bond the team even closer together, as Foale explained:

"We both realized what we were getting into. I was fortunate that Kaleri was an experienced Russian crew member and I think he felt he was lucky to have me (based on Foale's Mir experience, which included dealing with the aftermath of the collision with a Progress)."

Foale knew that Kaleri had worked in a two-person crew on Mir, while he had not, so he wanted to know up front what the differences would be in flying as a two-person, rather than a three-person crew – the pitfalls as well as the benefits. They agreed to bring any problems out in the open for discussion and as a result, Foale commented,

"In space we were very, very well fitted together and we never had a cross word. Losing Bill and Valery was a big deal. I had to deal with adjusting my psychology, my approach to the crew, and to the whole mission, which had completely changed from its original plan."

Foale, as station Commander, had originally assigned McArthur to be the American expert on the station and Tokarev to be the Russian one, while he would float between them and deal with the management requirements and control centers. But the reduction to a two-person crew meant that Foale now had to take on the role of American specialist aboard ISS as well as that of Commander. Now, his challenge to achieve all he had hoped for and deliver a successful residency had increased significantly, despite the reduction in mission objectives as a care-taker crew.

Ed Lu, who served as a crew member on Expedition 7, felt that his previous flight to the embryonic station on STS-106 in 2000, with his ISS commander Yuri Malenchenko, helped with training for the long ISS flight. Their ability to work together successfully was proven on the earlier Shuttle flight and made their time on station that much easier three years later. Another benefit of the STS-106 flight was that the crew fitted out the Zvezda Service Module shortly after it arrived in orbit:

"Zvezda is the main living compartment of the station, so we learned more about the station's systems than any other Shuttle crew before or after. In that sense, it was quite helpful for ISS expedition training, because much of ISS training was a review of our earlier Shuttle training."

The Russians have long stated that a cosmonaut cannot truly call themself a cosmonaut until they have made a space flight. For rookies whose first flight happens to be a long duration mission, the space station expedition is often seen as a tougher initiation. On ISS, a first flight as an expedition crew member can happen to astronauts from any of the partner nations. The real learning curve for living and working in space comes from doing precisely that, actually in orbit.

For Mike Fincke, who's first mission was ISS Expedition 9, the original plan was for him to fly a Shuttle mission before flying a long duration station expedition, but program scheduling in the wake of the Columbia accident changed this. So Fincke's first flight was a long duration ISS expedition:

> ***"We do get adequate training. In fact, it's pretty darn good training. I enjoyed my time aboard space station and even though I had not flown in space before, it was quite a good mission."***

The scale and intensity of pre-flight training for ISS is designed around the prior experience of each crew member and of the total expedition crew, as well as the scope and depth of the planned mission objectives. Whatever the mission, however, the very nature of ISS long duration training, with its international flavor and ever increasing range of scientific experiments and technical activities, means that no crew can possibly maintain peak proficiency across every field and discipline. There will inevitably be some on orbit training to brush up on some skills and procedures, especially in some of the less frequent operations or activities. Some of the activities in orbit may not even be started for months after launch, so there be a considerable time between training and operations. But this is not as overwhelming as it may seem, as ISS-8 Commander Mike Foale explained:

> ***"Inflight refresher training on the station is really not a big deal. It's 'maintaining your consistency' training. It's like flying an aircraft regularly to make sure you make good landings. We operate systems every so many weeks to remember all of the key factors in doing so."***

ISS Crew Assignments

The role of station Commander was initially intended to alternate between an American and a Russian. The ISS Commander has always been a veteran astronaut or cosmonaut. The Commander is responsible for managing day to day station activities. They coordinate the crew's objectives and, most importantly, are primarily responsible for the health and safety of the crew, to ensure that they meet expedition objectives within time lines and safety conditions. Station Commanders can also be Soyuz ferry vehicle trained. For her second residency, this time as Commander, US astronaut Peggy Whitson was presented with a pair of Kazakh riding whips by her Russian trainers, "to keep the men in line," on her crew. Clayton Anderson, who was already aboard ISS, was a little worried about this new 'management style' and was relieved to find she had left the whips on Earth. ISS-6 commander Ken Bowersox explained the role and responsibilities of an ISS commander in a 2006 interview:

> ***"The main job of an ISS commander is to help his crew to get the job done that the ground wants done. You are there as a facilitator, to provide the leadership communication with the ground and to make sure that the crew get what they need and the ground gets what they need."***

In 2010, Russian ISS-24 Commander Alexander Skvortsov gave his interpretation of the role of Expedition Commander on his blog:

> ***"The commander is responsible for the assignment and for the safety of the crew. Actually, this simple phase is completely determined by the terms of commanding duties to the ISS. It is impossible to compare the ISS commander and the commander of a military unit. Here (on station) we must take into account the specifics of our work in orbit. Each crew member has their own program of work, and, by and large, everyone knows what to do. In addition, as Commander, we are constantly in touch with the Earth (Mission Control), where the telemetry is gathered on the status of the station, and where experts analyze the data and make decisions. On Earth, experts are always on hand. The commander makes decisions in an emergency, in an acute shortage of time, when there is a serious threat to the crew or station."***

Skvortsov added that this was something he was glad he never had to put to the test. One or more crew members can be designated 'Flight Engineers' and they are chiefly responsible for science operations and maintenance on the station. On crews of six, the incoming crew all serve as 'Flight Engineers' until they take over the expedition. Flight Engineers can also serve as Soyuz Flight Engineers or Soyuz Commanders. Flight Engineer 1 is typically in charge of the Soyuz spacecraft systems and are usually either a Russian Air Force or civilian engineer cosmonaut.

Between 2005 and 2009, American astronauts journeyed to and from the station on Shuttle assembly missions, rotating the role of Flight Engineer-2. These astronauts were designated Shuttle Rotating Expedition Crew Members – or ShRECs. These crew members were assigned to a specific item of ISS hardware, to give them more familiarity with their installation and operations. For Peggy Whitson's ISS-16 command, she worked with Yuri Malenchenko as FE-1, visiting crew member Sheik Muszaphar Shuker from Malaysia (who returned with the ISS-15 crew members) and five ShRECs. Of these, Suni Williams, the first ShREC, was responsible for the P5 truss as her item of hardware, and Clayton Anderson focused on the S5 truss. Dan Tani focused on the P6

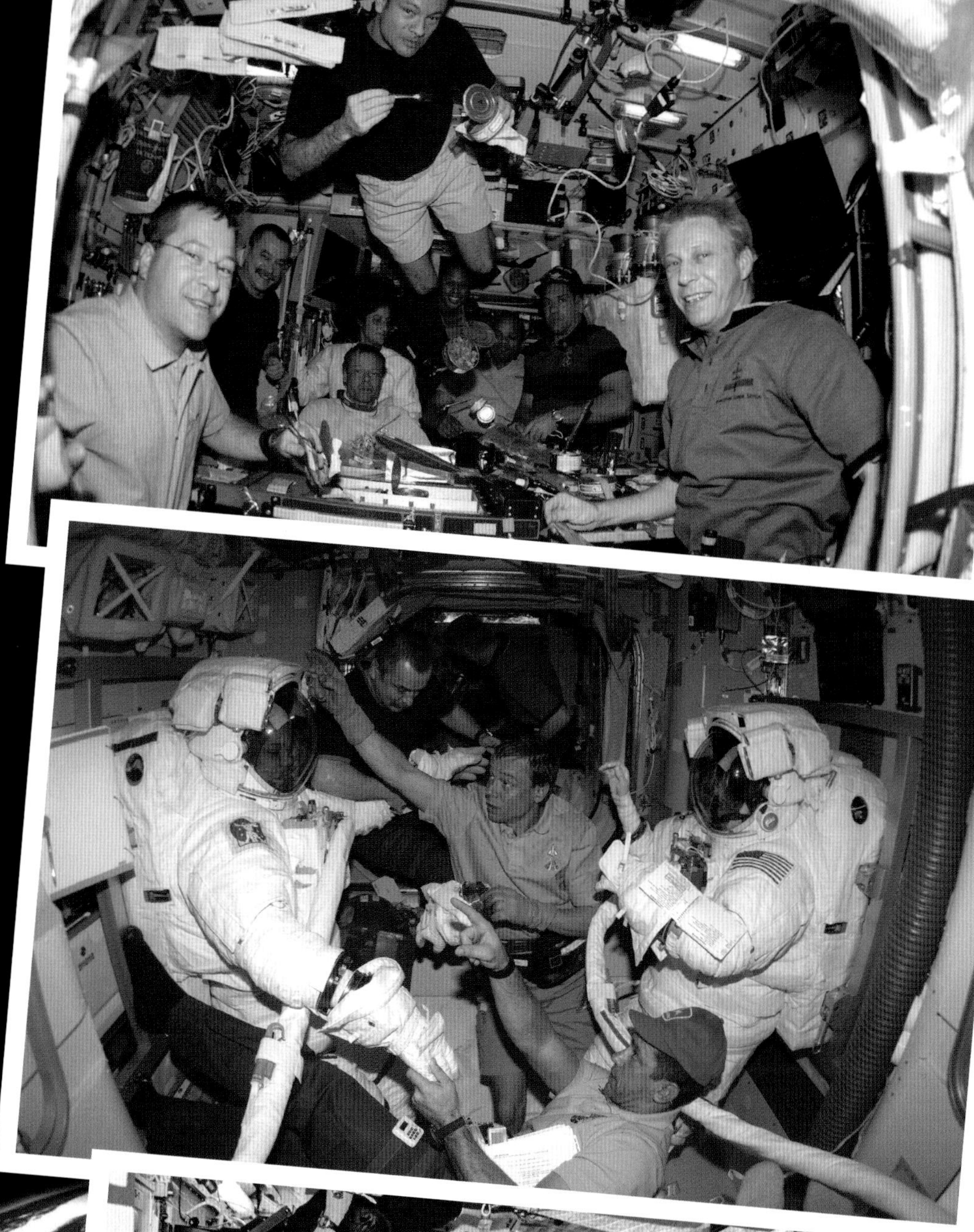

The STS-116 and Expedition 14 crewmembers share a meal in the Zvezda Service Module of the International Space Station. Pictured in the foreground are Nicholas J. M. Patrick (left), Michael E. López-Alegría (center top), Thomas Reiter (right). Pictured in the background are Mikhail Tyurin (left), Christer Fuglesang, Sunita L. Williams, Joan E. Higginbotham, Robert L. Curbeam, Jr., and William A. (Bill) Oefelein. Mark L. Polansky is out of frame.

As the mission's third spacewalk draws to a close, Robert L. Curbeam, Jr. (left) and Sunita L. Williams, get help as they remove their extravehicular mobility unit (EMU) spacesuits in the Quest Airlock of the International Space Station. William A. (Bill) Oefelein (bottom), Christer Fuglesang and Mikhail Tyurin, assisted Curbeam and Williams.

Carl E. Walz (lower left), plays host to some crewmates as he performs on a musical keyboard in the Destiny laboratory on the International Space Station (ISS). From the top, clockwise, are Rex J. Walheim, Jerry L. Ross, Ellen Ochoa, Lee M. E. Morin and Stephen N. Frick, along with Daniel W. Bursch (right foreground).

truss and Node 2; French ESA Astronaut Leopold Eyharts focused on the ESA Columbus module; and Garrett Reisman on the Kibo Japanese Module Logistics platform. The ISS-16 Expedition was affectionately known as 'Peggy and the seven dwarfs.'

Crewmembers often specialize in one or more systems. Additional training on station robotics qualifies a crew member to operate the Mobile Base System, the Station Canadarm2 and other robotic devices, in support of exterior operations. Japanese crew members can operate the KIBO RMS and several American crew members already had Shuttle RMS training and experience. On every crew, two or more crew members are trained to perform operations outside of the pressurized modules from either the American or Russian segments. To do so, the crew member must be certified to use either the American EMU or Russian Orlan pressure suits, or both.

To oversee the major NASA science objectives on each residency, an American crew member is designated the NASA Science Officer. This designation was first used by Peggy Whitson on Expedition 5 in 2002 and has been fulfilled by successive NASA astronauts since then. Flight Engineer 2 is often the science officer.

Don Pettit was assigned as the Science Officer for the 2003 Expedition 6 mission. This was prior to the Columbia accident and there was a lot of assembly and maintenance in the plan as well as science. Pettit shared the American segment science with Ken Bowersox, while Nicolai Budarin handled the Russian science program. The role was still new and not exactly defined, as Pettit explained:

> ***"Even though I had the title 'Science Officer', it was something that the staff at NASA came up with, one of those empty titles that was just bestowed upon somebody because it sounded good. I do have a science background, so if anybody was going to be science officer on station it was probably best that it was me, if only for that. I was in quarantine, getting ready for launch and I found out I was going to be the science officer. I didn't have two years prior warning and time to define my training and duties accordingly, I just found out ... 'Oh, I'm Science Officer. What's that?' Having said that, the title Science Officer on space station did legitimize a line of activities – if I was working on an experiment and I called down to the ground and said 'hey we would like a little extra time to do this project', the response would be 'Oh that's coming from the science officer. He's in charge of science, lets give him a little extra time.' So even though it started off as something of an empty title, we gradually adapted it and made it into whatever we wanted. We could use it when working with the ground help us get the timing and facilities needed for doing the science."***

From the beginning of the ISS, a full crew was planned to be 6 crew members. However, delays due to assembly and the Columbia accident meant that 'full' staffing could not be implemented until 2009, a decade after the start of station construction. Between Columbia, in February, 2003, and 2009, the reduced crew flew as 'caretakers', maintaining the station with as few as only two people. Without the Shuttle, adequate supplies and consumables could not be maintained for a full crew. After the solar arrays and final research modules arrived, and more focus could be placed on carrying logistics on the Shuttle and on Russian, European and Japanese unmanned spacecraft, and two Soyuz could be provided for ferrying crews between Earth and orbit, the full crew complement could grow to 6 and a more extensive science program could begin.

With a crew of three, each crew included one Russian, one American and one international crew member. Once the expedition crew size increased to six people, three crew members were Russian and focused their activities on maintaining systems and supporting payloads in the Russian segment. The US 'side' included US crew members from NASA, Japanese from JAXA, Europeans from ESA and Canadians from the CSA. European, Japanese or Canadian crew members were added as Soyuz third seat occupants, after the main assembly of the station was completed. They were primarily responsible for maintaining systems on the US side of the station and performing US, European, Japanese and Canadian experiments and hardware operations.

Originally the idea of launching on one system, either Shuttle or Soyuz, and returning on the other was fine, as long as the flight schedule followed the planned manifest without delay. In reality, however, it soon become apparent that alternating the two systems would not work as expected, given the launch restrictions of the more complex Shuttle, and the orbital lifetime constraints of the Soyuz.

Soyuz has an orbital life span of about 200 days. This means that a fresh Soyuz vehicle has to be sent up to the station every six months. During the assembly phase of the program, the expedition crew would be launched on the American Shuttle; the Russian expedition crew members only needed to focus on Soyuz landing skills as a contingency requirement. The visiting crews, flying for only the short duration of a taxi mission, could offer their third Soyuz seat to the European or Canadian partners so that their astronauts could gain operational experience aboard the station during a visiting mission prior to their inclusion as full members of an expedition crew after the assembly phase was completed. The seats could also be 'sold' to other space agencies, nations, and under commercial agreements. This did create some issues, especially with NASA, who initially refused access to the US segment by any 'space flight participant' who had not undergone stringent medical and preparation programs. It was pointed out that if NASA included such a 'passenger' on the Shuttle, they would not object to them visiting the Russian segment, so after the first visit of an SFP, the rules and criteria were amended to satisfy both of the major partners, as well as accommodating ESA and Canada, who were none too pleased with the loss of flight seat opportunities to fare paying individuals.

The instigation of two-person 'caretaker crews' following the loss of the Shuttle Columbia in 2003, called for a reassessment of visiting crews exchanging the Soyuz. Now, instead of flying a short visiting mission, the Commander and FE of the Soyuz would become the incoming expedition crew. After the handover period, the out-going station crew would return to Earth in the same Soyuz they had launched on, after a mission of four to six months. The third seat, when used, would be taken up by Russian or ESA crew members or Space Flight Participants flying a short, week-long mission before returning with the outgoing resident core crew members.

The Russian Soyuz has always been commanded by a Russian cosmonaut. On expeditions in which the Station Commander was Russian, he could also serve as the Soyuz Commander. There is always one Soyuz for every three crew members, and there is a Soyuz commander for every Soyuz. When the Shuttle Columbia was lost, and the Shuttle fleet grounded, the availability of the Soyuz meant that station operations could continue. Since the station was now solely dependent on supplies being delivered by Progress, and because most of the station's solar arrays were yet to be installed, it was decided to man the station with a reduced crew of only two people. These 'caretaker crews' could look after the station until the three- and then six-person crews could resume.

In 2006, ESA astronaut Thomas Reiter launched to ISS aboard the Shuttle to join the resident crew for an extended mission. His was the first of a series of flights that brought a third member of a resident crew up to join the other two, who had previously been launched by Soyuz. The third crew member was mainly a NASA astronaut, but occasionally an ESA one. The other two crew members (either two Russians or an American and a Russian) would launch on a Soyuz TMA, the same craft in which they would return at the end of their expedition. The pattern was changed once again with the advent of six-person crews.

The need to rotate the Soyuz craft, sometimes had crews overlapping, and at other times left only three people aboard the station between the departure of one expedition and the arrival of the next. This rotation made expeditions more fluid and difficult to define as distinct expedition crews beginning in 2009; therefore some crewmembers carried a designation as a crewmember split between two expeditions.

Assembly Crews

The Shuttle flight crews assigned to complete assembly or logistics missions during the construction of the station between 1998 and 2011 were known as Assembly Crews. The Assembly Crew included the Shuttle Commander, Pilot and Mission Specialists who operated the Remote Manipulator System (RMS) robot arm, conducted the space walks, and served as loadmaster for logistics transfers. Flight experience to ISS and to Mir on the Shuttle proved valuable to other areas of ISS crew operations, training, support roles and in management.

The decision to retire the Shuttle at the end of ISS assembly meant that the Soyuz would become the only vehicle for ferrying crews to and from the station for some time. This gave the program the chance to change the way the station was crewed once again, as two Soyuz could be docked to the station. Each expedition program changes with the departure of the outgoing main crew. The next crew could be launched to the ISS in advance of their formal take over and could assist the on-orbit team, overlapping the expeditions, supporting an expanded science program and giving the new arrivals time to adapt to life aboard the station prior to taking full control. The outgoing crew would help to bring the new crew up to speed and the longer hand over period would aid in the smooth running of station operations.

Two Genders – One Crew

In contrast to the early all-male space flight crews on either Russian or US missions, the International Space Station is a truly international and dual gender space complex. Although some have expressed reservations or concerns about flying mixed gender crews and having to overcome some difficult hurdles, female crew members have repeatedly shown their strengths and delivered on expectations, equal to and often surpassing their male colleagues.

After more than 50 years since the first female ventured into space, women now regularly crew the space station, and have served as Station Commander and Science Officers. Mixed gender crews would rather be recognized as 'the crew', not individualized by gender. Each member of any crew is selected, trained and assigned based on their skills, experience and talents. None of them, male or female, would be there if they couldn't do the job.

There have been physiological questions about female crew members flying on extended duration spaceflights, especially in relation to radiation levels and proposed long duration flights to deep space in the future. Peggy Whitson noted that there were no specific experiments which focused on women during her Expedition 5 in 2002 but:

> *"We do radiation monitoring for each individual crew member, plus we do area monitoring throughout the vehicle. There are dosimeters in various locations. Because females have a slightly higher radiation risk, I did make some extra effort. In my sleep station, I added water bags behind my sleeping bag on one side just for additional radiation protection."*

Aboard the station the officially recognized language is English, although in order to qualify for an expedition, each crew member must be able to speak and pass their training in Russian. Most Russians speak English as well as the Americans and vice versa. The language barrier has always been a hurdle to be overcome on any joint or multi-national mission involving the Russians. As far back as the early 1970s, the American Apollo-Soyuz crew had to learn Russian, which was something of a challenge. Tom Stafford, US commander of the Apollo Soyuz mission, never really lost his Oklahoma accent, prompting Alexei Leonov to call Stafford's Russian 'Oklahamski', a mixture of Russian and English. For ISS, there is a much longer period of joint training for each expedition to allow the crew members time to learn the languages properly. As new astronauts and cosmonauts are selected, Russian, English and Japanese language lessons are included in the training syllabus.

Some crew members also take a tour on the undersea habitat, Aquarius, the extreme environment research habitat off the coast of Florida near Key Largo. Known as the NASA Extreme Environment Mission operations (NEEMO), the facility parallels the kind of things that a station crew might have to address in orbit and is used as an analogue to ISS training.

Return to Earth

On orbit there are a couple of sessions where each crew goes inside the Soyuz to remind themselves how to strap themselves in the restraint couches, make sure that the Sokol suit fits properly (because the body changes slightly in orbit) and that they can still fit in the form fitting seat liner. They also conduct a few sessions through the residency powering up the Soyuz, studying the flight data file and communicating with Russian Mission Control.

The astronauts and cosmonauts, like most people on Earth, like to leave a memory of their presence in a certain place that they enjoyed. On ISS, this is usually done by attaching their mission emblem to a part of the station. Photos and mementos are part of committing a special journey to memory and experience once it is over and a trip to ISS, as out of this world as a journey can be, is no different.

ISS-27 Commander Dmitri Kondratyev added his thoughts about coming home in his blog over two months before his scheduled return:

"Yesterday we began advanced preparation for the return of our crew. Preparation includes repetition of time line descent action in an emergency situation, working with on board simulators to study the characteristics of the upcoming descent and much more. Although there remains a lot of time before May 16, the remaining weeks will be filled with a variety of tasks, loading a Progress cargo ship, the preparation for its undocking, preparations for the undocking of the Japanese cargo ship, the arrival of a new crew, the arrival of a Shuttle. Therefore, preparation for the descent is distributed over time to achieve greater efficiency and quality."

In the final days of her residence on ISS Expedition 5, Peggy Whitson mused about the prospect of returning to Earth:

"It was fun. I think that, at the end of my mission, the hardest thing was deciding that it was OK to leave, despite the fact that I wanted to see my family. It was some of the most fulfilling and gratifying time I have ever spent in my life, being on the station, so it was very difficult to leave. As my time aboard the station nears completion, I have a lot of mixed feelings about leaving. While of course I want to see all of my family and friends, it's hard to let go of the idea of living here."

Delays to her return mission, STS-111, delayed the arrival of new colleagues to the station, but it reinforced for Whitson that she did not find it hard to stay a little longer in space aboard the station. She sent an email to the ground with her ideas and opinions of what the crew should do in the intervening time before the Shuttle arrived. These feelings of excitement in continuing her work on the station surprised her a little, as she thought she was fully prepared to come home:

"Being here, living here, is something that I will probably spend the rest of my life striving to find just the right words for, to try and encompass and convey just a fraction of what makes our endeavors in space so special and essential. It is now time to go home. As much as I like being here, I realize that this is not home, and it's time to return."

Don Pettit wrote this towards the end of his first expedition aboard the station in 2003. He was looking forward to spending time with his young family and exploring new frontiers with them. Pettit explained that the feeling of being 'home' was directly proportional to the distance he had traveled, whether just locally for an evening out or returning from an international journey. For a space station crew returning to Earth in another country's spacecraft (cosmonauts on the Shuttle or astronauts on the Soyuz), home was still several thousand kilometers away from the landing site. But their first moments back on Earth – breathing in the natural air of Earth, feeling the sun's warmth on their skin, and feeling the soil beneath their feet means they are indeed 'home'.

Pettit and his Expedition 6 colleagues had originally been scheduled to return on a Shuttle flight, but the loss of Columbia meant their return would be on Soyuz, so there would be less room to bring back personal mementos for his family. It was difficult to decide what to bring home and what to leave behind. Pettit decided to bring back the three spoons he had used in space as a reminder of his flight on the station, and nothing from the personal items he had taken up with him. In thinking of his family, Pettit thought the three spoons would make excellent camping spoons:

"I can picture my boys sitting around a camp fire, eating beans out of the fire-charred can they were cooked in. As we chat about our world, our eyes will follow the sparks as they rise in the draft of hot air. Perhaps we will look at the stars and see the Space Station pass overhead".

John Phillips worked on his post-flight thoughts towards the end of Expedition 11 in 2005. He was trying to decide what were the most important things to report during his debriefing and what to show and tell the public during speaking engagements to try to give a realistic feel for what life is like in orbit. He edited numerous hours of video into a few manageable minutes and selected the best photos from the thousands the crew had taken. Phillips tried to avoid counting down to the time he was scheduled to come home, but he had started to think about some things that signaled the end of his stay, such as what kind of information he should pass on to the next resident crew. His packing was pretty simple, though. His allowance for returning on the Shuttle was two bags; one the size of an average suitcase, the other half that size. But as he was now coming home on a Soyuz, he packed a few personal items, secondary clothing and a few books and looked forward to seeing them one day after the Shuttle returned to flight. Phillips wrote:

"My allowance of items to carry down with me on Soyuz is essentially zero; just the spacesuit on my back, a wristwatch, and a radio dosimeter."

With the reduced return capability of the Soyuz, the crew were able to bring home some personal items but as Ken Bowersox recalled.:

"... there are a lot of personal items that just didn't come home with us and are still up on station. There were experiment samples that needed to be decided on. That was the hardest thing and we didn't make those decisions. Those calls were made by the ground. We made suggestions on how to package them to maximize the amount of items that we could bring home, but it was very difficult and some of our science had to stay on the station because of that. We were able to bring home some critical (top quality) science samples, but of course it would have been better if we could have been able to bring everything home on the Shuttle."

The end of his mission was a time for reflection for Greg Chamitoff on Expedition 17. He wrote:

"I suddenly find myself on the last night before the day I'll no longer be considered part of the crew on the Space Station. This is a very strange feeling indeed. As we fly around preparing every part of the Space Station for the arrival of Shuttle Endeavour tomorrow, I keep thinking about how I've really made this place my home for a long time. Every corner is familiar, every computer and panel and switch and cable is familiar, even if I didn't actually put it there myself. I have favorite places to work on different things, certain routes and handholds I use to fly around, and my stuff is scattered around the station as well. I came here with very little stuff but gradually, even here, I have accumulated all sorts of things that 'feel' like they are mine."

Chamitoff recalled one specific computer that he'd done most of his work on, or one still camera that he considered 'his':

"Everything from towels to office supplies, special food containers to tool sets, have become part of 'my stuff'. I have the same painful process to go through as I did on the way here – sort out which part of my stuff I'm going to take with me, and which I'm going to leave behind! Actually, as with returning from a vacation or a business trip, it's a lot easier to pack for the trip home. But I still have to put away and give up all the things that aren't actually mine, and that is what really makes me feel like my mission is truly coming to an end."

Expedition 24/25 cosmonaut Fyodor Yurchikhin blogged about the importance of undocking a returning Soyuz on time and what happened when Soyuz TMA-18, with the ISS 23/24 crew aboard, failed to undock on their first attempt on 24 September 2010. The crew was required to spend an extra night on station while the problem was investigated on the ground. When the undocking was canceled, there was a rush to get the biological samples back out of the Descent Module and into the freezer to protect them until the second attempted undocking the next day. They succeeded, but it showed the narrow window of time that the experiment samples could remain in the Soyuz from undocking to landing. Some weeks later, it was Yurchikhin himself who was preparing to return to Earth:

"There remains a matter of hours at the station. I ask myself whether I'll be back again."

That was in the future, but for now it was time to take the road home and for Yurchikhin a much anticipated long soak in a hot bath as soon as possible!

The Expedition 6 crew had launched on a Space Shuttle and they intended to return on a Shuttle before the Columbia was lost in February 2003. With the Shuttle fleet grounded they instead had to return home in the first Soyuz TMA, after their replacements had taken up residence on the station. Ken Bowersox explained:

"What was interesting is that in fact we were originally scheduled to come home on TMA-1. But the schedule was re-arranged so that another crew could bring it home, because there might be some software issues with a brand new vehicle like that. When the option to come home on Shuttle was lost we received plenty of good training from the specialists at Star City and Nicolai Budarin, with his experience, was able to get our team through."

The Expedition 12 crew, Valery Tokarev and Bill McArthur, had launched to the station aboard their Soyuz TMA with Space Flight

Participant Greg Olsen, with whom they had completed quite a bit of training. McArthur was extremely impressed with Olsen, his attitude and professionalism. Olsen was totally committed to the flight and to his role. Their return trip would be with Brazilian astronaut Marcos Pontes. Pontes was a professional test pilot and a NASA trained Mission Specialist. He was assigned to make a flight to ISS on Soyuz. Though the third seat crew member has minimal responsibility in operating the Soyuz, the Expedition 12 crew felt confident enough in both Olsen and Pontes to assign them some tasks on Soyuz during ascent and entry.

For Frank Culbertson, a former Shuttle Pilot and Commander, coming home in a vehicle he was not actually 'flying' was a different challenge. Culbertson felt pretty comfortable lying on his back on the mid-deck. While Culbertson understood the reason why he had to return on the mid-deck, the pilot in him still wanted to have an eye on the flight deck during landing:

> ***"Dom Gorie, the Commander of STS-108, was very kind to me. He gave me a small TV monitor to put on the bulkhead in front of me, though I actually put it on my knee. I could see the approach and runway as we turned around the final alignment circle. From the flight deck, Gorie played a video through the heads-up display and so I could actually see that we were above the clouds and he did not have the runway in sight. He gave me that monitor with the condition that I could have it as long as I didn't say anything. He didn't want any advice."***

Culbertson was not allowed to use the audio during the landing, just watch the events unfold. Gorie was clearly in command of the Shuttle.

Culbertson could definitely feel the gravity on his body after his long station mission, more so than on his two previous Shuttle flights. He had exercised really hard in the final weeks of the mission and after landing he was able to sit up after about 10 or 15 minutes. Culbertson even walked off the Orbiter under his own power. He recalled later:

> ***"It wasn't necessarily something I felt that I had to do, but since I felt up to it, I thought I'd give it a shot."***

For Leroy Chiao on Expedition 10, the most memorable part of the return was the landing and hatch opening:

> ***"The smell of the air was the main thing. I just got the smell of grass and mud again and it was very refreshing, but the whole Soyuz landing was very interesting. Having flown three times on the Shuttle, it was a great experience to fly and land on Soyuz."***

Coming back to Earth can be a jolt, in more than just the physical impact of a landing in the Soyuz. Expedition 7 astronaut Ed Lu came home to find that his credit card had expired, his phone had been disconnected and his car registration had run out. In February 2004, some four months after his return, Lu gave a public lecture in Honolulu, Hawaii. The former ISS inhabitant revealed:

> ***"You wouldn't believe how much mail and e-mail had piled up."***

He informed the audience that, in the first week home, he had over 3300 emails from friends and associates but no junk mail at all. At that time, he was just too busy trying to catch up with his bills and mail to think about a return to space any time soon. He subsequently decided not to fly in space again, but when asked about life aboard the station, he answered:

> ***"It's good. It's busy. It's kind of fun. I was wondering beforehand what it was going to be like. It was never boring."***

For many station crew members, the hardest part of the mission was the training and the time away from their families, spending hundreds of hours on aircraft flying across the world to different training locations. For Ed Lu, being away from home was hard. He was away for ten months for the mission. This included 2.5 months in Russia for final preparations, the six month mission itself, and another month post-flight in both Russia and the US for debriefings. Over the previous five years, Lu had completed 16 one-month trips to Russia, an added challenge for those who wished to complete an ISS expedition. For those with families, it is a very daunting, trying period of their lives, both for the crewman who will fly in space and for the loved ones they leave behind on Earth. The commitment for any mission comes not only from the crew member and their ground team, but also from the support of their family. Space flight is not just about flying the mission, but the dedication to the training, and time away from home, on Earth as well as in space.

As the station program expanded, trips to Canada, Europe and Japan added to the traveling time and duration away from home before even leaving the ground. For some space explorers this was too much. Many astronauts were not eager to make a long space station flight. But they had little choice. The retirement of Shuttle meant that their chances of short duration space flights were very limited.

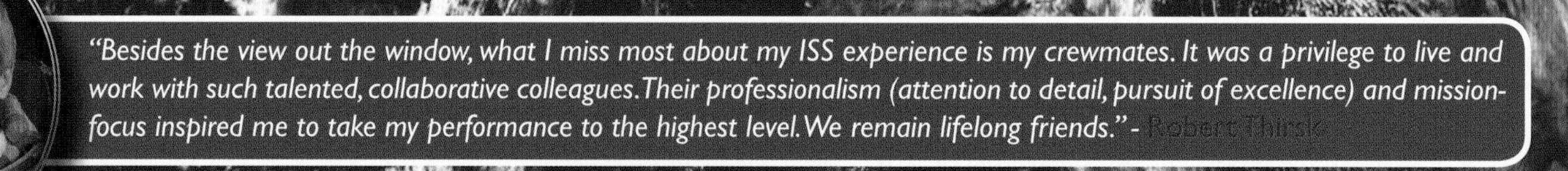

"Besides the view out the window, what I miss most about my ISS experience is my crewmates. It was a privilege to live and work with such talented, collaborative colleagues. Their professionalism (attention to detail, pursuit of excellence) and mission-focus inspired me to take my performance to the highest level. We remain lifelong friends." - Robert Thirsk

"I got this book, Rockets through Space [P.E. Cleator, 1936], on my twelfth birthday, in 1941, from my father. On the first page he says, "The fantasy of today is the reality of tomorrow". The author envisioned the construction of an artificial moon designed for the purpose of refueling spaceships. I was about 12 years old when I read this, and I said, "Boy, this is a pretty good thing to do." that's when I started getting interested in the Space Station." - Hans Mark

"We have established a presence up there. We have had humans living in space continuously for decades. I've got a picture of six people of diverse cultural backgrounds having a meal together on the International Space Station, working as one. We established that presence in space." - Robert D. Cabana

"I don't think there is anything unique that I have contributed. I am representative of many people, thousands of people, across the partnership, who have dedicated either their entire lives or much of their careers to making this thing work." - Jeffrey N. Williams

"Mankind needs something that leads to the unification of people. It is extremely important not to fight with each other in wars. Wars are expensive; we can't afford wars anymore. I think it's much smarter to spend money doing something like the International Space Station. It stops politicians from having an opportunity to make the wrong decisions." - Victor D. Blagov

"A frustration with the American space program is that human spaceflight has been a program that starts, comes to a complete halt, then we start over and do something different and we come to a halt, and we start over and do something different. Apollo, Skylab, Shuttle, ISS. I look at the Russian program and it has not had those big gaps. The Russian Program evolved. I wish the US could stay the course on human spaceflight and figure out how to use the previous program to go on to the next one. I hope that Space Station is part of that legacy." - Lynn F.H. Cline

"We have got to expand what we do. We have got to look into new ways and new frontiers. That is the basis for growth. You cannot find out anything you don't know unless you risk a little bit of time looking into new things. I don't know what the Space Station might produce. But, compared to the total national output, the total national economy, it is a small drop in a big bathtub. So we should do it. Man's curiosity is the big driver behind progress. Most leaders in any country don't have a great deal of curiosity. They have a great need to maintain the status quo. They are sitting where they want to be. But there will always be people with curiosity. The same way that we have got to support the arts, we have got to support the curious." - Max Faget

"I don't think a lot of people really appreciate what ISS does. ISS doesn't get a lot of publicity. The media rarely highlights its accomplishments. The ISS itself and all of the vehicles going to and from the ISS, you hear little about. It's in a black hole as far as the media and public are concerned. You cannot get a lot of support from the American public if they don't know what you're doing on the ISS. I don't think we do a good job of telling NASA's story about what we do on ISS and why it's important. The Canadian astronaut who was on the Space Station did a little singing, that was a highlight! Now, he's a rock star in Canada. It was good, and they should do more of it, but it wasn't planned by NASA. We need to have more of the good news and stories about ISS research accomplishments, to tell what the bang is that we are getting for the buck." - Daniel R. Mulville

"ISS enabled permanent habitation by humans in low-Earth orbit. ISS enabled exploration. ISS demonstrated in low-Earth orbit the ability to travel beyond low-Earth orbit; and ISS enabled the creation of commercial demand for access to space. ISS permitted the NASA budget to be focused towards exploration. I believe that once ISS is gone; we will be at that point where we've had humans in low-Earth orbit since the ISS program's inception, and we will continue to have humans in low-Earth orbit." - Michael T. Suffredini

"The Cold War no longer drives the space program. I think countries recognize now that by putting people in space, it looks good to everybody, especially when it is done internationally. It is good for the people who are participating and for other countries. They all admire what is happening. It gives people hope for humanity. Spaceflight is a shining beacon for the world. The greatest success of the ISS is the embodiment of a partnership between different countries doing something in space that is quite difficult." - Mike Foale

"I think we have now embarked upon an era where we will never again be without humans in space. It is a pretty momentous kind of change. It's also a major cultural change for the Space Center in the sense that previously we had worked 7 days a week, 24 hours a day for a couple of weeks at a time. Now it will be a permanent way of life, and that's a very difficult and different kind of a working environment." - Joseph Loftus

"The legacy of Station is as an international model of how we can do big things in space together. It is a powerful symbol to the world.... oriented towards collaboration, exploration and science. It's not about defense or competition. ISS is the world at its best." - Michael R. Barratt

"One legacy of the ISS is the framework it established for all these countries to work together successfully for the long term. Space Station negotiations were both my greatest accomplishment and my biggest challenge during 36 years with NASA. That was my biggest challenge on so many different levels. The fact that it was multilateral. Trying to blend a new partner in with established partners. Having the complexity of multilateral and bilateral negotiations. The changes in the program throughout the whole process of the negotiations." - Lynn F.H. Cline

"ISS has allowed all the international partners to learn how to work together. Over years continuously working together in orbit, Russians, Americans, Japanese, Canadians and Europeans are learning together. We learn by being out on the frontier. We are working in a Space Station continuously for years in a real environment; it is exploration that will help us go beyond Earth orbit." - Tom Cremins

"Typically we think of "space station" as they are all the same. Salyut was different from Skylab, which was different from Mir, which was different from ISS. Different environments made different demands on individuals that then drove different requirements for selection, training, support, and the type of people who flew. The contrast between ISS and Mir; got rid of a lot of the environmental issues that we had on Mir and solved a lot of the problems." - Albert W. Holland

"ISS has been a great project from the international partnerships perspective, and for trust, for engineering side, for science and for the social and cultural side. I think the Space Station's legacy is the trust between the partner agencies. When people ask me about the International Space Station, I say the science is important, but the relationships, the trust from one agency to the next, the transparency that we have with one another, I think that is the most significant part of the International Space Station. We will not have the Space Station forever. It is only a stepping-stone to help us build the next space project, and I think we will do it better for the next generation of explorers." - Sunita L. Williams

"The legacy of the ISS is the international partnership. I know there are other things that we do, globally, but this is the most technically difficult, complex thing that we've done as an international community. Even if you look at the construction; the engineering feat on orbit, most of these pieces had never met before they met in orbit. They never crossed borders here on Earth, but we assembled them in orbit. At seventeen thousand, five hundred miles an hour, we put these pieces together." - Peggy Whitson

"ISS is probably the largest international project ever attempted and successfully completed when you count up the number of people that have participated in its development—and now operation. ISS has proven that you can do a large, very complicated international program with partner nations. It proved that international programs are very difficult to run. But, in spite of the complexity and level of difficulty, it showed that you can do this. You can bring countries together, bring cultures together, oftentimes they have different objectives in mind, but together they do something incredible." - Kevin P. Chilton

"We [the Russians] had some doubts whether we would be able to build the ISS together. The station is a good illustration that we are capable of doing that and that we can achieve the result we want to achieve, together." - Sergei Krikalev

"Over the last forty years we have tended to be in the business of flying humans in space, for the sake of flying humans in space. The time has come where we must transition to having a purpose that is more related to other objectives, from commercial aspects, to scientific aspects, to preparing ourselves for the future and whatever that might be, and so there's a wide litany of possibilities, but no longer can we afford to just depend upon the glory of flying people in space as being a reason to spend the kind of money we do spend for human spaceflight." - Thomas Holloway

"International cooperation, with all the countries working together for a common goal is more valuable than all the science completed." - Don Pettit

"There is a plethora of results from ISS. We have shown that we can live and operate in space, 365 days a year, for years on end, which is phenomenal. We have shown that we can pull off a big, integrated, complex assembly of 30-plus spacecraft, over and over and over again. Ultimately, I think ISS will provide us some really groundbreaking science. I don't know what it will be, but I don't have much doubt that we'll find things up there that we wouldn't have found if we didn't have the Space Station." - Lauri N. Hansen

"The International Space Station is a symbol of what human beings can do when we work together constructively; if the former Soviet Union and the United States of America can put aside our differences, then the rest of the world can put aside their differences too. We can take the lessons that we have learned from space station and work together here on Earth." - Mike Fincke

"I think we are just beginning to see the results of the science that is going to be coming from the International Space Station. It takes time to gain the knowledge, and then actually prove what you've gained, back here, on Earth, in order to develop something that is a breakthrough. And I think you're going to see more and more, as time goes on—there's going to be more science. We're going to learn more. I think it's going to prove to be a great science laboratory, with many breakthroughs in the future. There are many, many more good things yet to come from the Space station." - Robert D. Cabana

"The international character of ISS is a big part of the legacy of the ISS. It shows that nations can come together and build something that is relatively untouched by geopolitical events and which keeps producing something and which you can look to and say, "we did that together and it is successful and it is a remarkable achievement." - Albert W. Holland

"The legacy of the Space Station is first of all to build the low-Earth orbit commercial market. We want to build economic activity and create jobs,... further knowledge, and solve problems here on the Earth… If we get to the end of the Space Station's useful life and we don't have any trophies in our trophy case that would really demonstrate that the Space Station was worth that $100 billion investment, then that would weaken our legacy. You can do things on the Space Station with astronauts up there that you can't do unmanned… Unmanned projects need to have a lot of redundancy, a lot of reliability. They have to be more complex and there's a lot more risk associated with them. The costs are extremely expensive. We try to communicate to users and say, "Hey, you've got an opportunity where you can ride on these commercial vehicles, get them up to Space Station, and solve your problems, advance your technologies." - Gregory H. Johnson

"Space Station is really about having humans see themselves with a home in the solar system and not just on the Earth. We are trying to determine if we can live and work productively beyond Earth. It's up to us to figure that out." - Bill Shepherd

"On ISS, we are doing scientific and engineering research. We're starting to find things that can affect people's lives on Earth in a direct way, through the life sciences experiments, through the technology. The technology of doing telemedicine… diagnosis via satellite communication from ultrasound...salmonella bacteria and looking at virulent strains which are grown on Space Station that are different than strains that are grown on Earth. Cold combustion—the idea that you can have combustion processes involving hydrocarbons that are occurring at less than half the normal combustion temperature. These advances that can be directly applied to life on Earth." - Donald R. Pettit

"Space Station has all the elements that will end up having to be a part of anything you do to go to the Moon or anything you do to stay on the Moon; the logistics chain to Earth orbit is a lot shorter than to the Moon. If something breaks on Station and it's life-threatening, you can get the crew out of there and get them back on the ground in forty-five minutes. If you need a part, you can get it up on the next vehicle that is going up. With the International Space Station, we are learning an enormous amount about how to keep a human-built piece of equipment functioning that has to sustain human life away from the surface of the Earth. Station gives us lessons every day it is up there. Thank God we have got a space station." - Neil Hutchinson

"The space stations that support long duration spaceflights are the foundation for future more remote missions. We have to learn to fly for a long time and to support life on board for a long time." - Sergei Krikalev

"The greatest legacy of ISS is international collaboration. I would really like to see us go to Mars one day. I don't think the US will ever go to Mars as a nation. I don't honestly think any single nation is going to have the resources to go to Mars. If humankind is really going to go beyond the Earth-Moon system, I think it has to be as humankind, not as the U.S. I don't think we have the resources to pull it off, or the will to pull it off at that expense. I think the greatest legacy ISS has left for us is setting the framework for how to pull a complicated, complex, difficult international collaboration together and make it work. If you think about it, no partners have walked out. ISS has lasted longer than many marriages. You look at the complexity of the relationships, and it's amazing that it's lasted, and really serves as an example. On ISS it was kind of, "We're going to do this, and we'd love to have you join us." For the next step, the "we are going to do this" has to be everybody. It's a different dynamic in the partnership. That, I think, is honestly its greatest legacy." - Lauri N. Hansen

"ISS is a stepping stone to a habitat on the Moon and to a Mars flight. I think we will have a lunar base one day and that this low-Earth orbit Space Station will have been good practice. The Space Station is a heavily ground-commanded platform. The fact that it's heavily ground-commanded takes away the crew's autonomy in some respects. It's a ground operated station, and the astronauts are just one part of the system that's operating up there. Mars, by necessity, we're going to go back to a less ground-commanded vehicle. There'll be more autonomy of the on-board crew; that will place new requirements on Mission Control; the "control" may be softened in that title for Mars." - Albert W. Holland

"I believe it is vitally important to have humans in space to do the challenging things that robots can't do or can't do efficiently. The only way we can send humans into space repeatedly for the long term, is to have an understanding of the effects of spaceflight on the human body; specifically weightlessness, radiation and isolation. The only way we're going to acquire that is by doing that kind of work on a Space Station. The International Space Station is a world-class research facility and it's incumbent on us to use it to the fullest extent, so we can then enable those future missions beyond low Earth orbit. Anybody that says we should have moved on, that the Space Station's a dead end, or it's a diversion, tells me that they are not thinking about people in space, and they're not thinking about usefulness in space. If the future of spaceflight involves people in space, then we need the ISS or something like it to do the basic work. We are doing our homework. We do not go on to the next grade until we've finished our homework here." - John B. Charles

"In human space flight we really want to get beyond low-Earth orbit. We have a plan for extended stays on the Moon and then that takes us to Mars. At the Moon we are days away. When we go to Mars, it is months away. It is not a forgiving environment. We are not technically ready to go to Mars, yet. So we better learn from Space Station, and then learn from the Moon so that we can be successful going to Mars. To go to Mars would require a spacecraft about the size of the Space Station. We would have to construct in orbit a spacecraft about the size of the Space Station and then make a six-month journey to Mars and then for about a week's stay, and then return back. I don't think we're ready to go to Mars. But we can use the Space Station to learn about long-duration space flight. Station is a transient. It makes sense if you look at it and then you look at that natural progression of stepping stones from Station to Moon to Mars." - William H. Gerstenmaier

"The Space Station is a superb engineering tool for developing the systems that we need to expand beyond low Earth orbit, to go out and establish human presence in the solar system. We've got a ways to go before we have an environmental control system that I'd feel comfortable relying upon while we are away from Earth for two years, knowing it had to work. The Space Station is an engineering test bed for developing those systems." - Robert D. Cabana

"As long as it doesn't cost too much for any one nation, the interest will be very strong to go to the moon and then to Mars. I believe it requires an international endeavor; it will not be possible for this nation to do it alone, because the political will, will not be there. But the will is there if it is together with other nations." - Mike Foale

"I think it's easy to argue that the Space Station is the most complex technological achievement of mankind, ever. One aspect is just in getting the pieces to orbit. On top of that is the international integration; the challenge of working with the Russians, from a technical, cultural and from language points of view. The technology and the scope of the technology that was integrated in this almost one-million-pound mass spacecraft assembled by way of 38 Shuttle launches, 180 EVAs, and more than 40 launches by the Russians. There were many challenges. It was a very large team of people doing many different aspects all of which had to be integrated." - Jeffrey N. Williams

"ISS is used for the science and engineering investigations required to enable human exploration beyond Earth orbit possible." - Don Pettit

"I think the most important role of the space station is preparing us for a mission to go back to the moon and then to Mars. I think we are learning about spacecraft design; how to make a much more robust spacecraft; how do we make a more maintainable spacecraft; how to make a spacecraft that can go to Mars and back. With this knowledge, we will be able to anticipate problems and we will have the wisdom to manifest the right spares on board and train the crews to do all the necessary maintenance without relying on ground control." - Leroy Chiao

"From an engineering perspective the building of the space station in orbit was the most complicated technological feat that we have ever achieved." - Peggy Whitson

"Many times people said: "It will never happen; there are too many EVAs; we'll never be successful, launching so many Shuttles and other country's rockets; we'll have problems when we lose a part, or when a whole segment doesn't make it up; none of that happened; the building of the ISS did not happen by accident. It was successful because a lot of people were so very dedicated and gave up so much of their time and their lives to make sure the program happened the way it was envisioned....Many times I took visiting dignitaries over to show them the scale model of what the completed International Space Station was going to look like. It wasn't a very big model, probably wingtip to wingtip it was maybe six feet. All of the modules, the truss and the solar arrays, and all the piece parts were displayed in the model. And while I was taking folks over to look at the model and explain to them how it was all going to come together and work, we hadn't launched a single piece of hardware yet. I kept looking at that thing going, "Man, we've got a long way to go." Then it seems like in the blink of an eye—not really, because there was a lot of work involved—but now we look up in the sky and it looks just like that model." - Kevin P. Chilton

"The decision to have a Space Station is not a technical decision; it is a political decision. I have been a little discouraged in seeing significant benefits from experiments. There is no question that there are numerous experiments that can be done: electrophoresis, crystal growth. They are great. But can you tie that to the investment? You could say, 'Well, on a cost benefit basis...' but we don't do it for that reason. The alternative, do we want to just opt out of spaceflight? That would be a significant decision. You might say, we continue the program for that reason, because we don't want to opt out of people operating in space." - Robert O. Piland

"At the end of the day, Space Station is a phenomenal achievement. No single thing on Space Station was that technically challenging. We weren't really pushing the envelope in terms of building a solar array, or building a lab module. The part that was technically challenging was the integration, making sure it could actually all come together in orbit. I look at where we were at the beginning, and we had all these assembly flights, and we had all these EVAs [Extravehicular Activities], and we all honestly thought there was no way this was really going to work the way we planned it. Every assembly stage was really a separate spacecraft. Each had all the systems and complications of a spacecraft. It was all assembled in orbit. Partway through, we threw all the cards up in the air. Yet it really did pretty much work the way it was planned." - Lauri N. Hansen

"We couldn't build the ISS today. It's kind of like a pyramid; could we build one of those now? The Station and the Shuttle were inextricably linked. One was designed to be built with the other. I think ISS will be the technological pinnacle for generations to come, because I don't see how we could ever build it again." - Charles Lundquist

"To be a member of the ISS team is an important thing for the Japanese people, the only Asian country to be involved in the ISS and now we co-operate with other Asian and African countries in developing their interests in space exploration, based upon our experiences in ISS. And so it is now a really broadening international co-operation on the ground as well as in space. By working in the ISS programme, global appreciation of Japanese technology and industry has changed." - Izumi Yoshizaki

"With Japanese astronauts now featuring on long missions, seeing them on TV, Japanese people including young people are beginning to see them as normal and not out of the ordinary. It's good that this has happened because it makes space not special; it is attainable as a career. For us we did not know what was involved with the programme and from our experiences with Kibo we have learnt how to work within the industry. From this, young people now have a career opportunity in the space programme. We were their pioneers." - Fumiya Tsutsui

"My most compelling memory as a kid was the picture of the Earth rising over the Moon. That gave us a whole new perspective on what the Earth was. Here's this little blue ball. As a kid I used to look at that and say all of us are in that picture. There are tremendous highs when years of work come together like when we launched the international partner modules and the Space Station was assembled. We, at NASA, are part of a thing that is bigger than ourselves. NASA has allowed us to rise above our day-to-day problems and our day-to-day crises and look at our world and our lives in a whole new perspective. We realize how small we are in the big scheme of things, how precious the Earth is." - William H. Gerstenmaier

"I have come to think of the space program as the means, over the very, very long term; maybe thousands of years or maybe a million years; to establish the human race elsewhere. We live on Earth; we think it is big but it's really a very small object. There is no backup. The Creator uniquely gave us the capability to do this. It seems almost divinely ordained that we should use this capability to ultimately preserve the race. It is up to us to focus and use our resources." - Fred Haise

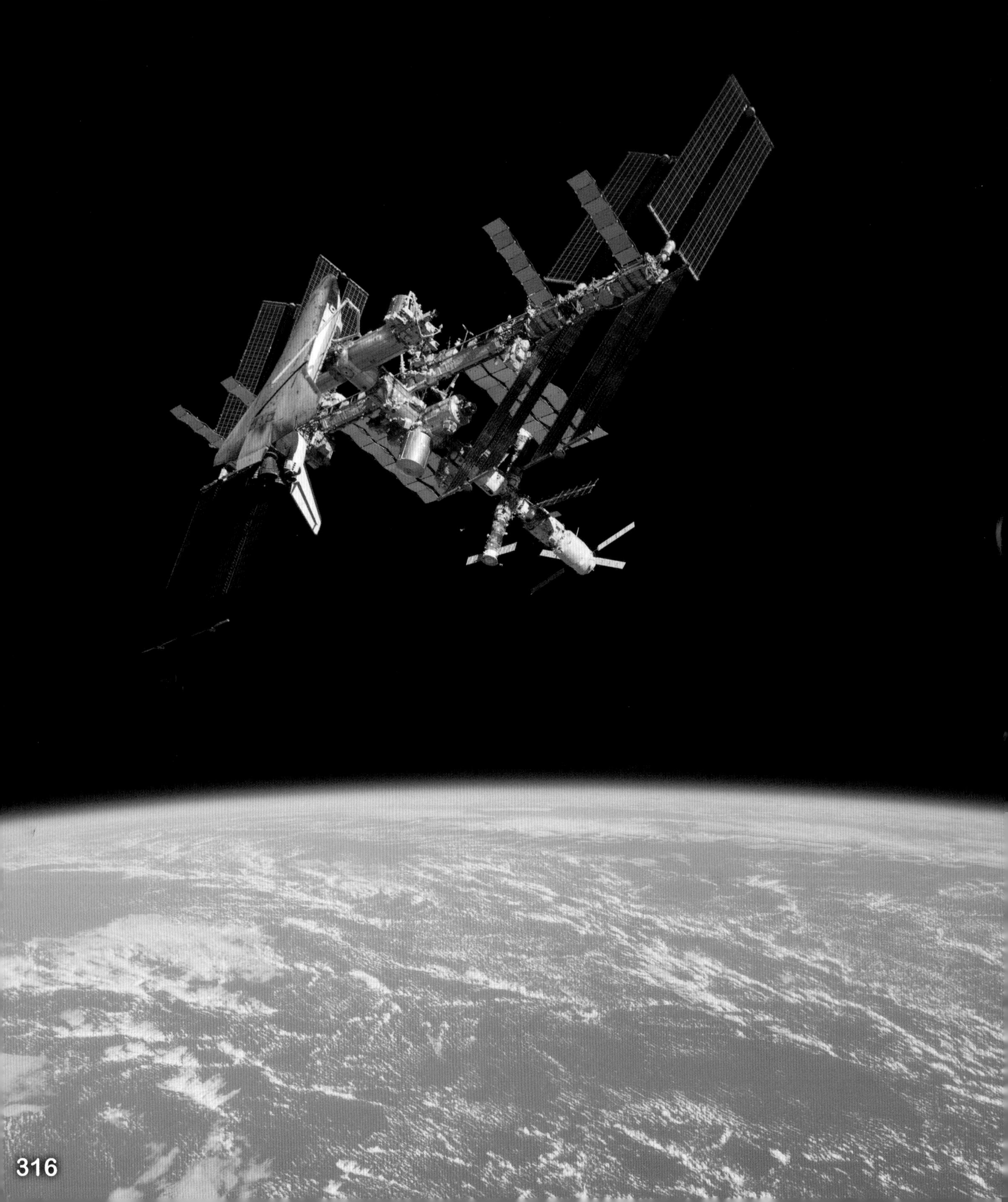

CONTRIBUTORS & COMMENTATORS

Isaac P. Abramov, Russia, lead designer, Soviet and Russian spacesuits, manager Zvezda Company
Anne Accola, NASA, Mission Planning and Analysis, developed many of the processes associated with crew operations on ISS.
Arnold D. Aldrich, 35 years with NASA, JSC, Headquarters, and 13 years with Lockheed Martin, started in flight operations, and moved into management positions in Skylab, Apollo, Shuttle and Space Station
Alexander P. Alexandrov, Russia, cosmonaut, flew on Mir in 1987, later Chief Cosmonaut, Energia, subsequently Department Chief for Flight Test, RSC Energia
Clayton C. Anderson, NASA, astronaut, aerospace engineer, flew on ISS Expedition 15/16, and Space Shuttle.
Michael P. Barrett, **NASA**, astronaut, physician, flew on ISS Expedition 19/20, and Soyuz and Space Shuttle.
Yuri M. Baturin, Russia Cosmonaut Expedition 2
Donald A. Beattie, NASA, Apollo Applications Program, Space Station Advisory Committee
James M. Beggs, NASA Administrator, 1981-1985, responsible for getting support of President Ronald Reagan for the Space Station Program
John Desmond Bernal, Irish molecular biologist and science historian, space colony conceptualizer
Jeff Bingham, Senior Science Advisor at the US Senate, focusing on NASA, space science, and space station
Victor D. Blagov, Russia, RSC Energia, involved in design of Vostok, and later spacecraft, including Mir Orbital Station. Later Chief of the Planning Department for Mir missions and Mission Director
Kenneth D. Bowersox, NASA, astronaut, pilot, commander, 5 Shuttle flights, 1 Soyuz return and ISS Expedition 6
Timothy Braithwaite, Canadian Space Agency manager and NASA Liaison, robotics flight controller for the Canadarm 2
Nicole Buckley, Canadian Space Agency Chief Scientist responsible for ISS utilization
Wernher von Braun, German/American engineer, responsible for German V-2, subsequent US Army military missiles, NASA MSFC Director, conceived designs for space stations, and popularized, built public support for the space program in the 1950s-60s
Randy H. Brinkley, NASA, ISS Program Manager, 1993-1999. Shuttle-Hubble Mission Manager
Robert D. Cabana, NASA, astronaut/pilot, 4 Shuttle missions, Deputy Manager, ISS, Director, Kennedy Space Center
Jeffrey A. Cardenas, NASA, JSC, Mir Operations Manager, Shuttle Mir Program
Robert E. Castle, NASA, JSC, flight director, lead flight director for several Shuttle Mir missions, ISS missions
John B. Charles, NASA, JSC, Chief Scientist for Human Research, during ISS and Shuttle-Mir
Gregory E. Chamitoff, NASA astronaut/engineer/flight controller, ISS expedition 17/18 and Space Shuttle
Robert Chesson, European Space Agency, Head of Space Flight Operations
Leroy Chiao, NASA astronaut/engineer, ISS expedition 10, Soyuz and Space Shuttle
Kevin P. Chilton, NASA, astronaut/engineer/pilot and commander, Space Shuttle/Mir
Jean-Loup Chretien, French cosmonaut, Chief CNES Astronaut, Soyuz, Salyut, Mir, Space Shuttle
Lynn Cline, NASA, Headquarters, lead negotiator for agreement for Russia to join ISS Program
Dandridge M. Cole, Senior Advanced Planning Specialist Martin Company, space colony conceptualizer,
Richard O. Covey, NASA astronaut/engineer/pilot and commander, Space Shuttle
Thomas Cremins, NASA Assoc. Administrator for Strategy and Plans. Established Shuttle-Mir Program.
Robert L. Crippen, NASA astronaut/engineer/pilot, MOL, Skylab, Space Shuttle
Walter Cugno, Thales Alenia Spazio, Node Project Manager, Director, Exploration and Science
Frank L. Culbertson, Jr. NASA, astronaut/engineer/pilot and commander, Shuttle-Mir Program Manager, Space Shuttle, ISS Expedition 3
Takao Doi, JAXA, astronaut/engineer, Space Shuttle, ISS
Bonnie J. Dunbar, NASA, astronaut/engineer, Space Shuttle, Spacelab, Mir
Freeman D'Vincent, Convair Astronautics, Space Vehicle Design Specialist
Kraft A. Ehricke, German/American engineer, Bell Aircraft and Convair Astronautics, futurist, space station conceptualizer
R. Bryan Erb, NASA Johnson Space Center, Canadian Space Agency, engineer, Assistant Director, CSA Space Station/ISS Program, Skylab
Kristen Facciol, Canadian Space Agency, engineer, ISS engineering support
Konstantin P. Feoktistov, Russia, cosmonaut, codesigner of Salyut space station
E. Michael Fincke, NASA, astronaut/ engineer, ISS Expedition 9/18, Soyuz, Space Shuttle
C. Michael Foale, NASA, astronaut/astrophysicist, ISS Expedition 8, Mir 23, Soyuz, Space Shuttle
Richard Fullerton, NASA, EVA project manager and working group co-chair during NASA-Mir.
Owen K. Garriott, NASA, astronaut/engineer, Skylab 3, Spacelab, Space Shuttle
William H. Gerstenmaier, NASA, Associate Administrator Exploration, ISS Program Manager
Mark S. Geyer, NASA, Director Johnson Space Center, Program Manager Orion, Project Manager FGB
Robert R. Gilruth, NASA, Director Johnson Space Center
Daniel S. Goldin, NASA Administrator, 1992-2001, responsible for Space Station Freedom becoming ISS Program under President William J Clinton
Walter W. Guy, NASA, Johnson Space Center, engineer, life support, crew systems
Chris Hadfield, Canadian Space Agency, astronaut/engineer/pilot, ISS Expedition 34/35, Mir, Soyuz, Space Shuttle
Fred W. Haise, Grumman Space Station Manager, NASA astronaut, Space Shuttle, Apollo
Laurie N. Hansen, NASA, Johnson Space Center, engineer, Engineering Director
Chikara Harada, head of JAXA Space Tracking and Communication Center
Henry W. Hartsfield, Jr. NASA astronaut/engineer/pilot, MOL, Skylab, Space Shuttle
John K. Hirasaki, NASA Johnson Space Center, engineer, Landing and Recovery Division, assessed use of Soyuz for space station.
John D. Hodge, NASA, Johnson Space Center, Headquarters, flight director, manager and organized program management for the Space Station Program.
Jeffrey A. Hoffman, NASA astronaut, astronomer, Space Shuttle
Albert W. Holland, NASA, Johnson Space Center, operational psychologist
Carolyn L. Huntoon, NASA physiologist, principal investigator, Spacelab, Mir, Dir. Johnson Space Center
Neil B. Hutchinson, NASA, Johnson Space Center, Headquarters, flight director, Manager Space Station Program
Thomas W. Holloway, NASA, Johnson Space Center, flight director, Manager, Space Shuttle, Director, Shuttle-Mir; Manager Space Station Program
John D. 'Denny' Holt, NASA, Johnson Space Center, manager, Operations Integration, Space Station Program
Dan Jacobs, NASA, Johnson Space Center, manager, international partners Space Station Program negotiator
Gary W. Johnson, NASA, Johnson Space Center, manager, spaceflight safety, Apollo, Apollo Soyuz, Shuttle-Mir, Space Station
Gregory H. Johnson, NASA, astronaut, pilot, Space Shuttle, President, Center for the Advancement of Science in Space (CASIS)
Shigeki Kamigaichi, JAXA, manager, public affairs, Space Station Program
Gary H. Kitmacher, NASA, Johnson Space Center, developer Cupola, Nodes, Stowage, author, manager, Space Shuttle, Spacehab, Shuttle-Mir, Space Station/ISS
Kenneth S. Kleinknecht, NASA, Johnson Space Center, Skylab Program Manager
Heinz Hermann Koelle, NASA, Marshall Space Flight Center, US Army, missile development, Saturn Program, ABMA Project Horizon, Advanced Projects Office, GfW
Richard H. Kohrs, NASA, Johnson Space Center, Headquarters, manager, Space Shuttle, Director, Space Station
Dmitri Y. Kondratyev, Russia, cosmonaut, pilot, ISS expedition 26/27, Soyuz
Vladimir V. Kovalyonok, Russia, cosmonaut, pilot, Salyut 6, Soyuz
Saunders B. Kramer, US engineer and scientist, Lockheed, conceived the Bucephalus Space Station
Sergei K. Krikalev, Russia, cosmonaut, manager, engineer, Salyut, Mir, ISS Expedition 1, 11, Soyuz, Space Shuttle
Philip A. Lapp, Engineer De Havilland Missile Division, Director/Co-Founder SPAR Aerospace, founder Canadian Astronautical Society
Daniele Laurini, European Space Agency, Cupola Project Manager
Valentin Lebedev, Russia cosmonaut, Salyut 7
William Leitch, Astronomer, scientist, Principal Queen's College, Kingston, Ontario, Canada
John "Mike" Lounge, NASA, astronaut, manager, Spacehab, Space Shuttle
Edward T. Lu, NASA, astronaut, physicist, ISS Expedition 7, Space Shuttle, Soyuz
William R. Lucas, NASA, Marshall Space Flight Center, chemist, metallurgist, Director, Marshall Spaceflight Center
Charles M. Lundquist, NASA, Johnson Space Center, manager, Orion, ISS
John MacNaughton, Vice-President & co-founder SPAR, principal advocate Canadarm program.
Sandy H. Magnus, NASA, astronaut, materials science engineering, ISS expedition 18, Space Shuttle
Hans M. Mark, NASA, Headquarters, Deputy Administrator,
Jeffrey Manber, space entrepreneur, ISS payloads, Mir/RSC Energia commercialization
Owen E. Maynard, NASA, Johnson Space Center, engineer, established early space station concepts and requirements
William S. McArthur, NASA, astronaut, pilot, ISS Expedition 12, Soyuz, Space Shuttle
Chiaki Mukai, JAXA, astronaut, doctor, first Japanese woman in space

Leonard S. Nicholson, NASA, Johnson Space Center, engineer, manager, Director, Space Shuttle Program
Alexander Nitsch, European Space Agency, engineer, ISS Increment Manager
Richard W. Nygren, NASA, Johnson Space Center, manager, ISS, Shuttle-Mir, Space Shuttle
Hermann J. Oberth, Austro-Hungary, physicist, established early space station concepts and requirements
Jean Olivier, ABMA Project Horizon, Senior Project Engineer NASA Launch Operations Center, Chief Engineer Hubble Space Telescope, Deputy Manager Chandra X-ray Telescope
Edward H. Olling, NASA, Johnson Space Center, manager, established early space station concepts and requirements
Bernardo Patti, European Space Agency, ISS and Exploration Program Manager
Donald R. Pettit, NASA, astronaut, chemical engineer, ISS Expedition 6, 30/31, Space Shuttle, Soyuz
Guido von Pirquet, Austrian, proposed space stations for refueling interplanetary vehicles.
Luther E. Powell, NASA, MSFC, Headquarters, engineer, Space Station Project Manager
William F. Readdy, NASA, astronaut, Associate Administrator, Space Flight, Space Shuttle, Mir
William D. 'Bill' Reeves, NASA, Johnson Space Center, Flight Director, Shuttle-Mir
Darrell C. Romick, Goodyear, Inc., engineer, concept of an orbiting space city.
Kent V. Rominger, NASA, astronaut, pilot, Space Shuttle, Spacelab, ISS
Harold W. Ross, British Interplanetary Society, engineer, developed concepts for spacecraft and space stations
Melanie W. Saunders, NASA, Johnson Space Center Associate Director, ISS international negotiations
Hans W. Schlegel, astronaut, European Space Agency, Space Shuttle, ISS
Emanuel Schnitzer, NASA Langley Research Center Head of Space Station Development
Carl B. Shelley, NASA, Johnson Space Center, engineer, Science and Technology Branch Chief, Deputy Manager, Space Station Projects
Kirk A. Shireman, Manager, ISS Program, NASA
Alexander A. Skvortsov, Russia, cosmonaut, ISS Expedition 23/24, 39/40, Soyuz
Charladean A. Smith, Canadian Space Agency, mission controller, ISS robotics
Thomas P. Stafford, NASA, astronaut, Apollo-Soyuz, helped establish Shuttle-Mir Program, ISS management processes
Ary A. Sternfeld, Poland, France, Soviet Union, engineer, established cosmonautics as a science
Michael T. Suffredini, NASA, Johnson Space Center, ISS Program Manager
Kathy D. Sullivan, NASA, astronaut, geologist, Space Shuttle
Svetlana Y. Savitskaya, Russia cosmonaut Salyut 7
Vladimir Syromyatnikov, Russia, engineer, Androgynous Peripheral Attach System designer
John J. "Tip" Talone, Director of ISS Processing for NASA at Kennedy Space Center
Alan Thirkettle, European Space Agency, ISS Program Manager
Robert F. Thompson, NASA, Skylab, Space Shuttle Program Manager
Fumiya Tsutsui, JAXA, Director, "Kibo" Japanese Experiment Module Project Team
Mikhail V. Tyurin, Russia, cosmonaut, ISS Expeditions 3, 14, 38/39, Soyuz, Space Shuttle
Chester A. Vaughn, NASA, Johnson Space Center , Headquarters, Chief Engineer, ISS Program
Suzan C. Voss, NASA, Johnson Space Center, technical manager, US Lab manager, ISS
James E. Webb, NASA Administrator
Peggy A. Whitson, NASA, astronaut, biochemist, ISS Expeditions 5, 16, 50/51, Soyuz, Space Shuttle
Dafydd R. 'Dave' Williams, astronaut, physician, Canadian Space Agency, Space Shuttle, ISS
Francis L. 'Frank' Williams, ABMA Project Horizon, NASA, Marshall Space Flight Center, Saturn program, Deputy Director Space Station Task Force
Jeffrey N. Williams, NASA, astronaut, pilot/engineer, ISS Expeditions 13, 21/22, 47/48, Soyuz, Space Shuttle
Sunita L. Williams, NASA, astronaut, engineer, ISS Expeditions 14/15, 32/33, Soyuz, Space Shuttle
Izumi Yoshizaki, JAXA, Spokesperson Public Affairs
Kenneth A. Young, NASA, Johnson Space Center, Chief, Mission Design Development Branch
Fyodor N. Yurchikhin, Russia, cosmonaut, ISS Expeditions 15, 36/37, 51/52, Soyuz, Space Shuttle

INDEX